Computational Intelligence

Leszek Rutkowski

Computational Intelligence

Methods and Techniques

 Springer

Prof. Leszek Rutkowski
Department of Computer Engineering
Technical University of Czestochowa
Armii Krajowej 36
42-200 Czestochowa
Poland
lrutko@kik.pcz.czest.pl

ISBN 978-3-642-09515-3   e-ISBN 978-3-540-76288-1

Originally published in Polish
METODY I TECHNIKI SZTUCZNEJ INTELIGENCJI
by Leszek Rutkowski, 2005 by Polish Scientific Publishers PWN

Copyright © by Wydawnictwo Naukowe PWN SA, Warszawa 2010

Copyright for the English edition © 2008 Springer-Verlag Berlin Heidelberg
Published by arrangement with Polish Scientific Publishers PWN, All rights reserved

Foreword

Publication of this book is a special event. This valuable title fills a serious gap in domestic science and technical literature. At the same time it introduces a reader to the most recent achievements in the quickly developing branch of knowledge which the computational intelligence has been for several years. The field, which is a subject of this book, is one of those important fields of science which enable to process information included in data and give their reasonable interpretation programmed by a user.

Recent decades have brought a stormy development of computer techniques and related computational methods. Together with their appearance and quick progress, theoretical and applied sciences developed as well, enabling the user to fully utilize newly created computational potential and to get knowledge out of increasing wealth of data. The development of computational intelligence is then strictly connected with the increase of available data as well as capabilities of their processing, mutually supportive factors. Without them the development of this field would be almost impossible, and its application practically marginal. That is why these techniques have especially developed in recent years.

The development of computational intelligence systems was inspired by observable and imitable aspects of intelligent activity of human being and nature. Nature when undertakes intelligent actions processes data in parallel regulating and adjusting these actions through feedback mechanisms. In such a system learning neural networks function. Another example can be optimization algorithms modeled based on natural selection processes or fuzzy logic systems reflecting vagueness, fuzziness, subjectivity or relativism of human being assessments.

Computational intelligence is a new branch of science and its development dates back to the 60s of the last century when the first algorithms of learning machines – forerunners of today's neural networks – were developed. Then, in the 70s foundations of set theory and fuzzy logic were created. In this early period of computational intelligence development genetic and evolutionary algorithms were introduced. In the 80s the bases for representation of knowledge using rough sets were also created. In recent years many hybrid techniques connecting learning systems with evolutionary and fuzzy ones were developed as well.

Developed theories of computational intelligence were quickly applied in many fields of engineering, data analysis, forecasting, in biomedicine and others. They are used in images and sounds processing and identifying, signals processing, multidimensional data visualization, steering of objects, analysis of lexicographic data, requesting systems in banking, diagnostic systems, expert systems and many other practical implementations.

The essence of the systems based on computational intelligence is to process and interpret data of various nature. These can be numerical, symbolic (e.g. language data of different degree of accuracy), binary, logical data or, for example, uncoded images read out directly on camera screen. The data can be formatted as numbers, that means single elements of vectors, as vectors or tables or as strings of elements or tables composed of them. They can also be composed of ordered sequences of elements or tables and contain elements described in a very inaccurate or even subjective manner.

The common feature of computational intelligence systems is that they process information in cases when presentation in the form of algorithms is difficult and they do it in connection with a symbolic representation of knowledge. These can be relations concerning an object known only based on a finite number of measurements of output and input state (activation). These can also be data binding the most probable diagnosis with a series of observed, often descriptive symptoms. In other cases these can be data characterizing sets in respect to some special features which are initially intangible to the user until they are derived from data and defined as dominant features. These systems have the capability to reconstruct behaviors observed in learning sequences, can form rules of inference and generalize knowledge in situations when they are expected to make prediction or to classify the object to one of previously observed categories.

This book is not only a valuable title on the publishing market, but is also a successful synthesis of computational intelligence methods in world literature. A special advantage of the book is that it contains many examples and illustrations of the methods described, which creates good opportunities to program the presented algorithms. This book should be recommended to engineers of various specialties, physicists, mathematicians, information technology specialists, economists and students of those or related specialties. It should give great satisfaction both to the author due to its publishing and to many readers who will use the techniques described in the book to solve practical issues they are interested in.

July 18, 2007
Jacek M. Żurada
Past President of IEEE Computational Intelligence Society

Contents

1
Introduction

The origins of artificial intelligence can be traced back to early centuries, even to the times of ancient philosophers, especially if we consider the philosophical aspects of this field of science. Less distant in time is the first half of the 19th century when a professor of the University of Cambridge, Charles Babbage, came up with an idea of the so-called "analytical machine" which could not only perform arithmetic operations of a certain type but was also capable of performing operations according to pre-defined instructions. What played an essential role in that project was a punched card which one hundred years later turned out to be a very important element of communication between man and computer. In 1950 Alan Turing came up with a test, the purpose of which was to check whether a given program is intelligent. Soon afterwards a number of works appeared and research projects were carried out in order to understand the natural language and solving complex problems. The ambition of scholars was to create a universal system named "General Problem Solver", which was supposed to solve problems in many areas. The project ended in failure, yet while it was in progress, the researchers had an opportunity to explore the complexity of the issue of artificial intelligence. The 60s and 70s of the last century are characterized by complete dominance of the so-called symbolic approach to solving various issues of artificial intelligence. Thus, decision tree induction methods and methods of predicate logic were used as well as, to a certain extent, classical probabilistic methods, which, however, gained greater significance later on, upon the development of Bayesian networks. A characteristic feature of that time was departure from the use of numerical calculations to solve the problems of artificial intelligence. The turning point in the development of artificial intelligence was publication of a book in 1986 in which Rumelhart and McClelland specified the method for learning of multilayer neural networks, which gave the possibility to solve the problems of, for instance, classifications that traditional methods could not handle. At the beginning of the nineties the concept of learning of neural networks was adopted to learning of fuzzy systems. In this way neuro-fuzzy structures were developed, moreover, a number of other combinations of neural networks, fuzzy systems and evolutionary algorithms were proposed. Today we have a separate branch of science defined in the English literature as Computational Intelligence. This term is understood as solving various problems of artificial intelligence with the use of computers to perform numerical calculations. Such computations are connected with application of the following techniques:

a) neural networks [242, 270],

b) fuzzy logic [94, 265],

c) evolutionary algorithms [57, 136],

d) rough sets [161, 163],

e) uncertain variables [18, 19],

f) probabilistic methods [1, 157].

Only those selected papers or monographs have been cited above, which present "soft computing" (soft techniques, [1, 108]). It must be emphasized that the subject of interest of computational intelligence covers not only individual techniques but also their various combinations [104]. There is an international society called IEEE Computational Intelligence Society, which organizes numerous conferences in the field of computational intelligence, moreover, it publishes three prestigious journals in this field, i.e. *IEEE Transactions on Neural Networks, IEEE Transactions on Fuzzy Systems and IEEE Transactions on Evolutionary Computation*. In Poland, the Polish section of this society exists. Methods of artificial intelligence and computational intelligence lie within the interests of the Polish Neural Networks Society, which organizes conferences called "Artificial Intelligence and Soft Computing" every two years. The purpose of those conferences is to integrate researchers who represent the traditional approach to the artificial intelligence methods and those who apply the methods of computational intelligence.

This book focuses on various techniques of computational intelligence, both single ones and those which form hybrid methods. Those techniques are today commonly applied to classical issues of artificial intelligence, e.g. to process speech and natural language, build expert systems and robots, search for information as well as for the needs of learning by machines. Below are specified the main threads of this book.

In Chapter 2 we briefly present the selected issues concerning artificial intelligence, beginning with the historic Turing test and the issue of the "Chinese room". This chapter contains introductory information on expert systems, robotics, issues of speech and natural language processing as well as heuristic methods. The second part of the chapter focuses on the importance of cognitivistics, i.e. science which attempts to understand the nature of the mind. Further, the chapter introduces the reader to the issues of intelligence of ants and ant algorithms, the field of science called "artificial life" as well as intelligent computer programs known as bots. In the conclusion of this chapter, we quote the opinions of well-known scientists on the perspectives of artificial intelligence and formulate conclusions which reflect the author's views on this topic.

The subsequent three chapters present methods of knowledge representation using various techniques, namely the rough sets, type-1 fuzzy sets and type-2 fuzzy sets.

Chapter 3 presents basic information on the subject of rough sets. The issue of approximation of set and family of sets is discussed therein. The second part of the chapter presents the issues of decision tables, and subsequently the LERS program is used to generate a rule base. The chapter in question, like the two subsequent ones, is richly illustrated with examples which make it easier for the reader to understand various definitions.

Chapter 4 presents basic terms and definitions of fuzzy sets theory. Then it discusses the issue of reasoning, i.e. reasoning on the basis of fuzzy antecedents. Moreover, the reader is introduced to the method for construction of fuzzy inference systems. The second part of the chapter contains numerous examples of applications of fuzzy sets in the issues of forecasting, planning and decision-making.

In Chapter 5, basic definitions concerning type-2 fuzzy sets are presented, operations on those sets are discussed, and subsequently type-2 fuzzy relations are discussed as well. Much attention has been given to the type-reduction method, i.e. a method of transformation of type-2 fuzzy sets into type-1 fuzzy sets. The last part of the chapter explains to the reader the issue of designing type-2 fuzzy inference systems.

Chapter 6 discusses artificial neural networks. This chapter first presents various mathematical models of a single neuron. Next the structure and functioning of multilayer neural networks are discussed. A number of algorithms for learning of those networks have been presented and the issue of choosing their architecture is given particular attention. In the subsequent paragraphs the reader is introduced to the idea of neural networks with feedback. The structure and functioning of the Hopfield, Hamming, Elman, RTRN and BAM networks are discussed. In the second part of the chapter we present the issue of self-organizing neural networks with competitive learning, ART networks, radial-basis function networks and probabilistic neural networks.

Chapter 7 discusses the family of evolutionary algorithms, in particular the classical genetic algorithm, evolutionary strategies and genetic programming. We also present advanced techniques used in evolutionary algorithms. The second part of the chapter discusses connections between evolutionary techniques and neural networks and fuzzy systems.

Chapter 8 presents various methods of data partitioning and algorithms of automatic data clustering. The definitions of hard, fuzzy and possibilistic partitions are provided. Subsequently distance measures applied in clustering methods are presented, which is followed by the discussion of the most popular data clustering algorithms, i.e. HCM algorithm, FCM algorithm, PCM algorithm, Gustafson-Kessel algorithm and FMLE algorithm. This chapter is finished with a presentation of known data clustering validity measures.

In Chapter 9 we present various neuro-fuzzy structures. Those structures are a multilayer (network) representation of a classical fuzzy system. To construct them, the Mamdani type inference and the logical-type inference were applied. Moreover, the so-called Takagi-Sugeno schema is discussed, where the consequents of rules are not fuzzy in nature but are functions of input variables. A characteristic feature of all structures is the possibility to enter weights reflecting the importance of both particular linguistic values in the antecedents of fuzzy rules and weights reflecting the importance of the entire rules. The concept of weighted triangular norms presented in Chapter 4 was used to build those structures. Those norms do not meet the boundary conditions of a classical t-norm and t-conorm, as the commonly applied Mamdani type inference rule does not meet the conditions of logical implication. This chapter illustrates that the application of weighted triangular norms leads to the construction of neuro-fuzzy structures characterized by a very low system operation error. In the second part of the chapter we present the algorithms for learning of all structures, and then we solve the issue of designing neuro-fuzzy systems which are characterized by a compromise between the system operation error and the number of parameters describing this system.

Chapter 10 presents the concepts of the so-called flexible neuro-fuzzy systems. Their characteristic feature is the possibility to find a method of inference (of Mamdani or logical type) as a result of the learning process. The execution of such systems will be possible thanks to specially constructed adjustable triangular norms which are presented in this chapter. Moreover, the following concepts have been used to build the neuro-fuzzy systems: the concept of soft triangular norms, parameterized triangular norms as well as weights used previously in Chapter 9 and describing the importance of particular rules and premises in those rules.

Some of the results presented in this book are based on the research conducted within the Professorial Grant (2005-2008) supported by the Foundation for Polish Science and Special Research Project (2006-2009) supported by Polish Ministry of Science and Higher Education.

The book is the result of, among other things, lectures given by its author in the last few years to graduate students of Politechnika Częstochowska and Academy of Humanities and Economics in Łódź, as well as to PhD students of the Systems Research Institute of the Polish Academy of Science.

The book is also the outcome of cooperation with colleagues from the Department of Computer Engineering of Politechnika Częstochowska, who learned the secrets of computational intelligence as my students back in their fourth year of studies. Thus, I would like to give heartfelt thanks for support to Krzysztof Cpałka, PhD. Eng., Robert Nowicki, PhD Eng., Rafał Scherer, PhD Eng., and Janusz Starczewski, PhD Eng. Moreover, I would like to thank the representatives of a slightly younger generation of scientists at the Department of Computer Engineering, namely Marcin Gabryel, PhD. Eng., Marcin Korytkowski, PhD. Eng., Agata Pokropińska, PhD. Eng., Łukasz Bartczuk, MSc. Eng. and Piotr Dziwiński, MSc. Eng. I also sincerely thank Ms Renata Marciniak, MSc., who took the trouble of preparing part of the drawings.

Special thanks to, Professor Janusz Kacprzyk, Professor Ryszard Tadeusiewicz and Professor Jacek Żurada, who were the first to learn about the idea of writing this book and made valuable comments.

2
Selected issues of artificial intelligence

2.1 Introduction

When considering the issues of artificial intelligence, we need to have a point of reference. This point of reference may be the definition of human intelligence. The literature contains many different definitions, but most of them come down to the conclusion that intelligence is the ability to adapt to new tasks and living conditions or a way in which humans process information and solve problems. Intelligence is also the ability to associate and to understand. It is influenced by both hereditary factors and by nurture. The most important processes and functions making up human intelligence are learning and using knowledge, ability to generalize, perception and cognitive abilities, e.g. ability to recognize a given object in any context. Moreover, we can list such elements as memorizing, setting and achieving objectives, ability to cooperate, formulation of conclusions, ability to analyze, creativity as well as conceptual and abstractive thinking. Intelligence is also related to such factors as self-consciousness, emotional and irrational states of human being.

The so-called man-made intelligent machines may be programmed to imitate only in a very limited scope, a few of above listed elements making up human intelligence. Thus, we have still a long way to go before we understand the functioning of the brain and are able to build its artificial counterpart. In this chapter, we shall briefly present the selected issues concerning artificial intelligence, beginning with the historical Turing test and the issue of the "Chinese room".

2.2   An outline of artificial intelligence history

Artificial intelligence (AI) is a term that stirs great interest as well as many controversies. The name was proposed for the first time by John McCarthy in 1956, when organizing a conference at the Dartmouth College on intelligent machines. The AI issues include, among others, the research for methods of solving problems. One of the examples is the research for chess algorithms. The logical reasoning is the second of many AI issues. It consists of building an algorithm imitating the way of inference occurring in the human brain. Another field of AI research is the processing of natural language, and in consequence, automatic translation of sentences from language to language, giving voice orders to machines and capturing information from voiced sentences and building knowledge bases based on this. AI researchers are faced with the challenge of creating software programs which learn by analogy and are able to perfect themselves. Predicting and forecasting of results and planning are also artificial intelligence domains. There is a large group of philosophers pondering over the problem of consciousness of an intelligent computer. The researchers also try to explore the processes of perception, i.e. vision, touch and hearing, and in consequence, to built electronic equivalents of these organs and apply them in robotics.

The literature presents different definitions of artificial intelligence:

a) Artificial intelligence is a science on machines performing tasks which require intelligence when performed by humans (M. Minsky).

b) Artificial intelligence is a domain of informatics concerning the methods and techniques of symbolic inference by a computer and symbolic representation of knowledge applied during such inference (E. Feigenbaum).

c) Artificial intelligence includes problem solving by methods modeled after natural activities and cognitive processes of humans using computer programs that simulate them (R. J. Schalkoff).

Even though AI is considered a domain of the informatics, it is a point of interest of researchers in other domains, like philosophers, psychologists, medical doctors and mathematicians. We may therefore firmly state that it is an interdisciplinary science, which aims to study human intelligence and implement it in machines. Knowing the definition of AI, we may ask the question: when is our program or machine intelligent? An attempt to answer this question was made in 1950 by the English mathematician Alan Turing. He is the creator of the so-called "Turing test" which is to decide whether the program is intelligent or not. This test consists of the idea that a man, using a keyboard and a computer screen, asks the same questions to the computer and to another person. If the interrogator is unable to differentiate the answers given by the computer from the answers given by a human, we can state that the computer (program) is intelligent. A known critic of the Turing idea was the American philosopher John Searle. He claimed that computers cannot be intelligent, because even though they are using symbols according to certain rules, they do not understand their meaning. To back up his thesis, the philosopher came up with the example known in the literature as the "Chinese Room". Let us assume that we have a closed room, in which there is a European who does not speak Chinese. He is given separate sheets of paper inscribed with Chinese symbols which tell a story. Our hero does not speak Chinese, but he notices a book on a shelf, written in a language he speaks and entitled *What to do when someone slips a paper with Chinese symbols under the door.* This book contains instructions how to make Chinese symbols correlated to the ones he received. To each question, the European prepares an answer according to the rules provided in the manual. Searle states that a man closed in the room in fact does not understand any of the information he is given, just like a computer executing a program. Therefore, there is an obvious difference between thinking and simulating thinking processes. According to Searle, even if we cannot distinguish between an answer given by a machine and a human, this does not mean that the machine is intelligent. Let us assume, however, that there is a machine which passed the Turing test. We can thus state that it is intelligent and able to think as such. In this case, Roger Penrose, the author of *The Emperor's New Mind,* wonders if it would be acceptable or rather reprehensible to use it for one's own purposes, not making note of its desires. Would selling of such machine, or cutting its power supply, which can be in such case considered as food, be ethical? In his book, Penrose describes one of the first AI devices. It was an electronic turtle built by Grey in the early 1950s. The device moved around the room using power from a battery. When the voltage dropped below a certain level, the turtle searched for the nearest socket and charged the batteries. Let us remark that such behavior is similar to the search and consumption of food by humans. Penrose goes further in his ideas and proposes to introduce a certain measure of "turtle's happiness" – the value from the interval, for instance −100 (the turtle is extremely unhappy) to +100 (extreme happiness). Let us also assume that our electronic pet may restore its power resources using solar energy. We can assume that the turtle is unhappy when it is in a dark place, where the sun does not reach (we can easily state that our artificial pet is hungry), and it really "enjoys" sunbathing. After such description of the device, few people could lock the turtle in a dark room, and still, it is just a machine, like a computer.

2.3   Expert systems

As one of many definitions states, the expert system is an "intelligent" computer program that applies knowledge and reasoning (inference) procedures in order to solve problems which require human experience (expert) acquired by many years of activity in a given domain. The general idea of expert systems consists in transposing the expert knowledge of a given domain to a knowledge base, designing an inference machine inferring on the basis of information possessed and adding a user interface used for communication.

The prototype of expert systems was the DENDRAL program, developed in the early 1960s at the Stanford University. Its task was to compute all possible configurations of a given set of atoms. The integral part of the program was a knowledge base containing chemistry laws and rules, which had been developed in chemical labs for decades. DENDRAL proved to be very helpful in solving issues for which analytical methods weren't developed.

At the same Stanford University in the 1970s, two other expert systems were created, which became a historical benchmark solutions in this domain. The PROSPECTOR system was designed to support geologists in defining the type of rock based on contents of different minerals. It facilitated the research for mineral deposits and estimation of deposit volume. PROSPECTOR was a conversation system using the rules obtained from specialists. Models of particular types of deposits contained from several dozens to several hundreds of rules constituting the knowledge base, which was separated from the inference mechanism. The use of PROSPECTOR proved to be a spectacular success as rich deposits of molybdenum have been discovered in Washington state (USA). The MYCIN system was designed to diagnose contagious diseases. The system was fed with data concerning the patient and the results of lab tests. The result of its operation was the diagnosis and recommendations for treatment in certain cases of blood infections. This system supported the decision-making process in case of incomplete data. In case of doubts the system provided the degree of certainty of its diagnosis and alternative solutions (diagnoses). On the basis of MYCIN system, the NEOMYCIN system was created and was used for training doctors.

It is worth to mention one of the largest projects in the history of artificial intelligence, known under the acronym CYC (the name is a fragment of the word encyclopedia) and developed in the USA. This system contained millions of rules (it was planned to have ultimately 100 million rules), which was supposed to give exceptional "intellectual" possibilities to a computer with appropriate software.

As has been mentioned before, the basic elements of an expert system are: a knowledge base, an inference machine and a user's interface. The knowledge base is made of a set of facts and rules. The rules are logical sentences which define some implications and lead to creation of new facts, which as a result allow to solve a given problem. The inference machine is a module which uses the knowledge base. This module may use different inference methods to solve the problem. The so-called shell expert systems are gaining on popularity – these are computer programs with a designed inference machine and an empty knowledge base. These programs are all equipped with special editor programs allowing to enter rules concerning a given problem the user wishes to solve. The issues of constructing expert systems are part of the so-called knowledge engineering. The scope of interest of specialists operating in the domain of knowledge engineering covers such issues as knowledge acquisition, its structuralization and processing as well as designing and selection of appropriate inference methods (inference machine) and designing appropriate interfaces between the computer and its user.

2.4   Robotics

The term "robot" appeared for the first time in 1920 in a play entitled "R.U.R" by the Czech author, Karel Čapek. The play presented a vision of inappropriate use of technology by humans. The story is developed around a factory which produces robots – slaves which are to replace humans in heavy tasks and difficult work. The industry producing robots is being developed and the machines are modernized and equipped with increasingly growing intelligence. A large demand allows to increase the number of robots built. Finally, they were used for military purposes, as soldiers. A time has come when the robots outnumbered their creators – humans. The play ends with a revolt of robots and with the end of human kind.

The dynamic development of robots was initiated by research conducted in the USA. In 1950s, robots developed to work in factories were created – among others, they assembled cars at a General Motors factory. Works were undertaken to build manipulating machines for nuclear industry and oceanographic exploration. Currently, the robots are small wonders of electronic engineering and their prices often exceed the prices of luxury cars. They are basically used everywhere. They perform all kinds of works, from insignificant and trivial, like bringing slippers or serving coffee, through works in difficult conditions in heavy industry, considered difficult and hard for people, to complicated surgical operations. In 2002, a robot steered by professor Louis Kavoussi from the distance of one thousand kilometers performed a surgical operation. The role of doctors supervising the work of the machine was limited to anesthetizing the patient. This way, the patient does not have to wait for the doctor to arrive, which significantly lowers the costs and the duration of the procedure. The da Vinci robot made by Intuitive Surgical imitates the movements of the surgeon's hands during operation, and at the same time eliminates the shaking. Moreover, it displays a large, magnified picture of the patient's heart. It facilitates the procedure, as the surgeon can precisely see the operated organ. The precision of robots causes a significant reduction of damages to the patient's tissues. Due to that, the patient may recover faster. Robots often replace people when difficult jobs e.g. disarming bombs must be performed.

For the last few years, the Japanese company Honda has presented subsequent versions of the ASIMO robot. Its creators claim that the robot speaks two languages – English and Japanese, and is able to hold a conversation. It smoothly moves up and down the stairs and avoids different obstacles. AIBO is another interesting robot. It was given the form of a metallic silver dog, which can play with a ball and pee, for instance. However, it has problems with avoiding obstacles and cannot climb or give a paw. One of the versions was able to recognize 75 voice commands. To learn the mode of operation of this toy, the user must go through a 150 – page manual.

The researchers ask themselves how intelligent a robot should be and what its intelligence should consist in. Two main approaches are emerging, referred to as the weak and the strong artificial intelligence hypothesis. *The weak hypothesis* of artificial intelligence assumes that an intelligent machine is able to simulate the human cognitive process, but it cannot experience any mental states by itself. It is possible that such machine succeeds in the Turing test. *The strong hypothesis* of artificial intelligence leads to the construction of machines that are able to reach cognitive mental states. This approach allows to build a machine which is conscious of its own existence, with real emotions and consciousness. Many research centers lead research on human brain and the entire nervous system of the human being. The understanding of rules functioning in nature will allow to build a "thinking robot". One of the examples is the "Dynamic Brain" robot, the creators of which (neurophysicists Stefan Schaal and Mitsuo Kawato) searched for the rules of learning and self-organization which enable a system to develop its own intelligence. This robot, by watching a film with a woman performing a Japanese folk dance, learned to dance. The project authors use the robot for research on the human brain functioning and interactions occurring between the brain and the human body.

The robot called Cog, created by Rodney Brooks, was supposed to have the intelligence of a six-year old child. The objective of this project was to study the issue of robot development, its physical personification and combination of sensory and motorical skills as well as social interactions. Cog imitated human reactions, was able to focus its vision on objects and extend its arms toward them. When moving, it corrected its actions. Its capabilities were developed towards the ability to recognize objects and living organisms. The opinion that a robot's skills may be very extensive was expressed by Cynthia Breazeal. She built the "Kismet" robot able to learn many behaviors. It was supposed to be able to communicate with people, understand their emotions and express its own emotions using "facial" expressions. The perspectives of robotics achieve unprecedented perspectives. For example, in the USA at the end of the 20th century, an extensive program of research in the area of molecular machines was initiated. One of the basic objectives of nanotechnology development are miniature robots – nanorobots, which may be helpful in supporting the immunological system, detecting bacteria, viruses and cancer cells.

2.5   Processing of speech and natural language

The obvious manner of communication between people is speech. Communicating with the computer or with other devices using spoken language may be a significant facilitation in the life of orally and aurally challenged or physically disabled people. The author of the "Chinese Room" concept, John Searle, answered the question concerning the most important achievements of AI: "I do not know the technological progresses in this domain well enough to give a specific answer. However, I have always been fascinated by achievements in the domain of natural language processing. I believe these works are worth real recognition". Research in the scope of speech and natural language processing covers the following issues:

a) speech synthesis,

b) automatic speech recognition,

c) natural language recognition,

d) automatic translation.

Speech synthesis may be considered equal to the attempt made by a computer to read a book. Speech synthesis has many applications, e.g. to learn foreign languages or to read information for blind people. The study of speech is not an easy task. This results from the fact that a person, when uttering words, intones them appropriately. In order to utter a given sentence, we have to understand its sense, and the computer is not conscious in this matter (as well as in all other issues). An interesting idea was the application by T.V. Raman, manager of IT specialist teams at Adobe, of different typefaces depending on how the computer should read a given text. For example, sentences written in italics are read more loudly. The program on which he worked is called ASTER (Audio System for Technical Readings). The application of ASTER for its author is basically a must, as he lost sight at the age of 14, but it is also used by his colleagues at Adobe, for whom it is a significant facilitation of their daily work. At first, creators of systems imitating human speech attempted to create devices modeled after the human speech organ. Unfortunately, the effects of operation of machines and programs based on formants were not satisfying and significantly differed from human speech. Therefore, such approach was abandoned, and the researchers started to use in algorithms pre-recorded fragments of speech, which were to be put together in an appropriate way. It appeared that it was a very good idea. This idea was applied later in various modifications and improvements of this method.

Another issue studied by AI researchers is automatic speech recognition. The development of this research will enable the communication with the computer, e.g. dictating texts, giving oral commands or voice user recognition (authorization). As we have already mentioned, people pronounce words in different manners (intonation, rate of speech, etc.), often at variance with the rules of grammar. The example of an operational system is the "Dragon Dictate". In the first phase of experiments, this program required a few hours of "adjustment" to the manner of speaking of the person who would dictate the text. Currently, this system is commercialized, just like its competitor, program called Angora. These systems utilize databases, in which words are placed together with their sound or phonemic representation. Based on comparisons, the system recognizes the word. The example application is, for instance, voice selection in cell phones, where each name has its voice label recorded by the user.

Another AI issue is natural language recognition. The problem comes down to retrieving essential data from sentences recorded in the form of text. The researchers create systems to capture knowledge from sentences and the computer should make the division of the sentence into parts of speech. This way, it is able to extract from the contents the objects (nouns), their qualities (adjectives) and relations between them. Earlier, the systems were prepared to work for specific branches of science and they included data from a given branch in a knowledge base. The example may be the Lunar system which answered the questions concerning rock samples brought from the moon.

The last AI issue is automatic translation. It consists in translating texts between various languages. Systems of this type are used by the European Union institutions. It should be noted that the problem may consist in different meaning of words depending on the context. Forty years ago in the USA, a report was formulated, which contained the statement: "No automatic translation of a general scientific text has ever succeeded, and there are no prospects for a quick progress in this scope." Currently, there are translation programs operating on PCs available on the market. However, there are still problems with translation of texts containing sentences from a narrow domain, e.g. different technical documents. The Transcend software may be used as an example. It operates on personal computers and is able to process several thousand words per minute. There are also systems which are able to translate spoken utterances (e.g. over the phone) in real time. Very fast, usually multiprocessor computers are used for this purpose.

2.6   Heuristics and research strategies

The word "heuristics" comes from the Greek words *heurisco*, which means to discover, to find. The easiest description of heuristics may be "creative solution of problems", both logical and mathematical, by way of experiment, trial-and-error method or by using analogies. Heuristic methods are applicable everywhere, where a solution of a problem requires large volumes of computation. Thanks to heuristics, we may eliminate some areas of the space searched. As a result, we may decrease computation costs, and at the same time speed up the discovery of solution. The literature does not provide any formal proofs for the correctness of operation of heuristic algorithms, but their efficiency is confirmed by simulations made. They are widely applied, among others, in expert systems, decision support systems and operation research. The defeat of the chess world champion by a computer was possible also thanks to heuristic techniques, which allowed to exclude variants not portending success. To understand what heuristics are, let us present a well known example in literature [70]. Let us assume that someone dropped a contact lens. Here are some possibilities for search:

1. Blind search – bending down and feeling around for the lens. Such search does not guarantee a positive result.

2. Methodical search – it consists in expanding the space of research methodically and in an organized way. It always guarantees the success, but is very time-consuming.

3. Analytical search – requires the solution of a mathematical equation concerning the fall of the contact lens, taking into consideration the air resistance, wind power, gravitation. It also guarantees the success, but is impractical.

4. Lazy search – consists in finding the nearest optician and purchasing a new lens.

5. Heuristic search – we define the approximate direction of the fall and we presume how far the lens could fall and then we search the selected area. It is the most natural behavior and we most often unconsciously select this method of proceeding.

In the example above, blind and heuristic search were referred to. We talk about blind search when we do not use information on the domain of the problem to be solved. In heuristic search, we use additional information on the space of states and we also are able to estimate the progress improving the efficiency of operation. The process of heuristic search is best presented in the form of tree or graph. In the literature, different strategies of graph search and defining the heuristic solution are considered.

Returning to chess: we may state that as soon as after just a few moves, there is such an unimaginable number of combinations that it is difficult to analyze them even using the best existing computer. Therefore, the current computer programs apply the techniques of artificial intelligence and in particular specially selected and well-developed heuristic methods. Thanks to it, chess computers are able to play a fair match with the best chess players of the world. Let us remind that in 1996, Garry Kasparov won by 4:2 the match with the first model of the Deep Blue computer. However, the following year he lost the match by 2.5:3.5 with the second model of this computer called Deep Blue II. Deep Blue II was a super-machine manufactured by IBM, with 32 nodes, and each of them was equipped with a board containing eight specialized chess processors. Therefore, each current move was analyzed simultaneously by 256 processors. Such processor capacity allowed to analyze 200 million positions on the board in one second. Additionally, this computer had a database containing all the openings from the last 100 years and a database with over a billion possible game ends. So, to win with a human being, incredible processing capacity was used. Next time, Kasparov met with a machine in 2003. This time, his opponent was a computer program called Deep Junior 7. It was written by two Israeli developers – Amir Ban and Shay Bushinsky. This program operated on a computer with 8 processors, much slower than Deep Blue. Its distinctive feature was a deeper knowledge of chess. The tournament, which lasted from January 26 to February 7 in New York, ended with a 3:3 draw. Deep Junior 7 analyzed 3 to 7 million positions of the chess board per second, and Garry Kasparov – a maximum of only 3. Only this comparison may be enough to convince that the computers are very far from the human way of thinking. However, humans are hampered by the fact that they tire quickly and are moreover led by emotions, which also impacts the game result.

2.7 Cognitivistics

Cognitivistics as a science exists for several decades now. In 1976, the quarterly magazine *Cognitive Science* was issued for the first time; it published the results of scientific research in that domain, and in 1979 the Cognitive Science Society, seated at the University of Michigan, was created. From that year, also scientific conferences have been organized, attracting researchers from all over the world. Apart from the name cognitivistics, we can come across other names like cognitive sciences or cognition science. Cognitivistics is the discipline of science which tries to understand the nature of the mind and which studies the phenomena concerning the mind. An essential issue of the cognitive sciences is the analysis of our method of perceiving the world and an attempt to understand what is going on in our minds when we perform basic mental operations. To this aim, studies on the functioning of the brain and models of its operation are used. The science uses scientific achievements of neurobiology and psychology. Cognitivistics is interdisciplinary by nature, therefore this science uses the methods and studies of other sciences, such as anthropology, psychophysics, artificial life, logic, linguistics, neurophysiology, philosophy, artificial intelligence and many other branches of science. We should firmly state that the interdisciplinarity is absolutely necessary for the development of cognitivistics. This science studies an extremely difficult research problem, which is the description of the functioning of the mind. It is obvious that the theories and methods developed only within one branch of science cannot lead to the solution of this problem. That is why any effective results may be obtained only by large research teams consisting of representatives of all the abovementioned disciplines. We should add that cognitivistics has a whole range of practical applications. For example, such domains as neurobiology, psychology and linguistics require the cooperation of appropriate specialists to develop methods of treating speech disorders after a cerebral hemorrhage. Another area of application are the cognitive models used to create computer software interfaces. One of the concepts provides for a possibility to create the image of associations we have in our minds on a computer screen.

A large challenge for the cognitive sciences is the creation of adequate brain models. Current models in the form of artificial neural networks are insufficient and have few things in common with their real equivalent. Moreover, the researchers in the area of cognitivistics will undoubtedly dwell for a long time on the issue of the so-called weak and strong artificial intelligence, which was already discussed in Subchapter 2.4.

2.8   Intelligence of ants

Ants are insects the survival of which depends mainly on cooperation. Many times, we have watched the anthill and dozens of ants wandering chaotically in search of food. When one of them managed to find its source, then other ants followed it shortly. The researchers were interested in the way ants find their way from the anthill to food. It turns out that ants usually choose the shortest route possible. The anthill was separated from the source of food and only two sticks were left – longer and shorter – as the only path. After a few minutes, it appeared that the ants started to return from the food source by the shorter way. The second experiment was made, during which only the longer stick was left. Of course, the ants immediately found their way, but when the shorter stick was added, they continued to walk the old way. After a closer study of their behavior, the researchers found out that an ant, during its march, leaves a trail behind it, in the form of a substance called pheromone, and thus creating a scent path. Its companion, when smelling such path, follows it, also leaving a trail. The next ant selects the path depending on the concentration of the pheromone in any given place. They then head where the largest number of their companions have passed. But what will happen when the source of food runs out? It appears that there is a solution to that as well. The pheromone volatilizes after some time, loosing its intensity. A path less frequented will just vanish after some time. Another behavior of ants attracted special interest from the researchers. Ants remove the bodies of their dead companions by piling them in stacks. It turns out that a small heap of bodies is enough for a "cemetery" to come into existence at this very spot. Also here, ants apply a very simple principle – they transport dead bodies to the place where there are already other bodies. This ant clustering may also be used in practice, for example in banking. The credit decision consists in reviewing customer details and defining whether he/she is creditworthy or not. Such factors as age, work, marital status, use of other bank services, etc. are taken into consideration. By imitating the behavior of ants, it is possible to create clusters of people with similar features. It turns out that dishonest customers are usually characterized by similar features. Verifying the customer will therefore consist in matching his/her data to an appropriate cluster and checking whether the customers classified therein are creditworthy. There are similar systems operating by this principle, but the advantage of the method described above is that the clusters are not defined top-down – they are created naturally.

Based on these observations, a certain type of algorithms has been created, called *ants algorithms*. The work of "artificial ants" applied in these algorithms differs a little from their living counterparts, and namely:

a) they live in an artificial discrete world, therefore they move on a completely different ground, for example between the graph vertexes;

b) their pheromone trail vanishes faster than in the real world;

c) the amount of pheromone secreted by an artificial ant depends on the quality of solution it obtained;

d) in most cases, the pheromone trail is updated only after the solution is generated.

These algorithms are used to solve difficult problems of combinatorial optimization, such as *the Traveling Salesman Problem*, for instance. The Traveling Salesman Problem is that he is supposed to visit a given number of cities by the shortest possible way. The cities are variously distant from one another, he cannot miss any of them and he cannot visit the same city twice. The task seems easy to solve using the algorithm checking all the variants. However, when we have just a dozen of cities, the number of possible ways increases to billions. Nevertheless, this may be perfectly solved using ants algorithms, i.e. using the work of "artificial ants".

Ants algorithms can be also used to solve discrete optimization problems, for instance to define vehicle routes, sequence sorting or defining routes in computer networks. They are applied in practice in telecommunications networks of France Telecom and British Telecommunications. Telephone exchanges are sometimes interconnected by low-capacity connections. When the load of the network increases, for example during TV phone contests, these connections get clogged. The solution is to use virtual agents, the work of which is based on ants behavior, thanks to which subsequent telephone exchanges may increase the capacity by avoiding overloaded segments of the network.

2.9 Artificial life

Artificial life is a relatively new branch of science. It was born in 1987 at the conference in Santa Fe in New Mexico (USA), where the term *Artificial Life* appeared for the first time. Christopher Langton, organizer of the said conference, defined artificial life as follows: "Artificial life is a domain of science dedicated to understanding life by attempting to extract the basic rules of dynamics which influence the biological phenomena. These phenomena are reproduced using different media – for example in computers – to be able to fully use new experimenting methods". It is a discipline of science which uses, among others, the achievements of biology, chemistry, physics, psychology, robotics and computer sciences. It deals with the simulation of life as we know it, but there are also works that study behaviors of organisms built on completely other basis than earthly creatures. However, the basis of this discipline of science is the definition of life. Unfortunately, researchers are not unanimous in this issue. The main reason why it is difficult to define this term is certainly the fact that we are dealing only with the forms of life present on Earth. The earliest, simple and well known example of artificial life is the *Game of Life*. The game was created in 1968 by the mathematician John Conway, who based its operation on cellular automatons. In this case, the environment is a two-dimensional grid of cells, the state of which may be described as full – representing a live cell, or empty – lack of live cell. The rules are very simple. A cell dies of loneliness or overcrowding, and a new one appears when it has exactly three neighbors. During the simulation, when visualizing it, we could observe quick, dynamic growth of cells, creating wonderful patterns, and just after it, declines of entire colonies, for instance. Initial, even only slightly differing settings of cells lead during simulation to very complicated and interesting patterns.

Another very interesting example of artificial life are *biomorphs*. Their creator is Richard Dawkins, a British zoologist. They were created in order to study the evolution of forms. Biomorphs are graphic shapes recorded in genotypes, the appearance of which reminds of living organisms. Dawkins applied simple genetic operations to obtain new shapes in subsequent generations. By starting the evolution from simple figures similar to trees, we can obtain shapes of insects and bugs. The simulation of growth of organisms was conducted using a formal description of development of so-called *L-systems*, proposed in 1968 by Aristid Lindenmayer. They were later used to describe and model plants growth. The effects were similar to the operation of fractals.

Tierra system is a virtual world created by the biologist Tom Ray. The environment is in this case a virtual computer, where programs-individuals live. These programs are written in a special, simple language reminding of an assembler. The evolution of individuals progresses using mutations, or a random exchange of one instruction, or recombination consisting in replacing a part of the code. The individuals compete with each other for resources, or the virtual machine memory, and for the time of processor use. Thus, they try to take as much space as possible, and use as much of their instructions as possible as compared to other individuals.

Framstick system is currently one of the most advanced artificial life projects. It has been conducted since 1997 by two Poles: Maciej Komosiński and Szymon Ulatowski. The simulations are performed in a virtual, three-dimensional world, where there are both land and water environment. The organisms (framsticks) are build of sticks, which additionally can play the role of receptors (having the sense of touch, balance or smell) or the function of the movement organs (using appropriate muscles). The organs of movement are steered with the help of the nervous system based on a neural network. Framsticks compete with each other for existence in the environment by fighting with each other and searching for food. Each of the individuals is described with a genotype, which is a special code describing its construction. Three types of chromosome notation are available: the first one – the simplest and describing directly the construction of framstick, the second has the form of a recurring notation, and the third one consists in notating the information on particular cell. The evaluation of adaptation is performed in the same environment and may take different forms. An individual may be assessed depending on its movement, size or resistance. The simulation may also take different forms, e.g. searching for the highest individual. Next generations of individuals are created by way of evolution, which consists of selection, mutation and crossover of genes. The evolution relates both to the external form of organism and the nervous system.

2.10   Bots

A bot is an automaton, a software tool, a program used most often to search and retrieve data. Intelligent bots can additionally make decisions based on knowledge acquired earlier. Currently, we may differentiate several kinds of bots, which due to their capabilities have been divided to:

1) chatterbots – these are automata for chats. They imitate natural language conversation, acquire information from the interlocutor;

2) searchbots – they maintain automatically databases, are used to search, index and collect data;

3) shoppingbots – these are automata help us when shopping over the Internet. They browse sites in search of given products, creating a report on price differences;

4) databots – automata to search data, solve problems; their construction is based on neural networks;

5) updatebots – are used to update data possessed by the user. They inform on changes in network resources;

6) infobots – programs automatically providing answers using e-mail. They are used to provide technical support and marketing information.

Among the bots, the greatest popularity and interest is shared by chatterbots, especially as a tool to analyze customer expectations in marketing. They are most often placed on web sites, used to promote products and help the navigation. For the companies that use them, they are a source of knowledge on customers, as during a conversation a lot of information can be obtained. We should watch whom we are chatting to, using the Internet, as it may be a chatterbot? Internet users find it difficult to identify the interlocutor, all the more as they do not expect to be chatting with an automaton. They are very surprised when they discover that it was only a computer program on the other side.

The first *chatterbots* appeared around 1966 as attempts to execute the CMC (*Computer Mediated Communication*) project, the objective of which was to initiate the human-computer communication. In 1968 Eliza was created – a program made only of 240 lines of code, simulating a conversation with a psychotherapeutist. The program was able to understand the structure of a sentence by analyzing the key words and formulate questions on this basis.

For many years, the developers specializing in writing intelligent chatting programs have been competing in a special contest. Several times in a row, the best developer was Richard Wallace, who created the bot called ALICE (*Artificial Linguistic Internet Computer Entity*). It is one of the best existing bots (www.alicebot.org). It owes its advantages to the users, with whom it talked, enriching its knowledge base. It is worth noting that even such a good program did not pass the Turing test as part of the mentioned competition. Different versions of ALICE program have commercial applications, e.g. to promote new products.

2.11 Perspectives of artificial intelligence development

For years, the researchers have been arguing on the perspectives of artificial intelligence development. Let us quote some statements of renown specialists in this domain.

Roger Penrose (professor of mathematics at Oxford University) in his book *The Emperor's New Mind* expresses the opinion that mind processes of humans differ fundamentally from computer operations. No machine working on the basis of calculations will be able to think and understand like we do. Processes in our brain are "non-computational". Moreover, Roger Penrose believes that human brain is completely explainable in the categories of physical world, but only the existing physical theories are incomplete to explain how our thinking processes proceed. Dr. Ray Kurzweil, studying, among others, the commercial application of artificial intelligence, claims that the disappearance of differences between machine and human is only a matter of time, as human brain in the last few centuries almost did not develop. Meanwhile, in the last twenty years, its electronic counterparts are developing at an incredible pace, and this trend is certainly going to hold in the years to come. Hans Moravec (director of the mobile robots laboratory at Carnegie Mellon University) thinks that "man is a complicated machine... and nothing more". He gained reputation with a daring statement that "in the future, human nervous system may be replaced by a more complex, artificial equivalent". He claims that sooner or later, Earth will be populated by products of our technology, more perfect and better managing the difficulties of life than sensitive and vulnerable representatives of *the homo sapiens* species. Kevin Warwick (professor of cybernetics at Reading University) claims that after we switch on the first powerful machine with intelligence comparable to human intelligence, most probably we will not be able to ever switch it off. We will set up a time bomb, ticking over human species, and we will be unable to disarm it. There will be no way to stop the march of the machines.

As we can see, research on artificial intelligence fascinates the scientists, but also stirs great controversies. When summing up the opinions and experiences of different researchers and eliminating extreme views, we may claim the following:

1) No machine made by humans so far ever managed to go outside of the set of rules programmed by man.

2) Artificial intelligent systems will not simulate exactly the functioning of human brain. It relates to hardware limitations and a large complexity of the brain.

3) Machines may pass the Turing test in a limited thematic scope (sports, chess, advisory system in economy or medicine).

4) In the future, it may seem to us that machines show signs of consciousness. However, machines will not be conscious in the philosophical sense.

5) In the perspective of several decades, intelligent machines will be our partners at work and home.

6) Computers will design next generations of computers and robots and will play a significant (dominant?) part in development of intelligence of inhabitants of Earth.

2.12 Notes

In this chapter, we have discussed, in a way accessible to the Reader, some selected issues of artificial intelligence. Detailed information can be found in many literature items on this subject. The most important directions of research in the scope of human intelligence were discussed in the monograph [143]. Historic and philosophic threads of discussion in the domain of artificial intelligence are contained in monographs [105,168]. Chess games between the machine and man have been discussed in articles [81, 124, 229]. The issues of expert systems are presented in monographs [5, 25, 89, 141, 146, 170, 179, 268]. The issues of robot construction and their applications discussed by authors of book [61] and [135]. The issues of speech and natural language processing have been discussed in works [45, 154, 228]. The idea of heuristic search and algorithms applied have been explained in works [62, 70]. The history and development of cognitivistics is presented in article [46]. Ants behavior and ants algorithms have been explained in study [14]. Different models of artificial life have been discussed in articles [114, 169]. Work [75] presents the applications of artificial intelligence in electrical power engineering, work [170] – in technical diagnostics, work [268] – in economics and management. Following monographs on artificial intelligence, classics of the world literature, are well worth recommending [129, 147, 184]. In monograph [245], the authors presented a very interesting approach to the issue of image recognition using artificial intelligence methods, and in particular they introduced the term of so-called "automatic understanding of images". In this chapter, we discussed only a few threads of the artificial intelligence issue. We refer the Reader interested for instance in multiagent intelligent systems to monographs [52, 109, 127, 255]. On the other hand, monograph [27] is a comprehensive study of learning systems and presents classic methods of artificial intelligence, like, for instance, the issue of decision trees induction and rules inductions as well as probabilistic methods and induction logical programming.

3

Methods of knowledge representation using rough sets

3.1 Introduction

In the physical world around us, it is impossible to find any two object (things) that are identical. By comparing any two objects, even if very similar, we will always be able to find differences between them, in particular if we consider a sufficiently large number of their features (attributes) with a sufficiently great accuracy. Of course, such a detailed description of the world is not always needed. If we decrease the precision of description, it may happen that some or even several objects that were distinguishable before become indiscernible. For example, all cities in Poland may be discernible with respect to the exact number of inhabitants. If we are interested in cities with the number of inhabitants within a given interval, e.g. from 100 to 300 thousand people, then some cities will be indiscernible with respect to the feature (attribute) "number of inhabitants". Moreover, in the description of any given object, we only consider a limited number of features, adequate to a given purpose. Quite often, we want to reduce that number to the necessary minimum. These are the problems dealt with by the theory of rough sets.

In order to facilitate further discussion, we shall introduce several notions and symbols. First, we shall define the universe of discourse U. It is the set of all objects which constitute the area of our interest. A single j-th element of this space will be denoted as x_j. Each object of the space U may be characterized using specific features. If it is a physical object, most certainly it has infinitely many features, however, we shall limit the selection to their specific subset. Let us denote the interesting set of object features of space U by the symbol Q. Let us denote the individual features by the symbol q appropriately indexed, e.g. q_i. What differentiates one object from another and makes other objects similar, these are the values of their features. Let us denote by V_q the set of values that the feature q can take. The value of feature q of the object x will be denoted as v_q^x. The vector of all object x features may be presented as $\mathbf{v}^x = \left[v_{q_1}^x, v_{q_2}^x, ..., v_{q_n}^x \right]$.

In this chapter the rough sets theory will be presented in the form of a series of definitions illustrated by examples. Table 3.1 will allow the reader an easier handling of them.

TABLE 3.1. List of definitions and examples

	Definition	Examples
3.1.	Information system	3.1, 3.2, 3.3
3.2.	Decision table	3.4
3.3.	Indiscernibility relation	
3.4.	Equivalence class	3.5, 3.6, 3.7
3.5.	Lower approximation of the set	3.8, 3.9, 3.10
3.6.	Upper approximation of the set	3.11, 3.12, 3.13
3.7.	Positive region of the set	3.8, 3.9, 3.10
3.8.	Boundary region of the set	3.14, 3.15, 3.16
3.9.	Negative region of the set	3.17, 3.18, 3.19
3.10.	$\widetilde{P}$-exactly set	3.20, 3.21, 3.22
3.11.	$\widetilde{P}$-definable set	3.20, 3.21, 3.22
3.12.	$\widetilde{P}$-accuracy of an approximation of the set	3.23, 3.24, 3.25
3.13.	Lower approximation of the family of sets	3.26
3.14.	Upper approximation of the family of sets	3.27
3.15.	Positive region of the family of sets	3.28
3.16.	Boundary region of the family of sets	3.29
3.17.	Negative region of the family of sets	3.30
3.18.	Quality of approximation of the family of sets	3.31
3.19.	Accuracy of approximation of the family of sets	3.32
3.20.	Dependence degree of attributes	3.33
3.21.	Deterministic rules	3.33
3.22.	Set of independent attributes	3.34
3.23.	Set of relatively independent attributes	3.35
3.24.	Reduct	3.36
3.25.	Relative reduct	3.36
3.26.	Indispensable attribute	3.37
3.27.	Core	3.38
3.28.	Normalized coefficient of attributes significance	3.39
3.29.	Error of approximate reduct	3.40

3.2   Basic terms

One of methods to present the information on objects characterized by the same set of features is the information system.

Definition 3.1
The information system is referred to an ordered 4-tuple $SI = \langle U, Q, V, f \rangle$ [1], where U is the set of objects, Q is the set of features (attributes), $V = \bigcup_{q \in Q} V_q$ is the set of all possible values of features, while $f : U \times Q \to V$ is called the information function. We can say that $v_q^x = f(x, q)$, of course $f(x, q) \in V_q$. The notation $v_q^x = f_x(q)$, which treats the information function as a family of functions, will be considered as equivalent. Then $f_x : Q \to V$.

Example 3.1
Let us consider a used car dealer. Currently, there are 10 cars. The universe of discourse U is therefore composed of 10 objects, which can be notated as

$$U = \{x_1, x_2, ..., x_{10}\}. \tag{3.1}$$

The car dealer notes in his documents four features of each car, which are usually referred to by customers during phone calls. These are: number of doors, horsepower, colour and make. Therefore, the set of features can be written as

$$\begin{aligned} Q &= \{q_1, q_2, q_3, q_4\} \\ &= \{\text{number of doors, horsepower, colour, make}\}. \end{aligned} \tag{3.2}$$

Based on the contents of Table 3.2 we can define the domains of particular features:

TABLE 3.2. Example of an information system

Object (U)	Number of doors (q_1)	Horsepower (q_2)	Colour (q_3)	Make (q_4)
x_1	2	60	blue	Opel
x_2	2	100	black	Nissan
x_3	2	200	black	Ferrari
x_4	2	200	red	Ferrari
x_5	2	200	red	Opel
x_6	3	100	red	Opel
x_7	3	100	red	Opel
x_8	3	200	black	Ferrari
x_9	4	100	blue	Nissan
x_{10}	4	100	blue	Nissan

$$V_{q_1} = \{2, 3, 4\}, \tag{3.3}$$

$$V_{q_2} = \{60, 100, 200\}, \tag{3.4}$$

$$V_{q_3} = \{black, bluc, red\}, \tag{3.5}$$

$$V_{q_4} = \{Ferrari, Nissan, Opel\}. \tag{3.6}$$

Example 3.2

Let us consider the set of real numbers in the interval U (see Fig. 3.1), where

$$U = [0, 10). \tag{3.7}$$

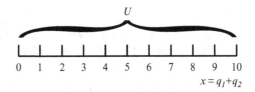

FIGURE 3.1. One-dimensional universe of discourse U

Let each element $x \in U$ be defined by two features making up the set of features

$$Q = \{q_1, q_2\}, \tag{3.8}$$

where q_1 is the integral part of the number x and q_2 is the decimal part of this number. Of course $x = \{q_1, q_2\}$. The information functions may be defined as follows

$$f_x(q_1) = \text{Ent}(x), \tag{3.9}$$

$$f_x(q_2) = x - \text{Ent}(x), \tag{3.10}$$

where function $\text{Ent}(\cdot)$ (*fr. entier*) means the integral part of the argument.

Knowing the definition of information functions, it is usually easy to define the domains of variability of particular features. In our example, they will be as follows:

$$V_{q_1} = \{0, 1, 2, 3, 4, 5, 6, 7, 8, 9\}, \tag{3.11}$$

$$V_{q_2} = [0; 1). \tag{3.12}$$

Example 3.3

In the following example, let us consider the space of pairs

$$U = \{\mathbf{x} = [x_1; x_2] \in [0; 10) \times [0; 10)\}. \tag{3.13}$$

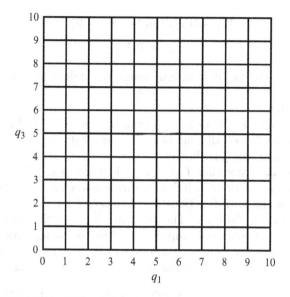

FIGURE 3.2. Two-dimensional universe of discourse U

The objects belonging to the space defined in this way may be interpreted as points located on a plane, as shown in Fig. 3.2. The most natural features of points are their coordinates x_1 and x_2. In our example, however, they will be defined otherwise. Let us define four features

$$Q = \{q_1, q_2, q_3, q_4\} \qquad (3.14)$$

where q_1 is the integral part of the first coordinate of point $\mathbf{x}$, q_2 is its decimal part, and q_3 and q_4 are the integral and the decimal part of the second coordinate of the point, respectively. The information functions will therefore be defined as follows:

$$f_{\mathbf{x}}(q_1) = \mathrm{Ent}\,(x_1), \qquad (3.15)$$

$$f_{\mathbf{x}}(q_2) = x_1 - \mathrm{Ent}\,(x_1), \qquad (3.16)$$

$$f_{\mathbf{x}}(q_3) = \mathrm{Ent}\,(x_2), \qquad (3.17)$$

$$f_{\mathbf{x}}(q_4) = x_2 - \mathrm{Ent}\,(x_2). \qquad (3.18)$$

Knowing the definition of information functions, it is usually easy to define the domains of variability of particular features. In this example, they will be as follows:

$$V_{q_1} = \{0, 1, 2, 3, 4, 5, 6, 7, 8, 9\}, \qquad (3.19)$$

$$V_{q_2} = [0; 1), \qquad (3.20)$$

$$V_{q_3} = \{0, 1, 2, 3, 4, 5, 6, 7, 8, 9\}, \tag{3.21}$$

$$V_{q_4} = [0; 1). \tag{3.22}$$

The special case of the information system is the decision table.

Definition 3.2

The decision table is the ordered 5-tuple $DT = \langle U, C, D, V, f \rangle$. The elements of the set C we call conditional features (attributes), and elements of D-decision features (attributes).

The information function f described in Definition 3.1 defines unambiguously the set of rules included in the decision table. In the notation, in the form of family of functions, the function $f_l : C \times D \to V$ defines l the decision rule of the table. The difference between the above definition and Definition 3.1 consists in separation of the set of features Q into two disjoint subsets C and D, complementary to Q. The decision tables are an alternative way of representing the information with relation to the rules:

$$R^l : \textbf{IF } c_1 = v_{c_1}^l \textbf{ AND } c_2 = v_{c_2}^l \textbf{ AND...AND } c_{n_c} = v_{c_{n_c}}^l \textbf{ THEN } d_1 = v_{d_1}^l$$
$$\textbf{AND } d_2 = v_{d_2}^l \textbf{ AND ... AND } d_{n_d} = v_{d_{n_d}}^l.$$

Example 3.4

Let us assume that basing on notes of the car dealer from Example 3.1, we shall build an expert system, which will define the car make based on information on the number of doors, horsepower and colour. We should divide the set Q (defined by formula (3.2)) into the set of conditional features

$$C = \{c_1, c_2, c_3\} = \{q_1, q_2.q_3\} \tag{3.23}$$
$$= \{\text{number of doors, horsepower, colour}\}$$

and a single-element set of decision features

$$D = \{d_1\} = \{q_4\} = \{\text{make}\}. \tag{3.24}$$

Information included in the information system presented in Table 3.2 will be used to build a decision table (Table 3.3). The description of each object of space U constitutes the basis to create a single rule.

The contents included in the decision table (Table 3.3) may also be presented in the form of rules:

$R^1 : \textbf{IF } c_1 = 2 \textbf{ AND } c_2 = 60 \textbf{ AND } c_3 = \text{blue} \textbf{THEN } d_1 = \text{Nissan}$
$R^2 : \textbf{IF } c_1 = 2 \textbf{ AND } c_2 = 100 \textbf{ AND } c_3 = \text{black} \textbf{ THEN } d_1 = \text{Nissan}$
$\dots$

$R^{10} : \textbf{IF } c_1 = 4 \textbf{ AND } c_2 = 100 \textbf{ AND } c_3 = \text{blue} \textbf{ THEN } d_1 = \text{Nissan}$

TABLE 3.3. Example of the decision table

Rule (l)	Number of doors (c_1)	Horsepower (c_2)	Colour (c_3)	Make (d_1)
1	2	60	blue	Opel
2	2	100	black	Nissan
3	2	200	black	Ferrari
4	2	200	red	Ferrari
5	2	200	red	Opel
6	3	100	red	Opel
7	3	100	red	Opel
8	3	200	black	Ferrari
9	4	100	blue	Nissan
10	4	100	blue	Nissan

Now we shall present two definitions that are very important in the rough sets theory. If given two objects $x_1, x_b \in U$ have the same values of all features q belonging to the set $P \subseteq Q$, which may be notated as $\forall_q \in P$, $f_{x_a}(q) = f_{x_b}(q)$, then we say that these objects are P-indiscernible or that they are to each other in P-*indiscernibility relation* $\left(x_a, \widetilde{P} x_b \right)$.

Definition 3.3
The P-*indiscernibility relation refers* to a $\widetilde{P}$ relation defined in the space $U \times U$ satisfying

$$x_a \widetilde{P} x_b \iff \forall_q \in P; \ f_{x_a}(q) = f_{x_b}(q), \tag{3.25}$$

where $x_a, x_b \in U$, $P \subseteq Q$.

It is easy to verify that the $\widetilde{P}$ relation is reflexive, symmetrical and transitive, and thus it is a relation of equivalence. The relation of equivalence divides a set in which it is defined, into a family of disjoint sets called equivalence classes of this relation.

Definition 3.4
The set of all objects $x \in U$ being in relation $\widetilde{P}$ we call *the equivalence class* of relation $\widetilde{P}$ in the space U. For each $x_a \in U$, there is exactly one such set denoted by the symbol $[x_a]_{\widetilde{P}}$, i.e.

$$[x_a]_{\widetilde{P}} = \left\{ x \in U : x_a \widetilde{P} x \right\}. \tag{3.26}$$

The family of all equivalence classes of the relation $\widetilde{P}$ in the space U (called the quotient of set U by relation $\widetilde{P}$) will be denoted using the symbol P^* or $U/\widetilde{P}$.

Example 3.5

Let us define the equivalence classes of relation C-indiscernibility $\widetilde{C}$ defined by the set of features C given by formula (3.23) for the information system defined in Example 3.1:

$$[x_1]_{\widetilde{C}} = \{x_1\}, \tag{3.27}$$

$$[x_2]_{\widetilde{C}} = \{x_2\}, \tag{3.28}$$

$$[x_3]_{\widetilde{C}} = \{x_3\}, \tag{3.29}$$

$$[x_4]_{\widetilde{C}} = [x_5]_{\widetilde{C}} = \{x_4, x_5\}, \tag{3.30}$$

$$[x_6]_{\widetilde{C}} = [x_7]_{\widetilde{C}} = \{x_6, x_7\}, \tag{3.31}$$

$$[x_8]_{\widetilde{C}} = \{x_8\}, \tag{3.32}$$

$$[x_9]_{\widetilde{C}} = [x_{10}]_{\widetilde{C}} = \{x_9, x_{10}\}. \tag{3.33}$$

We therefore can say that the objects x_4 and x_5 are C-indiscernible, similarly to x_6 and x_7 as well as x_9 and x_{10}. The family of above specified equivalence classes will be the set

$$c^* = \{\{x_1\}, \{x_2\}, \{x_3\}, \{x_4, x_5\}, \{x_6, x_7\}, \{x_8\}, \{x_9, x_{10}\}\}. \tag{3.34}$$

Example 3.6

For the set of features $Q = \{q_1, q_2\}$ defined in Example 3.2 all objects are discernible, i.e. there are infinitely many one-element equivalence classes of Q-indiscernibility relation and each element of space U forms its own class. It will be different when we divide the set Q into two features sets:

$$P_1 = \{q_1\}, \tag{3.35}$$

$$P_2 = \{q_2\}. \tag{3.36}$$

For P_1-indiscernibility relation, 10 equivalence classes are formed

$$[0]_{P_1} = [0; 1), \tag{3.37}$$

$$[1]_{P_1} = [1; 2), \tag{3.38}$$

$$\ldots$$

$$[9]_{P_1} = [9; 10). \tag{3.39}$$

Their family is the set

$$P_1^* = \{[0; 1); [1; 2); [2; 3); [3; 4); [4; 5); [5; 6); [6; 7); [7; 8); [8; 9); [9; 10)\}. \tag{3.40}$$

Figure 3.3 shows the exemplary equivalence class $[1]_{\widetilde{P_1}}$.

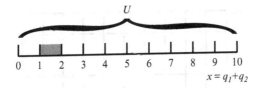

FIGURE 3.3. Example of equivalence class $[1]_{\tilde{P}_1}$

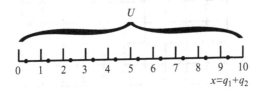

FIGURE 3.4. Example of equivalence class $[0.33]_{\tilde{P}_2}$

For P_2-indiscernibility relation, infinitely many ten-element equivalence classes are formed. These are sets of numbers from the space U with the same decimal part

$$[x]_{P_2} = \{\widehat{x} \in U : \widehat{x} - \mathrm{Ent}\,(\widehat{x}) = x - \mathrm{Ent}\,(x)\}. \qquad (3.41)$$

Their family is the set

$$P_2^* = \{[x]_{P_2} = \{\widehat{x} \in U : \widehat{x} - \mathrm{Ent}\,(\widehat{x}) = x - \mathrm{Ent}\,(x)\} : x \in [0;1)\} \qquad (3.42)$$
$$= \{[x]_{P_2} = \{\widehat{x} \in U : \widehat{x} - \mathrm{Ent}\,(\widehat{x}) = x\} : x \in [0;1)\}.$$

Figure 3.4 shows the exemplary equivalence class $[0.33]_{\tilde{P}_2}$.

Example 3.7
Like in Example 3.6, for the set of features Q defined by formula (3.14) in Example 3.3, all the objects are discernible, i.e. there are infinitely many one-element equivalence classes of Q-indiscernibility relation and each element of space U forms its own class. It will be different, if we consider the features set $P \subseteq Q$ given by

$$P = \{q_1, q_3\}. \qquad (3.43)$$

For the P-indiscernibility relation thus defined in the space U, we have 100 equivalence classes. The equivalence class of point $\mathbf{x} = (x_1, x_2)$ may be described as

$$[\mathbf{x}]_{\tilde{P}} = \{\widehat{\mathbf{x}} = (\widehat{x}_1, \widehat{x}_2) \in U : \mathrm{Ent}\,(\widehat{x}_1) = \mathrm{Ent}\,(x_1) \wedge \mathrm{Ent}\,(\widehat{x}_2) = \mathrm{Ent}\,(x_2)\}. \qquad (3.44)$$

Figure 3.5 presents the exemplary equivalence class.

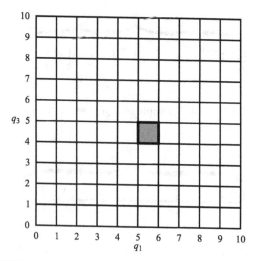

FIGURE 3.5. Example of equivalence class $[(5;4)]_{\tilde{P}}$

Their family is the set of all square fields visible in Figs. 3.2 and 3.5. We can describe this set as follows:

$$P^* = \left\{[\mathbf{x}]_{\tilde{P}} = \{\widehat{\mathbf{x}} = (\widehat{x}_1; \widehat{x}_2) \in U : \text{Ent}(\widehat{x}_1) = \text{Ent}(x_1) \wedge \text{Ent}(\widehat{x}_2)\right. \quad (3.45)$$

$$= \text{Ent}(x_2) : \mathbf{x} = (x_1; x_2); x_1; x_2 = 0; \ldots; 9\}$$

$$= \left\{[\mathbf{x}]_{\tilde{P}} = \{\widehat{\mathbf{x}} = (\widehat{x}_1; \widehat{x}_2) \in U : \text{Ent}(\widehat{x}_1) = x_1 \wedge \text{Ent}(\widehat{x}_2) = x_2\right\} :$$

$$\mathbf{x} = (x_1; x_2); x_1; x_2 = 0; \ldots; 9\}.$$

3.3   Set approximation

In the space U, certain sets X may exist. We can infer that particular objects $x \in U$ belong to sets X based on the knowledge of values of their features. The set of available features $P \subseteq Q$ is usually limited and the determination of membership of the object to a specific set may not be unequivocal. This situation is described by the terms of lower and upper approximation of set $X \subseteq U$.

Definition 3.5
The set $\underline{\widetilde{P}}X$ described as follows:

$$\underline{\widetilde{P}}X = \left\{x \in U : [x]_{\tilde{P}} \subseteq X\right\} \quad (3.46)$$

is called $\widetilde{P}$-lower approximation of the set $X \subseteq U$.

Therefore, the lower approximation of the set X is the set of the objects $x \in U$, with relation to which on the basis of values of features P, we can certainly state that they are elements of the set X.

Example 3.8

In the space U, defined by equation (3.1) in Example 3.1 there are three sets of car makes: Ferrari, Nissan and Opel (see Table 3.2). Let us mark them with letters X_F, X_N and X_O:

$$X_F = \{x_3, x_4, x_8\}, \tag{3.47}$$

$$X_N = \{x_2, x_9, x_{10}\}, \tag{3.48}$$

$$X_O = \{x_1, x_5, x_6, x_7\}. \tag{3.49}$$

We will infer the membership of various space objects based on based on the value of features of set C defined by notation (3.23). Applying directly Definition 3.5, let us determine $\widetilde{C}$-lower approximation of sets X_F, X_N and X_O. This definition says that the object $x \in U$ is an element of the lower approximation, if the whole equivalence class, to which it belongs, is a subset of the set X. Among the equivalence classes defined in Example 3.5, only classes $[x_3]_{\widetilde{C}}$ and $[x_8]_{\widetilde{C}}$ are the subsets of the set X_F, that is

$$\underline{\widetilde{C}}X_F = \{x_3\} \cup \{x_8\} = \{x_3, x_8\}. \tag{3.50}$$

The object x_4 does not belong to $\underline{\widetilde{C}}X_F$, even if it belongs to X_F, as the object x_5 with identical feature values from the set C, and therefore belonging to the same equivalence class, is not an element of X_F.

Sets $[x_2]_{\widetilde{C}}$ and $[x_9]_{\widetilde{C}} = [x_{10}]_{\widetilde{C}}$ are subsets of the set X_N, hence

$$\underline{\widetilde{C}}X_N = \{x_2\} \cup \{x_9, x_{10}\} = \{x_2, x_9, x_{10}\}. \tag{3.51}$$

Sets $[x_1]_{\widetilde{C}}$ and $[x_6]_{\widetilde{C}} = [x_7]_{\widetilde{C}}$ are subsets of the set X_O, hence

$$\underline{\widetilde{C}}X_O = \{x_1\} \cup \{x_6, x_7\} = \{x_1, x_6, x_7\}. \tag{3.52}$$

Example 3.9

Let us assume that in the space U defined in Example 3.2 there is a set X defined as follows:

$$X = [1, 75; \; 6, 50]. \tag{3.53}$$

Let us define the $\widetilde{P}_1$ and $\widetilde{P}_2$-lower approximation of this set. Four equivalence classes of P_1-indiscernibility relation (Example 3.6) belong entirely to the set X. Therefore, the $\widetilde{P}_1$-lower approximation will be their sum

$$\underline{\widetilde{P}_1}X = [2]_{P_1} \cup [3]_{P_1} \cup [4]_{P_1} \cup [5]_{P_1} = [2, 6), \tag{3.54}$$

which is illustrated in Fig. 3.6.

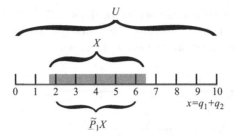

FIGURE 3.6. Lower approximation in one-dimensional universe of discourse

No equivalence class of the P_2-indiscernibility relation belongs entirely to the set X, therefore its $\tilde{P}_2$-lower approximation is an empty set, i.e.

$$\underline{\tilde{P}_2}X = \varnothing \tag{3.55}$$

Example 3.10
Let us in the space U, defined by notation (3.13), define the set X as shown in Fig. 3.7. This figure shows the marked equivalence classes making up the $\tilde{P}$-lower approximation of the set X. Among the 100 equivalence classes defined by formula (3.44), the lower approximation is made up by 25 equivalence classes – squares which are entirely subsets of the set X.

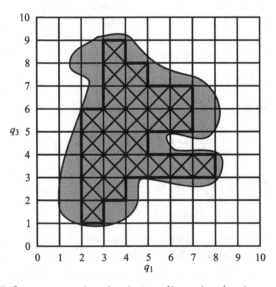

FIGURE 3.7. Lower approximation in two-dimensional universe of discourse

Definition 3.6
The set $\overline{\widetilde{P}}X$ described as follows:

$$\overline{\widetilde{P}}X = \{x \in U : [x]_{\widetilde{P}} \cap x \neq \varnothing\} \tag{3.56}$$

is called $\widetilde{P}$-*upper approximation* of the set $X \subseteq U$.

The upper approximation of the set X is the set of the objects $x \in U$, with relation to which, on the basis of values of features P, we can not certainly state that they are not elements of the set X.

Example 3.11
Applying directly Definition 3.6, let us determine $\widetilde{C}$-upper approximation of sets X_F, X_N, and X_O defined in Example 3.8. This definition says that the object $x \in X$ is an element of the upper approximation, if the whole equivalence class, to which it belongs, has a non-empty intersection with the set X. In other words, if at least one element of a given equivalence class belongs to the set X, then each element of this equivalence class belongs to the upper approximation of the set X. Among the equivalence classes defined in Example 3.5, elements of classes $[x_3]_{\widetilde{C}}$, $[x_4]_{\widetilde{C}}$, and $[x_8]_{\widetilde{C}}$ belong to the set X_F, hence

$$\overline{\widetilde{C}}X_F = \{x_3\} \cup \{x_4, x_5\} \cup \{x_8\} = \{x_3, x_4, x_5, x_8\}. \tag{3.57}$$

The object x_5 belongs to $\widetilde{C}$-upper approximation of the set X_F, even though it does not belong to X_F, as the object x_4 with identical values of features from set C, and therefore belonging to the same equivalence class, is an element of the set X_F. The set X_N contains elements from classes $[x_2]_{\widetilde{C}}$ and $[x_2]_{\widetilde{C}} = [x_{10}]_{\widetilde{C}}$, hence

$$\overline{\widetilde{C}}X_N = \{x_2\} \cup \{x_9, x_{10}\} = \{x_2, x_9, x_{10}\}. \tag{3.58}$$

The set X_O contains elements from classes $[x_1]_{\widetilde{C}}$, $[x_4]_{\widetilde{C}}$ and $[x_6]_{\widetilde{C}}$, so

$$\overline{\widetilde{C}}X_O = \{x_1\} \cup \{x_4, x_5\} \cup \{x_6, x_7\} = \{x_1, x_4, x_5, x_6, x_7\}. \tag{3.59}$$

Example 3.12
Let us determine $\widetilde{P}_1$ and $\widetilde{P}_2$-upper approximation of the set X defined in Example 3.9. Objects of six equivalence classes of P_1-indiscernibility relation belong to the set X. Therefore the $\widetilde{P}_1$-upper approximation will be their sum, i.e.

$$\overline{\widetilde{P}_1}X = [1]_{\widetilde{P}_1} \cup [2]_{\widetilde{P}_1} \cup [3]_{\widetilde{P}_1} \cup [4]_{\widetilde{P}_1} \cup [5]_{\widetilde{P}_1} \cup [6]_{\widetilde{P}_1} = [1, 7), \tag{3.60}$$

which is illustrated by Fig. 3.8.

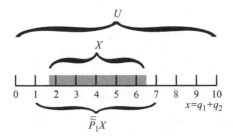

FIGURE 3.8. Upper approximation in one-dimensional universe of discourse

As the elements of all equivalence classes of the P_2-indiscernibility relation belong to the set X, so its $\widetilde{P}_2$-upper approximation is equal to the universe of discourse U, i.e.

$$\overline{\widetilde{P}_2}X = U. \tag{3.61}$$

Example 3.13

Figure 3.9 shows the marked equivalence classes included in the P-upper approximation of the set X described in the space U defined by formula (3.13).

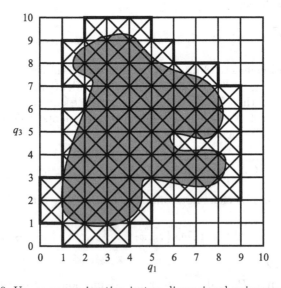

FIGURE 3.9. Upper approximation in two-dimensional universe of discourse

Definition 3.7

$\widetilde{P}$-*positive region* of the set X is defined as

$$\mathrm{Pos}_{\widetilde{P}}(X) = \underline{\widetilde{P}}X. \tag{3.62}$$

The positive region of the set X is equal to its lower approximation.

Definition 3.8

$\tilde{P}$-*boundary region* of the set X is defined as

$$\text{Bn}_{\tilde{P}}(X) = \overline{\tilde{P}}X \setminus \underline{\tilde{P}}X. \tag{3.63}$$

Example 3.14

By directly applying Definition 3.8, we shall find the boundary region of sets X_{F}, X_{N} and X_{O} defined in Example 3.8. We shall perform that by defining the difference of sets described in Examples 3.11 and 3.8. Therefore, we obtain

$$\text{Bn}_{\tilde{C}}(X_{\text{F}}) = \overline{\tilde{C}}X_{\text{F}} \setminus \underline{\tilde{C}}X_{\text{F}} \tag{3.64}$$
$$= \{x_3, x_4, x_5, x_8\} \setminus \{x_3, x_8\} = \{x_4, x_5\},$$

$$\text{Bn}_{\tilde{C}}(X_{\text{N}}) = \overline{\tilde{C}}X_{\text{N}} \setminus \underline{\tilde{C}}X_{\text{N}} = \varnothing, \tag{3.65}$$

$$\text{Bn}_{\tilde{C}}(X_{\text{O}}) = \overline{\tilde{C}}X_{\text{O}} \setminus \underline{\tilde{C}}X_{\text{N}} = \{x_4, x_5\}. \tag{3.66}$$

Example 3.15

Let us define the boundary region of set X defined in Example 3.9 for the set of features P_1 and P_2. In the first case, we have

$$\text{Bn}_{\tilde{P}_1}(X) = \overline{\tilde{P}_1}X \setminus \underline{\tilde{P}_1}X \tag{3.67}$$
$$= [1; 7) \setminus [2; 6) = [1; 2) \cup [6; 7),$$

which is illustrated by Fig. 3.10. In the second case

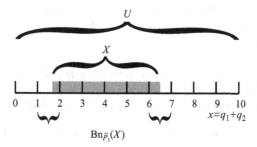

$$\text{Bn}_{\tilde{P}_1}(X)$$

FIGURE 3.10. Boundary region in one-dimensional universe of discourse

$$\text{Bn}_{\tilde{P}_2}(X) = \overline{\tilde{P}_2}X \setminus \underline{\tilde{P}_2}X \tag{3.68}$$
$$= U \setminus \varnothing = U.$$

Example 3.16

Figure 3.11 shows the marked equivalence classes included in the boundary region $\mathrm{Bn}_{\widetilde{P}}(X)$ described in the space U defined by formula (3.13).

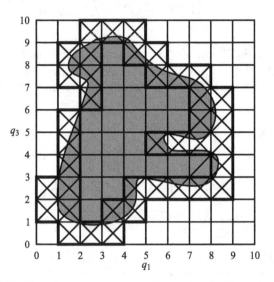

FIGURE 3.11. Boundary region in two-dimensional universe of discourse

Definition 3.9

$\widetilde{P}$-*negative region* of the set X is defined as

$$\mathrm{Neg}_{\widetilde{P}}(X) = U \setminus \overline{\widetilde{P}}X. \tag{3.69}$$

The negative region of the set X is the set of the objects $x \in U$, with relation to which, on the basis of values of features P, we can certainly state that they are not elements of the set X.

Example 3.17

According to Definition 3.9, we shall define the negative regions of the sets X_{F}, X_{N} and X_{O} considered in Example 3.8. By defining the complement of sets defined in Example 3.11 to the space U, we shall obtain

$$\mathrm{Neg}_{\widetilde{C}}(X_{\mathrm{F}}) = U \setminus \overline{\widetilde{C}}X = \{x_1, x_2, x_6, x_7, x_9, x_{10}\}, \tag{3.70}$$

$$\mathrm{Neg}_{\widetilde{C}}(X_{\mathrm{N}}) = U \setminus \overline{\widetilde{C}}X = \{x_1, x_3, x_4, x_5, x_6, x_7, x_8\}, \tag{3.71}$$

$$\mathrm{Neg}_{\widetilde{C}}(X_{\mathrm{O}}) = U \setminus \overline{\widetilde{C}}X = \{x_2, x_3, x_8, x_9, x_{10}\}. \tag{3.72}$$

Example 3.18

Let us define the boundary region of set X defined in Example 3.9 for the set of features P_1 and P_2. In the first case, we have

$$\mathrm{Neg}_{\widetilde{P}_1}(X) = U \setminus \overline{\widetilde{P}}_1 X \tag{3.73}$$
$$= U \setminus [1; 7) = [0; 1) \cup [7; 10),$$

which is illustrated by Fig. 3.12. In the second case, we have

$$\text{Neg}_{\widetilde{P}_2}(X) = U \setminus \overline{\widetilde{P}}_2 X \qquad (3.74)$$
$$= U \setminus U = \varnothing.$$

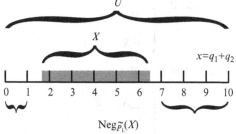

FIGURE 3.12. Negative region in one-dimensional universe of discourse

Example 3.19
Figure 3.13 shows the marked equivalence classes included in the negative region $\text{Neg}_{\widetilde{P}}(X)$, determined in the space U defined by formula (3.13).

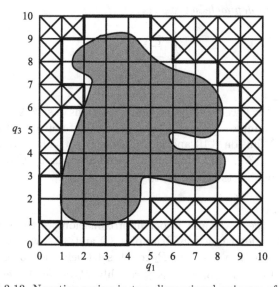

FIGURE 3.13. Negative region in two-dimensional universe of discourse

Definition 3.10
The set X is called a $\widetilde{P}$-*exactly set*, if its lower and upper approximation are equal

$$\underline{\widetilde{P}}X = \overline{\widetilde{P}}X \qquad (3.75)$$

and $\widetilde{P}$-rough set otherwise

$$\underline{\widetilde{P}}X \neq \overline{\widetilde{P}}X. \tag{3.76}$$

Definition 3.11
The set X is called
a) roughly $\widetilde{P}$-definable set, if

$$\begin{cases} \underline{\widetilde{P}}X \neq \varnothing \\ \overline{\widetilde{P}}X \neq U, \end{cases} \tag{3.77}$$

b) internally $\widetilde{P}$-non definable set, if

$$\begin{cases} \underline{\widetilde{P}}X = \varnothing \\ \overline{\widetilde{P}}X \neq U, \end{cases} \tag{3.78}$$

c) externally $\widetilde{P}$-non definable set, if

$$\begin{cases} \underline{\widetilde{P}}X \neq \varnothing \\ \overline{\widetilde{P}}X = U, \end{cases} \tag{3.79}$$

d) totally $\widetilde{P}$-non definable set, if

$$\begin{cases} \underline{\widetilde{P}}X = \varnothing \\ \overline{\widetilde{P}}X = U. \end{cases} \tag{3.80}$$

Example 3.20
By comparing the lower and upper approximations of sets X_F, X_N and X_O, described in Examples 3.8 and 3.11, we can easily notice that only the set X_N satisfies Definition 3.10 and is a $\widetilde{C}$-exactly set. The sets X_F and X_O satisfy equation (3.77) in Definition 3.11 and are roughly $\widetilde{C}$-definable sets, as well as $\widetilde{C}$-rough sets according to Definition 3.10.

Example 3.21
By comparing the lower and upper approximations of the set X defined in Example 3.9, determined in Examples 3.9 and 3.12, we can easily state that this set is both a $\widetilde{P}_1$- and $\widetilde{P}_2$-rough set (Definition 3.10). Moreover, according to Definition 3.11, it is a roughly $\widetilde{P}_1$-definable set and at the same time a totally $\widetilde{P}_2$-non definable set.

Example 3.22
By analyzing Figs. 3.7 and 3.9, we can state that the set X, defined in Example 3.10 and shown in Fig. 3.7, is a $\widetilde{P}$-rough set and at the same time a $\widetilde{P}$-definable set.

Definition 3.12

The value expressed by formula

$$\mu_{\tilde{P}}(X) = \frac{\overline{\overline{\underline{\underline{\tilde{P}X}}}}}{\overline{\overline{\tilde{P}X}}} \tag{3.81}$$

is called $\tilde{P}$-*accuracy* of approximation of the set X. The symbol $\overline{A}$ denotes the measure of the set A. In case of finite sets, we can use the cardinality as the measure, in case of continuous bounded sets, we can use such measures as the length of the interval, surface area, volume, etc.

Example 3.23

Let us determine $\tilde{C}$-accuracy of sets X_F, X_N and X_O, defined in Example 3.8. By applying formula (3.81), we obtain

$$\mu_{\tilde{C}}(X_F) = \frac{\overline{\overline{\underline{\underline{\tilde{C}X_F}}}}}{\overline{\overline{\underline{\underline{\tilde{C}X_F}}}}} = \frac{2}{4} = 0.5, \tag{3.82}$$

$$\mu_{\tilde{C}}(X_N) = \frac{\overline{\overline{\underline{\underline{\tilde{C}X_N}}}}}{\overline{\overline{\underline{\underline{\tilde{C}X_N}}}}} = \frac{3}{3} = 1, \tag{3.83}$$

$$\mu_{\tilde{C}}(X_O) = \frac{\overline{\overline{\underline{\underline{\tilde{C}X_O}}}}}{\overline{\overline{\underline{\underline{\tilde{C}X_O}}}}} = \frac{3}{5} = 0.6. \tag{3.84}$$

As we can see, $\tilde{C}$-accuracy of the approximation of the set X_N is 1, which confirms the previous observation that this set is a $\tilde{C}$-exact set. In other words, it is unambiguously defined by the features belonging to the set C given by formula (3.23).

Example 3.24

In case of continuous spaces of discourses, we can define the $\tilde{C}$-accuracy, using the length of appropriate intervals. Therefore, for the set X defined in Example 3.9, we have

$$\mu_{\tilde{P}_1}(X) = \frac{\overline{\overline{\underline{\underline{\tilde{P}_1X}}}}}{\overline{\overline{\underline{\underline{\tilde{P}_1X}}}}} = \frac{3}{3}, \tag{3.85}$$

$$\mu_{\tilde{P}_2}(X) = \frac{\overline{\overline{\underline{\underline{\tilde{P}_2X}}}}}{\overline{\overline{\underline{\underline{\tilde{P}_2X}}}}} = 0. \tag{3.86}$$

Example 3.25

In case of the set X, defined in Example 3.10, we can determine the $\tilde{P}$-accuracy of approximation using the surface area, as the measure. Based on Figs. 3.7 and 3.9 we have

$$\mu_{\tilde{P}}(X) = \frac{\overline{\overline{\tilde{P}X}}}{\overline{\overline{\tilde{P}X}}} = \frac{8}{21}. \tag{3.87}$$

3.4 Approximation of family of sets

Definitions 3.5 - 3.9 and 3.12 may be easily generalized for a certain family of sets of the space U. Let us denote the abovementioned family of sets by $X = \{X_1, X_2, ..., X_n\}$.

Definition 3.13

The set $\underline{\tilde{P}}X$ described as follows:

$$\underline{\tilde{P}}X = \left\{\underline{\tilde{P}}X_1, \tilde{P}X_2, ..., PX_n\right\} \tag{3.88}$$

is called $\tilde{P}$-*lower approximation* of the family of sets X.

Example 3.26

Let the elements of family of sets X be the sets X_F, X_N and X_O, defined in Example 3.8. We shall notate this as follows:

$$X = \{X_F, X_N, X_O\} \tag{3.89}$$
$$= \{\{x_3, x_4, x_8\}, \{x_2, x_9, x_{10}\}, \{x_1, x_5, x_6, x_7\}\}.$$

Using the sets determined in Example 3.8, according to Definition 3.13, we can write

$$\underline{\tilde{C}}X = \left\{\underline{\tilde{C}}X_F, \underline{\tilde{C}}X_N, \underline{\tilde{C}}X_O\right\} \tag{3.90}$$
$$= \{\{x_3, x_8\}, \{x_2, x_9, x_{10}\}, \{x_1, x_6, x_7\}\}.$$

Definition 3.14

The set $\overline{\tilde{P}}X$ described as follows:

$$\overline{\tilde{P}}X = \left\{\overline{\tilde{P}}X_1, \overline{\tilde{P}}X_2, ..., \overline{\tilde{P}}X_n\right\} \tag{3.91}$$

is *called $\tilde{P}$-upper approximation of family* of sets X.

Example 3.27
According to Definition 3.14, using the sets determined in Example 3.11, $\tilde{C}$-upper approximation of the family of sets X defined in Example 3.26, will be

$$\overline{\tilde{C}}X = \left\{ \overline{\tilde{C}}X_F, \overline{\tilde{C}}X_N, \overline{\tilde{C}}X_O \right\} \tag{3.92}$$

$$= \left\{ \{x_3, x_4, x_5, x_8\}, \{x_2, x_9, x_{10}\}, \{x_1, x_4, x_5, x_6, x_7\} \right\}.$$

Definition 3.15
$\tilde{P}$-*positive region of family* of the sets X is defined as

$$\mathrm{Pos}_{\tilde{P}}(X) = \bigcup_{X_i \in X} \mathrm{Pos}_{\tilde{P}}(X_i). \tag{3.93}$$

Example 3.28
The $\tilde{C}$-positive region of family of sets X, defined in Example 3.26, we can determine as follows:

$$\mathrm{Pos}_{\tilde{C}}(X) = \mathrm{Pos}_{\tilde{C}}(X_F) \cup \mathrm{Pos}_{\tilde{C}}(X_N) \cup \mathrm{Pos}_{\tilde{C}}(X_O) \tag{3.94}$$

$$= \{x_1, x_2, x_3, x_6, x_7, x_8, x_9, x_{10}\}.$$

As it can be inferred from the example, the term of the positive region of family of sets is not equal to the term of its lower approximation – by contrast with the terms of positive region and lower approximation of sets.

Definition 3.16
$\tilde{P}$-*boundary region of family* of the sets X is defined as

$$\mathrm{Bn}_{\tilde{P}}(X) = \bigcup_{X_i \in X} \mathrm{Bn}_{\tilde{P}}(X_i). \tag{3.95}$$

Example 3.29
According to Definition 3.16, $\tilde{C}$-boundary region of family of sets X, defined in Example 3.26, takes the form

$$\mathrm{Bn}_{\tilde{C}}(X) = \mathrm{Bn}_{\tilde{C}}(X_F) \cup \mathrm{Bn}_{\tilde{C}}(X_N) \cup \mathrm{Bn}_{\tilde{C}}(X_O) \tag{3.96}$$
$$= \{x_4, x_5\}.$$

Definition 3.17
$\tilde{P}$-*negative region of family* of the sets X is defined as

$$\mathrm{Neg}_{\tilde{P}}(X) = U \backslash \bigcup_{X_i \in X} \overline{\tilde{P}}X_i. \tag{3.97}$$

Example 3.30
$\widetilde{C}$-negative region of family of sets X, defined in Example 3.26, according to Definition 3.17 takes the form

$$\mathrm{Neg}_{\widetilde{C}}\left(X\right) = U \backslash \bigcup_{X_i c X} \overline{\widetilde{C}X_i} = \varnothing. \tag{3.98}$$

Definition 3.18
$\widetilde{P}$-*quality of approximation of family* of sets X is determined by the expression

$$\gamma_{\widetilde{P}}\left(X\right) = \frac{\overline{\overline{\mathrm{Pos}_{\widetilde{P}}\left(X\right)}}}{\overline{\overline{U}}}. \tag{3.99}$$

Example 3.31
$\widetilde{C}$-quality of approximation of family of sets X, defined in Example 3.26, is

$$\gamma_{\widetilde{C}}\left(X\right) = \frac{\overline{\overline{\mathrm{Pos}_{\widetilde{C}}\left(X\right)}}}{\overline{\overline{U}}} = \frac{8}{10}. \tag{3.100}$$

Definition 3.19
$\widetilde{P}$-*accuracy of approximation of family* of sets X is defined by

$$\beta_{\widetilde{P}}\left(X\right) = \frac{\overline{\overline{\mathrm{Pos}_{\widetilde{P}}\left(X\right)}}}{\sum_{X_i \epsilon X} \overline{\overline{\widetilde{P}X_i}}}. \tag{3.101}$$

Example 3.32
Using Definition 3.19 and notations (3.92) and (3.94), it is easy to check that $\widetilde{C}$-accuracy of approximation of family of sets X, defined in Example 3.26, is

$$\beta_{\widetilde{C}}\left(X\right) = \frac{\overline{\overline{\mathrm{Pos}_{\widetilde{C}}\left(X\right)}}}{\sum_{X_i \epsilon X} \overline{\overline{\widetilde{C}X_i}}} = \frac{8}{4+3+5} = \frac{2}{3}. \tag{3.102}$$

3.5   Analysis of decision tables

The theory of rough sets introduces the notion of dependency between features (attributes) of the information system. Thanks to that, we can check whether it is necessary to know the values of all features in order to unambiguously describe the object belonging to the set U.

Definition 3.20
Dependence degree of set of attributes P_2 on the set of attributes P_1, where $P_1, P_2 \subseteq Q$, is defined as follows:

$$k = \gamma_{\tilde{P}_1} (P_2^*), \tag{3.103}$$

where $\gamma_{\tilde{P}_1} (P_2^*)$ is determined pursuant to Definition 3.18.

The notation $P_1 \xrightarrow{k} P_2$ means that the set of attributes P_2 depends on the set of attributes P_1 to the degree $k < 1$. In case where $k = 1$, we shall simply write $P_1 \rightarrow P_2$.

The notion of dependence degree of attributes is used to define the correctness of construction of the decision table (Definition 3.2).

Definition 3.21
The rules of decision table are called *deterministic*, provided that for each pair of rules $l_a \neq l_b$ from the equality of values of all conditional attributes C, we can infer an equality of values of decision attributes D, i.e.

$$\forall_{\substack{l_a,l_b=1,\ldots,N \\ l_a \neq l_b}} : \forall_{c \in C} \; f_{l_a} (c) = f_{l_b} (c) \rightarrow \forall_{d \in D} \; f_{l_a} (d) = f_{l_b} (d). \tag{3.104}$$

If for a certain pair of rules $l_a \neq l_b$ the above condition is not met, i.e. the equality of values of all conditional attributes C does not result in the equality of values of decision attributes D, we shall call these rules as *non-deterministic*, i.e.

$$\exists_{\substack{l_a,l_b \\ l_a \neq l_b}} : \forall_{c \in C} \; f_{l_a} (c) = f_{l_b} (c) \rightarrow \exists_{d \in D} \; f_{l_a} (d) \neq f_{l_b} (d). \tag{3.105}$$

The decision table (Definition 3.2) is *well defined*, if all its rules are deterministic. Otherwise, we say that it is *not well defined*.

Let us notice that the decision table having a set of conditional attributes C and a set of decision attributes D is well defined, if the set of decision attributes depends on the set of conditional attributes to a degree which is equal to 1 $(C \rightarrow D)$, that is

$$\gamma_{\tilde{C}} (D^*) = 1. \tag{3.106}$$

The reason for the decision table to be not well defined is that it contains the so-called non-deterministic rules. The decision table that is not well defined may be "repaired" by removing the non-deterministic rules or expanding the set of conditional attributes C.

Example 3.33
Let us examine the dependence degree of the set of attributes D on the set of attributes C defined in Example 3.4. According to Definition 3.20, we have

$$k = \gamma_{\tilde{C}} (D^*) = \frac{\overline{\overline{Pos_{\tilde{C}} (D^*)}}}{\overline{\overline{U}}}. \tag{3.107}$$

Let us notice that the equivalence classes of D-indiscernibility relation are the sets X_F, X_N and X_O defined in Example 3.8. Therefore

$$D^* = \{X_F, X_N, X_O\}. \tag{3.108}$$

Based on Definition 3.15, using the lower approximations of sets X_F, X_N and X_O defined in Example 3.8, we obtain

$$\mathrm{Pos}_{\tilde{C}}(D^*) = \mathrm{Pos}_{\tilde{C}}(X_F) \cup \mathrm{Pos}_{\tilde{C}}(X_N) \cup \mathrm{Pos}_{\tilde{C}}(X_O) \tag{3.109}$$

$$= \underline{\tilde{C}}X_F \cup \underline{\tilde{C}}X_N \cup \underline{\tilde{C}}X_O$$

$$= \{x_3, x_8\} \cup \{x_2, x_9, x_{10}\} \cup \{x_1, x_6, x_7\}$$

$$= \{x_1, x_2, x_3, x_6, x_7, x_8, x_9, x_{10}\}.$$

By substituting the dependence (3.109) to formula (3.107), we obtain the result

$$k = \frac{\overline{\{x_1, x_2, x_3, x_6, x_7, x_8, x_9, x_{10}\}}}{\overline{\{x_1, x_2, x_3, x_4, x_5, x_6, x_7, x_8, x_9, x_{10}\}}} = \frac{8}{10}. \tag{3.110}$$

We can therefore state that the set of attributes D depends on the set of attributes C to the degree $k = 0.8$, which is notated as $C \xrightarrow{0.8} D$. The obtained value $k < 1$ informs us that the decision table given in Example 3.4 is not well defined. Based on the set of conditional attributes C, we cannot unambiguously infer on the membership of objects of the space U to the particular sets X_F, X_N and X_O, which are equivalence classes of the D-indiscernibility relation.

The used cars dealer from Example 3.1 should expand the set of conditional attributes C, if he wants to infer unambiguously on the car make on the basis of these attributes.

The non-deterministic rules in the decision table described in Example 3.4 are the rules 4 and 5. If they are removed, a not well-defined decision table (Table 3.3) is transformed into a well defined decision table (Table 3.4).

For the decision table thus defined, it is easy to check that

$$\gamma_{\tilde{C}}(D^*) = \frac{\overline{\{x_1, x_2, x_3, x_6, x_7, x_8, x_9, x_{10}\}}}{\overline{\{x_1, x_2, x_3, x_6, x_7, x_8, x_9, x_{10}\}}} = 1. \tag{3.111}$$

Hence, it is well defined.

TABLE 3.4. A well-defined decision table (after removing non-deterministic rules)

Rule (l)	Number of doors (c_1)	Horsepower (c_2)	Colour (c_3)	Make (d_1)
1	2	60	blue	Opel
2	2	100	black	Nissan
3	2	200	black	Ferrari
6	3	100	red	Opel
7	3	100	red	Opel
8	3	200	black	Ferrari
9	4	100	blue	Nissan
10	4	100	blue	Nissan

The second method used to create a well-defined table is to expand the sets of conditional attributes. The car dealer decided to add the type of fuel used, the type of upholstery and wheel rims, to the features considered so far. The new set of conditional attributes takes the following form

$$C = \{c_1, c_2, c_3, c_4, c_5, c_6\} \tag{3.112}$$
$$= \{\text{number of doors, horsepower, colour, fuel, upholstery, rims}\}.$$

The domains of new features are

$$V_{C_4} = \{\text{Diesel oil, Ethyl gasoline, gas}\}, \tag{3.113}$$
$$V_{C_5} = \{\text{woven fabric, leather}\}, \tag{3.114}$$
$$V_{C_6} = \{\text{steel, aluminium}\}. \tag{3.115}$$

Table 3.5 presents the decision table completed with new attributes and their values.

Let us determine for decision Table 3.5 the $\widetilde{C}$-quality and the $\widetilde{C}$-accuracy of approximation of family of sets D^*. The first step is to define the family of equivalence classes of the relation $\widetilde{C}$ in the space U. Each element of the space U has at least one different value of the feature, hence

$$C^* = \{\{x_1\}, \{x_2\}, \{x_3\}, \{x_4\}, \{x_5\}, \{x_6\}, \{x_7\}, \{x_8\}, \{x_9\}, \{x_{10}\}\}. \tag{3.116}$$

Therefore, the following positive regions of family of sets B^* are defined:

$$\text{Pos}_{\widetilde{C}}(X_F) = \underline{\widetilde{C}}X_F = \{x_3\} \cup \{x_4\} \cup \{x_8\} = \{x_3, x_4, x_8\} = X_F, \tag{3.117}$$

$$\text{Pos}_{\widetilde{C}}(X_N) = \underline{\widetilde{C}}X_N = \{x_2\} \cup \{x_9\} \cup \{x_{10}\} = \{x_2, x_9, x_{10}\} = X_N, \tag{3.118}$$

$$\text{Pos}_{\widetilde{C}}(X_O) = \underline{\widetilde{C}}X_O = \{x_1\} \cup \{x_5\} \cup \{x_6\} \cup \{x_7\} \tag{3.119}$$
$$= \{x_1, x_5, x_6, x_7\} = X_O.$$

TABLE 3.5. A well-defined decision table (after adding attributes)

Rule Number (l)	Number of doors (c_1)	Horsepower (c_2)	Colour (c_3)	Fuel (c_4)	Upholstery (c_5)	Rims c_6	Make (d_1)
1	2	60	blue	Ethyl gasoline	woven fabric	steel	Opel
2	2	100	black	Diesel oil	woven fabric	steel	Nissan
3	2	200	black	Ethyl gasoline	leather	Al	Ferrari
4	2	200	red	Ethyl gasoline	leather	Al	Ferrari
5	2	200	red	Ethyl gasoline	woven fabric	steel	Opel
6	3	100	red	Diesel oil	leather	steel	Opel
7	3	100	red	gas	woven fabric	steel	Opel
8	3	200	black	Ethyl gasoline	leather	Al	Ferrari
9	4	100	blue	gas	woven fabric	steel	Nissan
10	4	100	blue	Diesel oil	woven fabric	Al	Nissan

The $\widetilde{C}$-upper approximations of these sets may be defined similarly

$$\overline{\widetilde{C}}X_{\mathrm{F}} = \{x_3\} \cup \{x_4\} \cup \{x_8\} = \{x_3, x_4, x_8\} = X_{\mathrm{F}}, \quad (3.120)$$

$$\overline{\widetilde{C}}X_{\mathrm{N}} = \{x_2\} \cup \{x_9\} \cup \{x_{10}\} = \{x_2, x_9, x_{10}\} = X_{\mathrm{N}}, \quad (3.121)$$

$$\overline{\widetilde{C}}X_{\mathrm{O}} = \{x_1\} \cup \{x_5\} \cup \{x_6\} \cup \{x_7\} = \{x_1, x_5, x_6, x_7\} = X_{\mathrm{O}}. \quad (3.122)$$

$\widetilde{C}$-positive region of family of sets D^* takes the form

$$\mathrm{Pos}_{\widetilde{C}}(D^*) = \{x_3, x_4, x_8\} \cup \{x_2, x_9, x_{10}\} \cup \{x_1, x_5, x_6, x_7\} \quad (3.123)$$
$$= \{x_1, x_2, x_3, x_4, x_5, x_6, x_7, x_8 x_9, x_{10}\} = U.$$

Now, using Definition 3.18 and 3.19, we can directly determine:

$$\gamma_{\widetilde{C}}(D^*) = \frac{\overline{\overline{\mathrm{Pos}_{\widetilde{C}}(D^*)}}}{\overline{\overline{U}}} = \frac{\overline{\overline{U}}}{\overline{\overline{U}}} = \frac{10}{10} = 1, \quad (3.124)$$

$$\beta_{\widetilde{C}}(D^*) = \frac{\overline{\overline{\mathrm{Pos}_{\widetilde{C}}(D^*)}}}{\sum_{X_i \in D^*} \overline{\widetilde{C}X_i}} = \frac{10}{3+3+4} = 1. \quad (3.125)$$

On this basis, we can unambiguously state that the decision Table 3.5 is well defined.

Definition 3.22
The set of attributes $P_1 \subseteq Q$ is *independent* in a given information system, if for each $P_2 \subset P_1$ the inequality $\widetilde{P}_1 \neq \widetilde{P}_2$ occurs. Otherwise, the set P_1 is a *dependent one*.

Example 3.34
Let us consider data given in decision Table 3.5. It is easy to notice that the set C described by formula (3.112) is a dependent set. Exemplary subsets of the set C, $C_1 = \{c_1, c_2, c_3, c_5, c_6\}$ and $C_2 = \{c_1, c_2, c_3, c_4, c_5\}$, generate such quotient of the space U, as the set C (see. 3.116). The sets C_1 and C_2 are also dependent, as the sets $C_3 = \{c_1, c_3, c_5, c_6\} \subset C_1$ and $C_4 = \{c_1, c_3, c_4, c_5\} \subset C_2$ also generate the quotient of the space U described by formula (3.116). On the other hand, the sets C_3 and C_4 are independent sets.

Definition 3.23
The set of attributes $P_1 \subseteq Q$ is *independent with respect to* the set of attributes $P_2 \subseteq Q$ (P_2-*independent*), if for each $P_3 \subset P_1$ the following inequality holds

$$\text{Pos}_{\widetilde{P}_1}(P_2^*) \neq \text{Pos}_{\widetilde{P}_3}(P_2^*). \tag{3.126}$$

Otherwise, the set P_1 is P_2-*dependent*.

Example 3.35
The set C_3 is an independent set in a given information system (Definition 3.22). According to Definition 3.23, it is a D-dependent set. We shall demonstrate that the set C_3 together with its subset $C_5 = \{c_1, c_3, c_6\}$ does not meet condition (3.126), i.e.

$$\text{Pos}_{\widetilde{C}_3}(D^*) \neq \text{Pos}_{\widetilde{C}_5}(D^*). \tag{3.127}$$

Let us notice that

$$\begin{aligned}\text{Pos}_{\widetilde{C}_3}(D^*) &= \text{Pos}_{\widetilde{C}_3}(X_F) \cup \text{Pos}_{\widetilde{C}_3}(X_N) \cup \text{Pos}_{\widetilde{C}_3}(X_O) \\ &= X_F \cup X_N \cup X_O = U\end{aligned} \tag{3.128}$$

and

$$\begin{aligned}\text{Pos}_{\widetilde{C}_5}(D^*) &= \text{Pos}_{\widetilde{C}_5}(X_F) \cup \text{Pos}_{\widetilde{C}_5}(X_N) \cup \text{Pos}_{\widetilde{C}_5}(X_O) \\ &= X_F \cup X_N \cup X_O = U\end{aligned} \tag{3.129}$$

hence

$$\text{Pos}_{\widetilde{C}_3}(D^*) = \text{Pos}_{\widetilde{C}_5}(D^*). \tag{3.130}$$

Definition 3.24
Every independent set $P_2 \subset P_1$ for which $\tilde{P}_2 = \tilde{P}_1$ is called *the reduct* of the set of attributes $P_1 \subseteq Q$.

Definition 3.25
Every P_2-independent set $P_3 \subset P_1$ for which $\tilde{P}_3 = \tilde{P}_1$ is called *the relative reduct* of a set of attributes $P_1 \subseteq Q$ with respect to P_2 (the so-called P_2-reduct).

Example 3.36
If we return to the discussions in Examples 3.34 and 3.35; we can notice that the sets C_3 and C_4 presented therein are the reducts of the set C, whereas the set C_5 is the D-reduct of the set C_3 and C.

Definition 3.26
The attribute $p \in P_1$ is *indispensable from* P_1, if for $P_2 = P_1 \setminus \{p\}$, the equation $\tilde{P}_2 \neq \tilde{P}_1$ holds. Otherwise, the attribute p is *dispensable*.

Example 3.37
Using Definition 3.26, we shall check the indispensability of particular attributes $c \in C$ in the information system forming the decision Table 3.5. It is easy to check that

$$C^* = \{\{x_1\}, \{x_2\}, \{x_3\}, \{x_4\}, \{x_5\}, \tag{3.131}$$
$$\{x_6\}, \{x_7\}, \{x_8\}, \{x_9\}, \{x_{10}\}\},$$
$$(C \setminus \{c_1\})^* = \{\{x_1\}, \{x_2\}, \{x_3, x_8\}, \{x_4\}, \{x_5\}, \tag{3.132}$$
$$\{x_6\}, \{x_7\}, \{x_9\}, \{x_{10}\}\} \neq C^*,$$
$$(C \setminus \{c_2\})^* = \{\{x_1\}, \{x_2\}, \{x_3\}, \{x_4\}, \{x_5\}, \tag{3.133}$$
$$\{x_6\}, \{x_7\}, \{x_8\}, \{x_9\}, \{x_{10}\}\} = C^*,$$
$$(C \setminus \{c_3\})^* = \{\{x_1\}, \{x_2\}, \{x_3, x_4\}, \{x_5\}, \tag{3.134}$$
$$\{x_6\}, \{x_7\}, \{x_8\}, \{x_9\}, \{x_{10}\}\} \neq C^*,$$
$$(C \setminus \{c_4\})^* = \{\{x_1\}, \{x_2\}, \{x_3\}, \{x_4\}, \{x_5\}, \tag{3.135}$$
$$\{x_6\}, \{x_7\}, \{x_8\}, \{x_9\}, \{x_{10}\}\} = C^*,$$
$$(C \setminus \{c_5\})^* = \{\{x_1\}, \{x_2\}, \{x_3\}, \{x_4\}, \{x_5\}, \tag{3.136}$$
$$\{x_6\}, \{x_7\}, \{x_8\}, \{x_9\}, \{x_{10}\}\} = C^*,$$
$$(C \setminus \{c_6\})^* = \{\{x_1\}, \{x_2\}, \{x_3\}, \{x_4\}, \{x_5\}, \tag{3.137}$$
$$\{x_6\}, \{x_7\}, \{x_8\}, \{x_9\}, \{x_{10}\}\} = C^*.$$

As we can see, the attributes c_1 and c_3 are indispensable, while the attributes c_2, c_4, c_5 and c_6 are superfluous.

Definition 3.27
The set of all indispensable attributes from the set P is called a *core* of P, which is notated as follows:

$$\text{CORE}(P) = \left\{ p \in P : \tilde{P'} \neq \tilde{P}, P' = P \backslash \{p\} \right\}. \qquad (3.138)$$

Example 3.38
Using the results from Example 3.37, we can define the core of the set of attributes C as

$$\text{CORE}(C) = \{c_1, c_3\}. \qquad (3.139)$$

Definition 3.28
The normalized coefficient of significance of subset of the set of conditional attributes $C' \subset C$ is expressed by the following formula

$$\sigma_{(C,D)}(C') = \frac{\gamma_{\tilde{C}}(D^*) - \gamma_{\tilde{C}''}(D^*)}{\gamma_{\tilde{C}}(D^*)}, \qquad (3.140)$$

where $C'' = C \setminus C'$. Of course, in a special case the set C' may be a one-element set, then the coefficient (3.140) will express the significance of one conditional attribute.

The coefficient of significance plays an important role in the analysis of decision tables. The zero value obtained for a given subset of conditional attributes C indicates that this subset may be deleted from the set of conditional attributes without any detriment to the approximation of family of sets D^*.

Example 3.39
Let us determine the significance of an exemplary subset of the set of conditional attributes C defined by the notation (3.112). In Example 3.33, we have demonstrated (formula (3.124)), that $\tilde{C}$-quality of approximation of family of sets D^* for a well-defined decision table amounts to 1. For $C' = \{c_1\}$, we have $C'' = \{c_2, c_3, c_4, c_5, c_6\}$ and

$$\gamma_{\tilde{C}''}(D^*) = \frac{\overline{\overline{\text{Pos}_{\tilde{C}''}(D^*)}}}{\overline{\overline{U}}} \qquad (3.141)$$

$$= \frac{\overline{\overline{X_F \cup X_N \cup X_O}}}{\overline{\overline{U}}} = 1.$$

Hence

$$\sigma_{(C,D)}(\{c_1\}) = \frac{1-1}{1} = 0. \qquad (3.142)$$

Therefore, the attribute c_1 in the given decision table is insignificant, and due to that, its removal will not impact the quality of approximation of family of sets D^*.

For $C' = \{c_4, c_5, c_6\}$, we get $C'' = \{c_1, c_2, c_3\}$, hence

$$\gamma_{\tilde{C}''}(D^*) = \frac{\overline{\{x_3, x_8\} \cup \{x_2, x_9, x_{10}\} \cup \{x_1, x_6, x_7\}}}{\overline{\overline{U}}} = \frac{8}{10}, \qquad (3.143)$$

which, after substituting to formula (3.140), gives the value

$$\sigma_{(C,D)}(\{c_4, c_5, c_6\}) = \frac{1 - 0.8}{1} = 0.2. \qquad (3.144)$$

Based on the above discussion we see the attributes c_4, c_5 and c_6 added in Example 3.33 (Table 3.5) are of low significance.

Definition 3.29
Any given subset of the set of conditional attributes $C' \subset C$ is called a rough D-reduct of the set of attributes C, and *the approximation error* of this reduct is defined as follows:

$$\varepsilon_{(C,D)}(C') = \frac{\gamma_{\tilde{C}}(D^*) - \gamma_{\tilde{C}'}(D^*)}{\gamma_{\tilde{C}}(D^*)}. \qquad (3.145)$$

Example 3.40
Let us determine an approximation error of the set $C' = \{c_1, c_2, c_3\}$ which is the rough D-reduct of set of attributes C (decision Table 3.5). Using the result (3.143), we have

$$\varepsilon_{(C,D)}(\{c_1, c_2, c_3\}) = \frac{1 - 0.8}{1} = 0.2. \qquad (3.146)$$

3.6   Application of LERS software

LERS (*Learning from Examples based on Rough Sets*) software [67] has been created by RS Systems company. Its task is to generate the rule base, based on examples entered and to test the rule base generated or prepared independently. The data entered may be subject to some initial processing, among others by removing contradictions, eliminating or completing missing data and quantization of numerical values.

In order to present the capabilities of LERS software, let us consider two cases of data analysis. The first case, already discussed in the Example 3.1 – it is the case of the used car dealer. The second case – the problem of classification of Iris flowers, an example often used to illustrate and compare the performance of computational intelligence algorithms.

Example 3.41 (Cars in the parking lot)
Let the decision table describing the used car dealer from Example 3.1 have the form as in Table 3.5. In order to have a clear presentation of this example, it has been presented again in Table 3.6.

TABLE 3.6. Original decision table (before reduction)

Number of doors (c_1)	Horsepower (c_2)	Colour (c_3)	Fuel (c_4)	Upholstery (c_5)	Rims (c_6)	Make (d_1)
2	60	blue	Ethyl gasoline	woven fabric	steel	Opel
2	100	black	Diesel oil	woven fabric	steel	Nissan
2	200	black	Ethyl gasoline	leather	Al	Ferrari
2	200	red	Ethyl gasoline	leather	Al	Ferrari
2	200	red	Ethyl gasoline	woven fabric	steel	Opel
3	100	red	Diesel oil	leather	steel	Opel
3	100	red	gas	woven fabric	steel	Opel
3	200	black	Ethyl gasoline	leather	Al	Ferrari
4	100	blue	gas	woven fabric	steel	Nissan
4	100	blue	Diesel oil	woven fabric	Al	Nissan

In order to enter the data from Table 3.6 to LERS software, the following file must be prepared:

```
< a, a, a a a a d >
[doors horsepower]   colour   fuel              upholstery      rims    make
2        60          blue     Ethyl gasoline    woven fabric    steel   Opel
2        100         black    Diesel oil        woven fabric    steel   Nissan
2        200         black    Ethyl gasoline    leather         alum    Ferrari
2        200         red      Ethyl gasoline    leather         alum    Ferrari
2        200         red      Ethyl gasoline    woven fabric    steel   Opel
3        100         red      Diesel oil        leather         steel   Opel
3        100         red      gas               woven fabric    steel   Opel
3        200         black    Ethyl gasoline    leather         alum    Ferrari
4        100         blue     gas               woven fabric    steel   Nissan
4        100         blue     Diesel oil        woven fabric    alum    Nissan
```

In the first row of the file, the division to conditional attributes (a) and decision attributes (d) has been made. The second row contains the names of particular attributes. Based on data entered, LERS software generated 5 rules containing 8 conditions in total:

IF rims is steel **AND** colour is red **THEN** make is Opel
IF horsepower is 60 **THEN** make is Opel
IF doors is 4 **THEN** make is Nissan
IF colour is black **AND** fuel is Diesel oil **THEN** make is Nissan
IF horsepower is 200 **AND** upholstery is leather **THEN** make is Ferrari

TABLE 3.7. Decision table after removing redundant data

Number of doors (c_1)	Horsepower (c_2)	Colour (c_3)	Fuel (c_4)	Upholstery (c_5)	Rims (c_6)	Make (d_1)
	60					Opel
		black	Diesel oil			Nissan
	200			leather		Ferrari
	200			leather		Ferrari
		red			steel	Opel
		red			steel	Opel
		red			steel	Opel
	200			leather		Ferrari
4						Nissan
4						Nissan

The process of rules generation may be interpreted as removing redundant data from the decision table, which is shown in Table 3.7. The algorithm used for rules generation and removal of redundant data uses the rough sets theory.

By removing repeating entries from 3.7, we obtain Table 3.8, identical with the generated set of rules.

TABLE 3.8. Decision table obtained after reduction

Number of doors (c_1)	Horsepower (c_2)	Colour (c_3)	Fuel (c_4)	Upholstery (c_5)	Rims	Make (d_4)
		red			steel	Opel
	60					Opel
4						Nissan
		black	Diesel oil			Nissan
	200			leather		Ferrari

Example 3.42 (Classification of Iris flowers)

As it has been mentioned before, the problem of classification of Iris flowers is often used as an example to illustrate the performance of different types of computational intelligence algorithms. The task consists in determining the membership of flowers to one of three classes: Setosa, Virginica and Versicolor. The decision is made based on the value of four conditional attributes describing the dimensions (length and width) of the leaf and the flower petal.

We have 150 samples in our disposal, including 147 unique ones (not recurrent); 50 of them belongs to each of three classes. Table 3.9 presents the ranges of variability of particular attributes.

TABLE 3.9. Ranges of variability of attributes (classification of Iris flowers)

Attribute	Range	Number of unique values
p1	$\langle 4.3; 7.9 \rangle$	35
p2	$\langle 2.0; 4.4 \rangle$	23
p3	$\langle 1.0; 6.9 \rangle$	43
p4	$\langle 0.1; 2.5 \rangle$	22
Iris	Setosa, Virginica, Versicolor	3

The data have been divided into a learning and a testing part; 40 samples from each class have been selected randomly for the learning part and the remaining 30 samples have been used to create the testing part. Based on the contents of the learning sequence, the input file for LERS software has been prepared in the form:

```
< a  a  a  a  d >
[p1  p2  p3  p4 iris]
 4.4 2.9 1.4 0.2 Setosa
 4.8 3.0 1.4 0.1 Setosa
 5.4 3.4 1.7 0.2 Setosa
 ...
```

Based on data entered, LERS generated 34 rules containing 41 conditions in total:

IF p4 is 0.2 THEN iris is Setosa
IF p4 is 0.4 THEN iris is Setosa
IF p4 is 0.3 THEN iris is Setosa
IF p4 is 0.1 THEN iris is Setosa
IF p4 is 0.5 THEN iris is Setosa
IF p4 is 0.6 THEN iris is Setosa
IF p4 is 1.3 THEN iris is Versicolor
IF p4 is 1.5 AND p3 is 4.5 THEN iris is Versicolor

IF p4 is 1.0 THEN iris is Versicolor
IF p4 is 1.4 THEN iris is Versicolor
IF p4 is 1.5 AND p3 is 4.9 THEN iris is Versicolor
IF p4 is 1.2 THEN iris is Versicolor
IF p4 is 1.5 AND p1 is 5.9 THEN iris is Versicolor
IF p3 is 4.7 THEN iris is Versicolor
IF p4 is 1.1 THEN iris is Versicolor
IF p3 is 4.8 AND p1 is 5.9 THEN iris is Versicolor
IF p3 is 4.6 THEN iris is Versicolor
IF p4 is 1.6 AND p1 is 6.0 THEN iris is Versicolor
IF p4 is 2.1 THEN iris is Virginica
IF p4 is 2.3 THEN iris is Virginica
IF p3 is 5.5 THEN iris is Virginica
IF p4 is 2.0 THEN iris is Virginica
IF p1 is 7.3 THEN iris is Virginica
IF p3 is 6.0 THEN iris is Virginica
IF p3 is 5.1 THEN iris is Virginica
IF p3 is 5.8 THEN iris is Virginica
IF p3 is 6.1 THEN iris is Virginica
IF p4 is 2.4 THEN iris is Virginica
IF p4 is 1.8 AND p1 is 6.2 THEN iris is Virginica
IF p4 is 1.7 THEN iris is Virginica
IF p3 is 6.7 THEN iris is Virginica
IF p3 is 5.0 THEN iris is Virginica
IF p3 is 5.7 THEN iris is Virginica
IF p3 is 4.9 AND p2 is 2.7 THEN iris is Virginica

In the next step, data included in the learning sequence have been quantized so that the corresponding decision table remained still deterministic. LERS software defined the intervals given in Table 3.10 for particular conditional attributes.

The original input file has been replaced with the file presented below. Each value of the decision attribute has been replaced with an interval identifier it belongs to.

```
! Decision table produced by LERS (C version 1.0)
! First the attribute names list ...
!
[p1  p2  p3  p4  iris]
!
! Now comes the actual data. Please note that one example
! does NOT necessarily occupy one physical line
!
4.4..5.05  2.75..2.95  1..2.6  0.1..0.8  Setosa
4.4..5.05  2.95..3.05  1..2.6  0.1..0.8  Setosa
5.05..5.65  3.25..3.45  1..2.6  0.1..0.8  Setosa
...
```

TABLE 3.10. Result of quantization (classification of iris flowers)

Attribute	Range	Number of samples
	⟨4.4; 5.05⟩	26
	⟨5.05; 5.65⟩	26
p1	⟨5.65; 6.15⟩	22
	⟨6.15; 6.65⟩	22
	⟨6.65; 7.9⟩	24
	⟨2; 2.75⟩	24
	⟨2.75; 2.95⟩	18
p2	⟨2.95; 3.05⟩	21
	⟨3.05; 3.25⟩	21
	⟨3.25; 3.45⟩	17
	⟨3.45; 4.4⟩	19
	⟨1; 2.6⟩	40
p3	⟨2.6; 4.85⟩	40
	⟨4.85; 6.9⟩	40
	⟨0.1; 0.8⟩	40
p4	⟨0.8; 1.65⟩	42
	⟨1.65; 2.5⟩	38
	Setosa	40
Iris	Virginica	40
	Versicolor	40

Based on the file so prepared, LERS software generated 11 rules containing altogether 41 conditions:

IF p3 is > **THEN** iris is Setosa
IF p4 is <0.8; 1.65> **AND** p3 is <2.6; 4.85> **THEN** iris is Versicolor
IF p2 is <3.05; 3.25> **AND** p1 is <5.65; 6.15> **THEN** iris is Versicolor
IF p4 is <0.8; 1.65> **AND** p2 is <3.05; 3.25> **THEN** iris is Versicolor
IF p1 is <6.15; 6.65> **AND** p2 is <2; 2.75> **AND** p4 is <0.8; 1.65>
THEN iris is Versicolor
IF p3 is <4.85; 6.9> **AND** p4 is <1.65; 2.5> **THEN** iris is Virginica
IF p3 is <4.85; 6.9> **AND** p2 is <2.75; 2.95> **THEN** iris is Virginica
IF p1 is <5.65; 6.15> **AND** p3 is <4.85; 6.9> **THEN** iris is Virginica
IF p4 is <1.65; 2.5> **AND** p2 is <2.75; 2.95> **THEN** iris is Virginica
IF p2 is <2.95; 3.05> **AND** p1 is <6.65; 7.9> **THEN** iris is Virginica
IF p2 is <2; 2.75> **AND** p4 is <1.65; 2.5> **THEN** iris is Virginica

The rules obtained in the first and in the second trial have been used to classify the samples included in the testing set. The results of both experiments have been presented in Table 3.11. The first four columns contain the values of conditional attributes for test samples, the fifth column contains the correct result (decision attribute), the sixth column is the result obtained using the first set of rules, and the seventh column is the result obtained using the second set of rules.

By analyzing Table 3.11, one can notice, among others things, that the initial quantization of data, which resulted in the set of rules operating on intervals, leads to a more efficient inference system.

TABLE 3.11. Results of classification of iris flowers

p1	p2	p3	p4	pattern	classification 1	classification 2
5.0	3.6	1.4	0.2	Setosa	Setosa	Setosa
4.9	3.1	1.5	0.1	Setosa	Setosa	Setosa
4.3	3.0	1.1	0.1	Setosa	Setosa	Setosa
5.0	3.0	1.6	0.2	Setosa	Setosa	Setosa
5.5	4.2	1.4	0.2	Setosa	Setosa	Setosa
5.1	3.4	1.5	0.2	Setosa	Setosa	Setosa
5.1	3.8	1.5	0.3	Setosa	Setosa	Setosa
5.1	3.5	1.4	0.3	Setosa	Setosa	Setosa
4.6	3.1	1.5	0.2	Setosa	Setosa	Setosa
5.1	3.8	1.9	0.4	Setosa	Setosa	Setosa
5.1	2.5	3.0	1.1	Versicolor	Versicolor	Versicolor
6.1	2.8	4.7	1.2	Versicolor	Versicolor	Versicolor
6.0	2.7	5.1	1.6	Versicolor	???	Virginica
5.5	2.4	3.8	1.1	Versicolor	Versicolor	Versicolor
4.9	2.4	3.3	1.0	Versicolor	Versicolor	Versicolor
6.7	3.0	5.0	1.7	Versicolor	Virginica	Virginica
6.2	2.2	4.5	1.5	Versicolor	Versicolor	Versicolor
6.8	2.8	4.8	1.4	Versicolor	Versicolor	Versicolor
5.7	2.8	4.5	1.3	Versicolor	Versicolor	Versicolor
5.8	2.6	4.0	1.2	Versicolor	Versicolor	Versicolor
6.3	2.5	5.0	1.9	Virginica	Virginica	Virginica
6.1	3.0	4.9	1.8	Virginica	???	Virginica
6.3	2.9	5.6	1.8	Virginica	???	Virginica
6.7	3.1	5.6	2.4	Virginica	Virginica	Virginica
5.8	2.8	5.1	2.4	Virginica	Virginica	Virginica
6.1	2.6	5.6	1.4	Virginica	Versicolor	Virginica
6.4	2.7	5.3	1.9	Virginica	???	Virginica
6.9	3.1	5.4	2.1	Virginica	Virginica	Virginica
6.0	3.0	4.8	1.8	Virginica	???	???
6.4	2.8	5.6	2.2	Virginica	???	Virginica

3.7   Notes

The theory of rough sets was created by professor Zdzisław Pawlak [161 - 164]. The definitions provided in this chapter, as well as various applications of rough sets, are presented in a monograph [140], which is the first more comprehensive study on this subject in the Polish language. We refer the Reader interested in various aspects of rough sets to a rich set of publications [66, 67, 158, 177, 180, 233]. In Section 3.6, the LERS software has been used to generate the rules using the rough sets method. This software has been kindly made available for the purposes of this publication by professor Jerzy Grzymała-Busse of Kansas University, USA.

4
Methods of knowledge representation using type-1 fuzzy sets

4.1 Introduction

In everyday life, we come across phenomena and notions, the nature of which is ambiguous and imprecise. Using the classical theory of sets and bivalent logic, we are unable to formally describe such phenomena and notions. We are supported by the fuzzy sets theory, which in the last dozen of years has found many interesting applications.

In this chapter, we shall present, in a Reader friendly manner, the basic terms and definitions of fuzzy sets theory (points 4.2 – 4.7). We shall then discuss the issues of approximate reasoning, i.e. the reasoning on the basis of fuzzy antecedents (point 4.8). The next point relates to the problem of construction of fuzzy inference systems (point 4.9). The chapter is finalized by some examples of application of fuzzy sets in the issues of forecasting, planning and decision making.

4.2 Basic terms and definitions of fuzzy sets theory

Using the fuzzy sets, we can formally define imprecise and ambiguous notions, such as "high temperature", "young man", "average height" or "large city". Before providing the definition of a fuzzy set, we must determine the so-called *universe of discourse*. In case of an ambiguous term "a lot of money", a different sum will be considered to be large (in USD) if we limit the universe of discourse to [0; 1000], and a different one – if we assume the interval of $[0; 1000000]$. The universe of discourse, will be denoted by the letter $\mathbf{X}$. Let us remember that $\mathbf{X}$ is a non-fuzzy set.

Definition 4.1
The fuzzy set A in a given (non-empty) space $\mathbf{X}$, which is denoted as $A \subseteq \mathbf{X}$, is the set of pairs

$$A = \{(x, \mu_A(x)); x \in \mathbf{X}\}, \tag{4.1}$$

in which

$$\mu_A : \mathbf{X} \to [0, 1] \tag{4.2}$$

is the membership function of a fuzzy set A. This function assigns to each element $x \in \mathbf{X}$ its membership degree to the fuzzy set A, and we can distinguish 3 cases:

1) $\mu_A(x) = 1$ means the full membership of element x to the fuzzy set A, i.e. $x \in A$,

2) $\mu_A(x) = 0$ means the lack of membership of element x to the fuzzy set A, i.e. $x \notin A$,

3) $0 < \mu_A(x) < 1$ means a partial membership of element x to the fuzzy set A.

In the literature some authors use symbolic notations of fuzzy sets. If $\mathbf{X}$ is a space with a finite number of elements, $\mathbf{X} = \{x_1, ..., x_n\}$, then the fuzzy set $A \subseteq \mathbf{X}$ shall be notated as

$$A = \frac{\mu_A(x_1)}{x_1} + \frac{\mu_A(x_2)}{x_2} + \ldots + \frac{\mu_A(x_n)}{x_n} = \sum_{i=1}^{n} \frac{\mu_A(x_i)}{x_i}. \tag{4.3}$$

It should be reminded that the elements of $x_i \in \mathbf{X}$ may be not only numbers, but also persons, objects or other notions. Notation (4.3) has a symbolic character. The line of fraction does not symbolize the division, but means the assigning of membership degrees $\mu_A(x_1), \ldots, \mu_A(x_n)$ to particular elements $x_1, \ldots, x_n$. In other words, the notation

$$\frac{\mu_A(x_i)}{x_i} \quad i = 1, ..., n \tag{4.4}$$

shall mean the pair

$$(x_i, \mu_A(x_i)) \quad i = 1, ..., n. \tag{4.5}$$

Similarly, the "+" sign does not mean the addition, but the union of sets (4.5). It is worth noting that the non-fuzzy sets may be notated symbolically in a similar convention. For example, the set of school grades shall be symbolically noted as

$$D = 1 + 2 + 3 + 4 + 5 + 6, \tag{4.6}$$

which is equal to the notation

$$D = \{1, 2, 3, 4, 5, 6\}.\tag{4.7}$$

If $\mathbf{X}$ is a space with an infinite number of elements, then the fuzzy set $A \subseteq \mathbf{X}$ is notated symbolically as

$$A = \int_{\mathbf{X}} \frac{\mu_A(x)}{x}.\tag{4.8}$$

Example 4.1
Let us assume that $\mathbf{X} = \mathbf{N}$ is a set of natural numbers. We shall define the term of a natural numbers set "close to number 7". This can be achieved by defining the following fuzzy set $A \subseteq X$:

$$A = \frac{0.2}{4} + \frac{0.5}{5} + \frac{0.8}{6} + \frac{1}{7} + \frac{0.8}{8} + \frac{0.5}{9} + \frac{0.2}{10}.\tag{4.9}$$

Example 4.2
If $\mathbf{X} = \mathbf{R}$, where $\mathbf{R}$ is a set of real numbers, then the set of real numbers "close to number 7" shall be defined by the following membership function

$$\mu_A(x) = \frac{1}{1 + (x - 7)^2}.\tag{4.10}$$

Therefore, the fuzzy set of real numbers "close to number 7" shall be notated as

$$A = \int_{\mathbf{X}} \frac{\left[1 + (x - 7)^2\right]^{-1}}{x}.\tag{4.11}$$

Remark 4.1
The fuzzy sets of natural and real numbers "close to number 7" may be notated in many ways. For example, membership function (4.10) may be replaced by the formula

$$\mu_A(x) = \begin{cases} 1 - \sqrt{\dfrac{|x - 7|}{3}}, & \text{if } 4 \leq x \leq 10, \\ 0, & \text{otherwise.} \end{cases}\tag{4.12}$$

Figures 4.1a and 4.1b show two membership functions of the fuzzy set A of real numbers "close to number 7".

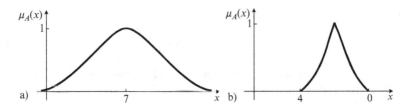

FIGURE 4.1. Illustration to Example 4.2 and Remark 4.1: membership functions of the fuzzy set A of real numbers "close to number 7"

Example 4.3

Let us formalize the imprecise notion "appropriate temperature of water in the Baltic Sea for swimming". Let us define the universe of discourse as the set $X = [15°, ..., 25°]$. The vacationer I, who prefers swimming in the water of $21°$, would define the following fuzzy set:

$$A = \frac{0.1}{16} + \frac{0.3}{17} + \frac{0.5}{18} + \frac{0.8}{19} + \frac{0.95}{20} + \frac{1}{21} + \frac{0.9}{22} \qquad (4.13)$$
$$+ \frac{0.8}{23} + \frac{0.75}{24} + \frac{0.7}{25}.$$

The vacationer II, preferring the temperature of $20°$, would give a different definition of this set

$$B = \frac{0.1}{15} + \frac{0.2}{16} + \frac{0.4}{17} + \frac{0.7}{18} + \frac{0.9}{19} + \frac{1}{20} + \frac{0.9}{21} \qquad (4.14)$$
$$+ \frac{0.85}{22} + \frac{0.8}{23} + \frac{0.75}{24} + \frac{0.7}{25}.$$

Using fuzzy sets A and B, we have formalized the imprecise notion "appropriate temperature of water in the Baltic Sea for bathing".

Remark 4.2

We should stress that the fuzzy sets theory describes the uncertainty in a different sense than the probability theory. Using the probability theory, we may define, for instance, the probability of casting 4, 5 or 6 while tossing the dice. Of course, this probability is 0.5. On the other hand, using fuzzy sets, we may describe the imprecise notion "casting a large number of pips". The appropriate fuzzy set may take the form

$$A = \frac{0.6}{4} + \frac{0.8}{5} + \frac{1}{6}$$

or

$$A = \frac{0.1}{3} + \frac{0.5}{4} + \frac{0.85}{5} + \frac{1}{6}.$$

The only similarity between the fuzzy sets theory and the probability theory is the fact that both the fuzzy set membership function and the probability take the values in the interval $[0, 1]$.

In some applications, the standard forms of membership function are used. Below we shall specify these functions and will present their graphic representations.

1. The *singleton* function shall be defined as follows:

$$\mu_A(x) = \left\{ \begin{array}{ll} 1, & \text{if} \quad x = \overline{x}, \\ 0, & \text{if} \quad x \neq \overline{x}. \end{array} \right. \tag{4.15}$$

The singleton is a specific membership function, as it takes the value 1 only in a single point of the universe of discussion, belonging fully to the fuzzy set. In other points, it takes the value of 0. This membership function characterizes a single-element fuzzy set. The only element having the full membership to the fuzzy set A is the point $\overline{x}$. The singleton membership function is mainly used to perform fuzzification operation applied in fuzzy inference systems.

2. *Gaussian* membership function (Fig. 4.2) is described by the formula

$$\mu_A(x) = \exp\left(-\left(\frac{x - \overline{x}}{\sigma} \right)^2 \right), \tag{4.16}$$

where $\overline{x}$ is the middle and σ defines the width of the Gaussian curve. It is the most common membership function.

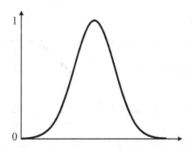

FIGURE 4.2. Gaussian membership function

3. *Bell* membership function (Fig. 4.3) takes the form of

$$\mu(x; a, b, c) = \frac{1}{1 + \left| \dfrac{x - c}{a} \right|^{2b}}, \tag{4.17}$$

where the parameter a defines its width, the parameter b its slopes, and the parameter c its center.

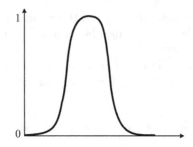

FIGURE 4.3. Bell membership function

4. Membership function of *class* s (Fig. 4.4) is defined as

$$
s\left(x;a,b,c\right)=\begin{cases}0 & \text{for} \quad x\le a, \\ 2\left(\dfrac{x-a}{c-a}\right)^{2} & \text{for} \quad a<x\le b, \\ 1-2\left(\dfrac{x-c}{c-a}\right)^{2} & \text{for} \quad b<x\le c, \\ 1 & \text{for} \quad x>c.\end{cases}
\tag{4.18}
$$

where $b=\left(a+c\right)/2$. The graph of the membership function belonging to this class takes a graphic form reminding of the letter "s", and its shape depends on the selection of the a, b and c parameters. In the point $x=b=\left(a+c\right)/2$ the membership function of class s takes the value of 0.5.

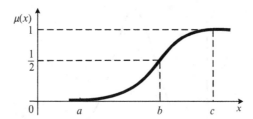

FIGURE 4.4. Membership function of class s

5. The membership function of *class* π (Fig. 4.5) is defined by the membership function of class s

$$
\pi\left(x;b,c\right)=\begin{cases}s\left(x;c-b,\ c-b/2,c\right) & \text{for} \quad x\le c, \\ 1-s\left(x;c,c+b/2,c+b\right) & \text{for} \quad x>c.\end{cases}
\tag{4.19}
$$

The membership function of class π takes the zero values for $x\ge c+b$ and $x\le c-b$. In points $x=c\pm b/2$ its value is 0.5.

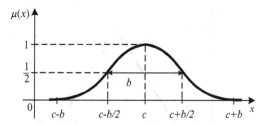

FIGURE 4.5. Membership function of class π

6. The membership function of *class* γ (Fig. 4.6) is given by the formula

$$\gamma(x; a, b) = \begin{cases} 0 & \text{for} \quad x \le a, \\ \dfrac{x - a}{b - a} & \text{for} \quad a < x \le b, \\ 1 & \text{for} \quad a > b. \end{cases} \tag{4.20}$$

The Reader will easily notice the analogies between the shapes of the membership function of class s and γ.

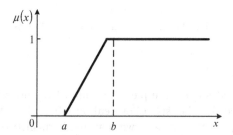

FIGURE 4.6. Membership function of class γ

7. The membership function of *class* t (Fig. 4.7) is defined as follows:

$$t(x; a, b, c) = \begin{cases} 0 & \text{for} \quad x \le a, \\ \dfrac{x - a}{b - a} & \text{for} \quad a < x \le b, \\ \dfrac{c - x}{c - b} & \text{for} \quad b < x \le c, \\ 0 & \text{for} \quad x > c. \end{cases} \tag{4.21}$$

In some applications, the membership function of class t may be alternative to the function of class π.

8. The membership function of *class* L (Fig. 4.8) is defined by the formula

$$L(x; a, b) = \begin{cases} 1 & \text{for} \quad x \le a, \\ \dfrac{b - x}{b - a} & \text{for} \quad a < x \le b, \\ 0 & \text{for} \quad a > b. \end{cases} \tag{4.22}$$

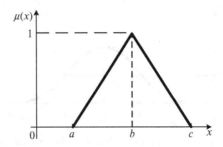

FIGURE 4.7. Membership function of class t

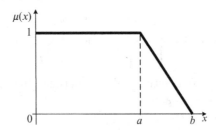

FIGURE 4.8. Membership function of class L

Remark 4.3

Above, we provided some examples of standard membership function for fuzzy sets defined in the space of real numbers i.e. $\mathbf{X} \subset \mathbf{R}$. When $\mathbf{X} \subset \mathbf{R}^n$, $\mathbf{x} = [x_1, ..., x_n]^T$, $n > 1$, we may distinguish two cases. The first case occurs when we assume the independence of particular variables x_i, $i = 1, ..., n$. Then the multidimensional membership functions are created by applying the definition of Cartesian product of fuzzy sets (Definition 4.14) and using standard membership functions of one variable. In case the variables x_i are dependent, we apply the multidimensional membership function. Below, three examples of such functions are specified:

1. The membership function of *class* Π (Fig. 4.9) is defined as follows:

$$\mu_A(\mathbf{x}) = \begin{cases} 1 - 2 \cdot \left(\dfrac{\|\mathbf{x} - \overline{\mathbf{x}}\|}{\alpha} \right)^2 & \text{for} \quad \|\mathbf{x} - \overline{\mathbf{x}}\| \leq \dfrac{1}{2}\alpha, \\ 2 \cdot \left(1 - \dfrac{\|\mathbf{x} - \overline{x}\|}{\alpha} \right)^2 & \text{for} \quad \dfrac{1}{2}\alpha < \|\mathbf{x} - \overline{\mathbf{x}}\| \leq \alpha, \\ 0 & \text{for} \quad \|\mathbf{x} - \overline{\mathbf{x}}\| > \alpha, \end{cases} \qquad (4.23)$$

when $\overline{\mathbf{x}}$ is the center of the membership function, $\alpha > 0$ is the parameter defining its spread.

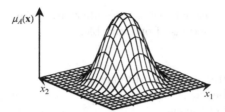

FIGURE 4.9. Two-dimensional membership function of class II

2. *The radial* membership function (Fig. 4.10) takes the form

$$\mu_A\left(\mathbf{x}\right) = e^{\frac{\|\mathbf{x}-\overline{\mathbf{x}}\|^2}{2\cdot\sigma^2}}, \tag{4.24}$$

where $\overline{x}$ is the center, and the value of the parameter σ influences the shape of this function.

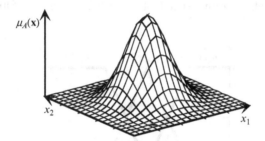

FIGURE 4.10. Two-dimensional radial membership function

3. *The ellipsoidal* membership function (Fig. 4.11) is defined as follows:

$$\mu_A\left(\mathbf{x}\right) = \exp\left(-\frac{\left(\mathbf{x}-\overline{\mathbf{x}}\right)^T \mathbf{Q}^{-1}\left(\mathbf{x}-\overline{\mathbf{x}}\right)}{\alpha}\right), \tag{4.25}$$

where $\overline{\mathbf{x}}$ is the center, $\alpha > 0$ is the parameter defining the spread of this function, and $\mathbf{Q}$ is the so-called covariance matrix. By modifying this matrix, we may model the shape of this function.

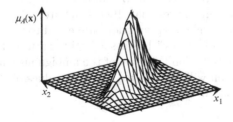

FIGURE 4.11. Two-dimensional elipsoidal membership function

We shall provide below two examples illustrating the application of standard membership functions of one variable.

Example 4.4

Let us consider three imprecise statements:
1) "low speed of the car",
2) "medium speed of the car",
3) "high speed of the car".

We shall assume the interval $[0, x_{max}]$ as the universe of discourse $\mathbf{X}$, where x_{max} is the maximum speed. Figure 4.12 illustrates fuzzy sets A, B, C corresponding to the above statements. Let us notice that the membership function of the set A is of the L type, of the set B is of the t type, and of the set C is the class γ. In the fixed point $x = 40$ km/h the membership function of the fuzzy set "low speed of the car" takes the value 0.5, i.e. $\mu_A = (40) = 0.5$. The same value is taken by the membership function of the fuzzy set "medium speed of the car", i.e. $\mu_B = (40) = 0.5$ and $\mu_C = (40) = 0$.

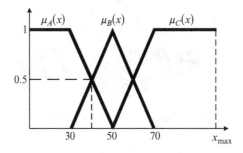

FIGURE 4.12. Illustration to Example 4.4

Example 4.5

Figure 4.13 illustrates the membership function of the fuzzy set "a lot of money" (in USD). It is the function of the class s, and $\mathbf{X} = [0; 100000]$, $a = 1000$, $c = 10000$. Therefore, we can certainly consider amounts exceeding 10000 USD as "large", as then, the values of the membership function are equal to 1. Amounts lower than 1000 USD are not "large", as the values of the relevant membership functions are equal to 0. Such definition of the fuzzy set "a lot of money" has a subjective nature. The Reader may have his/her own opinion on the issue of the ambiguous statement "a lot of money". This opinion will be reflected by other values of the parameters a and c of the function of class s.

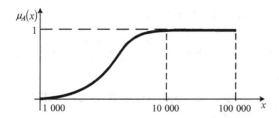

FIGURE 4.13. Illustration to Example 4.5

Definition 4.2
The set of elements of the universe $\mathbf{X}$, for which $\mu_A(x) > 0$, is called *the support of a fuzzy set A* and is denoted as supp A (*support*), i.e.

$$\text{supp } A = \{x \in \mathbf{X}; \ \mu_A(x) > 0\}. \tag{4.26}$$

Definition 4.3
The height of a fuzzy set A shall be denoted as $h(A)$ and defined as

$$h(A) = \sup_{x \in \mathbf{X}} \mu_A(x). \tag{4.27}$$

Example 4.6
If $X = \{1, 2, 3, 4, 5\}$ and

$$A = \frac{0.2}{1} + \frac{0.4}{2} + \frac{0.7}{4}, \tag{4.28}$$

then supp $A = \{1, 2, 4\}$.
   If $\mathbf{X} = \{1, 2, 3, 4\}$ and

$$A = \frac{0.3}{2} + \frac{0.8}{3} + \frac{0.5}{4}, \tag{4.29}$$

then $h(A) = 0.8$.

Definition 4.4
The fuzzy set A is called *normal* if and only if $h(A) = 1$. If the fuzzy set A is not normal, it can be normalized using the transformation

$$\mu_{A_{nor}}(x) = \frac{\mu_A(x)}{h(A)}, \tag{4.30}$$

where $h(A)$ is the height of this set.

Example 4.7
The fuzzy set

$$A = \frac{0.1}{2} + \frac{0.5}{4} + \frac{0.3}{6} \tag{4.31}$$

after normalizing takes the form

$$A_{nor} = \frac{0.2}{2} + \frac{1}{4} + \frac{0.6}{6}.$$ (4.32)

Definition 4.5
The fuzzy set A is called *empty* which shall be notated $A = \varnothing$, if and only if $\mu_A(x) = 0$ for each $x \in \mathbf{X}$.

Definition 4.6
The fuzzy set A is *included* in the fuzzy set B, which shall be notated $A \subset B$, if and only if

$$\mu_A(x) \le \mu_B(x)$$ (4.33)

for each $x \in \mathbf{X}$. An example of *inclusion* of the fuzzy set A in the fuzzy set B is illustrated by Fig. 4.14.

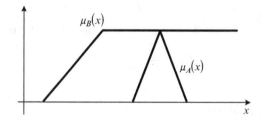

FIGURE 4.14. Inclusion of the fuzzy set A in the fuzzy set B

Definition 4.7
The fuzzy set A is *equal* to the fuzzy set B, which shall be notated $A = B$, if and only if

$$\mu_A(x) = \mu_B(x)$$ (4.34)

for each $x \in \mathbf{X}$. The above definition, similarly to Definition 4.6, is not "flexible", as it does not consider the case when the values of the membership functions $\mu_A(x)$ and $\mu_B(x)$ are almost equal. Then we can introduce the term of equality degree of fuzzy sets A and B as for example

$$E(A = B) = 1 - \max_{x \in T} |\mu_A(x) - \mu_B(x)|,$$ (4.35)

where $T = \{x \in X : \mu_A(x) \ne \mu_B(x)\}$. Different definitions of the inclusion degree and the equality degree of fuzzy sets have been presented in detail in monograph [94].

Definition 4.8

α-*cut* of the fuzzy set $A \subseteq \mathbf{X}$, notated as A_α is called the following non-fuzzy set:

$$A_\alpha = \{x \in \mathbf{X} : \mu_A(x) \geq \alpha\}, \quad \forall_{\alpha \in [0,1]}, \tag{4.36}$$

or the set defined by the characteristic function

$$\chi_{A_\alpha}(x) = \begin{cases} 1 & \text{for} \quad \mu_A(x) \geq \alpha, \\ 0 & \text{for} \quad \mu_A(x) < \alpha. \end{cases} \tag{4.37}$$

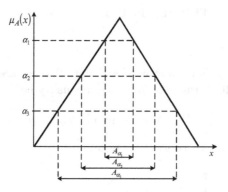

FIGURE 4.15. Illustration of α-cuts of the fuzzy set A

The definition of α-cut of a fuzzy set is illustrated by Fig. 4.15. It can be easily noted that there is the following implication:

$$\alpha_2 < \alpha_1 \implies A_{\alpha_1} \subset A_{\alpha_2}. \tag{4.38}$$

Example 4.8

Let us consider the fuzzy set $A \subseteq \mathbf{X}$

$$A = \frac{0.1}{2} + \frac{0.3}{4} + \frac{0.7}{5} + \frac{0.8}{8} + \frac{1}{10}, \tag{4.39}$$

while $\mathbf{X} = \{1, ..., 10\}$. According to Definition 4.8 particular α-cuts are defined as follows:

$$A_0 = \mathbf{X} = \{1, ..., 10\},$$
$$A_{0.1} = \{2, 4, 5, 8, 10\},$$
$$A_{0.3} = \{4, 5, 8, 10\},$$
$$A_{0.7} = \{5, 8, 10\},$$
$$A_{0.8} = \{8, 10\},$$
$$A_1 = \{10\}.$$

Definition 4.9

The fuzzy set $A \subseteq \mathbf{R}$ is *convex* if and only if for any $x_1, x_2 \in \mathbf{R}$ and $\lambda \in [0,1]$ the following occurs

$$\mu_A[\lambda x_1 + (1-\lambda)x_2] \geq \min\{\mu_A(x_1), \mu_A(x_2)\}. \tag{4.40}$$

Figure 4.16 illustrates an example of a convex fuzzy set.

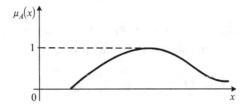

FIGURE 4.16. Convex fuzzy set A

Definition 4.10

The fuzzy set $A \subseteq \mathbf{R}$ is *concave* if and only if there are such points x_1, $x_2 \in \mathbf{R}$ and $\lambda \in [0, 1]$, that the following inequality holds

$$\mu_A \left[\lambda x_1 + (1 - \lambda) x_2 \right] < \min \left\{ \mu_A \left(x_1 \right), \mu_A \left(x_2 \right) \right\}. \tag{4.41}$$

Figure 4.17 illustrates a concave fuzzy set.

FIGURE 4.17. Concave fuzzy set A

4.3   Operations on fuzzy sets

In this subchapter, we shall present the basic operations on fuzzy sets, both set operations and algebraic operations.

Definition 4.11

The intersection of fuzzy sets A, $B \subseteq \mathbf{X}$ is the fuzzy set $A \cap B$ with the membership function

$$\mu_{A \cap B} \left(x \right) = \min \left(\mu_A \left(x \right), \mu_B \left(x \right) \right) \tag{4.42}$$

for each $x \in \mathbf{X}$. This operation has been presented graphically in Fig. 4.18. The intersection of fuzzy sets $A_1, A_2, ..., A_n$ is defined by the membership function

$$\mu_{A_1 \cap A_2 \cdots \cap A_n}(x) = \min\left[\mu_{A_1}(x), \mu_{A_2}(x), ..., \mu_{A_n}(x)\right] \qquad (4.43)$$

for each $x \in \mathbf{X}$.

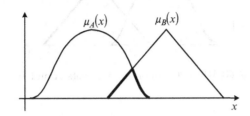

FIGURE 4.18. Intersection of fuzzy sets A and B

Remark 4.4

Apart form the definition of *intersection* of fuzzy sets, the literature also contains the definition of *algebraic product* of these sets. The algebraic product of fuzzy sets A and B is the fuzzy set $C = A \cdot B$ defined as follows:

$$C = \{(x, \mu_A(x) \cdot \mu_B(x)) \mid x \in \mathbf{X}\}. \qquad (4.44)$$

The operation of algebraic product is illustrated by Fig. 4.19.

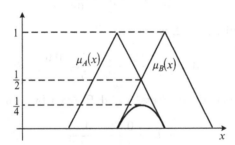

FIGURE 4.19. Algebraic product of fuzzy sets A and B

Definition 4.12

The union of fuzzy sets A and B is the fuzzy set $A \cup B$ defined by the membership function

$$\mu_{A \cup B}(x) = \max\left(\mu_A(x), \mu_B(x)\right) \qquad (4.45)$$

for each $x \in \mathbf{X}$.

This operation is illustrated by Fig. 4.20. The membership function of the union of fuzzy sets $A_1, A_2, ..., A_n$ is expressed by the formula

$$\mu_{A_1 \cup A_2 \cup \cdots \cup A_n}(x) = \max\left[\mu_{A_1}(x), \mu_{A_2}(x), ..., \mu_{A_n}(x)\right] \qquad (4.46)$$

for each $x \in \mathbf{X}$.

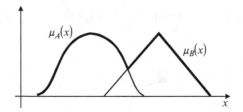

FIGURE 4.20. Union of fuzzy sets A and B

Example 4.9

Let us assume that $\mathbf{X} = \{1, 2, 3, 4, 5, 6, 7\}$ and

$$A = \frac{0.9}{3} + \frac{1}{4} + \frac{0.6}{6},$$ (4.47)

$$B = \frac{0.7}{3} + \frac{1}{5} + \frac{0.4}{6}.$$ (4.48)

According to Definition 4.11, we have

$$A \cap B = \frac{0.7}{3} + \frac{0.4}{6}.$$ (4.49)

By virtue of Definition 4.12, we have

$$A \cup B = \frac{0.9}{3} + \frac{1}{4} + \frac{1}{5} + \frac{0.6}{6}$$ (4.50)

whereas the product of fuzzy sets A and B given by formula (4.44) takes the form of

$$A \cdot B = \frac{0.63}{3} + \frac{0.24}{6}.$$ (4.51)

In the literature a very useful is the so-called *decomposition theorem*. It allows to represent any fuzzy set A in the form of union of fuzzy sets generated by α-cuts of the set A.

Theorem 4.1

Any fuzzy set $A \subseteq \mathbf{X}$ may be presented in the form

$$A = \bigcup_{\alpha \in [0,1]} \alpha A_\alpha,$$ (4.52)

where αA_α means a fuzzy set, to the elements of which the following membership degrees have been assigned:

$$\mu_{\alpha A_\alpha}(x) = \begin{cases} \alpha & \text{for} \quad x \in A_\alpha, \\ 0 & \text{for} \quad x \notin A_\alpha. \end{cases}$$ (4.53)

Example 4.10

We shall decompose fuzzy set (4.39). In accordance with formula (4.52), we have

$$A = \left(\frac{0.1}{2} + \frac{0.1}{4} + \frac{0.1}{5} + \frac{0.1}{8} + \frac{0.1}{10} \right) \cup \left(\frac{0.3}{4} + \frac{0.3}{5} + \frac{0.3}{8} + \frac{0.3}{10} \right)$$

$$\cup \left(\frac{0.7}{5} + \frac{0.7}{8} + \frac{0.7}{10} \right) \cup \left(\frac{0.8}{8} + \frac{0.8}{10} \right) \cup \frac{1}{10} \qquad (4.54)$$

$$= \frac{0.1}{2} + \frac{0.3}{4} + \frac{0.7}{5} + \frac{0.8}{8} + \frac{1}{10}.$$

Remark 4.5

Definitions 4.11 and 4.12 are not the only known in literature definitions of intersection and union of fuzzy sets. Instead of the equalities (4.42) and (4.45) repeated below,

$$\begin{cases} \mu_{A \cap B}(x) = \min(\mu_A(x), \mu_B(x)) \\ \mu_{A \cup B}(x) = \max(\mu_A(x), \mu_B(x)) \end{cases}$$

we can find alternative definitions using the terms of the so-called *t-norm* and *t-conorm*. Therefore, operation (4.42) is an example of the operation of *t*-norm (intersection operation), and operation (4.45) is an example of the operation of *t*-conorm (union operation). Subchapter 4.6 will present the formal definitions of *t*-norm and *t*-conorm and more general definitions of intersection and union of fuzzy sets.

Remark 4.6

The literature contains attempts to analytically find the "best" operations of intersection and union of fuzzy sets. For example, Bellman and Giertz [8] defined and solved the problem of finding two such functions f and g

$$f, g : [0, 1] \times [0, 1] \to [0, 1]$$

that

$$\mu_{A \cap B}(x) = f(\mu_A(x), \mu_B(x)), \qquad (4.55)$$

$$\mu_{A \cup B}(x) = g(\mu_A(x), \mu_B(x)). \qquad (4.56)$$

The authors of the above mentioned publication imposed many conditions on functions f and g, and then they demonstrated that these conditions are met only by operation (4.42) and (4.45) This does not mean that operation (4.42) and (4.45) are adequate in all applications, e.g. if

$$\mu_A(x) < \mu_B(x), \quad \forall_{x \subseteq \mathbf{X}}, \qquad (4.57)$$

then as a result of operation (4.42), we will have

$$\mu_{A \cap B}(x) = \mu_A(x) \qquad (4.58)$$

regardless of $\mu_B(x)$. In other words, the membership function of the fuzzy set B has no influence whatsoever on the definition of the intersection of fuzzy sets A and B. This fact is illustrated by Fig. 4.21. In such case, it seems more reasonable to apply, for instance, formula (4.44) as the intersection operation. Then the intersection of two fuzzy sets will be identical to the product of these sets (see Remark 4.4).

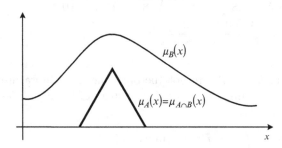

FIGURE 4.21. Intersection of fuzzy sets A and B when $\mu_A(x) < \mu_B(x)$

Definition 4.13

The complement of a fuzzy set $A \subseteq \mathbf{X}$ is the fuzzy set $\widehat{A}$ with the membership function

$$\mu_{\widehat{A}}(x) = 1 - \mu_A(x) \tag{4.59}$$

for each $x \in \mathbf{X}$. The complement operation is illustrated by Fig. 4.22.

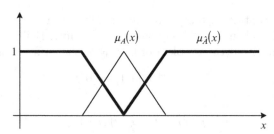

FIGURE 4.22. Complement of fuzzy set A

Example 4.11

Let us assume that $\mathbf{X} = \{1, 2, 3, 4, 5, 6\}$ and

$$A = \frac{0.3}{2} + \frac{1}{3} + \frac{0.7}{5} + \frac{0.9}{6}. \tag{4.60}$$

According to Definition 4.13, the complement of the set A is the set

$$\widehat{A} = \frac{1}{1} + \frac{0.7}{2} + \frac{1}{4} + \frac{0.3}{5} + \frac{0.1}{6}. \tag{4.61}$$

Let us notice that

$$A \cap \widehat{A} = \frac{0.3}{2} + \frac{0.3}{5} + \frac{0.1}{6} \neq \emptyset \tag{4.62}$$

and

$$A \cup \widehat{A} = \frac{1}{1} + \frac{0.7}{2} + \frac{1}{3} + \frac{1}{4} + \frac{0.7}{5} + \frac{0.9}{6} \neq \mathbf{X}. \tag{4.63}$$

We may demonstrate that the above presented operations on fuzzy sets (Definitions 4.11 – 4.13) have the properties of commutativity, associativity and distributivity and moreover de Morgan's laws and absorption laws occur. However, in case of fuzzy sets, the laws of crisp sets are not met, i.e.

$$A \cap \widehat{A} \neq \emptyset, \tag{4.64}$$

$$A \cup \widehat{A} \neq \mathbf{X}. \tag{4.65}$$

This fact is illustrated by Fig. 4.23 and Example 4.11. It is worth noting that the membership function of the intersection of fuzzy sets A and $\widehat{A}$ meets the inequality (see [42]):

$$\mu_{A \cap \widehat{A}}(x) = \min \left(\mu_A(x), \mu_{\widehat{A}}(x) \right) \leq \frac{1}{2}. \tag{4.66}$$

Similarly, in case of the union, we have

$$\mu_{A \cup \widehat{A}}(x) = \max \left(\mu_A(x), \mu_{\widehat{A}}(x) \right) \geq \frac{1}{2}. \tag{4.67}$$

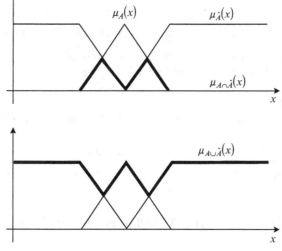

FIGURE 4.23. Fuzzy sets $A \cap \widehat{A}$ and $A \cup \widehat{A}$

Definition 4.14

The Cartesian product of fuzzy sets $A \subseteq \mathbf{X}$ and $B \subseteq \mathbf{Y}$ is notated as $A \times B$ and defined as

$$\mu_{A \times B}(x, y) = \min(\mu_A(x), \mu_B(y)) \tag{4.68}$$

or

$$\mu_{A \times B}(x, y) = \mu_A(x)\, \mu_B(y) \tag{4.69}$$

for each $x \in \mathbf{X}$ and $y \in \mathbf{Y}$. The Cartesian product of fuzzy sets $A_1 \subseteq \mathbf{X}_1, A_2 \subseteq \mathbf{X}_2, ..., A_n \subseteq \mathbf{X}_n$ is notated as $A_1 \times A_2 \times ... \times A_n$ and defined as

$$\mu_{A_1 \times A_2 \times ... \times A_n}(x_1, x_2, ..., x_n) = \min(\mu_{A_1}(x_1), \mu_{A_2}(x_2), ..., \mu_{A_n}(x_n)) \tag{4.70}$$

or

$$\mu_{A_1 \times A_2 \times ... \times A_n}(x_1, x_2, ..., x_n) = \mu_{A_1}(x_1)\, \mu_{A_2}(x_2), ..., \mu_{A_n}(x_n) \tag{4.71}$$

for each $x_1 \in \mathbf{X}_1, x_2 \in \mathbf{X}_2, ..., x_n \in \mathbf{X}_n$.

Example 4.12

Let us assume that $\mathbf{X} = \{2, 4\}$, $\mathbf{Y} = \{2, 4, 6\}$ and

$$A = \frac{0.5}{2} + \frac{0.9}{4}, \tag{4.72}$$

$$B = \frac{0.3}{2} + \frac{0.7}{4} + \frac{0.1}{6}. \tag{4.73}$$

By applying Definition 4.14 of the Cartesian product of sets A and B, we obtain

$$A \times B = \frac{0.3}{(2, 2)} + \frac{0.5}{(2, 4)} + \frac{0.1}{(2, 6)} + \frac{0.3}{(4, 2)} + \frac{0.7}{(4, 4)} + \frac{0.1}{(4, 6)}. \tag{4.74}$$

The following algebraic operations on fuzzy sets play a significant role in the semantics of linguistic variables (Subchapter 4.8).

Definition 4.15

The concentration of a fuzzy set $A \subseteq \mathbf{X}$ shall be notated as $CON(A)$ and defined as

$$\mu_{CON(A)}(x) = (\mu_A(x))^2 \tag{4.75}$$

for each $x \in \mathbf{X}$.

Definition 4.16

The dilation of a fuzzy set $A \subseteq \mathbf{X}$ shall be notated as $DIL(A)$ and defined as

$$\mu_{DIL(A)}(x) = (\mu_A(x))^{0.5} \tag{4.76}$$

for each $x \in \mathbf{X}$.

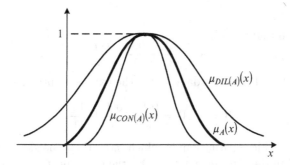

FIGURE 4.24. The operation of concentration and dilation of fuzzy sets

The operation of concentration and dilation of fuzzy sets is illustrated by Fig. 4.24.

Example 4.13
If $\mathbf{X} = \{1, 2, 3, 4\}$ and

$$A = \frac{0.4}{2} + \frac{0.7}{3} + \frac{1}{4}, \tag{4.77}$$

then according to Definitions 4.15 and 4.16, we have

$$CON(A) = \frac{0.16}{2} + \frac{0.49}{3} + \frac{1}{4}, \tag{4.78}$$

$$DIL(A) = \frac{0.63}{2} + \frac{0.84}{3} + \frac{1}{4}. \tag{4.79}$$

4.4 The extension principle

The extension principle allows to extend different mathematical operations from non-fuzzy sets to fuzzy sets. Let us consider a non-fuzzy mapping f of the space $\mathbf{X}$ in the space $\mathbf{Y}$

$$f : \mathbf{X} \to \mathbf{Y}. \tag{4.80}$$

Let A be a given fuzzy set defined in the space $\mathbf{X}$, i.e. $A \in \mathbf{X}$. If the fuzzy set A has the form (4.3), i.e.

$$A = \frac{\mu_A(x_1)}{x_1} + \frac{\mu_A(x_2)}{x_2} + \cdots + \frac{\mu_A(x_n)}{x_n}$$

and the mapping f is one-to-one, then the extension principle says that the fuzzy set B induced by this mapping and defined in the space $\mathbf{Y}$ takes the form

$$B = f(A) = \frac{\mu_A(x_1)}{f(x_1)} + \frac{\mu_A(x_2)}{f(x_2)} + \cdots + \frac{\mu_A(x_n)}{f(x_n)}. \tag{4.81}$$

Example 4.14

Let us assume that

$$A = \frac{0.1}{3} + \frac{0.4}{2} + \frac{0.7}{5} \tag{4.82}$$

and $f(x) = 2x + 1$. In accordance with the extension principle, we have

$$B = f(A) = \frac{0.1}{7} + \frac{0.4}{5} + \frac{0.7}{11}. \tag{4.83}$$

Now, let us consider a situation, in which more than one element of the set **X** is mapped in the same element $y \in Y$ (the mapping f is not one-to-one). Then the membership degree of the element y to the fuzzy set $B = f(A)$ is equal to the maximum degree from the membership degrees of elements of set **X**, which are mapped in the same element y. The following example shall illustrate this case of the extension principle.

Example 4.15

If

$$A = \frac{0.3}{-2} + \frac{0.5}{3} + \frac{0.7}{2} \tag{4.84}$$

and $f(x) = x^2$, then the fuzzy set B induced by the mapping f is equal to

$$B = f(A) = \frac{0.5}{9} + \frac{0.7}{4}, \tag{4.85}$$

as $\max\{0.3;\ 0.7\} = 0.7$. Let us notate by $f^{-1}(y)$ the set of the elements $x \in X$ which are mapped in the element $y \in Y$ by transformation of f. If $f^{-1}(y)$ is an empty set, i.e. $f^{-1}(y) = \varnothing$, then the membership degree of the element y to the fuzzy set B is equal to zero. The above discussion and the example that illustrate it allow to formulate the following extension principle.

Extension principle I

If we have a given non-fuzzy mapping (4.80) and some fuzzy set $A \subseteq \mathbf{X}$, then the extension principle says that the fuzzy set B induced by this mapping has the form of

$$B = f(A) = \{(y, \mu_B(y)) \mid y = f(x),\ x \in \mathbf{X}\}, \tag{4.86}$$

where

$$\mu_B(y) = \begin{cases} \sup\limits_{x \in f^{-1}(y)} \mu_A(x), & \text{if } f^{-1}(y) \neq \varnothing, \\ 0, & \text{if } f^{-1}(y) = \varnothing. \end{cases} \tag{4.87}$$

The extension principle I includes both the case of the space **X** with a finite number of elements (set B is defined by formula (4.81)), and with an infinite number of elements. In the latter case, the fuzzy set B induced by the mapping f may be notated as

$$B = f(A) = \int_y \frac{\mu_A(y)}{f(x)}. \tag{4.88}$$

In some applications (e.g. fuzzy numbers, Subchapter 4.5), a different form of the extension principle is useful.

Extension principle II
Let $\mathbf{X}$ be the Cartesian product of non-fuzzy sets $\mathbf{X}_1 \times \mathbf{X}_2 \times \ldots \times \mathbf{X}_n$. If we have a given non-fuzzy mapping

$$f : \mathbf{X}_1 \times \mathbf{X}_2 \times \ldots \times \mathbf{X}_n \to \mathbf{Y} \tag{4.89}$$

and given fuzzy sets $A_1 \subseteq \mathbf{X}_1$, $A_2 \subseteq \mathbf{X}_2, \ldots, A_n \subseteq \mathbf{X}_n$, then the extension principle says that the fuzzy set B induced by the mapping f has the form

$$B = f(A_1, \ldots, A_n) = \{(y, \mu_B(y)) \mid y = f(x_1, \ldots, x_n), (x_1, \ldots, x_n) \in \mathbf{X}\}, \tag{4.90}$$

while

$$\mu_B(y) = \begin{cases} \sup\limits_{\substack{(x_1,\ldots,x_n) \\ \in f^{-1}(y)}} \min\{\mu_{A_1}(x_1), \ldots, \mu_{A_n}(x_n)\}, & \text{if } f^{-1}(y) \neq \varnothing, \\ 0, & \text{if } f^{-1}(y) = \varnothing. \end{cases} \tag{4.91}$$

In formula (4.91), the minimum operation may be replaced with an algebraic product or, more generally, with the so-called t-norm (Subchapter 4.6). The following three examples illustrate the fact that the extension principle allows to extend the arithmetical operations onto fuzzy sets.

Example 4.16
Let us consider the function

$$f(x_1, x_2) = \frac{x_1 x_2}{(x_1 + x_2)}. \tag{4.92}$$

Let us determine a fuzzy set $B = f(A_1, A_2)$ induced by mapping (4.92).
In accordance with formula (4.90), we have

$$B = f(A_1, A_2) = \int\limits_{x_1 \in \mathbf{X}_1} \int\limits_{x_2 \in \mathbf{X}_2} \sup\limits_{\substack{(x_1,\ldots,x_n) \\ \in f^{-1}(y)}} \min(\mu_{A_i}(x_1), \mu_{A_2}(x_2)) \left| \frac{x_1 x_2}{x_1 + x_2} \right. .$$

Example 4.17
Let us assume that $\mathbf{X}$ is the Cartesian product of sets $\mathbf{X}_1 = \mathbf{X}_2 = \{1, 2, 3, 4, 5, 6\}$. Let A_1 be the fuzzy set of numbers "close to number 2"

$$A_1 = \frac{0.7}{1} + \frac{1}{2} + \frac{0.8}{3} \tag{4.93}$$

and let A_2 be the fuzzy set of numbers "close to number 4"

$$A_2 = \frac{0.8}{3} + \frac{1}{4} + \frac{0.9}{5}. \tag{4.94}$$

If

$$y = f(x_1, x_2) = x_1 x_2, \tag{4.95}$$

then the set $B = f(A_1, A_2)$ induced by mapping (4.95) will be a fuzzy set of numbers "close to number 8", while $B \subseteq Y = \{1, 2, ..., 36\}$. By virtue of the extension principle II, we have

$$B = f(A_1, A_2) = \sum_{i,j=1}^{3} \left[\min\left(\mu_{A_1}\left(x_1^{(i)}\right), \mu_{A_2}\left(x_2^{(j)}\right) \right) \right] / x_1^{(i)} x_2^{(j)}$$

$$= \frac{\min(0.7; 0.8)}{3} + \frac{\min(0.7; 1)}{4} + \frac{\min(0.7; 0.9)}{5}$$

$$+ \frac{\min(1; 0.8)}{6} + \frac{\min(1; 1)}{8} + \frac{\min(1; 0.9)}{10} + \frac{\min(0.8; 0.8)}{9} \tag{4.96}$$

$$+ \frac{\min(0.8; 1)}{12} + \frac{\min(0.8; 0.9)}{15}$$

$$= \frac{0.7}{3} + \frac{0.7}{4} + \frac{0.7}{5} + \frac{0.8}{6} + \frac{1}{8} + \frac{0.8}{9} + \frac{0.9}{10} + \frac{0.8}{12} + \frac{0.8}{15}.$$

The following example illustrates the case when the element $y = f\left(x_1^{(i)}, x_2^{(j)}\right)$ takes the same value for different values of elements $x_1^{(i)}$ and $x_2^{(j)}$.

Example 4.18
Let us assume that $\mathbf{X}$ is the Cartesian product of sets $\mathbf{X}_1 = \mathbf{X}_2 = \{1, 2, 3, 4\}$. We shall define the following fuzzy set A_1 of numbers "close to number 2"

$$A_1 = \frac{0.7}{1} + \frac{1}{2} + \frac{0.8}{3} \tag{4.97}$$

and the fuzzy set A_2 of numbers "close to number 3"

$$A_2 = \frac{0.8}{2} + \frac{1}{3} + \frac{0.6}{4}. \tag{4.98}$$

Currently, the set $B = f(A_1, A_2)$ induced by mapping (4.95) will be a fuzzy set of numbers "close to number 6", while $B \subseteq Y = \{1, 2, ..., 16\}$. In accordance with the extension principle II, we have

$$B = f(A_1, A_2) = \frac{\min(0.7; 0.8)}{2} + \frac{\min(0.7; 1)}{3}$$
$$+ \frac{\max[\min(0.7; 0.6); \min(1; 0.8)]}{4}$$
$$+ \frac{\max[\min(1; 1); \min(0.8; 0.8)]}{6} \qquad (4.99)$$
$$+ \frac{\min(1; 0.6)}{8} + \frac{\min(0.8; 1)}{9} + \frac{\min(0.8; 0.6)}{12}$$
$$= \frac{0.7}{2} + \frac{0.7}{3} + \frac{0.8}{4} + \frac{1}{6} + \frac{0.6}{8} + \frac{0.8}{9} + \frac{0.6}{12}.$$

4.5 Fuzzy numbers

In the fuzzy sets theory, we can differentiate the fuzzy sets defined on the axis of real numbers. For example, fuzzy sets of numbers "close to number 7" (Fig. 4.25) are defined in the set $\mathbf{R}$, and in addition, are normal and convex and have a continuous membership function.

Such fuzzy sets are called fuzzy numbers. Below, we will present the definition of a fuzzy number.

Definition 4.17
A fuzzy set A defined on the set of real numbers, $A \subseteq \mathbf{R}$, the membership function of which

$$\mu_A : \mathbf{R} \to [0, 1]$$

meets the conditions:
1) $\sup_{x \in \mathbf{R}} \mu_A(x) = 1$, i.e. the fuzzy set A is normal,
2) $\mu_A[\lambda x_1 + (1 - \lambda) x_2] \geq \min\{\mu_A(x_1), \mu_A(x_2)\}$, i.e. the set A is convex,
3) $\mu_A(x)$ is a continuous function by intervals, is called a *fuzzy number*.

Figure 4.25 illustrates an example of fuzzy numbers. The theory of fuzzy numbers distinguishes the positive and negative fuzzy numbers.

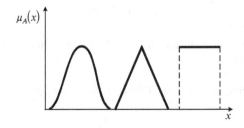

FIGURE 4.25. Examples of fuzzy numbers

Definition 4.18

The fuzzy number $A \subseteq \mathbf{R}$ is *positive*, if $\mu_A(x) = 0$ for all $x < 0$.
The fuzzy number $A \subseteq \mathbf{R}$ is *negative*, if $\mu_A(x) = 0$ for all $x > 0$.

Figure 4.26 illustrates an example of the positive fuzzy number, negative fuzzy number and a number that is neither positive not negative.

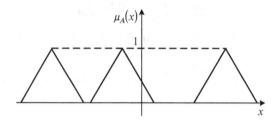

FIGURE 4.26. Examples of the positive fuzzy number, negative fuzzy number and a number that is neither positive nor negative

The Reader, who has studied Subchapter 4.4, will have no difficulties in defining the basic arithmetical operations on fuzzy numbers. We shall define these operations using the extension principle, which allows to formulate the definition of adding, subtracting, multiplying and dividing two fuzzy numbers $A_1, A_2 \subseteq \mathbf{R}$.

Definition 4.19 is the consequence of extension principle II, in which mapping (4.89) takes the form

$$y = f(x_1, x_2) = \begin{cases} x_1 + x_2 \text{ in case of adding fuzzy numbers } A_1 \text{ and } A_2 \\ x_1 - x_2 \text{ in case of subtracting fuzzy numbers } A_1 \text{ and } A_2 \\ x_1 \cdot x_2 \text{ in case of multiplying fuzzy numbers } A_1 \text{ and } A_2 \\ x_1 : x_2 \text{ in case of dividing fuzzy numbers } A_1 \text{ and } A_2 \end{cases}$$

Definition 4.19

The basic arithmetic operations on fuzzy numbers $A_1, A_2 \subseteq \mathbf{R}$ shall be defined as follows:

a) Adding two fuzzy numbers A_1 and A_2 shall be notated

$$A_1 \oplus A_2 \stackrel{\text{def}}{=} B, \tag{4.100}$$

while the membership function of sum (4.100) is defined by formula (4.91) taking the form of

$$\mu_B(y) = \sup_{\substack{x_1, x_2 \\ y = x_1 + x_2}} \min\{\mu_{A_1}(x_1), \mu_{A_2}(x_2)\}. \tag{4.101}$$

b) *Subtracting two fuzzy numbers A_1 and A_2 shall be notated*

$$A_1 \ominus A_2 \stackrel{\text{def}}{=} B, \tag{4.102}$$

while the membership function of difference (4.102) is defined by formula (4.91) taking the form of

$$\mu_B(y) = \sup_{\substack{x_1,x_2 \\ y=x_1-x_2}} \min\{\mu_{A_1}(x_1), \mu_{A_2}(x_2)\}. \tag{4.103}$$

c) *Multiplication of two fuzzy numbers A_1 and A_2 shall be notated*

$$A_1 \odot A_2 \stackrel{\text{def}}{=} B, \tag{4.104}$$

while the membership function of product (4.104) is defined by formula (4.91) taking the form of

$$\mu_B(y) = \sup_{\substack{x_1,x_2 \\ y=x_1 \cdot x_2}} \min\{\mu_{A_1}(x_1), \mu_{A_2}(x_2)\}. \tag{4.105}$$

d) *Division of two fuzzy numbers A_1 and A_2 shall be notated*

$$A_1 \oslash A_2 \stackrel{\text{def}}{=} B, \tag{4.106}$$

while the membership function of quotient (4.106) is defined by formula (4.91) taking the form of

$$\mu_B(y) = \sup_{\substack{x_1,x_2 \\ y=x_1 : x_2}} \min\{\mu_{A_1}(x_1), \mu_{A_2}(x_2)\}. \tag{4.107}$$

Although from the application perspective, we are interested in fuzzy numbers having continuous membership functions, we shall consider a discrete case to illustrate the above definition.

Example 4.19
Let us add and multiply two fuzzy numbers of the following form

$$A_1 = \frac{0.7}{2} + \frac{1}{3} + \frac{0.6}{4}, \tag{4.108}$$

$$A_2 = \frac{0.8}{3} + \frac{1}{4} + \frac{0.5}{6}. \tag{4.109}$$

In accordance with formula (4.101), we have

$$A_1 \oplus A_2 = \frac{\min(0.7;\ 0.8)}{5} + \frac{\max\{\min(0.7;\ 1),\min(1;\ 0.8)\}}{6}$$
$$+ \frac{\max\{\min(1;\ 1),\min(0.6;\ 0.8)\}}{7}$$
$$+ \frac{\max\{\min(0.7;\ 0.5),\min(0.6;\ 1)\}}{8} \tag{4.110}$$
$$+ \frac{\min(1;\ 0.5)}{9} + \frac{\min(0.6;\ 0.5)}{10}$$
$$= \frac{0.7}{5} + \frac{0.8}{6} + \frac{1}{7} + \frac{0.6}{8} + \frac{0.5}{9} + \frac{0.5}{10}.$$

On the basis of formula (4.105), we obtain

$$A_1 \odot A_2 = \frac{\min(0.7;\ 0.8)}{6} + \frac{\min(0.7;\ 1)}{8} + \frac{\min(1;\ 0.8)}{9}$$
$$+ \frac{\max\{\min(0.7;\ 0.5),\min(1;\ 1),\min(0.6;\ 0.8)\}}{12} \tag{4.111}$$
$$+ \frac{\min(0.6;\ 1)}{16} + \frac{\min(1;\ 0.5)}{18} + \frac{\min(0.6;\ 0.5)}{24}$$
$$= \frac{0.7}{6} + \frac{0.7}{8} + \frac{0.8}{9} + \frac{1}{12} + \frac{0.6}{16} + \frac{0.5}{18} + \frac{0.5}{24}.$$

Arithmetical operations on fuzzy number do not necessarily result in a fuzzy number. This problem is eliminated when we perform operations on fuzzy numbers with continuous membership functions, which is stated in the following theorem.

Theorem 4.2 (Dubois and Prade [42])
If the fuzzy numbers A_1 and A_2 have continuous membership functions, then arithmetical operations of adding, subtracting, multiplying and dividing result in fuzzy numbers.

We have discussed the basic binary operations on fuzzy numbers. Unary operations are performed also using the extension principle. If f is the mapping

$$f : \mathbf{R} \rightarrow \mathbf{R} \tag{4.112}$$

and $A \subseteq \mathbf{R}$, $y = f(x)$, then according to formula (4.87) we have

$$\mu_B(y) = \sup_{\substack{x \\ y=f(x)}} \mu_A(x), \tag{4.113}$$

where $B = f(A)$.

Below, we shall present some examples of unary operations on fuzzy numbers.

1) Reversal of sign operation

As a result of operation $f(x) = -x$ we obtain a fuzzy number which is opposite to the fuzzy number $A \subseteq \mathbf{R}$. This number shall be notated $-A \subseteq \mathbf{R}$, and its membership function is equal to

$$\mu_{-A}(x) = \mu_A(-x). \tag{4.114}$$

The fuzzy numbers A and $-A$ are symmetrical about the x axis.

2) Inverse operation

As a result of operation $f(x) = x^{-1}, x \neq 0$, we obtain a fuzzy number which is inverse to the fuzzy number $A \subseteq \mathbf{R}$. This number shall be notated $A^{-1} \subseteq \mathbf{R}$, and its membership function is equal to

$$\mu_{A^{-1}}(x) = \mu_A(x^{-1}). \tag{4.115}$$

Let us assume that A is a positive or a negative fuzzy number. If the fuzzy number A is neither positive nor negative, then the fuzzy set $B = f(A) = A^{-1}$ is not convex, and therefore B is not a fuzzy number.

3) Scaling operation

As a result of operation $f(x) = \lambda x, \lambda \neq 0$, we obtain a fuzzy number which is scaled in relation to the fuzzy number $A \subseteq \mathbf{R}$. This number shall be notated $\lambda A \subseteq \mathbf{R}$, and its membership function is equal to

$$\mu_{\lambda A}(x) = \mu_A(x\lambda^{-1}). \tag{4.116}$$

4) Exponent operation

As a result of operation $f(x) = e^x, x > 0$, we obtain the power of the fuzzy number $A \subseteq \mathbf{R}$. This number shall be notated $e^A \subseteq \mathbf{R}$, and its membership function is equal to

$$\mu_{e^A}(x) = \begin{cases} \mu_A(\log x) & \text{for} \quad x > 0, \\ 0 & \text{for} \quad x < 0, \end{cases} \tag{4.117}$$

and therefore e^A is a positive fuzzy number.

5) Absolute value operation

The absolute value of a fuzzy number $A \subseteq \mathbf{R}$ shall be notated as $|A| \subseteq \mathbf{R}$ and defined as

$$\mu_{|A|}(x) = \begin{cases} \max(\mu_A(x), \mu_A(-x)) & \text{for} \quad x \geq 0, \\ 0 & \text{for} \quad x < 0. \end{cases} \tag{4.118}$$

Of course, $|A|$ is a positive fuzzy number.

Example 4.20

If

$$A = \frac{0.7}{1} + \frac{1}{2} + \frac{0.6}{5}, \tag{4.119}$$

then the fuzzy number $-A$ has the form

$$-A = \frac{0.6}{-5} + \frac{1}{-2} + \frac{0.7}{-1}, \tag{4.120}$$

while the fuzzy number A^{-1} shall be notated

$$A^{-1} = \frac{0.6}{0.2} + \frac{1}{0.5} + \frac{0.7}{1}. \tag{4.121}$$

Using Definition 4.19, it is easy to check that in the above example

$$A + (-A) \neq \frac{1}{0} \tag{4.122}$$

and

$$A \cdot A^{-1} \neq \frac{1}{1}. \tag{4.123}$$

Therefore, the fuzzy numbers are characterized by a lack of opposite and inverse fuzzy number with relation to adding and multiplication. This fact makes impossible to use, for instance, the elimination method to solve equations with fuzzy numbers.

Arithmetical operations on fuzzy numbers call for rather complicated computations. That is why Dubois and Prade [41] have proposed a certain particular representation of fuzzy numbers. This representation shows the fuzzy numbers using 3 parameters, which much simplifies arithmetic operations. Let L and P be the functions mapping

$$(-\infty, \infty) \rightarrow [0, 1] \tag{4.124}$$

and meeting the conditions
1) $L(-x) = L(x)$ and $P(-x) = P(x)$,
2) $L(0) = 1$ and $P(0) = 1$,
3) L and P are nonincreasing functions in the interval $[0, +\infty)$.

We may give the following examples of function L:

$$L(x) = P(x) = e^{-|x|^p} \quad p > 0, \tag{4.125}$$

$$L(x) = P(x) = \frac{1}{1 + |x|^p} \quad p > 0, \tag{4.126}$$

$$L(x) = P(x) = \max(0, 1 - |x|^p) \quad p > 0, \tag{4.127}$$

$$L(x) = P(x) = \begin{cases} 1 & \text{for} \quad x \in [-1, 1], \\ 0 & \text{for} \quad x \notin [-1, 1]. \end{cases} \tag{4.128}$$

Below, we present the definition of a fuzzy number of the $L - P$ type.

Definition 4.20
The fuzzy number $A \subseteq \mathbf{R}$ is *a fuzzy number of the $L - P$ type* if and only if its membership function has the form of

$$\mu_A(x) = \begin{cases} L\left(\dfrac{m-x}{\alpha}\right), & \text{if} \quad x \leq m, \\ P\left(\dfrac{x-m}{\beta}\right), & \text{if} \quad x \geq m, \end{cases} \tag{4.129}$$

where m is a real number, called the average value of the fuzzy number A ($\mu_A(m) = 1$), α-positive real number, called the left-sided spread, β-positive real number, called the right-sided spread. Let us notice that if the α and β spreads increase, then number A becomes "more" fuzzy. The fuzzy number of the $L - P$ type may be notated in the short form as

$$A = (m_A, \alpha_A, \beta_A)_{LP}. \tag{4.130}$$

Example 4.21
The fuzzy number "more or less 9" may be notated in the form

$$A = (9, 3, 3)_{LP}. \tag{4.131}$$

The membership function of this number is illustrated in Fig. 4.27, and

$$L(x) = P(x) = \frac{1}{1 + x^2}. \tag{4.132}$$

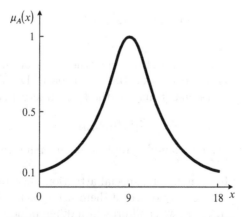

FIGURE 4.27. Illustration to Example 4.21

Arithmetic operations on fuzzy numbers $L - P$ come down to operations on three parameters. The fuzzy number opposite to fuzzy number (4.130) is equal

$$-A = (-m_A, \alpha, \beta)_{LP} .$$ (4.133)

The sum of fuzzy numbers $A = (m_A, \alpha_A, \beta_A)_{LP}$ and $B = (m_B, \alpha_B, \beta_B)_{LP}$ has the form of

$$A \oplus B = (m_A + m_B, \alpha_A + \alpha_B, \beta_A + \beta_B)_{LP} .$$ (4.134)

Other arithmetic operations (e.g. multiplication and division) on fuzzy numbers of the $L - P$ type are more complicated, and their result is an approximation.

The membership function $\mu_A(x)$ of the fuzzy number of the $L - P$ type takes the value 1 only in point $x = m$. Now, we shall modify Definition 4.20 so that $\mu_A(x) = 1$ not only in a single point $x = m$, but in all points of the interval $[m_1, m_2]$, while $m_1 < m_2$ and $m_1, m_2 \in \mathbf{R}$. We will then obtain the definition of the so-called flat fuzzy number. This definition may be applied to modeling fuzzy intervals.

Definition 4.21
A *flat* fuzzy number of the $L - P$ type is called the fuzzy number with the membership function

$$\mu_A(x) = \begin{cases} L\left(\dfrac{m_1 - x}{\alpha}\right), & \text{if} \quad x \leq m_1, \\ 1, & \text{if} \quad m_1 \leq x \leq m_2, \\ P\left(\dfrac{x - m_2}{\beta}\right), & \text{if} \quad x \geq m_2. \end{cases}$$ (4.135)

A flat fuzzy number A may be identified with the fuzzy interval A of the form

$$A = (m_1, m_2, \alpha, \beta)_{LP} .$$ (4.136)

Example 4.22
Let us consider the imprecise statement "the price of the motorbike in this store varies from approx. 3,000 USD to 6,000 USD". The appropriate formalization of this statement may be the fuzzy interval A of the form

$$A = (3, 6, \alpha, \beta)_{LP} .$$ (4.137)

Figure 4.28 illustrates an example graph of membership function of fuzzy interval (4.137).

In the theory of fuzzy numbers, special attention must be paid to triangular fuzzy numbers. We may describe them using membership functions of class t (4.21). At present, we shall present a different description of these numbers. A *triangular* fuzzy number A is defined on the interval $[a_1, a_2]$,

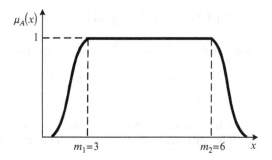

FIGURE 4.28. Illustration to Example 4.22

and its membership function takes the value equal to 1 in the point a_M. Therefore, the triangular fuzzy number may be notated as

$$A = (a_1, a_M, a_2). \qquad (4.138)$$

In many applications, there is the need to perform the so-called *defuzzification* of a triangular fuzzy number. As a result of defuzzification, the knowledge described by the fuzzy number is presented as a real number. Below, we shall present four methods of defuzzification of a triangular fuzzy number [13]:

$$y^{(1)} = a_M,$$

$$y^{(2)} = \frac{a_1 + a_M + a_2}{3},$$

$$y^{(3)} = \frac{a_1 + 2a_M + a_2}{4},$$

$$y^{(4)} = \frac{a_1 + 4a_M + a_2}{6},$$

Let us notice that the values a_1 and a_2 do not influence the determination of value of a defuzzified $y^{(1)}$. It is worth noting that the sum of two triangular fuzzy numbers $A_1 = \left(a_1^{(1)}, a_M^{(1)}, a_2^{(1)}\right)$ and $A_2 = \left(a_1^{(2)}, a_M^{(2)}, a_2^{(2)}\right)$ is also a triangular number [171]

$$A_1 + A_2 = \left(a_1^{(1)}, a_M^{(1)}, a_2^{(1)}\right) + \left(a_1^{(2)}, a_M^{(2)}, a_2^{(2)}\right) \qquad (4.139)$$

$$= \left(a_1^{(1)} + a_1^{(2)}, a_M^{(1)} + a_M^{(2)}, a_2^{(1)} + a_2^{(2)}\right).$$

This is also true for $n > 2$ triangular numbers.

4.6   Triangular norms and negations

In point 4.3, we have defined the operations of intersection and union of fuzzy sets as

$$\mu_{A \cap B}(x) = \min(\mu_A(x), \mu_B(x)),$$
$$\mu_{A \cup B}(x) = \max(\mu_A(x), \mu_B(x)).$$

At the same time, we have stressed that these are not the only definitions of such operations. The *intersection* of fuzzy sets may be more generally defined as

$$\mu_{A \cap B}(x) = T(\mu_A(x), \mu_B(x)), \tag{4.140}$$

where the function T is the so-called t-norm. Therefore, $\min(\mu_A(x), \mu_B(x)) = T(\mu_A(x), \mu_B(x))$ is an example of operation of the t-norm. Similarly, the union of fuzzy sets is defined as follows:

$$\mu_{A \cup B}(x) = S(\mu_A(x), \mu_B(x)), \tag{4.141}$$

where the function S is the so-called t-conorm. In this case, $\max(\mu_A(x), \mu_B(x)) = S(\mu_A(x), \mu_B(x))$ is an example of the t-conorm. It is worth noting that the t-norms and the t-conorms belong to the so-called triangular norms. These norms will be applied several times further in this book, not only to define the operations of intersection and union of fuzzy sets.

The Reader has been presented with the examples of operation of a t-norm and a t-conorm, and now we shall present their formal definitions.

Definition 4.22
The function of two variables T

$$T : [0, 1] \times [0, 1] \rightarrow [0, 1] \tag{4.142}$$

is called a t-norm, if
(i) function T is nondecreasing with relation to both arguments

$$T(a, c) \leq T(b, d) \quad \text{for} \quad a \leq b, \ c \leq d \tag{4.143}$$

(ii) function T satisfies the condition of commutativity

$$T(a, b) = T(b, a) \tag{4.144}$$

(iii) function T satisfies the condition of associativity

$$T(T(a, b), c) = T(a, T(b, c)) \tag{4.145}$$

(iv) function T satisfies the boundary condition

$$T(a, 1) = a, \tag{4.146}$$

where $a, b, c, d \in [0, 1]$.

From the assumptions it follows that

$$T(a, 0) = T(0, a) \leq T(0, 1) = 0. \tag{4.147}$$

Therefore, the second boundary condition takes the form

$$T(a, 0) = 0. \tag{4.148}$$

Further in this chapter, we shall notate the operation of t-norm on arguments a and b in the following way

$$T(a, b) = a \overset{T}{*} b. \tag{4.149}$$

If, for instance, a and b are identified with the membership functions of fuzzy sets A and B, then we shall notate equality (4.140) as

$$\mu_{A \cap B}(x) = T(\mu_A(x), \mu_B(x)) = \mu_A(x) \overset{T}{*} \mu_A(x). \tag{4.150}$$

Using property (4.145), the definition of t-norm may be generalized for the case of a t-norm of multiple variables

$$\underset{i=1}{\overset{n}{T}} \{a_i\} = T\left\{ \underset{i=1}{\overset{n-1}{T}} \{a_i\}, a_n \right\} = T\{a_1, a_2, ..., a_n\} = T\{\mathbf{a}\} \tag{4.151}$$
$$= a_1 \overset{T}{*} a_2 \overset{T}{*} ... \overset{T}{*} a_n.$$

Below, we present the definition of triangular norm with weights $0 \leq w_i \leq 1$, $i = 1, ..., n$.

Definition 4.23
A *weighted* t-norm shall be denoted as T^* and defined as follows:

$$T^*\{a_1, ..., a_n; w_1, ..., w_n\} = \underset{i=1}{\overset{n}{T}} \{1 - w_i(1 - a_i)\}, \tag{4.152}$$

where T is any t-norm, and the weights meet the condition $0 \leq w_i \leq 1$, $i = 1, ..., n$.

Let us assume that $w_i = 1$, $i = 1, ..., n$. Then the weighted t-norm T^* is reduced to the t-norm T. In Chapters 9 and 10 the weights w_i are interpreted as degrees of truth of the antecedens of fuzzy rules or degrees of truth of particular rules in the so-called logical model. It can be easily checked that the weighted t-norm (similarly to the weighted t-conorm – Definition 4.25) does not meet the boundary conditions of a classic t-norm. However, as we will demonstrate in Chapters 9 and 10, the application of this concept allows to design neuro-fuzzy structures characterized by a high accuracy.

Definition 4.24

The function of two variables S

$$S : [0, 1] \times [0, 1] \rightarrow [0, 1] \qquad (4.153)$$

is called a *t-conorm*, if it is nondecreasing with relation to both arguments, meets the condition of commutativity and associativity, and the following boundary condition is met:

$$S(a, 0) = a. \qquad (4.154)$$

From the assumptions and condition (4.154) we get:

$$S(a, 1) = S(1, a) \geq S(1, 0) = 1. \qquad (4.155)$$

Therefore, the second boundary condition takes the form

$$S(a, 1) = 1. \qquad (4.156)$$

The operation of *t*-conorm on arguments a and b will be notated in the following way

$$S(a, b) = a \overset{S}{*} b. \qquad (4.157)$$

Using the property of associativity, the above definition may be generalized for the case of a *t*-conorm of multiple variables

$$\overset{n}{\underset{i=1}{S}} \{a_i\} = S\left\{ \overset{n-1}{\underset{i=1}{S}} \{a_i\}, a_n \right\} = S\{a_1, a_2, ..., a_n\} = S\{\mathbf{a}\} \qquad (4.158)$$

$$= a_1 \overset{S}{*} a_2 \overset{S}{*} ... \overset{S}{*} a_n.$$

Definition 4.25

A *weighted* *t*-conorm shall be notated as S^* and defined as follows:

$$S^*\{a_1, ..., a_n; w_1, ..., w_n\} = \overset{n}{\underset{i=1}{S}} \{w_i a_i\}. \qquad (4.159)$$

In Chapters 9 and 10 the weights w_i are interpreted as degrees of truth of particular rules in the so-called Mamdani model.

Definition 4.26

Functions T and S, meeting the conditions

$$\overset{n}{\underset{i=1}{S}} \{a_i\} = 1 - \overset{n}{\underset{i=1}{T}} \{1 - a_i\}, \qquad (4.160)$$

$$\mathop{T}_{i=1}^{n} \{a_i\} = 1 - \mathop{S}_{i=1}^{n} \{1 - a_i\}, \tag{4.161}$$

are called *dual* triangular norms.

Definition 4.27
We say that a pair of dual triangular operators T and S has an *Archimedean property*, if

$$T\{a,a\} < a < S\{a,a\} \tag{4.162}$$

for each $a \in (0,1)$.

Definition 4.28
We say that a pair of dual continuous triangular operators T and S is *of np (nilpotent) type*, if for a given sequence of arguments $a_i \in (0,1)$, $i = 1, 2, ...,$ there is an index n such as

$$T\{a_1, a_2, ..., a_n\} = 0, \tag{4.163}$$

$$S\{a_1, a_2, ..., a_n\} = 1. \tag{4.164}$$

Definition 4.29
We say that a pair of dual continuous triangular operators T and S is of *st (strict) type*, if

$$T\{a_1, a_2, \ldots, a_n\} > 0, \tag{4.165}$$

$$S\{a_1, a_2, \ldots, a_n\} < 1 \tag{4.166}$$

for $0 < a_i < 1$, $i = 1, ..., n$, $n \geq 2$ and $a_1 = a_2 = ... = a_n$.

Below, examples of t-norms and corresponding t-conorms have been presented. Their 2-dimensional representations are depicted in Fig. 4.29–4.32.

Example 4.23 (Triangular norms of the *min/max* type)
The triangular norms of the min/max type, called the Zadeh triangular norms, are described by the following dependencies:

$$T_M\{a_1, a_2\} = \min\{a_1, a_2\}, \tag{4.167}$$

$$S_M\{a_1, a_2\} = \max\{a_1, a_2\}, \tag{4.168}$$

$$T_M\{a_1, a_2, ..., a_n\} = \min_{i=1,...,n}\{a_i\}, \tag{4.169}$$

$$S_M\{a_1, a_2, ..., a_n\} = \max_{i=1,...,n}\{a_i\}. \tag{4.170}$$

Min/max triangular norms are dual but are not Archimedean.

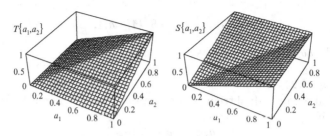

FIGURE 4.29. Triangular norms described by formulas (4.167) and (4.168)

Example 4.24 (Algebraic triangular norms)
Algebraic triangular norms are described by the following formulas:

$$T_P \{a_1, a_2\} = a_1 a_2, \tag{4.171}$$

$$S_P \{a_1, a_2\} = a_1 + a_2 - a_1 a_2, \tag{4.172}$$

$$T_P \{a_1, a_2, ..., a_n\} = \prod_{i=1}^{n} a_i, \tag{4.173}$$

$$S_P \{a_1, a_2, ..., a_n\} = 1 - \prod_{i=1}^{n} (1 - a_i). \tag{4.174}$$

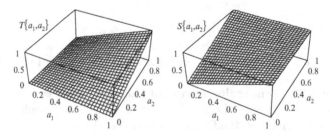

FIGURE 4.30. Triangular norms described by formulas (4.171) and (4.172)

Algebraic triangular norms are dual triangular norms of *the strict* type.

Example 4.25 (Łukasiewicz triangular norms)
Łukasiewicz triangular norms are described by the following dependencies:

$$T_L \{a_1, a_2\} = \max \{a_1 + a_2 - 1, 0\}, \tag{4.175}$$

$$S_L \{a_1, a_2\} = \min \{a_1 + a_2, 1\}, \tag{4.176}$$

$$T_L \{a_1, a_2, ..., a_n\} = \max \left\{ \sum_{i=1}^{n} a_i - (n - 1), 0 \right\}, \tag{4.177}$$

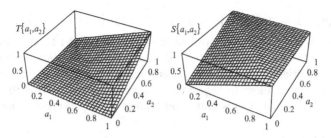

FIGURE 4.31. Triangular norms described by formulas (4.175) and (4.176)

$$S_L\{a_1, a_2, ..., a_n\} = \min\left\{\sum_{i=1}^{n} a_i, 1\right\}. \tag{4.178}$$

Łukasiewicz triangular norms are dual triangular norms of *the nilpotent* type.

Example 4.26 (Boundary triangular norms)
Boundary triangular norms are described by the following formulas:

$$T_D\{a_1, a_2\} = \begin{cases} 0, & \text{if } S_M\{a_1, a_2\} < 1, \\ T_M\{a_1, a_2\}, & \text{if } S_M\{a_1, a_2\} = 1. \end{cases} \tag{4.179}$$

$$S_D\{a_1, a_2\} = \begin{cases} 1, & \text{if } T_M\{a_1, a_2\} > 0, \\ S_M\{a_1, a_2\}, & \text{if } T_M\{a_1, a_2\} = 0. \end{cases} \tag{4.180}$$

$$T_D\{a_1, a_2, ..., a_n\} = \begin{cases} 0, & \text{if } S_M\{a_1, a_2, ..., a_n\} < 1, \\ T_M\{a_1, a_2, ..., a_n\}, & \text{if } S_M\{a_1, a_2, ..., a_n\} = 1. \end{cases} \tag{4.181}$$

$$S_D\{a_1, a_2, ..., a_n\} = \begin{cases} 1, & \text{if } T_M\{a_1, a_2, ..., a_n\} > 0, \\ S_M\{a_1, a_2, ..., a_n\}, & \text{if } T_M\{a_1, a_2, ..., a_n\} = 0, \end{cases} \tag{4.182}$$

It is worth reminding that all the triangular norms satisfy the following inequalities:

$$S_M\{a_1, a_2, ..., a_n\} \leq S\{a_1, a_2, ..., a_n\} \leq S_D\{a_1, a_2, ..., a_n\}, \tag{4.183}$$

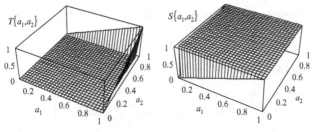

FIGURE 4.32. Triangular norms described by formulas (4.179) and (4.180)

$$T_D\{a_1, a_2, ..., a_n\} \leq T\{a_1, a_2, ..., a_n\} \leq T_M\{a_1, a_2, ..., a_n\}. \qquad (4.184)$$

The generators of triangular norms are directly related to the triangular norms. Functions called generators of triangular norms allow to define the properties of triangular norms, e.g. distinguish the nilpotent or strict type triangular norms, allow to generalize binary triangular norms to multidimensional triangular norms and allow to derive new triangular norms.

Definition 4.30

The function of the form
$$t_{\text{mul}} : [0,1] \rightarrow [t_{\text{mul}}(0), 1] \qquad (4.185)$$

is called a *multiplicative generator of Archimedean t-norm*

$$T\{a_1, a_2, ..., a_n\} = t_{\text{mul}}^{-1}\left(\left(\prod_{i=1}^{n} t_{\text{mul}}(a_i)\right) \vee t_{\text{mul}}(0)\right) \qquad (4.186)$$

for $a_1, a_2, ..., a_n \in [0,1]$. For Archimedean t-norm of st (*strict*) type, we have

$$t_{\text{mul}}(0) = 0. \qquad (4.187)$$

Therefore, formula (4.186) takes the form

$$T\{a_1, a_2, ..., a_n\} = t_{\text{mul}}^{-1}\left(\prod_{i=1}^{n} t_{\text{mul}}(a_i)\right) \qquad (4.188)$$

for $a_1, a_2, ..., a_n \in [0,1]$.

Definition 4.31

The function of the form

$$t_{\text{add}} : [0,1] \rightarrow [0, t_{\text{add}}(0)] \qquad (4.189)$$

is called an *additive generator of Archimedean t-norm*

$$T\{a_1, a_2, ..., a_n\} = t_{\text{add}}^{-1}\left(\left(\sum_{i=1}^{n} t_{\text{add}}(a_i)\right) \wedge t_{\text{add}}(0)\right) \qquad (4.190)$$

for $a_1, a_2, ..., a_n \in [0,1]$.

For Archimedean t-norm of st (*strict*) type, we have

$$t_{\text{add}}(0) = \infty. \qquad (4.191)$$

Then formula (4.190) takes the form

$$T\{a_1, a_2, ..., a_n\} = t_{\text{add}}^{-1}\left(\sum_{i=1}^{n} t_{\text{add}}(a_i)\right) \qquad (4.192)$$

for $a_1, a_2, ..., a_n \in [0,1]$.

Definition 4.32

The function of the form

$$s_{\text{mul}} : [0, 1] \to [s_{\text{mul}} (1) , 1] \tag{4.193}$$

is called a *multiplicative generator of Archimedean t-conorm*

$$S\{a_1, a_2, ..., a_n\} = s_{\text{mul}}^{-1} \left(\left(\prod_{i=1}^{n} s_{\text{mul}} (a_i) \right) \vee s_{\text{mul}} (1) \right) \tag{4.194}$$

for $a_1, a_2, \ldots, a_n \in [0, 1]$.

For Archimedean t-conorm of st (*strict*) type, we have

$$s_{\text{mul}} (1) = 0. \tag{4.195}$$

Therefore, formula (4.194) takes the form

$$S\{a_1, a_2, ..., a_n\} = s_{\text{mul}}^{-1} \left(\prod_{i=1}^{n} s_{\text{mul}} (a_i) \right) \tag{4.196}$$

for $a_1, a_2, \ldots, a_n \in [0, 1]$.

Definition 4.33

The function of the form

$$s_{\text{add}} : [0, 1] \to [0, s_{\text{add}} (1)] \tag{4.197}$$

is called an *additive generator of Archimedean t-conorm (nilpotent)*

$$S\{a_1, a_2, ..., a_n\} = s_{\text{add}}^{-1} \left(\left(\sum_{i=1}^{n} s_{\text{add}} (a_i) \right) \wedge s_{\text{add}} (1) \right) \tag{4.198}$$

for $a_1, a_2, \ldots, a_n \in [0, 1]$.

For Archimedean t-conorm of st (*strict*) type, we have

$$s_{\text{add}} (1) = \infty. \tag{4.199}$$

Therefore, formula (4.198) takes the form

$$S\{a_1, a_2, ..., a_n\} = s_{\text{add}}^{-1} \left(\sum_{i=1}^{n} s_{\text{add}} (a_i) \right) \tag{4.200}$$

for $a_1, a_2, \ldots, a_n \in [0, 1]$.

The relation of a multiplicative and additive generator of Archimedean t-norm is as follows:

$$t_{\text{mul}} (a) = \exp (-t_{\text{add}} (a)), \tag{4.201}$$

$$t_{\text{add}}(a) = -\ln(t_{\text{mul}}(a)). \tag{4.202}$$

The relation of a multiplicative and additive generator of dual triangular norms is following:

$$s_{\text{mul}}(a) = t_{\text{mul}}(1-a), \tag{4.203}$$

$$s_{\text{add}}(a) = t_{\text{add}}(1-a). \tag{4.204}$$

Remark 4.7

Let f be any multiplicative generator of Archimedean t-norm T. Then t-norm T is of the nilpotent type if and only if $f(0) > 0$, and is of the strict type if and only if, $f(0) = 0$. This result has been presented in monograph [144]. If however f is any additive generator of Archimedean t-norm T, then t-norm T is of the nilpotent type if and only if $f(0) < \infty$, and is of the strict type if and only if $f(0) = \infty$. This result has been presented in monograph [111].

Example 4.27

Let us consider a multiplicative generator of Archimedean t-norm given by the formula

$$t_{\text{mul}}(a) = a^p, \quad p > 0. \tag{4.205}$$

Using dependencies (4.201) – (4.204), we may determine an additive generator of the Archimedean t-norm and a multiplicative and additive generator of Archimedean t-conorm as follows:

$$t_{\text{add}}(a) = -\ln(t_{\text{mul}}(a)) = -p\ln(a), \tag{4.206}$$

$$s_{\text{mul}}(a) = t_{\text{mul}}(1-a) = (1-a)^p, \tag{4.207}$$

$$s_{\text{add}}(a) = t_{\text{add}}(1-a) = -p\ln(1-a). \tag{4.208}$$

If we want to use the abovementioned generators to generate triangular norms, we should define their inverse functions

$$t_{\text{mul}}^{-1}(a) = a^{\frac{1}{p}}, \tag{4.209}$$

$$t_{\text{add}}^{-1}(a) = \exp(-a)^{\frac{1}{p}}, \tag{4.210}$$

$$s_{\text{mul}}^{-1}(a) = 1 - a^{\frac{1}{p}}, \tag{4.211}$$

$$s_{\text{add}}^{-1}(a) = 1 - \exp(-a)^{\frac{1}{p}}. \tag{4.212}$$

Let us note that for the generator given by formula (4.205), we have

$$t_{\mathrm{mul}}(0) = s_{\mathrm{mul}}(1) = 0. \tag{4.213}$$

Boundary conditions indicate that the generators used are related to triangular norms of the strict type. That is why we shall use dependencies (4.188), (4.192), (4.196) or (4.200), to determine these norms. Therefore

$$\begin{aligned} T\{a_1, a_2\} &= t_{\mathrm{mul}}^{-1}(t_{\mathrm{mul}}(a_1)\, t_{\mathrm{mul}}(a_2)) \tag{4.214} \\ &= t_{\mathrm{add}}^{-1}(t_{\mathrm{add}}(a_1) + t_{\mathrm{add}}(a_2)) \\ &= a_1 a_2 \\ &= T_P\{a_1, a_2\} \end{aligned}$$

and

$$\begin{aligned} S\{a_1, a_2\} &= s_{\mathrm{mul}}^{-1}(s_{\mathrm{mul}}(a_1)\, s_{\mathrm{mul}}(a_2)) \tag{4.215} \\ &= s_{\mathrm{add}}^{-1}(s_{\mathrm{add}}(a_1) + s_{\mathrm{add}}(a_2)) \\ &= a_1 + a_2 - a_1 a_2 \\ &= S_P\{a_1, a_2\}. \end{aligned}$$

Example 4.28

Let us consider a multiplicative generator of Archimedean t-norm

$$t_{\mathrm{mul}}(a) = \exp(a - 1). \tag{4.216}$$

Using dependencies (4.201) – (4.204), we may determine an additive generator of the Archimedean t-norm and a multiplicative and additive generator of Archimedean t-conorm as follows:

$$t_{\mathrm{add}}(a) = -\ln(t_{\mathrm{mul}}(a)) = 1 - a, \tag{4.217}$$

$$s_{\mathrm{mul}}(a) = t_{\mathrm{mul}}(1 - a) = \exp(-a), \tag{4.218}$$

$$s_{\mathrm{add}}(a) = t_{\mathrm{add}}(1 - a) = a. \tag{4.219}$$

If we want to use the abovementioned generators to generate triangular norms, we should define their inverse functions

$$t_{\mathrm{mul}}^{-1}(a) = 1 + \ln(a), \tag{4.220}$$

$$t_{\mathrm{add}}^{-1}(a) = 1 - a, \tag{4.221}$$

$$s_{\mathrm{mul}}^{-1}(a) = -\ln(a), \tag{4.222}$$

$$s_{\mathrm{add}}^{-1}(a) = a. \tag{4.223}$$

Let us remark that for the generator given by formula (4.216), we have

$$t_{\text{mul}}(0) = s_{\text{mul}}(1) = \frac{1}{e}. \tag{4.224}$$

Boundary conditions indicate that the generators used are related to triangular norms of the *nilpotent* type. That is why we shall use dependencies (4.186), (4.190), (4.194) or (4.198) to determine these norms. Then

$$\begin{aligned}
T\{a_1, a_2\} &= t_{\text{mul}}^{-1}\left((t_{\text{mul}}(a_1)\, t_{\text{mul}}(a_2)) \vee t_{\text{mul}}(0)\right) \tag{4.225}\\
&= t_{\text{add}}^{-1}\left((t_{\text{add}}(a_1) + t_{\text{add}}(a_2)) \wedge t_{\text{add}}(0)\right)\\
&= \max\{a_1 + a_2 - 1, 0\}\\
&= T_L\{a_1, a_2\}
\end{aligned}$$

and

$$\begin{aligned}
S\{a_1, a_2\} &= s_{\text{mul}}^{-1}\left((s_{\text{mul}}(a_1)\, s_{\text{mul}}(a_2)) \vee s_{\text{mul}}(1)\right) \tag{4.226}\\
&= s_{\text{add}}^{-1}\left((s_{\text{add}}(a_1) + s_{\text{add}}(a_2)) \wedge s_{\text{add}}(1)\right)\\
&= \min\{a_1 + a_2, 1\}\\
&= S_L\{a_1, a_2\}.
\end{aligned}$$

In Subchapter 4.3, we have defined the basic operations on fuzzy sets (intersection, union and Cartesian product), using the min/max. type operators. Now, we shall generalize these definitions, using any triangular norms.

Definition 4.34
The intersection of n fuzzy sets $A_1, A_2, ..., A_n \subseteq \mathbf{X}$ is the fuzzy set $A = A_1 \cap A_2 \cap ... \cap A_n = \bigcap_{i=1}^{n} A_i$ defined by the membership function

$$\mu_A(\mathbf{x}) = \mathop{T}_{i=1}^{n} \mu_{A_i}(\mathbf{x}). \tag{4.227}$$

Definition 4.35
The union of fuzzy sets $A_1, A_2, ..., A_n \subseteq \mathbf{X}$ is the fuzzy set $A = A_1 \cup A_2 \cup ... \cup A_n = \bigcup_{i=1}^{n} A_i$ defined by the membership function

$$\mu_A(\mathbf{x}) = \mathop{S}_{i=1}^{n} \mu_{A_i}(\mathbf{x}). \tag{4.228}$$

Definition 4.36
The Cartesian product of n fuzzy sets $A_1 \subseteq \mathbf{X}_1, A_2 \subseteq \mathbf{X}_2, ..., A_n \subseteq \mathbf{X}_n$, notated as $A_1 \times A_2 \times ... \times A_n$, is defined by

$$\begin{aligned}
\mu_{A_1 \times A_2 \times ... \times A_n}(\mathbf{x}_1, \mathbf{x}_2, ..., \mathbf{x}_n) &= T\{\mu_{A_1}(\mathbf{x}_1), \mu_{A_2}(\mathbf{x}_2) \tag{4.229}\\
..., \mu_{A_n}(\mathbf{x}_n)\} &= \mathop{T}_{i=1}^{n}\{\mu_{A_i}(\mathbf{x}_i)\}.
\end{aligned}$$

Negations are extensions of a logical contradiction. They have the role of contradiction operators in fuzzy linguistic models.

Definition 4.37

(i) A nonincreasing function $N : [0,1] \rightarrow [0,1]$ is called a *negation*, if $N(0) = 1$ and $N(1) = 0$.

(ii) Negation $N : [0,1] \rightarrow [0,1]$ is called a *st (strict) type* negation, if it is continuous and decreasing.

(iii) A negation of *st (strict)* type is called *strong type* negation, if it is involution, i.e. $N(N(a)) = a$.

Below, we have specified a list of selected negations.

Example 4.29

Zadeh negation is described below by the following dependency

$$N(a) = 1 - a. \tag{4.230}$$

It may be easily noticed that Zadeh negation is a *strong* type negation.

Example 4.30

Yager negation is described by the following dependency

$$N(a) = (1 - a^p)^{\frac{1}{p}}, \quad p > 0. \tag{4.231}$$

Yager negation is a *strict* type negation.

Example 4.31

Sugeno negation is described below by the following dependency

$$N(a) = \frac{1 - a}{1 + pa}, \quad p > -1. \tag{4.232}$$

Sugeno negation is a *strong* type negation.

Remark 4.8

If t-norm, t-conorm and negation N of the *strong* type meet the dependencies

$$\mathop{S}_{i=1}^{n} \{a_i\} = N^{-1}\left(\mathop{T}_{i=1}^{n} \{N(a_i)\}\right) \tag{4.233}$$

and

$$\mathop{T}_{i=1}^{n} \{a_i\} = N^{-1}\left(\mathop{S}_{i=1}^{n} \{N(a_i)\}\right), \tag{4.234}$$

then T, S and N form the co-called De Morgan triple and we say that t-norm T and t-conorm S are mutually N-dual. It should be noted that triangular norms in Definition 4.26 are dual when applying the Zadeh negation $N(a) = 1 - a$.

4.7 Fuzzy relations and their properties

One of the basic notions of the fuzzy sets theory is the notion of a fuzzy relation. Fuzzy relations allow to formalize the imprecise formulations like "x is almost equal to y" or "x is much greater than y". We shall present the definitions of fuzzy relation and composition of fuzzy relations. From the perspective of applications, Definition 4.40 concerning the composition of a fuzzy set and fuzzy relation is particularly important.

Definition 4.38
The fuzzy relation R between two non-empty (non-fuzzy) sets $\mathbf{X}$ and $\mathbf{Y}$ is called the fuzzy set determined on the Cartesian product $\mathbf{X} \times \mathbf{Y}$, i.e.

$$R \subseteq \mathbf{X} \times \mathbf{Y} = \{(x, y) : x \in \mathbf{X}, \, y \in \mathbf{Y}\}. \tag{4.235}$$

In other words, the fuzzy relation is a set of pairs

$$R = \{((x, y), \mu_R(x, y))\}, \quad \forall_{x \in \mathbf{X}} \forall_{y \in \mathbf{Y}}, \tag{4.236}$$

where

$$\mu_R : \mathbf{X} \times \mathbf{Y} \to [0, 1] \tag{4.237}$$

is the membership function. To each pair (x, y), $x \in \mathbf{X}$, $y \in \mathbf{Y}$ this function assigns membership degree $\mu_R(x, y)$, which is interpreted as strength of relation between elements $x \in \mathbf{X}$ and $y \in \mathbf{Y}$. According to the convention we adopted (Subchapter 4.2), the fuzzy relation may be notated in the form

$$R = \sum_{\mathbf{X} \times \mathbf{Y}} \frac{\mu_R(x, y)}{(x, y)} \tag{4.238}$$

or

$$R = \int_{\mathbf{X} \times \mathbf{Y}} \frac{\mu_R(x, y)}{(x, y)}. \tag{4.239}$$

Example 4.32
We shall apply Definition 4.38 to formalize an imprecise statement "y is more or less equal to x". Let $\mathbf{X} = \{3, 4, 5\}$ and $\mathbf{Y} = \{4, 5, 6\}$. Relation R may be defined as follows:

$$R = \frac{1}{(4, 4)} + \frac{1}{(5, 5)} + \frac{0.8}{(3, 4)} + \frac{0.8}{(4, 5)} + \frac{0.8}{(5, 4)} \tag{4.240}$$
$$+ \frac{0.8}{(5, 6)} + \frac{0.6}{(3, 5)} + \frac{0.6}{(4, 6)} + \frac{0.4}{(3, 6)}.$$

Therefore the membership function $\mu_R(x, y)$ of relation R has the form

$$\mu_R(x, y) = \begin{cases} 1 & \text{if} \quad x = y, \\ 0.8 & \text{if} \quad |x - y| = 1, \\ 0.6 & \text{if} \quad |x - y| = 2, \\ 0.4 & \text{if} \quad |x - y| = 3. \end{cases} \tag{4.241}$$

Relation R may also be notated using the matrix

$$
\begin{array}{c}
 & \begin{array}{ccc} y_1 & y_2 & y_3 \end{array} \\
\begin{array}{c} x_1 \\ x_2 \\ x_3 \end{array} & \left[\begin{array}{ccc} 0.8 & 0.6 & 0.4 \\ 1 & 0.8 & 0.6 \\ 0.8 & 1 & 0.8 \end{array} \right]
\end{array}
\tag{4.242}
$$

where $x_1 = 3$, $x_2 = 4$, $x_3 = 5$, and $y_1 = 4$, $y_2 = 5$, $y_3 = 6$.

Example 4.33
Let $\mathbf{X} = \mathbf{Y} = [0, 120]$ be the human lifespan. Then relation R with the membership function

$$
\mu_R(x, y) = \begin{cases} 0 & \text{if} \quad x - y \leq 0, \\[2mm] \dfrac{x - y}{30} & \text{if} \quad 0 < x - y < 30, \\[2mm] 1 & \text{if} \quad x - y \geq 30 \end{cases}
\tag{4.243}
$$

represents the imprecise proposition "a person of age x is much older than a person of age y". We should stress that the fuzzy relation R is a fuzzy set and therefore the definitions of intersection, union and complement stated in Subchapter 4.3 remain valid, i.e.

$$
\mu_{R \cap S}(x, y) = \min(\mu_R(x, y), \mu_S(x, y)),
\tag{4.244}
$$

$$
\mu_{R \cup S}(x, y) = \max(\mu_R(x, y), \mu_S(x, y)),
\tag{4.245}
$$

$$
\mu_{\widehat{R}}(x, y) = 1 - \mu_R(x, y).
\tag{4.246}
$$

In the fuzzy sets theory, an important role is given to the notion of composition of two fuzzy relations. Let us consider three non-fuzzy sets $\mathbf{X}$, $\mathbf{Y}$, $\mathbf{Z}$ and two fuzzy relations $R \subseteq \mathbf{X} \times \mathbf{Y}$ and $S \subseteq \mathbf{Y} \times \mathbf{Z}$ with membership functions $\mu_R(x, y)$ and $\mu_S(y, z)$, respectively.

Definition 4.39
Composition of sup-T type of fuzzy relations $R \subseteq \mathbf{X} \times \mathbf{Y}$ and $S \subseteq \mathbf{Y} \times \mathbf{Z}$ is called a fuzzy relation $R \circ S \subseteq \mathbf{X} \times \mathbf{Z}$ with the membership function

$$
\mu_{R \circ S}(x, z) = \sup_{y \in \mathbf{Y}} \left\{ \mu_R(x, y) \overset{T}{*} \mu_S(y, z) \right\}.
\tag{4.247}
$$

A specific form of the membership function $\mu_{R \circ S}(x, z)$ of the composition $R \circ S$ depends on the adopted t-norm in formula (4.247). If we take min as a t-norm, i.e. $T(a, b) = \min(a, b)$, then equality (4.247) may be notated as follows:

$$
\mu_{R \circ S}(x, z) = \sup_{y \in \mathbf{Y}} \left\{ \min[\mu_R(x, y), \mu_S(y, z)] \right\}.
\tag{4.248}
$$

Formula (4.248) is known in the literature as the sup-min type composition. If the set $\mathbf{Y}$ has an infinite number of elements, then the composition sup-min comes down to the max-min type composition of the form

$$\mu_{R \circ S}(x, z) = \max_{y \in \mathbf{Y}} \{\min[\mu_R(x, y), \mu_S(y, z)]\}. \tag{4.249}$$

Example 4.34

Let us assume that relations R and S are represented by matrices

$$R = \begin{bmatrix} 0.2 & 0.5 \\ 0.6 & 1 \end{bmatrix}, \quad S = \begin{bmatrix} 0.3 & 0.6 & 0.8 \\ 0.7 & 0.9 & 0.4 \end{bmatrix}, \tag{4.250}$$

while $\mathbf{X} = \{x_1, x_2\}$, $\mathbf{Y} = \{y_1, y_2\}$, $\mathbf{Z} = \{z_1, z_2, z_3\}$. The max-min type composition of the relations R and S has the form

$$Q = R \circ S = \begin{bmatrix} 0.2 & 0.5 \\ 0.6 & 1 \end{bmatrix} \circ \begin{bmatrix} 0.3 & 0.6 & 0.8 \\ 0.7 & 0.9 & 0.4 \end{bmatrix} \tag{4.251}$$

$$= \begin{bmatrix} q_{11} & q_{12} & q_{13} \\ q_{21} & q_{22} & q_{23} \end{bmatrix},$$

where

$$q_{11} = \max[\min(0.2; 0.3), \min(0.5; 0.7)] = 0.5,$$
$$q_{12} = \max[\min(0.2; 0.6), \min(0.5; 0.9)] = 0.5,$$
$$q_{13} = \max[\min(0.2; 0.8), \min(0.5; 0.4)] = 0.4,$$
$$q_{21} = \max[\min(0.6; 0.3), \min(1; 0.7)] = 0.7,$$
$$q_{22} = \max[\min(0.6; 0.6), \min(1; 0.9)] = 0.9,$$
$$q_{23} = \max[\min(0.6; 0.8), \min(1; 0.4)] = 0.6.$$

Therefore

$$Q = \begin{bmatrix} 0.5 & 0.5 & 0.4 \\ 0.7 & 0.9 & 0.6 \end{bmatrix}. \tag{4.252}$$

Table 4.1 lists the basic properties of fuzzy relations, where I means the unitary matrix and O the zero matrix.

TABLE 4.1. Basic properties of fuzzy relations

1	$R \circ I = I \circ R = R$
2	$R \circ O = O \circ R = O$
3	$(R \circ S) \circ T = R \circ (S \circ T)$
4	$R^m \circ R^n = R^{m+n}$
5	$(R^m)^n = R^{mn}$
6	$R \circ (S \cup T) = (R \circ S) \cup (R \circ T)$
7	$R \circ (S \cap T) \subseteq (R \circ S) \cap (R \circ T)$
8	$S \subset T \rightarrow R \circ S \subset R \circ T$

As we have already mentioned, the composition of a fuzzy set with a fuzzy relation is particularly important. The composition of this type will be used repeatedly further in this chapter. Let us consider a fuzzy set $A \subseteq \mathbf{X}$ and a fuzzy relation $R \subseteq \mathbf{X} \times \mathbf{Y}$ with membership functions $\mu_A(x)$ and $\mu_R(x, y)$, respectively.

Definition 4.40
Composition of a fuzzy set $A \subseteq \mathbf{X}$ and a fuzzy relation $R \subseteq \mathbf{X} \times \mathbf{Y}$ shall be notated $A \circ R$ and defined as a fuzzy set $B \subseteq \mathbf{Y}$

$$B = A \circ R \tag{4.253}$$

with the membership function

$$\mu_B(y) = \sup_{x \in \mathbf{X}} \left\{ \mu_A(x) \overset{T}{*} \mu_R(x, y) \right\}. \tag{4.254}$$

A specific form of formula (4.254) depends on the chosen t-norm (see Table 4.1) and on properties of set $\mathbf{X}$. Below, we shall present 4 cases:

1) If $T(a, b) = \min(a, b)$, then we obtain a composition of the sup-min type

$$\mu_B(y) = \sup_{x \in \mathbf{X}} \{\min[\mu_A(x), \mu_R(x, y)]\}. \tag{4.255}$$

2) If $T(a, b) = \min(a, b)$, and $\mathbf{X}$ is a set with a finite number of elements, then we obtain the composition of the max-min type

$$\mu_B(y) = \max_{x \in \mathbf{X}} \{\min[\mu_A(x), \mu_R(x, y)]\}. \tag{4.256}$$

3) If $T(a, b) = a \cdot b$, then we obtain a composition of the sup-product type

$$\mu_B(y) = \sup_{x \in \mathbf{X}} \{\mu_A(x) \cdot \mu_R(x, y)\}. \tag{4.257}$$

4) If $T(a, b) = a \cdot b$, and $\mathbf{X}$ is a set with a finite number of elements, then we obtain the composition of the max-product type

$$\mu_B(y) = \max_{x \in \mathbf{X}} \{\mu_A(x) \cdot \mu_R(x, y)\}. \tag{4.258}$$

Example 4.35
Let us assume that $\mathbf{X} = \{x_1, x_2, x_3\}$ and $\mathbf{Y} = \{y_1, y_2\}$, fuzzy set A has the form

$$A = \frac{0.4}{x_1} + \frac{1}{x_2} + \frac{0.6}{x_3}, \tag{4.259}$$

and the relation R represents the matrix

$$R = \begin{array}{c} \\ x_1 \\ x_2 \\ x_3 \end{array} \begin{array}{cc} y_1 & y_2 \\ \left[\begin{array}{cc} 0.5 & 0.7 \\ 0.2 & 1 \\ 0.9 & 0.3 \end{array}\right] \end{array}. \tag{4.260}$$

Composition $A \circ R$ of the max-min type is determined according to formula (4.256). The result of the composition is the fuzzy set B of the form

$$B = \frac{\mu_B(y_1)}{y_1} + \frac{\mu_B(y_2)}{y_2}, \qquad (4.261)$$

while

$$\mu_B(y_1) = \max\{\min(0.4;\ 0.5), \min(1;\ 0.2), \min(0.6;\ 0.9)\} \qquad (4.262)$$
$$= 0.6,$$

$$\mu_B(y_2) = \max\{\min(0.4;\ 0.7), \min(1;\ 1), \min(0.6;\ 0.3)\} \qquad (4.263)$$
$$= 1.$$

Therefore

$$B = \frac{0.6}{y_1} + \frac{1}{y_2}. \qquad (4.264)$$

Composition $A \circ R$ may also be defined as the projection of the intersection of the cylindrical extension $ce(A)$ of the fuzzy set A and fuzzy relation R, to space $\mathbf{Y}$, i.e.

$$B = A \circ R = \text{proj}\,\{ce(A) \cap R\} \text{ to } \mathbf{Y}, \qquad (4.265)$$

where the operations of cylindrical extension and projection have been defined below.

Definition 4.41
Cylindrical extension of set $A \subseteq \mathbf{X}$ to set $ce(A) \subseteq \mathbf{X} \times \mathbf{Y}$ is defined as follows:

$$ce(A) = \int_{\mathbf{X} \times \mathbf{Y}} \frac{\mu_A(x)}{(x,y)}. \qquad (4.266)$$

Definition 4.42
Projection of fuzzy set $A \subseteq \mathbf{X} \times \mathbf{Y}$ to space $\mathbf{Y}$ is called an operation defined as follows:

$$\text{proj}\,A \text{ to } \mathbf{Y} = \int_{y \times \mathbf{Y}} \frac{\sup_{x \in \mathbf{X}} \mu_A(x,y)}{y}. \qquad (4.267)$$

4.8   Approximate reasoning

4.8.1   Basic rules of inference in binary logic

In traditional (binary) logic, we infer about the truth of some sentences based on the truth of some other sentences. This reasoning shall be notated in the form of a schema: above the horizontal line, we shall put all the sentences, on the basis of which we make the reasoning, below the line, we shall put the inference. The schema of correct inference has such property that if all sentences above the horizontal line are true, then also the sentence below the line is true, as true sentences may lead only to a true inference. In this point capital letters A and B symbolize the sentences rather than fuzzy sets. Let A and B be the sentences, while the notation $A = 1(B = 1)$ means that the logical value $A(B)$ is true and the notation $A = 0(B = 0)$ means, that the logical value of the sentence $A(B)$ is falsehood. We shall present below two rules of inference used in binary logic.

Definition 4.43
The reasoning rule modus ponens is defined by the following reasoning schema:

Premise	A
Implication	$A \rightarrow B$

Inference	B

$$(4.268)$$

Example 4.36
Let sentence A have the form "John is a driver", and sentence B – "John has got a driving license". Pursuant to the rule of modus ponens, if $A = 1$, then also $B = 1$, as if it is true that "John is a driver", then it is also true that "John has got a driving license". In other words, the truth of the premise and implication (sentences above the line) result in the truth of the inference (sentence below the line).

Definition 4.44
The reasoning rule modus tollens is defined by the following reasoning schema:

Premise	$\overline{B}$
Implication	$A \rightarrow B$

Inference	$\overline{A}$

$$(4.269)$$

Example 4.37
When continuing Example 4.36 we understand that if "John does not have a driving license", or $B = 0$ $\left(\overline{B} = 1 \right)$, then "John is not a driver", or $A = 0$ $\left(\overline{A} = 1 \right)$. Also in this example, the truth of the premise and the implication lead to the truth of the inference.

We have presented only the two inference rules in the binary logic, which will be generalized to the fuzzy case. Of course, in the binary logic, a series

of other inference rules is known. We shall refer the interested Reader to a
rich literature in this subject (e.g. [234]).

4.8.2 Basic rules of inference in fuzzy logic

Currently, we shall extend the basic inference rules in binary logic to a
fuzzy case. Let us assume that the sentences occurring in the modus ponens (4.268) and modus tollens (4.269) rules are characterized by some
fuzzy sets. In this way, we shall obtain a generalized modus ponens inference rule and a generalized modus tollens inference rule.

4.8.2.1 A generalized fuzzy modus ponens inference rule

Definition 4.45
A generalized (fuzzy) modus ponens inference rule is defined by the following reasoning schema:

Premise	x is A'
Implication	**IF** x is A **THEN** y is B
Inference	y is B'

$$(4.270)$$

where $A, A' \subseteq \mathbf{X}$ and $B, B' \subseteq \mathbf{Y}$ are fuzzy sets, and x and y are the
so-called linguistic variables.

According to the above definition, *linguistic variables* are variables which
take as values words or sentences uttered in the natural language.
Examples may be provided by such statements as "low speed", "temper-
ate temperature" or "young person". These statements may be formalized
by assigning some fuzzy sets to them. It should be stressed that linguistic variables may, apart from word values, take numerical values just like
ordinary mathematical variables. The following example illustrates the generalized (fuzzy) modus ponens inference rule and presents the notion of the
linguistic variable.

Example 4.38
Let us consider the following reasoning schema

Premise	The car speed is high
Implication	If the car speed is very high, then the noise level is high
Inference	The noise level in the car is medium-high

$$(4.271)$$

In the schema above, the premise, implication and inference are imprecise statements. We shall distinguish the following linguistic variables: x – car speed, y – noise level. The set

$$T_1 = \{\text{"low", "medium", "high", "very high"}\}$$

is a set of values of the linguistic variable x. Similarly, the set

$$T_2 = \{\text{"low", "medium", "medium-high", "high"}\}$$

is a set of values of the linguistic variable y.

Each element of the set T_1 and T_2 may be assigned with an appropriate fuzzy set. When analyzing the reasoning schema (4.270) and (4.271), we obtain the following fuzzy sets:

$$A = \text{"very high speed of the car"},$$

$$A' = \text{"high speed of the car"},$$

and

$$B = \text{"high noise level"},$$

$$B' = \text{"medium-high noise level"}.$$

The Reader may propose membership functions for these fuzzy sets, like it has been presented in Fig. 4.12. Let us discuss the difference between non-fuzzy rule (4.268) and fuzzy rule (4.270). In both cases the implication is of the same form $A \to B$, where A and B are sentences (rule (268)) or fuzzy sets (rule (270)). However, the sentence A in the implication of the non-fuzzy rule is also included in the premise of this rule. On the other hand, the premise of the fuzzy rule does not concern the fuzzy set A, but relates to a certain fuzzy set A', which may be in a sense close to A, but not necessarily $A = A'$. In Example 4.38, the fuzzy set $A = $ "very high speed of the car" is not equal to the fuzzy set $A' = $ "high speed of the car". As a result, the inferences of schemas (4.268) and (4.270) are different. The inference of the fuzzy rule relates to a certain fuzzy set B', which is defined by the composition of the fuzzy set A' and a fuzzy implication $A \to B$, i.e.

$$B' = A' \circ (A \to B). \tag{4.272}$$

The fuzzy implication $A \to B$ is equivalent to a certain fuzzy relation $R \in \mathbf{X} \times \mathbf{Y}$ with the membership function $\mu_R(x, y)$. Therefore, the membership function of the fuzzy set B' may be determined using formula (4.254), which shall be notated as

$$\mu_{B'}(y) = \sup_{x \in \mathbf{X}} \left\{ \mu_{A'}(x) \overset{T}{*} \mu_{A \to B}(x, y) \right\}, \tag{4.273}$$

while $\mu_{A \rightarrow B}(x, y) = \mu_R(x, y)$. In the special case, when t-norm is of the min type, formula (4.273) takes the form

$$\mu_{B'}(y) = \sup_{x \in \mathbf{X}} \{ \min [\mu_{A'}(x), \mu_{A \rightarrow B}(x, y)] \}. \qquad (4.274)$$

The Reader will easily notice that if

$$A' = A, \quad \text{then} \quad B' = B, \qquad (4.275)$$

and the generalized fuzzy modus ponens rule of inference (4.270) is reduced to the modus ponens rule (4.268) discussed in point 4.8 1.

Let us now assume that there is an implication $A \rightarrow B$ in schema (4.270), while the fuzzy set A' (premise) is equal in turn:
1) $A' = A$,
2) $A' =$ "very A", while $\mu_{A'}(x) = \mu_A^2(x)$,
3) $A' =$ "more or less A", while $\mu_{A'}(x) = \mu_A^{1/2}(x)$,
4) $A' =$ "not A", while $\mu_{A'}(x) = 1 - \mu_A(x)$.

The fuzzy set "very A" is defined through the operation of concentration (4.75), the fuzzy set "more or less A" is defined through operation of dilation (4.76), and the fuzzy set "not A" is defined through the operation of complement (4.59). Table 4.2 presents (see [58]) the obvious relations that may exist between fuzzy sets A' and B'. Relation 1 is a *modus ponens* scheme (4.268), relations 2b and 3b occur, when there is no significant relation between A' and B', relation 4a means, that from the premise x is "not A" we cannot infer about y.

TABLE 4.2. Intuitive relations between the premises and inferences of the generalized modus ponens rule

Relation	Premise x is A'	Inference y is B'
1	x is A	y is B
2a	x is "very A"	y is "very B"
2b	x is "very A"	y is B
3a	x is "more or less A"	y is "more or less B"
3b	x is "more or less A"	y is B
4a	x is "not A"	y is undefined
4b	x is "not A"	y is "not B"

4.8.2.2. Generalized fuzzy modus tollens inference rule

Definition 4.46

A generalized (fuzzy) modus tollens inference rule is defined by the following reasoning schema:

Premise	y is B'
Implication	**IF** x is A **THEN** y is B

Inference	x is A'

$$(4.276)$$

where A, $A' \subseteq \mathbf{X}$ and B, $B' \subseteq \mathbf{Y}$ are fuzzy sets, and x and y are the linguistic variables.

Example 4.39

This example refers to Example 4.38, and at the same time the description which follows schema (4.271) remains valid.

Premise	The noise level in the car is medium-high
Implication	If the car speed is very high, then the noise level is high

Inference	The car speed is high

$$(4.277)$$

The fuzzy set A' in the inference of scheme (4.276) is defined through the composition of relations

$$A' = (A \rightarrow B) \circ B', \tag{4.278}$$

while

$$\mu_{A'}(x) = \sup_{y \in \mathbf{Y}} \left\{ \mu_{A \rightarrow B}(x, y) \overset{T}{*} \mu_{B'}(y) \right\}. \tag{4.279}$$

If t-norm is of the min type, then formula (4.279) takes the form

$$\mu_{A'}(x) = \sup_{y \in \mathbf{Y}} \left\{ \min \left[\mu_{A \rightarrow B}(x, y), \mu_{B'}(y) \right] \right\}. \tag{4.280}$$

If

$$A' = \overline{A} \quad \text{and} \quad B' = \overline{B}, \tag{4.281}$$

then the generalized fuzzy modus tollens rule of inference (4.276) is reduced to the modus tollens rule discussed in point 4.8.1. Table 4.3 shows the obvious relations [58] between the premises and inferences of the generalized modus tollens rule.

TABLE 4.3. Intuitive relations between the premises and inferences of the generalized modus tollens rule

Relation	Premise y is B'	Inference x is A'
1	y is "not B"	x is "not A"
2	y is "not very B"	x is "not very A"
3	y is "more or less B"	x is "more or less A"
4a	y is B	x is undefined
4b	y is B	x is A

4.8.3 Inference rules for the Mamdani model

In the previous point, we have discussed the generalized fuzzy modus ponens and modus tollens schemas of inference. Membership functions (4.273) and (4.279) in the inferences of these schemes depend on membership function $\mu_{A \to B}(x, y)$ of the fuzzy implication $A \to B$, which is equal to a certain fuzzy relation $R \subseteq \mathbf{X} \times \mathbf{Y}$. We shall present different methods of determining the function $\mu_{A \to B}(x, y)$ based on the knowledge of the membership function $\mu_A(x)$ and $\mu_B(y)$. In case of the Mamdani model, the membership functions $\mu_{A \to B}(x, y)$ shall be determined as follows:

$$\mu_{A \to B}(x, y) = T\ (\mu_A(x), \mu_B(y)), \tag{4.282}$$

where T is any t-norm. We may interpret function T in formula (4.282) as the correlation function between the antecedens and consequences in fuzzy rules. Most often, the minimum type rule defined below is applied:

- **Minimum type rule**

$$\mu_{A \to B}(x, y) = \mu_R(x, y) = \mu_A(x) \wedge \mu_B(y) \tag{4.283}$$
$$= \min[\mu_A(x), \mu_B(y)].$$

Another known rule is the product type rule (also referred to as the Larsen rule):

- **Product type rule (Larsen)**

$$\mu_{A \to B}(x, y) = \mu_R(x, y) = \mu_A(x) \cdot \mu_B(y). \tag{4.284}$$

It should be stressed that Mamdani type rules are not implications in the logical meaning, which can be easily demonstrated when analyzing Table 4.4.

TABLE 4.4. Illustration of the Mamdani type rules

$\mu_A(x)$	$\mu_B(y)$	$\min[\mu_A(x), \mu_B(y)]$	$\mu_A(x)\mu_B(y)$	$\mu_{A\to B}(x,y)$
0	0	0	0	1
0	1	0	0	1
1	0	0	0	0
1	1	1	1	1

Remark 4.9

Although the Mamdani rules are not implications in the logical sense, we may encounter in literature an erroneous interpretation of these rules as fuzzy implications. In monograph [134] Mendel refers to the Mamdani rules as "engineering implications", to differentiate them from fuzzy implications meeting the conditions of Definition 4.47 presented in the following point.

4.8.4 Inference rules for the logical model

We shall present different inference rules for the logical model using the fuzzy implication definition [55].

Definition 4.47

A *fuzzy implication* is the function $I : [0,1]^2 \to [0,1]$ meeting the following conditions:

a) if $a_1 \le a_3$, then $I(a_1, a_2) \ge I(a_3, a_2)$ for all $a_1, a_2, a_3 \in [0,1]$,

b) if $a_2 \le a_3$, then $I(a_1, a_2) \le I(a_1, a_3)$ for all $a_1, a_2, a_3 \in [0,1]$,

c) $I(0, a_2) = 1$ for all $a_2 \in [0,1]$,

d) $I(a_1, 1) = 1$ for all $a_1 \in [0,1]$,

e) $I(1, 0) = 0$.

Table 4.5 presents the most commonly applied fuzzy implications meeting all or some requirements of Definition 4.47.

Some of the fuzzy implications in Table 4.5 belong to special implication groups:

a) *S-implications* defined as follows:

$$I(a, b) = S\{1 - a, b\}.$$

Examples of S-implication are implications 1, 2, 3 and 4 in Table 4.5. They meet all the conditions of Definition 4.47.

b) *R-implications* defined as follows:

$$I(a, b) = \sup_z \{z \mid T\{a, z\} \le b\}, \quad a, b \in [0,1].$$

Examples of R-implication are implications 6 and 7 in Table 4.5. They meet all the conditions of Definition 4.47.

TABLE 4.5. Fuzzy implications

No.	Name	Implication
1	Kleene-Dienes (binary)	$\max\{1-a, b\}$
2	Łukasiewicz	$\min\{1, 1-a+b\}$
3	Reichenbach	$1-a+a\cdot b$
4	Fodor	$\begin{cases} 1, & \text{if } a \le b \\ \max\{1-a, b\}, & \text{if } a > b \end{cases}$
5	Rescher	$\begin{cases} 1, & \text{if } a \le b \\ 0, & \text{if } a > b \end{cases}$
6	Goguen	$\begin{cases} 1, & \text{if } a = 0 \\ \min\left\{1, \dfrac{b}{a}\right\}, & \text{if } a > 0 \end{cases}$
7	Gödel	$\begin{cases} 1, & \text{if } a \le b \\ b, & \text{if } a > b \end{cases}$
8	Yager	$\begin{cases} 1, & \text{if } a = b \\ b^a, & \text{if } a > b \end{cases}$
9	Zadeh	$\max\{\min\{a, b\}, 1-a\}$
10	Willmott	$\min\left\{ \begin{array}{l} \max\{1-a, b\} \\ \max\{a, 1-b, \min\{1-a, b\}\} \end{array} \right\}$
11	Dubois-Prade	$\begin{cases} 1-a, & \text{if } b = 0 \\ b, & \text{if } a = 1 \\ 1, & \text{otherwise} \end{cases}$

c) *Q-implications* defined as follows:

$$I(a, b) = S\{N(a), T\{a, b\}\}, \quad a, b \in [0, 1],$$

where $N(a)$ is a negation operator. An example of Q-implication is the Zadeh implication, which does not meet the condition a) and d) of Definition 4.47.

Using Table 4.5, the membership functions $\mu_{A\to B}(x, y)$ shall be determined in case of a logical model as follows:

- **Binary implication (Kleene-Dienes)**

$$\mu_{A\to B}(x, y) = \max[1 - \mu_A(x), \mu_B(y)]. \tag{4.285}$$

- **Łukasiewicz implication**

$$\mu_{A\to B}(x, y) = \min[1.1 - \mu_A(x) + \mu_B(y)]. \tag{4.286}$$

- **Reichenbach implication**

$$\mu_{A\to B}(x, y) = 1.1 - \mu_A(x) + \mu_A(x) \cdot \mu_B(y). \tag{4.287}$$

- **Fodor implication**

$$\mu_{A \to B}(x, y) = \begin{cases} 1, & \text{if } \mu_A(x) \leq \mu_B(y), \\ \max\{1 - \mu_A(x), \mu_B(y)\}, & \text{if } \mu_A(x) > \mu_B(y). \end{cases}$$
(4.288)

- **Rescher implication**

$$\mu_{A \to B}(x, y) = \begin{cases} 1, & \text{if } \mu_A(x) \leq \mu_B(y), \\ 0, & \text{if } \mu_A(x) > \mu_B(y). \end{cases}$$
(4.289)

- **Goguen implication**

$$\mu_{A \to B}(x, y) = \begin{cases} \min\left[1, \dfrac{\mu_B(y)}{\mu_A(x)}\right], & \text{if } \mu_A(x) > 0, \\ 1, & \text{if } \mu_A(x) = 0. \end{cases}$$
(4.290)

- **Gödel implication**

$$\mu_{A \to B}(x, y) = \begin{cases} 1, & \text{if } \mu_A(x) \leq \mu_B(y), \\ \mu_B(y), & \text{if } \mu_A(x) > \mu_B(y). \end{cases}$$
(4.291)

- **Yager implication**

$$\mu_{A \to B}(x, y) = \begin{cases} 1, & \text{if } \mu_A(x) = 0, \\ \mu_B(y)^{\mu_A(x)}, & \text{if } \mu_A(x) > 0. \end{cases}$$
(4.292)

- **Zadeh implication**

$$\mu_{A \to B}(x, y) = \max\{\min[\mu_A(x), \mu_B(y)], 1 - \mu_A(x)\}.$$
(4.293)

- **Willmott implication**

$$\mu_{A \to B}(x, y) = \min \left\{ \begin{array}{l} \max\{1 - \mu_A(x), \ \mu_B(y)\}, \\ \max\{\mu_A(x), 1 - \mu_B(y), \\ \min\{1 - \mu_A(x), \mu_B(y)\}\}. \end{array} \right\}$$
(4.294)

- **Dubois-Prade implication**

$$\mu_{A \to B}(x, y) = \begin{cases} 1 - \mu_A(x), & \text{if } \mu_B(y) = 0 \\ \mu_B(y), & \text{if } \mu_A(x) = 1 \\ 1, & \text{otherwise} \end{cases}$$
(4.295)

It may be easily checked that in case of the logical model, particular inference rules are implications in the logical meaning. This fact is illustrated by Tables 4.6, 4.7 and 4.8 for the binary, Łukasiewicz and Reichenbach implications.

TABLE 4.6. Illustration of operation of binary implication

$\mu_A(x)$	$\mu_B(y)$	$1 - \mu_A(x)$	$\max[1 - \mu_A(x),\ \mu_B(y)]$	$\mu_{A \to B}(x, y)$
0	0	1	1	1
0	1	1	1	1
1	0	0	0	0
1	1	0	1	1

TABLE 4.7. Illustration of operation of Łukasiewicz implication

$\mu_A(x)$	$\mu_B(y)$	$1 - \mu_A(x) + \mu_B(y)$	$\min[1.1 - \mu_A(x) + \mu_B(y)]$	$\mu_{A \to B}(x, y)$
0	0	1	1	1
0	1	2	1	1
1	0	0	0	0
1	1	1	1	1

TABLE 4.8. Illustration of operation of Reichenbach implication

$\mu_A(x)$	$\mu_B(y)$	$1 - \mu_A(x) + \mu_A(x) \cdot \mu_B(y)$	$1.1 - \mu_A(x) + \mu_A(x) \cdot \mu_B(y)$	$\mu_{A \to B}(x, y)$
0	0	1	1	1
0	1	1	1	1
1	0	0	0	0
1	1	1	1	1

4.9   Fuzzy inference systems

In many issues concerning the technological processes control, it will be necessary to determine a model of the considered process. The knowledge of the model allows to select the appropriate controller. However, often it is very difficult to find an appropriate model, it is a problem which sometimes requires different simplifying assumptions. The application of the fuzzy sets theory to control technological processes does not require any knowledge of models of these processes. It is enough to formulate rules of procedure in the form of sentences like: **IF** ... **THEN**. Similarly, classification tasks may be solved. The approach using rules of the **IF** ... **THEN** type allows to solve a classification problem without the knowledge of probability densities of particular classes. Fuzzy control systems and classifiers are particular cases of fuzzy inference systems. Figure 4.33 illustrates a typical schema of such a system. It consists of the following elements:

1) rules base,
2) fuzzification block,
3) inference block,
4) defuzzification block.

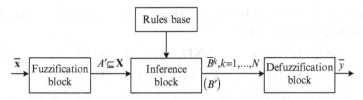

FIGURE 4.33. Block diagram of fuzzy inference system

Further in this chapter, we will assume that the inference block uses the Mamdani type model, in which the antecedents and consequents of rules are combined using a t-norm operation. In Chapter 9, we shall present in detail the neuro-fuzzy structures built using both the Mamdani and the logical model. Now, we will discuss particular elements of the fuzzy inference system.

4.9.1   Rules base

The rules base, sometimes called a *linguistic model,* is a set of fuzzy rules $R^{(k)}$, $k = 1, ..., N$, of the form

$$R^{(k)} : \textbf{IF } x_1 \text{ is } A_1^k \textbf{ AND } x_2 \text{ is } A_2^k \textbf{ AND...AND} \qquad (4.296)$$
$$x_n \text{ is } A_n^k \textbf{ THEN } y_1 \text{ is } B_1^k \textbf{ AND } y_2 \text{ is } B_2^k \textbf{ AND...AND } y_m \text{ is } B_m^k,$$

where N is the number of fuzzy rules, A_i^k – fuzzy sets such as

$$A_i^k \subseteq \mathbf{X}_i \subset \mathbf{R}, \quad i = 1, ..., n, \qquad (4.297)$$

B_j^k – fuzzy sets such as

$$B_j^k \subseteq \mathbf{Y}_j \subset \mathbf{R}, \quad j = 1, ..., m, \qquad (4.298)$$

$x_1, x_2, ..., x_n$ – input variables of the linguistic model, while

$$[x_1, x_2, ..., x_n]^T = \mathbf{x} \in \mathbf{X}_1 \times \mathbf{X}_2 \times ... \times \mathbf{X}_n, \qquad (4.299)$$

$y_1, y_2, ..., y_m$ – output variables of the linguistic model, while

$$[y_1, y_2, ..., y_m]^T = \mathbf{y} \in \mathbf{Y}_1 \times \mathbf{Y}_2 \times ... \times \mathbf{Y}_m. \qquad (4.300)$$

Symbols $\mathbf{X}_i$, $i = 1, ..., n$, and $\mathbf{Y}_j$, $j = 1, ..., m$, denote the spaces of input and output variables, respectively.

Further in our discussion, we shall assume that particular rules $R^{(k)}$, $k = 1, ..., N$, are related to each other using a logical operator "or". Moreover, we shall assume that outputs $y_1, y_2, ..., y_m$ are mutually independent.

Therefore, without losing the generality, we shall consider fuzzy rules with a scalar output of the form

$$R^{(k)} : \textbf{IF} \quad x_1 \text{ is } A_1^k \textbf{ AND } x_2 \text{ is } A_2^k \textbf{ AND} \qquad (4.301)$$

$$\ldots \textbf{AND} \quad x_n \text{ is } A_n^k \textbf{ THEN } y \text{ is } B^k,$$

where $B^k \subseteq \textbf{Y} \subset \textbf{R}$ and $k = 1, ..., N$. Let us notice that any rule of form (4.301) consists of the part **IF**, called *antecedent*, and a part **THEN** called *consequent*. The antecedent of the rule contains a set of conditions, and the consequent contains the inference. Variables $\textbf{x} = [x_1, x_2, ..., x_n]^T$ and y may take both linguistic values defined in words, like "small", "average" and "high", and numerical values. Let us denote

$$\textbf{X} = \textbf{X}_1 \times \textbf{X}_2 \times \ldots \times \textbf{X}_n \qquad (4.302)$$

and

$$A^k = A_1^k \times A_2^k \times \ldots \times A_n^k. \qquad (4.303)$$

Applying the above notations, rule (4.301) may be presented in the form

$$R^{(k)} : A^k \rightarrow B^k, \quad k = 1, \ldots, N. \qquad (4.304)$$

Let us notice that the rule $R^{(k)}$ may be interpreted as a fuzzy relation defined on the set $\textbf{X} \times \textbf{Y}$ i.e. $R^{(k)} \subseteq \textbf{X} \times \textbf{Y}$ is a fuzzy set with membership function

$$\mu_{R^{(k)}} (\textbf{x}, y) = \mu_{A^k \rightarrow B^k} (\textbf{x}, y). \qquad (4.305)$$

When designing fuzzy controllers, it should be decided whether the number of rules is sufficient, whether they are consistent and whether there are interactions between particular rules. These issues have been discussed in detail in works [33], [40] and [165].

4.9.2 Fuzzification block

A control system with fuzzy logic operates on fuzzy sets. That is why a specific value $\overline{\textbf{x}} = [\overline{x}_1, \overline{x}_2, ..., \overline{x}_n]^T \in \textbf{X}$ of an input signal of the fuzzy controller is subject to a fuzzification operation, as a result of which it is mapped into a fuzzy set $A' \subseteq \textbf{X} = \textbf{X}_1 \times \textbf{X}_2 \times \ldots \times \textbf{X}_n$. Most often in case of control problems, singleton type fuzzification is applied as follows

$$\mu_{A'} (\textbf{x}) = \delta (\textbf{x} - \overline{\textbf{x}}) = \begin{cases} 1, & \text{if } \textbf{x} = \overline{\textbf{x}}, \\ 0, & \text{if } \textbf{x} \neq \overline{\textbf{x}}. \end{cases} \qquad (4.306)$$

The fuzzy set A' is an input of the inference block. If the input signal is measured together with the interference (noise), then the fuzzy set A' may be defined using the membership function

$$\mu_{A'} (\textbf{x}) = \exp \left[-\frac{(\textbf{x} - \overline{\textbf{x}})^T (\textbf{x} - \overline{\textbf{x}})}{\delta} \right], \qquad (4.307)$$

where $\delta > 0$.

4.9.3   Inference block

Let us assume that at the input to the inference block we have a fuzzy set $A' \subseteq \mathbf{X} = \mathbf{X}_1 \times \mathbf{X}_2 \times \ldots \times \mathbf{X}_n$. We shall find an appropriate fuzzy set at the output of this block. Let us consider two cases to which different defuzzification methods will correspond.

Case 1. At the output of the inference block we obtain N fuzzy sets $\overline{B}^k \subseteq \mathbf{Y}$ according to the generalized fuzzy *modus ponens* inference rule. The fuzzy set $\overline{B}^k$ is determined by the composition of the fuzzy set A' and the relation $R^{(k)}$, i.e.

$$\overline{B}^{(k)} = A' \circ \left(A^k \rightarrow B^k \right), \quad k = 1, \ldots, N. \tag{4.308}$$

Using Definition 4.39, we shall determine the membership function of the fuzzy set $\overline{B}^k$ as follows

$$\mu_{\overline{B}^k}(y) = \sup_{\mathbf{x} \in \mathbf{X}} \left[\mu_{A'}(\mathbf{x}) \overset{T}{*} \mu_{A^k \rightarrow B^k}(\mathbf{x}, y) \right]. \tag{4.309}$$

A specific form of the function $\mu_{\overline{B}^k}(y)$ depends on the chosen t-norm (Subchapter 4.6), inference rule (points 4.8.3 and 4.8.4) and on the method of defining the Cartesian product of fuzzy sets (Definition 4.36). We should note that in case of singleton type fuzzification (4.306), formula (4.309) takes the form

$$\mu_{\overline{B}^k}(y) = \mu_{A^k \rightarrow B^k}(\overline{\mathbf{x}}, y). \tag{4.310}$$

Example 4.40

If $n = 2$, t-norm is of the min type, the fuzzy inference is defined by a rule of the min type and the Cartesian product of fuzzy sets is defined by formula (4.68), then formula (4.309) takes the form

$$\mu_{\overline{B}^k}(y) = \sup_{\mathbf{x} \in \mathbf{X}} \left[\min \left(\mu_{A'}(\mathbf{x}), \mu_{A^k \rightarrow B^k}(\mathbf{x}, y) \right) \right] \tag{4.311}$$

$$= \sup_{\mathbf{x} \in \mathbf{X}} \left\{ \min \left[\mu_{A'}(\mathbf{x}), \min \left(\mu_{A'}(\mathbf{x}), \mu_{B^k}(y) \right) \right] \right\}$$

$$= \sup_{\substack{x_1 \in \mathbf{X}_1, \\ x_2 \in \mathbf{X}_2}} \left\{ \min \left[\mu_{A'_1}(x_1), \mu_{A'_2}(x_2), \mu_{A^k_1}(x_1), \mu_{A^k_2}(x_2), \mu_{B^k}(y) \right] \right\}.$$

The last equality results from the fact that

$$\mu_{A^k}(\mathbf{x}) = \mu_{A^k_1 \times A^k_2}(x_1, x_2) = \min \left[\mu_{A^k_1}(x_1), \mu_{A^k_2}(x_2) \right] \tag{4.312}$$

and

$$\mu_{A'}(\mathbf{x}) = \mu_{A'_1 \times A'_2}(x_1, x_2) = \min \left[\mu_{A'_1}(x_1), \mu_{A'_2}(x_2) \right] \tag{4.313}$$

Example 4.41

If $n = 2$, t-norm is of the product type, the fuzzy inference is defined by a rule of the product type and the Cartesian product of fuzzy sets is defined by formula (4.69), then formula (4.309) takes the form

$$\mu_{\overline{B}^k}(y) = \sup_{x \in X}\{\mu_{A'}(\mathbf{x}) \cdot \mu_{A^k \to B^k}(\mathbf{x}, y)\} \tag{4.314}$$

$$= \sup_{x \in X}\{\mu_{A'}(\mathbf{x}) \cdot \mu_{A^k}(\mathbf{x}) \cdot \mu_{B^k}(y)\}$$

$$= \sup_{\substack{x_1 \in X_1, \\ x_2 \in X_2}} \left\{\mu_{A'_1}(x_1) \cdot \mu_{A'_2}(x_2) \cdot \mu_{A_1^k}(x_1) \cdot \mu_{A_2^k}(x_2) \cdot \mu_{B^k}(y)\right\}.$$

Case 2. At the output of the inference block, we obtain one fuzzy set $B' \subseteq \mathbf{Y}$, defined by the formula

$$B' = \bigcup_{k=1}^{N} A' \circ R^{(k)} = \bigcup_{k=1}^{N} A' \circ (A^k \to B^k). \tag{4.315}$$

Applying Definition 4.35, we obtain the membership function of a fuzzy set B'

$$\mu_{B'}(y) = \overset{N}{\underset{k=1}{S}} \mu_{\overline{B}^k}(y), \tag{4.316}$$

while the membership function $\mu_{\overline{B}^k}(y)$ is given by formula (4.309).

Example 4.42

Let us consider a fuzzy inference system with a rules base:

$$R^{(1)} : \textbf{IF } x_1 \text{ is } A_1^1 \textbf{ AND } x_2 \text{ is } A_2^1 \textbf{ THEN } y \text{ is } B^1, \tag{4.317}$$

$$R^{(2)} : \textbf{IF } x_1 \text{ is } A_1^2 \textbf{ AND } x_2 \text{ is } A_2^2 \textbf{ THEN } y \text{ is } B^2. \tag{4.318}$$

At the input of the controller, signal $\overline{\mathbf{x}} = [\overline{x}_1, \overline{x}_2]^T$ was given. As a result of singleton type fuzzification at the input to the inference block, we obtain fuzzy sets A'_1 and A'_2, while

$$\mu_{A'_1}(x_1) = \delta(x_1 - \overline{x}_1), \quad \mu_{A'_2}(x_2) = \delta(x_2 - \overline{x}_2). \tag{4.319}$$

We shall determine the output signal $\overline{y}$ of a fuzzy controller. As the t-norm the minimum operation is chosen. On the basis of formula (4.309), we have

$$\mu_{\overline{B}^k}(y) = \sup_{\substack{x_1 \in X_1, \\ x_2 \in X_2}} \left[\min\left(\mu_{A'_1 \times A'_2}(x_1, x_2), \mu_{R^{(k)}}(x_1, x_2, y)\right)\right]. \tag{4.320}$$

Additionally, we shall assume that

$$\mu_{A'_1 \times A'_2}(x_1, x_2) = \min\left[\mu_{A'_1}(x_1), \mu_{A'_2}(x_2)\right] \tag{4.321}$$

$$= \min\left[\delta(x_1 - \overline{x}_1), \delta(x_2 - \overline{x}_2)\right].$$

Therefore

$$\mu_{\overline{B}^k}(y) = \sup_{x_1 \in \mathbf{X}_1, x_2 \in \mathbf{X}_2} \left[\min\left(\delta(x_1 - \overline{x}_1), \delta(x_2 - \overline{x}_2), \mu_{R^{(k)}}(x_1, x_2, y)\right) \right]$$

$$= \mu_{R^{(k)}}(\overline{x}_1, \overline{x}_2, y) \tag{4.322}$$

and

$$\mu_{R^{(k)}}(\overline{x}_1, \overline{x}_2, y) = \mu_{A_1^k \times A_2^k \to B^k}(\overline{x}, \overline{x}_2, y). \tag{4.323}$$

In case the minimum (Mamdani) type rule is applied, we obtain

$$\mu_{A_1^k \times A_2^k \to B^k}(\overline{x}_1, \overline{x}_2, y) = \min\left[\mu_{A_1^k \times A_2^k}(\overline{x}_1, \overline{x}_2), \mu_{B^k}(y)\right]. \tag{4.324}$$

Moreover,

$$\mu_{A_1^k \times A_2^k}(\overline{x}_1, \overline{x}_2) = \min\left[\mu_{A_1^k}(\overline{x}_1), \mu_{A_2^k}(\overline{x}_2)\right]. \tag{4.325}$$

In consequence,

$$\mu_{\overline{B}^k}(y) = \min\left\{\min\left[\mu_{A_1^k}(\overline{x}_1), \mu_{A_2^k}(\overline{x}_2)\right], \mu_{B^k}(y)\right\} \tag{4.326}$$

$$= \min\left[\mu_{A_1^k}(\overline{x}_1), \mu_{A_2^k}(\overline{x}_2), \mu_{B^k}(y)\right]$$

and

$$\mu_{B'}(y) = \max_{k=1,2}\left\{\min\left[\mu_{A_1^k}(\overline{x}_1), \mu_{A_2^k}(\overline{x}_2), \mu_{B^k}(y)\right]\right\}. \tag{4.327}$$

Figure 4.34 illustrates the graphic interpretation of the fuzzy inference.

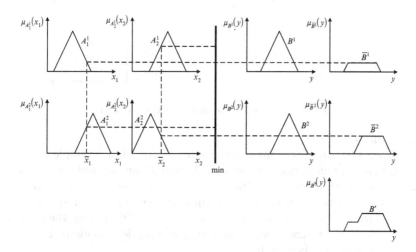

FIGURE 4.34. Illustration to Example 4.42

Example 4.43

In this example we will repeat the reasoning made in Example 4.42, but instead of rule (4.324), we shall apply the product (Larsen) type rule, i.e.

$$\mu_{A_1^k \times A_2^k \to B^k} (\overline{x}_1, \overline{x}_2, y) = \mu_{A_1^k \times A_2^k} (\overline{x}_1, \overline{x}_2) \cdot \mu_{B^k} (y). \qquad (4.328)$$

As a result of aggregation of rules 1 and 2, we will obtain the fuzzy set B' with the membership function

$$\mu_{B'} (y) = \max_{k=1,2} \left\{ \mu_{B^k} (y) \min \left[\mu_{A_1^k} (\overline{x}_1), \mu_{A_2^k} (\overline{x}_2) \right] \right\}. \qquad (4.329)$$

In this case,

$$\mu_{\overline{B}^k} (y) = \mu_{B^k} (y) \cdot \min \left[\mu_{A_1^k} (\overline{x}_1), \mu_{A_2^k} (\overline{x}_2) \right]. \qquad (4.330)$$

The graphic interpretation of the fuzzy inference is illustrated in Fig. 4.35.

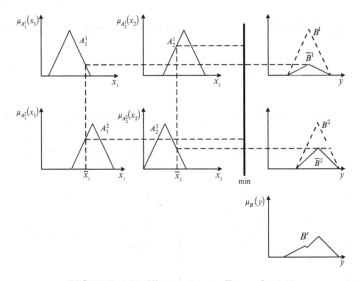

FIGURE 4.35. Illustration to Example 4.43

Example 4.44

Let us consider the fuzzy inference system described in Example 4.42 assuming that input (numerical) signals $\overline{x}_1$ and $\overline{x}_2$ are subject to fuzzification, as a result of which at the input of the inference block we obtain fuzzy sets A_1' and A_2' with membership functions $\mu_{A_1'} (x_1)$ and $\mu_{A_2'} (x_2)$. In other words, we withdraw from assumption (4.319) limiting the class of sets A_1' and A_2' to fuzzy singletons. Other assumptions made in Example 4.42 remain valid. Then we have

$$\mu_{\overline{B}^k}(y) = \sup_{\mathbf{x}\in\mathbf{X}}\left[\mu_{A'}(\mathbf{x}) \overset{T}{*} \mu_{R^{(k)}}(\mathbf{x},y)\right] \tag{4.331}$$

$$= \sup_{\substack{x_1\in\mathbf{X}_1,\\ x_2\in\mathbf{X}_2}} \min\{\min\left[\mu_{A_1'}(x_1),\mu_{A_2'}(x_2)\right], \mu_{R^{(k)}}(x_1,x_2,y)\}.$$

The min operation shall be notated with the symbol $\wedge$. Therefore

$$\mu_{\overline{B}^k}(y) = \sup_{\substack{x_1\in\mathbf{X}_1,\\ x_2\in\mathbf{X}_2}}\left[\mu_{A_1'}(x_1)\wedge\mu_{A_2'}(x_2)\wedge\mu_{R^{(k)}}(x_1,x_2,y)\right]$$

$$= \sup_{\substack{x_1\in\mathbf{X}_1,\\ x_2\in\mathbf{X}_2}}\left\{\mu_{A_1'}(x_1)\wedge\mu_{A_2'}(x_2)\wedge\left[\left(\mu_{A_1^k}(x_1)\wedge\mu_{A_2^k}(x_2)\right)\wedge\mu_{B^k}(y)\right]\right\}$$

$$= \left\{\sup_{\substack{x_1\in\mathbf{X}_1,\\ x_2\in\mathbf{X}_2}}\left[\mu_{A_1'}(x_1)\wedge\mu_{A_2'}(x_2)\wedge\left(\mu_{A_1^k}(x_1)\wedge\mu_{A_2^k}(x_2)\right)\right]\right\} \tag{4.332}$$

$$\wedge\,\mu_{B^k}(y)$$

$$= \left\{\sup_{x_1\in\mathbf{X}_1}\left[\mu_{A_1'}(x_1)\wedge\mu_{A_1^k}(x_1)\right]\wedge\sup_{x_2\in\mathbf{X}_2}\left[\mu_{A_2'}(x_2)\wedge\mu_{A_2^k}(x_2)\right]\right\}$$

In consequence, $\wedge\,\mu_{B^k}(y)$.

$$\mu_{B'}(y) = \max_{k=1,2}\left\{\left(\sup_{x_1\in\mathbf{X}_1}\left[\mu_{A_1'}(x_1)\wedge\mu_{A_1^k}(x_1)\right]\right.\right. \tag{4.333}$$

$$\left.\left.\wedge\sup_{x_2\in\mathbf{X}_2}\left[\mu_{A_2'}(x_2)\wedge\mu_{A_2^k}(x_2)\right]\right)\wedge\mu_{B^k}(y)\right\}.$$

Figure 4.36 illustrates the graphic interpretation of the fuzzy inference.

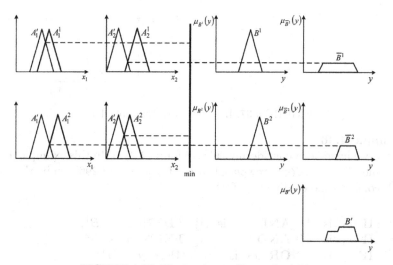

FIGURE 4.36. Illustration to Example 4.44

Example 4.45

In Example 4.44 we assumed that the t-norm, the Cartesian product and the inference rule are defined using the minimum operation. Now, we shall replace the minimum operation by product. In accordance with formula (4.309), we have

$$\mu_{\bar{B}^k}(y) = \sup_{\substack{x_1 \in \mathbf{X}_1, \\ x_2 \in \mathbf{X}_2}} [\mu_{A'}(x_1, x_2) \cdot \mu_{R^{(k)}}(x_1, x_2, y)]$$

$$= \sup_{\substack{x_1 \in \mathbf{X}_1, \\ x_2 \in \mathbf{X}_2}} \left[\mu_{A_1'}(x_1) \, \mu_{A_2'}(x_2) \, \mu_{A_1^k \times A_2^k}(x_1, x_2) \, \mu_{B^k}(y) \right] \qquad (4.334)$$

$$= \sup_{x_1 \in \mathbf{X}_1} \left[\mu_{A_1'}(x_1) \, \mu_{A_1^k}(x_1) \right] \sup_{x_2 \in \mathbf{X}_2} \left[\mu_{A_2'}(x_2) \, \mu_{A_2^k}(x_2) \right] \mu_{B^k}(y).$$

The membership function of fuzzy set B' shall be determined based on dependencies (4.334) and (4.316).

Figure 4.37 illustrates the graphic interpretation of the fuzzy inference.

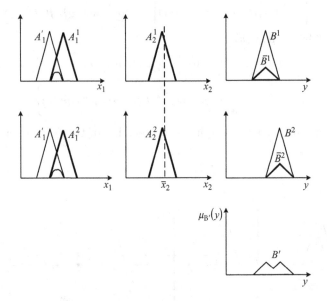

FIGURE 4.37. Illustration to Example 4.45

Example 4.46

So far, we have been considering rules of form (4.296). In this example, first two rules $R^{(1)}$ and $R^{(2)}$ are special cases of notation (4.296), while the rule $R^{(3)}$ contains the conjunction **OR**:

$R^{(1)}$: **IF** x_1 is A_1^1 **AND** x_2 is A_2^1 **THEN** y is B^1,
$R^{(2)}$: **IF** x_1 is A_1^2 **AND** x_2 is A_2^2 **THEN** y is B^2,
$R^{(3)}$: **IF** x_1 is A_1^3 **OR** x_3 is A_2^3 **THEN** y is B^3.

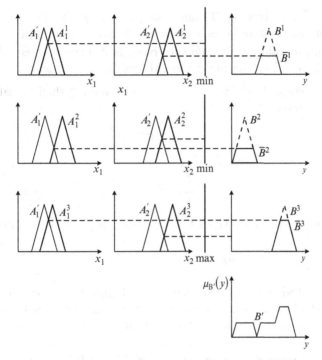

FIGURE 4.38. Illustration to Example 4.46

Figure 4.38 shows a graphic interpretation of a fuzzy inference with the assumption that the t-norm, Cartesian product and the inference rule are of the min type. The above problem may be solved in an alternative way. To do this, we shall notice that the rule $R^{(3)}$ may be notated as two rules $\overline{R}^{(3)}$ and $\overline{R}^{(4)}$:

$\overline{R}^{(3)}$: **IF** x_1 is A_1^3 **THEN** y is B^3,
$\overline{R}^{(4)}$: **IF** x_1 is A_2^3 **THEN** y is B^3.

We have obtained rules $R^{(1)}$, $R^{(2)}$, $\overline{R}^{(3)}$ and $\overline{R}^{(4)}$, which are special cases of notation (4.296). The Reader will easily derive analytical form of the membership function of set B', using the results of Example 4.44 as a template.

4.9.4   Defuzzification block

The output value of the inference block is either N fuzzy sets $\overline{B}^k$ with membership functions $\mu_{\overline{B}^k}(y)$, $k = 1, 2, \ldots, N$, or a single fuzzy set B' with membership function $\mu_{B'}(y)$. Now we consider a problem of mapping fuzzy sets $\overline{B}^k$ (or fuzzy set B') into a single value $\overline{y} \in \mathbf{Y}$. This mapping is called d*efuzzification* and it is made in the defuzzification block.

If the output value of the inference block is N fuzzy sets $\overline{B}^k$, then the value $\overline{y} \in \mathbf{Y}$ may be determined using the following methods:

1. *Center average defuzzification method.* The value $\overline{y}$ shall be calculated using the formula

$$\overline{y} = \frac{\sum\limits_{k=1}^{N} \mu_{\overline{B}^k}\left(\overline{y}^k\right) \overline{y}^k}{\sum\limits_{k=1}^{N} \mu_{\overline{B}^k}\left(\overline{y}^k\right)}, \tag{4.335}$$

where $\overline{y}^k$ is the point in which the function $\mu_{B^k}(y)$ takes the maximum value, i.e.

$$\mu_{B^k}\left(\overline{y}^k\right) = \max_{y} \mu_{B^k}(y). \tag{4.336}$$

Point $\overline{y}^k$ is called *center* of the fuzzy set B^k. Figure 4.39 shows the concept of this method for $N = 2$. Let us notice that value $\overline{y}$ does not depend on the shape and support of the membership function $\mu_{B^k}(y)$.

2. *Center of sums defuzzification method.* The value $\overline{y}$ is computed as follows:

$$\overline{y} = \frac{\int_{\mathbf{Y}} y \sum\limits_{k=1}^{N} \mu_{\overline{B}^k}(y)\,\mathrm{d}y}{\int_{\mathbf{Y}} \sum\limits_{k=1}^{N} \mu_{\overline{B}^k}(y)\,\mathrm{d}y}. \tag{4.337}$$

If the output value of the inference block is a single fuzzy set B', then the value $\overline{y}$ may be determined using the following methods:

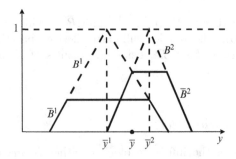

FIGURE 4.39. Illustration of the center average defuzzification method

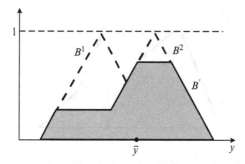

FIGURE 4.40. Illustration of the center of gravity method

3. *Center of gravity method* (or *center of area method*). The value $\bar{y}$ is calculated as *the center of gravity* of the membership function $\mu_{B'}(y)$, i.e.

$$\bar{y} = \frac{\int_Y y\mu_{B'}(y)\,dy}{\int_Y \mu_{B'}(y)\,dy} = \frac{\int_Y y\,S_{k=1}^N\,\mu_{\bar{B}^k}(y)}{\int_Y S_{k=1}^N\,\mu_{\bar{B}^k}(y)}, \qquad (4.338)$$

assuming that both integrals in the above formula exist. In a discrete case, the above formula takes the form

$$\bar{y} = \frac{\sum_{k=1}^N \mu_{B'}\left(\bar{y}^k\right)\bar{y}^k}{\sum_{k=1}^N \mu_{B'}\left(\bar{y}^k\right)}. \qquad (4.339)$$

Figure 4.40 shows the method of value $\bar{y}$ determination using the center of gravity method.

4. *Maximum membership function method.* The value $\bar{y}$ is computed according to the formula

$$\mu_{B'}(\bar{y}) = \sup_{y\in Y}\mu_{B'}(y) \qquad (4.340)$$

assuming that $\mu_{B'}(y)$ is a unimodal function. This method does not consider the shape of the membership function, which is illustrated in Fig. 4.41.

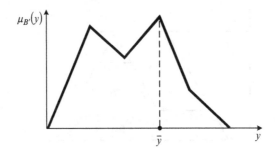

FIGURE 4.41. Illustration of the maximum membership function method

4.10   Application of fuzzy sets

4.10.1   Fuzzy Delphi method

Fuzzy Delphi method is the generalization of the classic Delphi technique concerning the long-term forecasting. The classic Delphi technique has been developed in the 1960s by Rand corporation in Santa Monica, California. The name of the method originates from ancient Greece, namely from the Delphi oracle, famous for predicting the future. The essence of the Delphi technique may be described as follows [13]:

- Highly qualified experts in a given discipline express independently their opinions on a certain event in the domain of science, technology or business. These opinions may be related to forecasts concerning the market, economy, technological progress, etc.

- Opinions of the experts are identified with data which are subjective in nature and are analyzed using statistical methods. Their average value is determined and the results are analyzed by the company's management board.

- In case of results that are not satisfactory for the company's management board, experts are requested once again to provide their opinions. At the same time, experts are provided with results of the previous round of inquiries.

- The process is repeated until a solution is obtained that is reasonable for the management board. In practice, usually two or three rounds are sufficient.

In long-term forecasting, we are dealing with imprecise and incomplete data. Decisions made by experts are subjective and depend mainly on their individual concepts. That is why it seems appropriate to present data using fuzzy numbers. It is particularly appropriate to use triangular numbers, as they may be easily constructed using three specific values: the smallest, the largest and the most probable (in the common meaning of this word). The analysis is based on the fuzzified average rather than on the real average. The fuzzy Delphi method consists of the following steps:

Step 1. Experts E_i, $i = 1, ..., n$, express their opinion on a certain event, e.g. the lowest $a_1^{(i)}$, the most probable $a_M^{(i)}$, and the highest $a_2^{(i)}$ price of Euro. Information provided by experts $E_{(i)}$ are presented by the company's management board in the form of fuzzy triangular numbers

$$A_i = \left(a_1^{(i)}, a_M^{(i)}, a_2^{(i)}\right), \quad i = 1, ..., n. \tag{4.341}$$

Step 2. The average is computed

$$A_{\text{aver}} = (m_1, m_M, m_2) = \left(\frac{1}{n}\sum_{i=1}^{n} a_1^{(i)}, \frac{1}{n}\sum_{i=1}^{n} a_M^{(i)}, \frac{1}{n}\sum_{i=1}^{n} a_2^{(i)}\right). \tag{4.342}$$

Step 3. Each expert E_i expresses his/her opinion again, taking into account the received averages from the previous round of inquiries and new fuzzy numbers are created.

$$B_i = \left(b_1^{(i)}, b_M^{(i)}, b_2^{(i)}\right), \quad i = 1, ..., n. \tag{4.343}$$

The process is repeated from step 2. The average B_{aver} is calculated according to formula (4.342), only that now $a_1^{(i)}, a_M^{(i)}, a_2^{(i)}$ are replaced respectively by $b_1^{(i)}, b_M^{(i)}, b_2^{(i)}$. If needed, the following triangular numbers $C_i = \left(c_1^{(i)}, c_M^{(i)}, c_2^{(i)}\right)$ are generated and their average $C_{\text{aver.}}$ is computed. The process is repeated until two subsequent sufficiently close averages $(A_{\text{aver}}, B_{\text{aver}}, C_{\text{aver}}, ...)$ are obtained.

Step 4. Later, if new relevant information is obtained concerning a given problem, the above procedure may be repeated.

The fuzzy Delphi method is a typical multi-expert forecasting procedure used to connect different views and opinions. We shall now present two examples of application of the fuzzy Delphi method using triangular fuzzy numbers.

Example 4.47. Estimate values of the Euro exchange rate in July next year

A group of 16 experts E_i, $i = 1, ..., 16$, was requested to express their opinions on the value of the Euro exchange rate (EUR/PLN) in July next year, using the fuzzy Delhi method. It is assumed that the experts' opinions have the same weight. Experts' forecasts, represented by triangular numbers A_i, $i = 1, ..., 16$ (4.341) have been placed in Table 4.9.

TABLE 4.9. Experts' forecasts (first inquiry)

Expert	Fuzzy number	Lowest value	Most probable value	Highest value
E_1	A_1	$a_1^{(1)} = 3.5882$	$a_M^{(1)} = 4.2062$	$a_2^{(1)} = 4.5060$
E_2	A_2	$a_1^{(2)} = 3.9854$	$a_M^{(2)} = 4.2070$	$a_2^{(2)} - 4.6020$
E_3	A_3	$a_1^{(3)} = 3.4868$	$a_M^{(3)} = 4.2071$	$a_2^{(3)} = 4.8524$
E_4	A_4	$a_1^{(4)} = 3.9201$	$a_M^{(4)} = 4.2065$	$a_2^{(4)} = 4.7925$
E_5	A_5	$a_1^{(5)} = 4.0012$	$a_M^{(5)} = 4.2080$	$a_2^{(5)} = 4.5900$
E_6	A_6	$a_1^{(6)} = 3.8724$	$a_M^{(6)} = 4.2063$	$a_2^{(6)} = 4.9825$
E_7	A_7	$a_1^{(7)} = 3.7760$	$a_M^{(7)} = 4.2062$	$a_2^{(7)} = 4.8250$
E_8	A_8	$a_1^{(8)} = 3.8925$	$a_M^{(8)} = 4.2065$	$a_2^{(8)} = 4.6872$
E_9	A_9	$a_1^{(9)} = 3.6823$	$a_M^{(9)} = 4.2085$	$a_2^{(9)} = 4.6257$
E_{10}	A_{10}	$a_1^{(10)} = 4.0010$	$a_M^{(10)} = 4.2067$	$a_2^{(10)} = 4.6889$
E_{11}	A_{11}	$a_1^{(11)} = 3.8926$	$a_M^{(11)} = 4.2051$	$a_2^{(11)} = 4.9820$
E_{12}	A_{12}	$a_1^{(12)} = 3.5868$	$a_M^{(12)} = 4.2061$	$a_2^{(12)} = 4.9560$
E_{13}	A_{13}	$a_1^{(13)} = 3.8101$	$a_M^{(13)} = 4.2055$	$a_2^{(13)} = 4.9920$
E_{14}	A_{14}	$a_1^{(14)} = 3.7865$	$a_M^{(14)} = 4.2082$	$a_2^{(14)} = 5.0101$
E_{15}	A_{15}	$a_1^{(15)} = 3.7826$	$a_M^{(15)} = 4.2069$	$a_2^{(15)} = 4.9840$
E_{16}	A_{16}	$a_1^{(16)} = 3.7824$	$a_M^{(16)} = 4.2067$	$a_2^{(16)} = 4.7805$

To obtain the average A_{aver}, first we shall sum up the numbers in the last three columns of Table 4.9

$$\sum_{i=1}^{16} a_1^{(i)} = 60.8469; \quad \sum_{i=1}^{16} a_M^{(i)} = 67.3075; \quad \sum_{i=1}^{16} a_2^{(i)} = 76.8568,$$

and then we use formula (4.342)

$$A_{\text{aver}} = \left(\frac{60.8469}{16}, \frac{67.3075}{16}, \frac{76.8568}{16} \right) = (3.80293; \ 4.20671; \ 4.80355).$$

The approximate result has the form

$$A_{\text{aver}}^p = (3.8029; \ 4.2067; \ 4.8036).$$

It may be easily noticed that, for instance, opinions of experts E_7 and E_{16} are close to the average A_{aver}^p, while the opinions of experts E_1 and E_{12} differ from it significantly. Let us assume that the company's management board decided to repeat the inquiry addressed to the experts, who receive the results of the previous round of inquiries. The experts propose new

TABLE 4.10. Experts' forecasts (second inquiry)

Expert	Fuzzy number	Lowest value	Most probable value	Highest value
E_1	B_1	$b_1^{(1)} = 3.6892$	$b_M^{(1)} = 4.2060$	$b_2^{(1)} = 4.7892$
E_2	B_2	$b_1^{(2)} = 3.8026$	$b_M^{(2)} = 4.2072$	$b_2^{(2)} = 4.8020$
E_3	B_3	$b_1^{(3)} = 3.7956$	$b_M^{(3)} = 4.2060$	$b_2^{(3)} = 4.8024$
E_4	B_4	$b_1^{(4)} = 3.8026$	$b_M^{(4)} = 4.2064$	$b_2^{(4)} = 4.7824$
E_5	B_5	$b_1^{(5)} = 3.9217$	$b_M^{(5)} = 4.2050$	$b_2^{(5)} = 4.7986$
E_6	B_6	$b_1^{(6)} = 3.8056$	$b_M^{(6)} = 4.2077$	$b_2^{(6)} = 4.8008$
E_7	B_7	$b_1^{(7)} = 3.7856$	$b_M^{(7)} = 4.2066$	$b_2^{(7)} = 4.8125$
E_8	B_8	$b_1^{(8)} = 3.7985$	$b_M^{(8)} = 4.2067$	$b_2^{(8)} = 4.7892$
E_9	B_9	$b_1^{(9)} = 3.8006$	$b_M^{(9)} = 4.2079$	$b_2^{(9)} = 4.9254$
E_{10}	B_{10}	$b_1^{(10)} = 3.9121$	$b_M^{(10)} = 4.2067$	$b_2^{(10)} = 4.7986$
E_{11}	B_{11}	$b_1^{(11)} = 3.8564$	$b_M^{(11)} = 4.2066$	$b_2^{(11)} = 4.7891$
E_{12}	B_{12}	$b_1^{(12)} = 3.7859$	$b_M^{(12)} = 4.2070$	$b_2^{(12)} = 4.7682$
E_{13}	B_{13}	$b_1^{(13)} = 3.8026$	$b_M^{(13)} = 4.2065$	$b_2^{(13)} = 4.7851$
E_{14}	B_{14}	$b_1^{(14)} = 3.7998$	$b_M^{(14)} = 4.2070$	$b_2^{(14)} = 4.8102$
E_{15}	B_{15}	$b_1^{(15)} = 3.7548$	$b_M^{(15)} = 4.2067$	$b_2^{(15)} = 4.7986$
E_{16}	B_{16}	$b_1^{(16)} = 3.7266$	$b_M^{(16)} = 4.2066$	$b_2^{(16)} = 4.7256$

forecasts of Euro exchange rate, which are transformed by the company's board onto triangular numbers B_i. Experts' new forecasts have been presented in Table 4.10.

By reapplying formula (4.342), we shall determine

$$B_{\text{aver}} = (3.80258;\ 4.206663;\ 4.798619).$$

The approximate result is as follows:

$$B_{\text{aver}}^p = (3.8026;\ 4.2067;\ 4.7986).$$

Now, the company's management board is satisfied, as the averages A_{aver}^p and B_{aver}^p are very close, and in consequence the algorithm is stopped and the triangular number B_{aver}^p is considered as the inference connecting the experts' opinions. This result is interpreted as follows: the estimated values of the Euro exchange rate in July next year are within the interval [3.8026; 4.7986], while the forecasted price of the Euro is PLN 4.2067. This forecast has been obtained by defuzzifying the fuzzy triangular number $B_{\text{aver}}^p = (3.8026;\ 4.2067;\ 4.7986)$.

4.10.2   Weighted fuzzy Delphi method

In many domains of our lives (e.g. economy, finance, management), the knowledge, experience and expertise of a certain group of experts are often more valued than the knowledge and expertise of other experts. It is expressed using weights w_i assigned to experts. We shall now describe the weighted fuzzy Delphi method. Let us assume that the competence of expert E_i, $i = 1, ..., 16$, is reflected by the weight w_i, $i = 1, ..., 16$, $w_i + ... + w_n = 1$. Next steps in the fuzzy Delphi method will be subject to slight changes, namely: in step 2 instead of triangular average A_{aver} weighted average A_{aver}^w appears. It is also the case of step 3, where instead of arithmetic averages, we have weighted averages.

Example 4.48

Let us return to Example 4.47, where 16 experts presented their opinions expressed using triangular numbers A_i placed in Table 4.9. Let us assume now that the competences of experts E_{10}, E_{16} are estimated as the highest (weight 0.13), competences of experts E_4, E_8 and E_{15} have the weight of 0.1 and of other experts 0.04; the sum of all weights is equal to 1. Table 4.11 illustrates the weighted forecasts of experts.

Having summed up the values in the last line of Table 4.11, we obtain a weighted triangular average

TABLE 4.11. Weighted forecasts of experts

Expert	w_i	$w_i \times a_1^{(i)}$	$w_i \times a_M^{(i)}$	$w_i \times a_2^{(i)}$
E_1	0.04	0.1435	0.1682	0.1802
E_2	0.04	0.1594	0.1683	0.1841
E_3	0.04	0.1395	0.1683	0.1941
E_4	0.1	0.3920	0.4207	0.4793
E_5	0.04	0.1600	0.1683	0.1836
E_6	0.04	0.1549	0.1683	0.1993
E_7	0.04	0.1510	0.1682	0.1930
E_8	0.1	0.3893	0.4207	0.4687
E_9	0.04	0.1473	0.1683	0.1850
E_{10}	0.13	0.5201	0.5469	0.6096
E_{11}	0.04	0.1557	0.1682	0.1993
E_{12}	0.04	0.1435	0.1682	0.1982
E_{13}	0.04	0.1524	0.1682	0.1997
E_{14}	0.04	0.1515	0.1683	0.2004
E_{15}	0.1	0.3783	0.4207	0.4984
E_{16}	0.13	0.4917	0.5469	0.6215
Sum:	1	3.830094	4.2067	4.79434

$$A^w_{\text{aver}} = (3.830094;\ 4.2067;\ 4.79434).$$

The approximate result is as follows:

$$A^{wp}_{\text{aver}} = (3.8301;\ 4.2067;\ 4.7943).$$

The result obtained is almost the same as in Example 4.47. As a result of defuzzification of the weighted average A^{wp}_{aver} we obtain the value of 4.2067.

4.10.3 Fuzzy PERT method

Planning of the order of actions is a complicated undertaking that requires taking into account many actions which are to be performed in the process of design of new product or technology. In the second half of 1950s in the USA, two new methods were proposed to be used in organizing large, complex production or construction enterprises in which many cooperating parties and contractors participate. The literature knows these methods under the name of PERT (*Project Evaluation and Review Technique*) and *Critical Path Method* (CPM). PERT and CPM techniques are similar and often used together as one method. For the first time, the PERT method was used in the USA in 1957 with relation to the construction of nuclear submarines and Polaris rockets. On the other hand, CPM was used more or less at the same time in research facilities of Remington Rand and DuPont companies working in the planning strategy in chemical factories. The PERT method consists in creating a model of the network of activities aiming to achieve the expected objective, taking into account the duration of each of them (the shortest, the longest and the most probable). The sequence of operations, of which each subsequent one depends on the execution of the previous one, creates the so-called critical path determining the longest time of task execution. Its shortening may be obtained by excluding from the critical path any activities that may be performed in parallel, and by hastening the execution of the remaining ones. A developed form of this method takes into account also the costs of execution of particular activities (stages) of the task. The PERT method allows to prepare a schedule of works which optimizes the time and costs of task execution, allowing for a smooth cooperation of all its participants, elimination of downtimes and so-called bottlenecks. We shall present the operation of the PERT method on a simplified example of designing production process for household appliances. The project of a given appliance requires that particular components be designed, manufactured and assembled and the ready product be tested. In our example, the project is made of ten different actions A, B, C, D, E, F, G, H and I. The required completion time for each activity has been presented in the last column of Table 4.12. It was estimated by managers responsible for particular activities.

TABLE 4.12. Weighted forecasts of experts

	Type of activity	Preceding activities	Parallel activities	Subsequent activities	Required completion time (days)
A	Designing of mechanical parts	–	–	B, C	30
B	Designing of electrical installation	A	C, E	D	30
C	Production of mechanical parts	A	B	E	28
D	Production of electrical parts	B	E, G	F	28
E	Assembly of mechanical parts	C	B, D	G	25
F	Assembly of electrical parts	D, E	–	H	20
G	Assembly of electronic parts	E	–	H	12
H	Start up of a new product	G, F	–	I	8
I	Tests	H	–	–	10

At first, let us build a network model of planning which takes into account the data included in Table 4.12. This model has been presented in Fig. 4.42. Each activity is represented by a rectangle in the middle of which the symbol of the activity type is placed together with the number of days necessary for its completion. A network model of planning shows the sequential relations between activities. A *critical path* is defined as the sequence of activities ranged from the first to the last in the Project, that requires the longest completion time. As a consequence, the total required time of project execution is identical to the time needed to complete the activities in the critical path.

The network planning model allows to set a critical path which in Fig. 4.42 has been represented using blocks connected by arrows which link activities A, B, D, F, H, I. Therefore the whole project completion time amounts to: $30 + 30 + 28 + 20 + 8 + 10 = 126$ days. In Fig. 4.42,

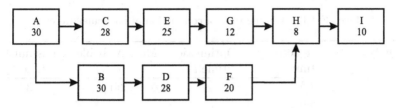

FIGURE 4.42. A network model of planning

TABLE 4.13. Estimated time of task A completion

Expert	T_i^A	Optimistic time	Most probable time	Pessimistic time
E_1	T_1^A	28	29	33
E_2	T_2^A	27	30	32
E_3	T_3^A	27	31	34
Sum:	$\sum_{i=1}^{3} T_i^A$	82	90	99

we may notice that activities C, E and G are not on the critical path. Therefore, their completion may take longer than $28 + 25 + 12 = 65$ days, however, the delay may not be greater than 13 days so as not to cause the extension of the critical path.

In our example, the time of each kind of activity will be estimated by three experts. The experts' task will be to estimate an optimistic, most probable (in the common meaning of this word) and a pessimistic time of completion of tasks A, B, ...,I. Experts' opinions have been notated as triangular fuzzy numbers $T_i^A, T_i^B, ..., T_i^I$, $i = 1, 2, 3$. Table 4.13 illustrates the experts' opinions concerning task A completion time.

Average time of task A completion is represented by a fuzzy triangular number of the form

$$T_{ave}^A = \left(\frac{82}{3}, \frac{90}{3}, \frac{99}{3}\right) \approx (27.3; \ 30; \ 33).$$

shall obtain the actual task completion time by defuzzifying the triangular number T_{aver}^A. Depending on the method of defuzzification used (see Subchapter 4.5), we will obtain the following results:

$$y^{(1)} = 30,$$

$$y^{(2)} = \frac{27.3 + 30 + 33}{3} = 30.1,$$

$$y^{(3)} = \frac{27.3 + 2(30) + 33}{4} = 30.075,$$

$$y^{(4)} = \frac{27.3 + 4(30) + 33}{6} = 30.05.$$

TABLE 4.14. Average time of completion of particular tasks

Activity	Average activity time	Optimistic time t_1	Most probable time t_M	Pessimistic time t_2
A	T^A_{aver}	27	30	33
B	T^B_{aver}	28	30	32
C	T^C_{aver}	24	27	31
D	T^D_{aver}	24	26	29
E	T^E_{aver}	22	25	27
F	T^F_{aver}	17	20	23
G	T^G_{aver}	10	12	14
H	T^H_{aver}	6	8	11
I	T^I_{aver}	7	9	12

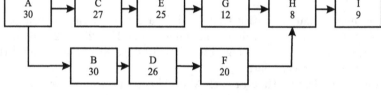

FIGURE 4.43. A modified network model of planning

Similarly, eight other three expert groups expressed their opinions concerning the estimation of required completion time for particular tasks. Table 4.14 presents the average times $T^B_{aver}, \ldots, T^I_{aver}$ (the time T^A_{aver} is also included). Each triangular number in Table 4.14, representing the average time of completion of a given time, is defuzzified (max operation) in order to obtain the actual time of completion of this task.

Figure 4.43 illustrates the network planning model taking into account the experts' opinion.

Total time needed to complete the project is

$$T = T^A_{aver} + T^B_{aver} + T^D_{aver} + T^F_{aver} + T^H_{aver} + T^I_{aver} = (109, 123, 140).$$

As a consequence, the project duration varies from 109 to 140 days, while the most probable time is, according to experts, 123 days.

4.10.4   Decision making in a fuzzy environment

The fuzzy sets theory allows to make decisions in the so-called *fuzzy environment*, which is made of fuzzy objectives, fuzzy constraints and a fuzzy decision. Let us consider a certain set of options (also referred to as choices or variants) notated using $X_{op} = \{x\}$. *A fuzzy objective* is defined as a fuzzy set G defined in the set of options X_{op}. *The fuzzy set G* is described by the membership function $\mu_G : X_{op} \to [0,1]$. The function $\mu_G(x) \in [0,1]$ for a given x defines the membership degree of option $x \in X_{op}$ to the fuzzy set G (fuzzy objective). A *fuzzy constraint* is defined as a fuzzy set C also defined in the set of options X_{op}. The fuzzy set C is described by the membership function $\mu_C : X_{op} \to [0,1]$. The function $\mu_C(x) \in [0,1]$ for a given x defines the membership degree of option $x \in X_{op}$ to the fuzzy set C (fuzzy constraint). Let us consider the task of determining a decision, achieving at the same time the fuzzy objective G and meeting the fuzzy constraint C. A *fuzzy decision D* is a fuzzy set created as a result of intersection of the fuzzy objective and fuzzy constraint:

$$D = G \cap C, \tag{4.344}$$

while

$$\mu_D(x) = T\{\mu_G(x), \mu_C(x)\} \tag{4.345}$$

for each $x \in X$. It should be noted that notation (4.345) suggests the following interpretation of the decision making task in a fuzzy environment: "reach G and meet C". The specific form of formula (4.345) depends on the t-norm adopted.

The above considerations can be easily generalized to the case of many objectives and constraints. Let us assume that we have $n > 1$ fuzzy objectives, $G_1, ..., G_n$, and $m > 1$ fuzzy constraints, $C_1, ..., C_m$, and all are defined as fuzzy sets in the set of options X_{op}. A fuzzy decision is determined as follows:

$$D = G_1 \cap ... \cap G_n \cap C_1 \cap ... \cap C_m, \tag{4.346}$$

while

$$\mu_D(x) = T\{\mu_{G_1}(x), ..., \mu_{G_n}(x), \mu_{C_1}(x), ..., \mu_{C_m}(x)\} \tag{4.347}$$

for each $x \in X_{op}$. A maximization decision is the option $x^* \in X$, such as

$$\mu_D(x^*) = \max_{x \in X} \mu_D(x). \tag{4.348}$$

The above considerations shall be illustrated on examples of specific applications (see [13]) of decision making in a fuzzy environment:
a) division of dividend,
b) employment policy,
c) housing policy for low income families,
d) assessment of students,
e) college selection strategy,
f) setting the price of a new product.
In all examples, we shall determine a fuzzy decision of the minimum type, i.e. the t-norm in formula (4.345) is defined by the min operation.

Example 4.49. Division of dividend

The general shareholders' meeting, having approved the company's balance sheet, is considering the amount of dividend for one share. The amount of the dividend is a linguistic variable taking on two values: *attractive dividend* and *moderate dividend*. The linguistic value *attractive dividend* is an objective described by the fuzzy set G, defined on the set of options $X_{op} = \{x : 0 < x \leq 70\}$, where option x is expressed in USD. The membership function $\mu_G(x)$ is increasing. The linguistic value *moderate dividend* is a constraint described by the fuzzy set C, defined on the set of options X_{op}, with a decreasing membership function $\mu_C(x)$. Let us assume that the fuzzy set G *attractive dividend* has the form of

$$\mu_G(x) = \begin{cases} 0 & \text{for} \quad 0 < x \leq 10, \\ \dfrac{1}{40}x - \dfrac{1}{4} & \text{for} \quad 10 \leq x \leq 50, \\ 1 & \text{for} \quad 50 \leq x \leq 70, \end{cases}$$

while the fuzzy set C *moderate dividend* is given on X_{op} as follows:

$$\mu_C(x) = \begin{cases} 1 & \text{for} \quad 0 < x \leq 20, \\ -\dfrac{1}{40}x + \dfrac{3}{2} & \text{for} \quad 20 \leq x \leq 60, \\ 0 & \text{for} \quad 60 \leq x \leq 70. \end{cases}$$

The fuzzy decision of the minimum type has the form

$$\mu_D(x) = \min(\mu_G(x), \mu_C(x)),$$

which is illustrated in Fig. 4.44. The point of intersection of function $\mu_G(x) = \dfrac{1}{40}x - \dfrac{1}{4}$ and $\mu_C(x) = -\dfrac{1}{40}x + \dfrac{3}{2}$ is $(35; 0.625)$. Therefore $x^* = 35$ and the amount of the dividend to be paid is 35 USD.

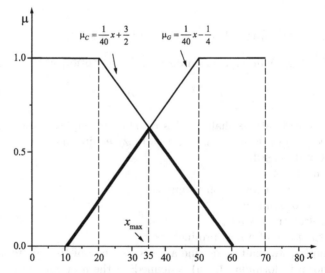

FIGURE 4.44. Illustration to Example 4.49

Example 4.50. Employment policy

Let us consider a company which organized a competition for the position of director's assistant. During the interview, the candidates were asked about their qualifications, experience, knowledge in a given domain, etc. In our example, the following criteria for assessing the candidates were set:

G_1 – experience,
G_2 – computer literacy,
G_3 – young age,
G_4 – foreign language.

The company's objective is to hire for the position of the director's assistant the best candidate who will accept the remuneration offered by the company. Individual candidates x_k, $k = 1, ..., 4$, are evaluated from the point of view of meeting the objectives G_1, G_2, G_3 and G_4 and meeting the constraint C. The result of this evaluation are fuzzy sets defined on the set $X_{op} = \{x_1, x_2, x_3, x_4\}$

$$G_1 = \frac{0.7}{x_1} + \frac{0.2}{x_2} + \frac{0.5}{x_3} + \frac{0.3}{x_4},$$

$$G_2 = \frac{0.8}{x_1} + \frac{0.8}{x_2} + \frac{0.5}{x_3} + \frac{0.2}{x_4},$$

$$G_3 = \frac{0.7}{x_1} + \frac{0.8}{x_2} + \frac{0.4}{x_3} + \frac{0.5}{x_4},$$

$$G_4 = \frac{0.5}{x_1} + \frac{0.6}{x_2} + \frac{0.7}{x_3} + \frac{0.8}{x_4}.$$

The constraint C is understood as the willingness of the candidates to accept the remuneration offered by the company and according to the company's views, is as follows:

$$C = \frac{0.3}{x_1} + \frac{0.4}{x_2} + \frac{0.6}{x_3} + \frac{0.9}{x_4}.$$

The decision D is determined on the basis of the following formula:

$$D = G_1 \cap G_2 \cap G_3 \cap G_4 \cap C.$$

As a result of some simple computations, we obtain

$$D = \frac{0.3}{x_1} + \frac{0.2}{x_2} + \frac{0.4}{x_3} + \frac{0.2}{x_4}.$$

Candidate No. 3 is characterized by the greatest membership degree equal to 0.4, and hence, he is the best candidate for the position offered by the company. The decision model for recruitment policy presented in our example may be applied in similar situations.

Example 4.51. Housing policy for low income families

The city council intends to implement a housing policy for low income families living in old apartment houses located on large plots. Three alternative projects are being considered:

x_1 – renovation and management of the buildings,
x_2 – Social Building Program,
x_3 – preferential credit to purchase new apartaments.

The set of options is $X_{op} = \{x_1\, x_2, x_3\}$. These projects will require a partial or total relocation of families. The city council, having studied the experts' opinions, proposed three objectives and two constraints described by the fuzzy sets defined on X_{op}. They are as follows:

$$\text{"Improvement of the living standards"} = G_1 = \frac{0.7}{x_1} + \frac{0.8}{x_2} + \frac{0.9}{x_3},$$

$$\text{"Greater number of apartaments in the same area"} = G_2 = \frac{0}{x_1} + \frac{0.9}{x_2} + \frac{0.9}{x_3},$$

$$\text{"Better housing conditions"} = G_3 = \frac{0.3}{x_1} + \frac{0.7}{x_2} + \frac{0.8}{x_3},$$

$$\text{"Reasonable cost"} = C_1 = \frac{0.8}{x_1} + \frac{0.6}{x_2} + \frac{0.3}{x_3},$$

$$\text{"Short realization time"} = C_2 = \frac{0.9}{x_1} + \frac{0.2}{x_2} + \frac{0.7}{x_3}.$$

We shall determine the fuzzy decision as follows:

$$D = G_1 \cap G_2 \cap G_3 \cap C_1 \cap C_2.$$

For t-norm of the minimum type, we obtain

$$D = \frac{0}{x_1} + \frac{0.2}{x_2} + \frac{0.3}{x_3}.$$

Project x_3 with the highest membership degree 0.3 appeared to be the best solution.

Example 4.52. Assessment of students

A company offered a summer internship for the students who obtained the best results in science (electronics, informatics, mathematics) and in languages (English, German). The word *the best* is a linguistic value, which was described separately for science subjects (NS) and languages (NJ) and presented in Fig. 4.45, assuming that the interval of marks is $[2, 5]$.

The membership functions of fuzzy sets NS and NJ are the following:

$$\mu_{NS}(x) = \begin{cases} 0 & \text{for} \quad 2 \le x \le 4.3; \\ \dfrac{x - 4.3}{0.5} & \text{for} \quad 4.3 \le x \le 4.8; \\ 1 & \text{for} \quad 4.8 \le x \le 5, \end{cases} \qquad (4.349)$$

and

$$\mu_{NJ}(x) = \begin{cases} 0 & \text{for} \quad 2 \leq x \leq 4.2; \\ \dfrac{x - 4.2}{0.4} & \text{for} \quad 4.2 \leq x \leq 4.6; \\ 1 & \text{for} \quad 4.6 \leq x \leq 5. \end{cases} \qquad (4.350)$$

The students who obtained in science subjects the average of 4.8 and higher, are assigned the membership degree equal to 1. In case of languages, the analogous value is 4.6. In our example, six students (x_1 = Kate, x_2 = Margaret, x_3 = Ann, x_4 = Tom, x_5 = Jack, x_6 = Michael) are applying for the internship. The set of options is $X_{op} = \{x_1, x_2, x_3, x_4, x_5, x_6\}$. Table 4.15 contains the average notes of students in particular subjects.

By substituting the average of students' marks in science subjects to formula (4.349), we obtain membership degrees to the fuzzy set NS. Similarly, by substituting the average of students' marks in foreign languages to formula (4.350), we obtain membership degrees to the fuzzy set NJ.

The next step is to create fuzzy sets corresponding to the data included in Table 4.16.

$$\text{"The best in electronics"} = G_1 = \frac{1}{x_1} + \frac{0.2}{x_2} + \frac{1}{x_3} + \frac{0.4}{x_4} + \frac{1}{x_5} + \frac{1}{x_6},$$

$$\text{"The best in informatics"} = G_2 = \frac{1}{x_1} + \frac{0.8}{x_2} + \frac{1}{x_3} + \frac{1}{x_4} + \frac{0.6}{x_5} + \frac{0.4}{x_6},$$

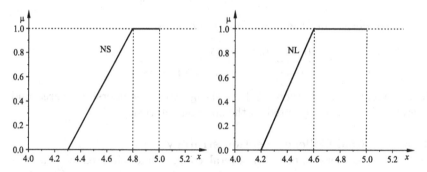

FIGURE 4.45. Membership functions of fuzzy sets NS and NJ

TABLE 4.15. Average notes of students in particular subjects

Student	Electronics	Informatics	Mathematics	English	German
Kate (x_1)	4.8	5.0	4.7	4.3	4.7
Margaret (x_2)	4.4	4.7	4.8	4.4	4.4
Ann (x_3)	4.9	4.9	4.6	4.7	4.3
Tom (x_4)	4.5	4.8	4.9	5.0	4.5
Jack (x_5)	5.0	4.6	4.7	4.4	5.0
Michael (x_6)	4.9	4.5	5.0	4.5	4.4

TABLE 4.16. Values of membership degrees to fuzzy sets NS and NJ

Student	Electronics	Informatics	Mathematics	English	German
Kate (x_1)	1	1	0.8	0.25	1
Margaret (x_2)	0.2	0.8	1	0.5	0.5
Ann (x_3)	1	1	0.6	1	0.25
Tom (x_4)	0.4	1	1	1	0.75
Jack (x_5)	1	0.6	0.8	0.5	1
Michael (x_6)	1	0.4	1	0.75	0.5

$$\text{``The best in mathematics''} = G_3 = \frac{0.8}{x_1} + \frac{1}{x_2} + \frac{0.6}{x_3} + \frac{1}{x_4} + \frac{0.8}{x_5} + \frac{1}{x_6},$$

$$\text{``Perfect in English''} = G_4 = \frac{0.25}{x_1} + \frac{0.5}{x_2} + \frac{1}{x_3} + \frac{1}{x_4} + \frac{0.5}{x_5} + \frac{0.75}{x_6},$$

$$\text{``Perfect in German''} = G_5 = \frac{1}{x_1} + \frac{0.5}{x_2} + \frac{0.25}{x_3} + \frac{0.75}{x_4} + \frac{1}{x_5} + \frac{0.5}{x_6}.$$

By substituting the data to formula (4.346), we obtain

$$D = G_1 \cap G_2 \cap G_3 \cap G_4 \cap G_5.$$

The fuzzy decision of the minimum type has the form

$$D = \frac{0.25}{x_1} + \frac{0.2}{x_2} + \frac{0.25}{x_3} + \frac{0.4}{x_4} + \frac{0.5}{x_5} + \frac{0.4}{x_6}.$$

The student x_5 is characterized by the greatest membership degree and therefore he will be accepted to the summer internship.

Example 4.53. College selection strategy
A talented student applied to several colleges and after exams, he was accepted to 4 of the colleges that make up the set of options $X_{op} = \{x_1, x_2, x_3, x_4\}$. Now, he must make a decision concerning the college he will go to. The objective of our student is to learn at a renown college (i.e. at the top of the ranking of the best colleges). At the same time, the future college student would like that some conditions be met, and namely: the school should be located not far from his place of residence; it should have a program of international exchange; it should have good technical back-office facilities and after graduating from it, the student wants to have high odds to find a job. These constraints were notated using fuzzy sets:

$$\text{``Not far from the place of residence''} = C_1 = \frac{0.8}{x_1} + \frac{0.9}{x_2} + \frac{0.4}{x_3} + \frac{0.5}{x_4},$$

"International exchange program" $= C_2 = \dfrac{0.2}{x_1} + \dfrac{0.2}{x_2} + \dfrac{0.9}{x_3} + \dfrac{0.6}{x_4}$,

"Good technical back-office facilities at the college (equipment of rooms, labs, etc.)" $= C_3 = \dfrac{0.5}{x_1} + \dfrac{0.3}{x_2} + \dfrac{0.6}{x_3} + \dfrac{0.7}{x_4}$,

"High odds to find a job"$= C_4 = \dfrac{0.6}{x_1} + \dfrac{0.5}{x_2} + \dfrac{0.7}{x_3} + \dfrac{0.7}{x_4}$.

In Table 4.17, particular colleges were assigned membership degrees (where x_2 with the membership degree equal to 1 is the college which ranked first in the ranking, etc.)

TABLE 4.17. Colleges together with the membership degrees assigned

College	x_1	x_2	x_3	x_4
The membership degree of the place in ranking	0.75	1	0.25	0.5

Using the data included in Table 4.17, we shall create the fuzzy set G describing the objective

$$G = \frac{0.75}{x_1} + \frac{1}{x_2} + \frac{0.25}{x_3} + \frac{0.5}{x_4}.$$

Using formula (4.346), we shall obtain the following fuzzy decision:

$$D = G \cap C_1 \cap C_2 \cap C_3 \cap C_4.$$

The fuzzy decision of the minimum type has the form

$$D = \frac{0.2}{x_1} + \frac{0.2}{x_2} + \frac{0.25}{x_3} + \frac{0.5}{x_4}.$$

The greatest membership degree is 0.5, therefore our student will select the college x_4.

Example 4.54. Determination of price of a new product
Determination of the price for a new product being launched is a complicated process. It requires a joint effort from experts in such domains as finance, management, marketing and sales. The experts' task is to define the product price. Typical requirements concerning the new product price definition are as follows:
W_1 – the product should have a *low price*,
W_2 – the product should have a *high price*,
W_3 – the product should have a *price close to the competitive price*,
W_4 – the product should have a *price close to twice the manufacturing cost*.

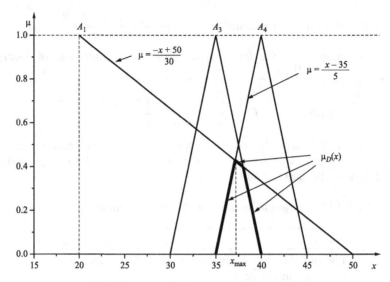

FIGURE 4.46. Illustration to Example 4.54

Let us consider a pricing model made of three requirements W_1, W_3, W_4. Let us assume that the competitive price is 35 USD and the double manufacturing cost is 40 USD. The set of option X_{op} is within the interval $[20, 60]$, which means that the product price should be included in this interval. This model has been presented in Fig. 4.46. The linguistic values in particular requirements are described by fuzzy sets as follows: requirement W_1 is expressed by the fuzzy triangular number A_1 (*low price*), requirements W_3 and W_4 have been presented using fuzzy triangular numbers A_3 (*close to the competitive price*) and A_4 (*close to twice the manufacturing cost*). The membership functions of triangular fuzzy numbers A_1, A_3 and A_4 are the following:

$$\mu_{A_1}(x) = \begin{cases} \dfrac{-x+50}{30} & \text{for } 20 \leq x \leq 50, \\[2mm] 0 & \text{otherwise;} \end{cases}$$

$$\mu_{A_3}(x) = \begin{cases} \dfrac{x-30}{5} & \text{for } 30 \leq x \leq 35, \\[2mm] \dfrac{-x+40}{5} & \text{for } 35 \leq x \leq 40, \\[2mm] 0 & \text{otherwise;} \end{cases}$$

$$\mu_{A_4}(x) = \begin{cases} \dfrac{x-35}{5} & \text{for } 35 \leq x \leq 40, \\[2mm] \dfrac{-x+45}{5} & \text{for } 40 \leq x \leq 45, \\[2mm] 0 & \text{otherwise.} \end{cases}$$

The fuzzy decision D of the minimum type has the form

$$\mu_D(x) = \min\left(\mu_{A_1}(x), \mu_{A_3}(x), \mu_{A_4}(x)\right).$$

By finding the intersection point of lines $\mu = \frac{-x+50}{30}$ and $\mu = \frac{x-35}{5}$, we obtain the decision $x^* = 37.14$ interpreted as the product price. The experts accept this price as recommended. We may observe in Fig. 4.46 that the triangular fuzzy number A_3 (*close to the competitive price*) impacts the fuzzy decision D, but does not influence the maximizing decision x^*. Only the triangular fuzzy number A_4 (*close to twice the manufacturing cost*) and A_1 (*low price*) influence the value of x^*.

Example 4.55. Definition of price of a new product
Let us continue Example 4.54, modifying the requirement W_1 as follows: W_1- the product should have a *very low price*. Other requirements W_2, W_3 and W_4 are not changed.

According to formula (4.75) the membership function of *very* A_1 has the form

$$\mu_{veryA_1}(x) = (\mu_{A_1}(x))^2 = \begin{cases} \left(\dfrac{-x+50}{30}\right)^2 & \text{for } 20 \leq x \leq 50, \\[2mm] 0 & \text{otherwise.} \end{cases}$$

It is a parabola defined in the interval $[20, 50]$, shown in Fig. 4.47.

The fuzzy decision D is determined as follows:

$$\mu_D(x) = \min\left(\mu_{veryA_1}(x), \mu_{A_3}(x), \mu_{A_4}(x)\right).$$

To define x^*, we should determine the intersection point of the function $\mu = \left(\frac{-x+50}{30}\right)^2$ and $\mu = \frac{x-35}{5}$. We will obtain a quadratic equation $x^2 - 280x + 8800 = 0$, the solutions of which are 36.075 and 243.925. The solution within the interval $[35,40]$, $x^* = 36.075 \approx 36$, provides the suggested product price. The modifier *very* provides a greater focus on *low price*, that is why we obtained 36, i.e. a price lower than 37.14 obtained in the previous example. Like in Example 4.54, triangular number A_3 (*price close to the competitive one*) impacts the fuzzy decision D, but does not influence the value of x^*.

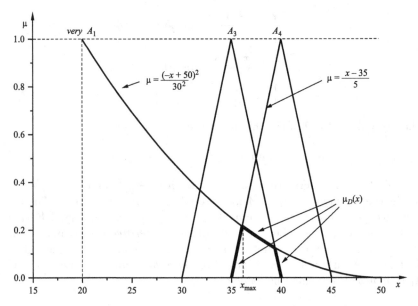

FIGURE 4.47. Illustration to Example 4.55

Example 4.56. Definition of price of a new product
Now, we shall modify in Example 4.54 the requirement W_1 as follows:
W_1 – the product should have a *rather low price*. Other requirements
W_2, W_3 and W_4 are not changed.

According to formula (4.76) the membership function of *rather* A_1 has
the form

$$\mu_{rather A_1}(x) = (\mu_{A_1}(x))^{\frac{1}{2}} = \begin{cases} \left(\dfrac{-x+50}{30}\right)^{\frac{1}{2}} & \text{for } 20 \leq x \leq 50, \\ 0 & \text{otherwise,} \end{cases}$$

or it is a parabola defined in the interval $[20, 50]$, shown in Fig. 4.48.

Figure 4.48 indicates that requirement W_1 (*rather low price*) does not
influence the making of the fuzzy decision D characterized by the membership function $\mu_D(x)$, while $x^* = 37.5$.

Let us notice that the pricing models in Examples 4.54 and 4.55 lead
to decisions that respect *the low price* and *the price close to twice the
manufacturing costs* without considering *the competitive price*. A company
with such pricing strategy may create a favorable market for the competition, but the company may be threatened by losses or elimination from
the market.

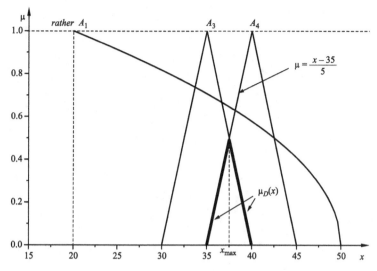

FIGURE 4.48. Illustration to Example 4.56

4.11   Notes

The fuzzy set theory has been introduced in 1965 by Lotfi Zadeh [265]. The foundations of this theory and its many applications have been presented in several monographs [17, 33, 79, 94, 118, 130, 165-167, 269]. An extensive monograph by Piegat [171] is particularly worth mentioning. The issues of designing fuzzy inference system for controlling purposes are dealt with by monographs [3, 33, 39, 40, 253]. The excellent monograph by Kacprzyk [95] on decision making in fuzzy conditions and on multi-stage fuzzy controlling is also well worth recommending. In monograph [13], the authors provided many examples of application of fuzzy sets in economy and management. Similar examples we presented in Subchapter 4.10. Various generators of triangular norms and their properties have been discussed in monographs [111, 144]. In works [43, 44, 87, 150, 180], their authors discuss many issues concerning the combination of the fuzzy sets theory with the rough sets theory. In monographs [116] and [119], the authors presented some applications of fuzzy sets in industrial processes diagnostic issues. Work [252] provides the method of generation of fuzzy rules based on a learning sequence. In monograph [21], the authors present different types of applied operators, for instance, for fuzzy rules aggregation.

5
Methods of knowledge representation using type-2 fuzzy sets

5.1   Introduction

The fuzzy sets, discussed in the previous chapter, are called type-1 fuzzy sets. They are characterized by the membership function, while the value of this function for a given element x is called the grade of membership of this element to a fuzzy set. In case of type-1 fuzzy sets, the membership grade is a real number taking values in the interval $[0, 1]$. This chapter will present another concept of a fuzzy description of uncertainty. According to this concept, the membership grade is not a number any more, but it has a fuzzy character. Figure 5.1 shows a graphic illustration of type-1 fuzzy sets $A_1, ..., A_5$ and corresponding type-2 fuzzy sets $\widetilde{A}_1, ..., \widetilde{A}_5$. It should be noted that in case of type-2 fuzzy sets, for any given element x, we cannot speak of an unambiguously specified value of the membership function. In other words, the membership grade is not a number, as in case of type-1 fuzzy sets.

In subsequent points of this chapter, basic definitions concerning type-2 fuzzy sets will be presented and operations on these sets will be discussed. Then type-2 fuzzy relations and methods of transformation of type-2 fuzzy sets into type-1 fuzzy sets will be introduced.

In the last part of this chapter, the theory of type-2 fuzzy sets will serve for the construction of the fuzzy inference system. Particular blocks of such system will be discussed in details, including type-2 fuzzification, type-2 rules base, type-2 inference mechanisms and the two-stage defuzzification consisting of type-reduction and defuzzification.

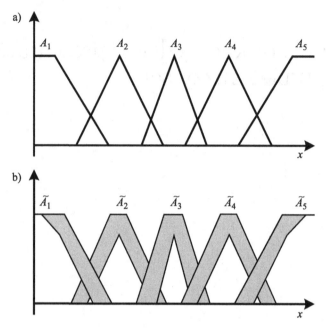

FIGURE 5.1. Graphic illustrations of type-1 fuzzy sets and corresponding type-2 fuzzy sets

5.2   Basic definitions

Definition 5.1
Type-2 fuzzy set $\widetilde{A}$ defined on a universe of discourse X, which is denoted as $\widetilde{A} \subseteq X$, is a set of pairs

$$\{x, \mu_{\widetilde{A}}(x)\}, \tag{5.1}$$

where x is an element of a fuzzy set, and its grade of membership $\mu_{\widetilde{A}}(x)$ in the fuzzy set $\widetilde{A}$ is a type-1 fuzzy set defined in the interval $J_x \subset [0,1]$, i.e.

$$\mu_{\widetilde{A}}(x) = \int_{u \in J_x} f_x(u)/u. \tag{5.2}$$

Function $f_x : [0,1] \rightarrow [0,1]$ will be called *the secondary membership function*, and its value $f_x(u)$ will be called *the secondary grade* or *secondary membership*. Of course, u is an argument of the secondary membership function. The interval J_x, being a domain of the secondary membership function f_x, is called *the primary membership* of element x. The fuzzy set $\widetilde{A}$ may be notated, in the notation of fuzzy sets, as follows:

$$\widetilde{A} = \int_{x \in X} \mu_{\widetilde{A}}(x)/x \tag{5.3}$$

or

$$\tilde{A} = \int \mu_{\tilde{A}}(x)/x = \int_{x \in X} \left[\int_{u \in J_x} f_x(u)/u \right]/x, \quad J_x \subseteq [0, 1]. \qquad (5.4)$$

Example 5.1

Fig. 5.2a depicts the method of construction of type-2 fuzzy sets. For a given element x_1 we get the interval $J_{x_1} = [0.4, 0.7]$ being a domain of the secondary membership function f_{x_1}. Figures 5.2b, 5.2c and 5.2d show exemplary secondary membership functions of triangular, interval and Gaussian types with a finite support [171].

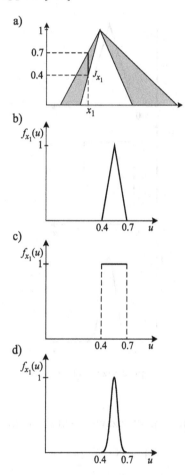

FIGURE 5.2. Illustration of type-2 fuzzy set and secondary membership functions for $J_{x_1} = [0.4; 0.7]$

Figure 5.3 depicts the same type-2 fuzzy set, but another element x_2 is chosen, $x_2 \in X$, as well as a corresponding membership grade being a type-1 fuzzy set (of a triangular, interval or Gaussian type with a finite support) defined on the interval $J_{x_2} = [0.1, 0.6]$.

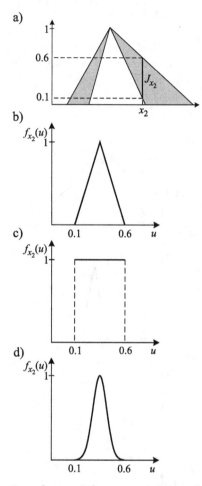

FIGURE 5.3. Illustration of type-2 fuzzy set and secondary membership functions for $J_{x_2} = [0.1; 0.6]$

In a discrete case, the type-2 fuzzy set will be defined in a similar way, i.e.

$$\tilde{A} = \sum_{x \in X} \mu_{\tilde{A}}(x) / x \tag{5.5}$$

and

$$\mu_{\tilde{A}}(x) = \sum_{u \in J_x} f_x(u) / u. \tag{5.6}$$

Let us assume that the set X has been discretized and takes R values $x_1, ..., x_R$. Moreover, the intervals J_x corresponding to these values, have been discretized and each of them takes M_i values, $i = 1, ..., R$. We can then note

$$\tilde{A} = \sum_{x \in X} \left[\sum_{u \in J_x} f_x(u)/u \right] /x = \sum_{i=1}^{R} \left[\sum_{u \in J_{x_i}} f_{x_i}(u)/u \right] /x_i \qquad (5.7)$$

$$= \left[\sum_{k=1}^{M_1} f_{x_i}(u_{1k})/u_{1k} \right] /x_1 + ... + \left[\sum_{k=1}^{M_R} f_{x_R}(u_{Rk})/u_{Rk} \right] /x_R.$$

Remark 5.1

The fuzzy membership grade can take two characteristic and extreme forms of the type-1 fuzzy set:

$\mu_{\tilde{A}}(x) = 1/1$ meaning a full membership of element x to the fuzzy set $\tilde{A}$,

$\mu_{\tilde{A}}(x) = 1/0$ meaning the lack of membership of element x to the fuzzy set $\tilde{A}$,

Example 5.2

Let us assume that $X = \{1, 2, 3\}$ and $J_{x_1} = \{0.2, 0.5, 0.7\}$, $J_{x_2} = \{0.5, 1\}$, $J_{x_3} = \{0.1, 0.3, 0.5\}$. If we assign appropriate grades of secondary membership to particular elements of sets $J_{x_1}, J_{x_2}, J_{x_3}$ we may define the following type-2 fuzzy set:

$$\tilde{A} = (0.5/0.2 + 1/0.5 + 0.5/0.7)/1 + (0.5/0.5 + 1/1)/2 \qquad (5.8)$$
$$+ (0.5/0.1 + 1/0.3 + 0.5/0.5)/3.$$

Definition 5.2

Let us assume that each secondary membership function f_x of a type-2 fuzzy set takes value 1 only for one element $u \in J_x$. Then the union of elements u forms a so-called *principal membership function*, i.e.

$$\mu_{A_g}(x) = \int_{x \in X} u/x, \quad \text{where} \quad f_x(u) = 1. \qquad (5.9)$$

The principal membership function defines the appropriate type-1 fuzzy set denoted as A_g.

Remark 5.2

In case where the secondary membership function f_x is an interval function, then the principal membership function will be determined as a union of all the elements u being mid-points of the primary membership $J_x, x \in X$.

Example 5.3

We are going to discuss a type-2 fuzzy set given by formula (5.8). Upon the basis of Definition 5.2 we may determine the following fuzzy set A_g:

$$A_g = \frac{0.5}{1} + \frac{1}{2} + \frac{0.3}{3}. \qquad (5.10)$$

5.3 Footprint of uncertainty

The type-2 fuzzy set may be described using the notion of the footprint of uncertainty.

Definition 5.3
Let us assume that $J_x \subset [0,1]$ means the primary membership of element x. *The footprint of uncertainty (FOU)* of a type-2 fuzzy set $\tilde{A} \subseteq X$ will be a bounded region consisting of all the points of primary membership of elements x, i.e.

$$FOU\left(\tilde{A}\right) = \bigcup_{x \in X} J_x \qquad (5.11)$$

Example 5.4
Let us discuss the family of membership functions of the type-1, fuzzy set which is described by the Gaussian function with the assumption that a standard deviation σ changes in the interval $[\sigma_1, \sigma_2]$, i.e.:

$$\mu_A(x) = N(m, \sigma; x) = \exp\left[-\frac{1}{2}\left(\frac{x-m}{\sigma}\right)^2\right], \quad \sigma \in [\sigma_1, \sigma_2]. \qquad (5.12)$$

The family of membership functions (5.12) forms a type-2 fuzzy set. A full description of this set would require to define the secondary membership function for each point x and the corresponding interval J_x. Figure 5.4 shows the footprint of uncertainty of the discussed type-2 fuzzy set.

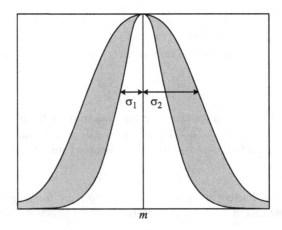

FIGURE 5.4. Footprint of uncertainty of a type-2 fuzzy set: $\sigma \in [\sigma_1, \sigma_2]$

Example 5.5

Let us discuss the family of membership functions of the type-1 fuzzy set which is described by the Gaussian function with the assumption that the average value m changes in interval $[m_1, m_2]$, i.e.

$$\mu_A(x) = \exp\left[-\frac{1}{2}\left(\frac{x-m}{\sigma}\right)^2\right], \quad m \in [m_1, m_2]. \qquad (5.13)$$

The family of membership functions (5.13) forms a type-2 fuzzy set. As in the previous example, a full description of this set would require to define the secondary membership function for each point x and the corresponding interval J_x. Figure 5.5 shows the footprint of uncertainty of the discussed type-2 fuzzy set.

Let us assume that $J_x = \left[\underline{J_x}, \overline{J_x}\right], \quad x \in X.$

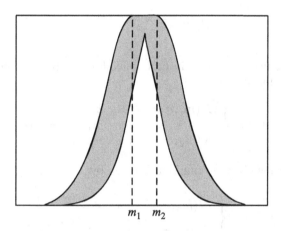

FIGURE 5.5. Footprint of uncertainty of a type-2 fuzzy set: $m \in [m_1, m_2]$

Definition 5.4

The upper membership function (UMF) is the membership function of the type-1 fuzzy set defined by:

$$\overline{\mu}_{\tilde{A}}(x) = UMF(\tilde{A}) = \bigcup_{x \in X} \overline{J_x} \quad \forall x \in X. \qquad (5.14)$$

Definition 5.5
The lower membership function (LMF) is the membership function of the type-1 fuzzy set defined by:

$$\underline{\mu}_{\tilde{A}}(x) = LMF(\tilde{A}) = \bigcup_{x \in X} J_x \quad \forall x \in X. \qquad (5.15)$$

Example 5.6
We are going to determine the footprint of uncertainty for the type-2 fuzzy set given in Example 5.4. It is easy to notice that the upper membership function takes the form

$$\overline{\mu}_{\tilde{A}}(x) = N(m, \sigma_2; x), \qquad (5.16)$$

and the lower membership function is given by

$$\underline{\mu}_{\tilde{A}}(x) = N(m, \sigma_1; x). \qquad (5.17)$$

Example 5.7
For the type-2 fuzzy set given in Example 5.5 the upper membership function takes the form

$$\overline{\mu}_{\tilde{A}}(x) = \begin{cases} N(m_1, \sigma; x) & \text{for} \quad x < m_1, \\ 1 & \text{for} \quad m_1 \leq x \leq m_2, \\ N(m_2, \sigma; x) & \text{for} \quad x > m_2, \end{cases} \qquad (5.18)$$

and the lower membership function is given by

$$\underline{\mu}_{\tilde{A}}(x) = \begin{cases} N(m_2, \sigma; x) & \text{for} \quad x \leq \dfrac{m_1 + m_2}{2}, \\ N(m_1, \sigma; x) & \text{for} \quad x > \dfrac{m_1 + m_2}{2}. \end{cases} \qquad (5.19)$$

5.4   Embedded fuzzy sets

In type-2 fuzzy sets we can distinguish between so-called embedded type-1 and embedded type-2 fuzzy sets.

Definition 5.6
From each interval J_x, $x \in X$, we will select only one element $\theta \in J_x$.
The embedded type-2 set in set $\tilde{A}$ is set $\tilde{A}_o$

$$\tilde{A}_o = \int_{x \in X} [f_x(\theta)/\theta]/x \quad \theta \in J_x \subseteq U = [0, 1]. \qquad (5.20)$$

Of course, there is an uncountable number of embedded sets $\tilde{A}_\mathrm{o}$ in set $\tilde{A}$. In a discrete case, the embedded set $\tilde{A}_0$ is defined as follows:

$$\tilde{A}_\mathrm{o} = \sum_{i=1}^{R} \left[f_{x_i}\left(\theta_i\right)/\theta_i \right]/x_i \quad \theta_i \in J_{x_i} \subseteq U = [0,1]. \qquad (5.21)$$

It is easy to notice that there are $\prod_{i=1}^{R} M_i$ embedded fuzzy sets $\tilde{A}_\mathrm{o}$ in set $\tilde{A}$.

Example 5.8

Let us assume that

$$\tilde{A} = (0.5/0.2 + 1/0.5 + 0.5/0.7)/2 + (0.3/0.5 + 1/1)/3 \qquad (5.22)$$
$$+ (0.5/0.1 + 1/0.3 + 0.5/0.5)/4.$$

Then one of the 18 embedded fuzzy sets $\tilde{A}_\mathrm{o}$ takes the form

$$\widetilde{A_\mathrm{o}} = (0.5/0.7)/2 + (0.3/0.5)/3 + (1/0.3)/4. \qquad (5.23)$$

Each embedded type-2, fuzzy set $\widetilde{A_\mathrm{o}}$ is connected with an embedded type-1 fuzzy set denoted as A_o.

Definition 5.7

The embedded type-1 set is defined as follows:

$$A_\mathrm{o} = \int_{x \in X} \theta/x \quad \theta \in J_x \subseteq U = [0,1]. \qquad (5.24)$$

There is an uncountable number of embedded fuzzy sets A_o. In a discrete case, formula (5.24) becomes

$$A_\mathrm{o} = \sum_{i=1}^{R} \theta_i/x_i \quad \theta_i \in J_{x_i} \subseteq U = [0,1]. \qquad (5.25)$$

The number of all sets A_o is $\prod_{i=1}^{R} M_i$.

A particular case of an embedded type-1 set is fuzzy set A_g defined by the principal membership function given by formula (5.9). Furthermore, it should be noted that embedded fuzzy set A_o looses all the information about secondary grades. Thus, upon the basis of the family of embedded sets A_o, it is not possible to reconstruct the type-2 fuzzy set, but only its footprint of uncertainty. However, the notion of embedded set will turn to be especially useful when discussing the fast algorithm of type-reduction presented further in this chapter.

Example 5.9
Figure 5.6 shows three different embedded type-1 sets.

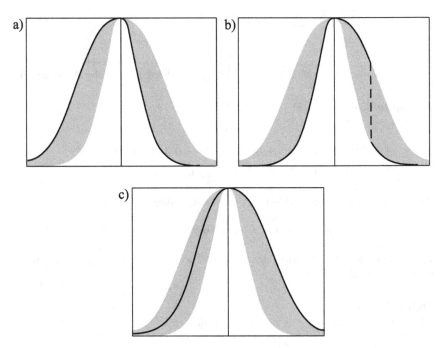

FIGURE 5.6. Embedded type-1 fuzzy sets

Example 5.10
Let us discuss a type-2 fuzzy set given by

$$\widetilde{A} = (0.6/0.3 + 1/0.7)/3 + (0.4/0.4 + 1/1)/5. \tag{5.26}$$

We can distinguish four embedded fuzzy sets A_o in set $\widetilde{A}$:

$$\begin{aligned}
A_o &= 0.3/3 + 0.4/5, \\
A_o &= 0.7/3 + 0.4/5, \\
A_o &= 0.3/3 + 1/5, \\
A_o &= 0.7/3 + 1/5.
\end{aligned} \tag{5.27}$$

5.5 Basic operations on type-2 fuzzy sets

The extension principle (Subchapter 4.4) allows to extend operations on type-1 fuzzy sets to operations on type-2 sets.

We are going to discuss two type-2 fuzzy sets, $\widetilde{A}$ and $\widetilde{B}$, defined as follows:

$$\widetilde{A} = \int_{x \in X} \left(\int_{u \in J_x^u} f_x(u)/u \right)/x \qquad (5.28)$$

and

$$\widetilde{B} = \int_{x \in X} \left(\int_{v \in J_x^v} g_x(v)/v \right)/x, \qquad (5.29)$$

where $J_x^u, J_x^v \subset [0,1]$. The sum of sets $\widetilde{A}$ and $\widetilde{B}$ is a type-2 fuzzy set notated as $\widetilde{A} \cup \widetilde{B}$ and defined as follows:

$$\mu_{\widetilde{A} \cup \widetilde{B}} = \int_{w \in J_x^w} h_x(w)/w = \phi\left(\mu_{\widetilde{A}}(x), \mu_{\widetilde{B}}(x)\right) \qquad (5.30)$$

$$= \phi\left(\int_{u \in J_x^u} f_x(u)/u, \int_{v \in J_x^v} g_x(v)/v \right).$$

In this case, the extended ϕ function is any t-conorm, but its arguments are not common numbers, but type-1 fuzzy sets $\mu_{\widetilde{A}}(x)$ and $\mu_{\widetilde{B}}(x)$ for the given $x \in X$. In accordance with the extension principle

$$\phi\left(\int_{u \in J_x^u} f_x(u)/u, \int_{v \in J_x^v} g_x(v)/v \right) = \int_{u \in J_x^u} \int_{v \in J_x^v} f_x(u) \overset{T}{*} g_x(v)/\phi(u,v).$$
$$(5.31)$$

After substituting t-conorm in place of function ϕ, the sum of type-2 fuzzy sets is given by the fuzzy membership function

$$\mu_{\widetilde{A} \cup \widetilde{B}}(x) = \int_{u \in J_x^u} \int_{v \in J_x^v} f_x(u) \overset{T}{*} g_x(v)/u \overset{S}{*} v. \qquad (5.32)$$

The formula above allows to determine the sum of type-2 fuzzy sets for each value of x. The membership function of the resulting set is the highest value of expression $f_x(u) \overset{T}{*} g_x(v)$ for all the pairs (u,v), which give the same element $w = u \overset{S}{*} v$ as a result.

Example 5.11

This example will explain in details the manner of determining the sum of type-2 fuzzy sets. We assume the minimum operation as the t-norm, and the maximum operation as the t-conorm. We are going to discuss two type-2 fuzzy sets, $\widetilde{A}$ and $\widetilde{B}$, defined as follows:

$$\widetilde{A} = (0.5/0.2 + 1/0.5 + 0.5/0.7)/1 + (0.5/0.5 + 1/1)/2 \qquad (5.33)$$
$$+ (0.5/0.1 + 1/0.3 + 0.5/0.5)/3$$

and

$$\tilde{B} = (1/0)\,/1 + (0.5/0.5 + 1/0.8)\,/2 + (1/0.6 + 0.5/1)\,3 \qquad (5.34)$$

In accordance to formula (5.32) for $x = 1$ we have

$$\mu_{\tilde{A} \cup \tilde{B}}\,(1) = \frac{0.5 \wedge 1}{0.2 \vee 0} + \frac{1 \wedge 1}{0.5 \vee 0} + \frac{0.5 \wedge 1}{0.7 \vee 0} \qquad (5.35)$$
$$= \frac{0.5}{0.2} + \frac{1}{0.5} + \frac{0.5}{0.7}.$$

For $x = 2$ we obtain

$$\mu_{\tilde{A} \cup \tilde{B}}\,(2) = \frac{0.5 \wedge 0.5}{0.5 \vee 0.5} + \frac{0.5 \wedge 1}{0.5 \vee 0.8} + \frac{\max\,(1 \wedge 0.5, 1 \wedge 1)}{1} \qquad (5.36)$$
$$= \frac{0.5}{0.5} + \frac{0.5}{0.8} + \frac{1}{1}.$$

For $x = 3$ we have

$$\mu_{\tilde{A} \cup \tilde{B}}(3) = \frac{\max\,(0.5 \wedge 1, 1 \wedge 1, 0.5 \wedge 1)}{0.6} + \frac{\max\,(0.5 \wedge 0.5, 1 \wedge 0.5, 0.5 \wedge 0.5)}{1}$$
$$= \frac{1}{0.6} + \frac{0.5}{1}. \qquad (5.37)$$

Thus, the sum of sets $\tilde{A}$ and $\tilde{B}$ is

$$\tilde{A} \cup \tilde{B} = (0.5/0.2 + 1/0.5 + 0.5/0.7)/1 + (0.5/0.5 + 0.5/0.8 + 1/1)/2 \qquad (5.38)$$
$$+ (1/0.6 + 0.5/1)/3.$$

The intersection of sets $\tilde{A}$ and $\tilde{B}$ is a type-2 fuzzy set with the fuzzy membership function given by the following formula

$$\mu_{\tilde{A} \cap \tilde{B}}\,(x) = \int_{w \in J_x^w} h_x\,(w)\,/w = \phi\,(\mu_{\tilde{A}}\,(x), \mu_{\tilde{B}}\,(x)) \qquad (5.39)$$

$$= \phi\left(\int_{u \in J_x^u} f_x\,(u)\,/u, \int_{v \in J_x^v} g_x\,(v)\,/v\right),$$

where the extended function ϕ is any t-norm this time. The arguments of the function ϕ are type-1 fuzzy sets, i.e, $\mu_{\tilde{A}}\,(x)$ and $\mu_{\tilde{B}}\,(x)$. Thus, the intersection of type-2 fuzzy sets is specified as follows:

$$\mu_{\tilde{A} \cap \tilde{B}}\,(x) = \int_{u \in J_x^u} \int_{v \in J_x^v} f_x\,(u) \overset{T^*}{*} g_x\,(v)\,/u \overset{T}{*} v. \qquad (5.40)$$

In formula (5.40), the t-norm aggregating secondary memberships has been denoted by T^*, and its form can be selected irrespectively of the selection of the extended t-norm T. Also in this case the membership function of the resulting set is the highest value of the expression $f_x(u) \overset{T^*}{*} g_x(v)$ for all the pairs (u, v), which bring the same element $w = u \overset{T}{*} v$ as a result.

Example 5.12
We are going to determine the intersection of the type-2 fuzzy sets discussed in Example 5.11. We assume the minimum operation as t-norm T^* and T. In accordance with formula (5.40), for $x = 1$ and $x = 2$, we obtain

$$\mu_{\tilde{A} \cap \tilde{B}}(1) = \frac{\max(0.5 \wedge 1, 1 \wedge 1, 0.5 \wedge 1)}{0} = \frac{1}{0} \tag{5.41}$$

and

$$\mu_{\tilde{A} \cap \tilde{B}}(2) = \frac{\max(0.5 \wedge 0.5, 0.5 \wedge 1, 1 \wedge 0.5)}{0.5} + \frac{1 \wedge 1}{1 \wedge 0.8} \tag{5.42}$$
$$= \frac{0.5}{0.5} + \frac{1}{0.8}.$$

A complement of the type-2 fuzzy set is a type-2 fuzzy set with the fuzzy membership function given by the formula

$$\mu_{\widehat{\tilde{A}}}(x) = \phi\left(\mu_{\tilde{A}}(x)\right) \tag{5.43}$$
$$= \int_{u \in J_x^u} f_x(u) / (1 - u).$$

Example 5.13
Let us discuss a type-2 fuzzy set defined by the following formula:

$$\mu_{\tilde{A}}(x) = (0.4/0.6 + 1/0.7)/9. \tag{5.44}$$

In accordance with formula (5.43), we have

$$\mu_{\widehat{\tilde{A}}}(x) = (0.4/0.4 + 1/0.3)/9. \tag{5.45}$$

Remark 5.3
Sum (5.32) and intersection (5.40) of type-2 fuzzy sets may be treated as a result of applying the operator of the extended t-norm $\tilde{T}$ and extended t-conorm $\tilde{S}$. These operators may also be discussed in the context of type-1 fuzzy sets defined in the interval $[0, 1]$. We are going to discuss two such sets

$$F = \int_{u \in J^u} f(u) / u \quad \text{and} \quad G = \int_{v \in J^v} g(v) / v. \tag{5.46}$$

The operator of the extended t-norm, whose arguments and resulting value are type-1 fuzzy sets defined within the universe of discourse $[0, 1]$ is given by

$$F \overset{\tilde{T}}{*} G = \int_{u \in J^u} \int_{v \in J^v} g(u) \overset{T^*}{*} f(v) / u \overset{T}{*} v. \tag{5.47}$$

An analogic result may be obtained in a discrete case. We are going to discuss two type-1 sets

$$F = \sum_{u \in J^u} f(u)/u \quad \text{and} \quad G = \sum_{v \in J^v} g(v)/v \qquad (5.48)$$

The operator of the extended t-norm is given by the formula

$$\mu_{\tilde{A}} \overset{\tilde{T}}{*} \mu_{\tilde{B}} = \sum_{u \in J^u} \sum_{v \in J^v} \left(f(u) \overset{T^*}{*} g(v) \right) / u \overset{T}{*} v, \qquad (5.49)$$

and the operator of the extended t-conorm takes the form

$$\mu_{\tilde{A}} \overset{\tilde{S}}{*} \mu_{\tilde{B}} = \sum_{u \in J^u} \sum_{v \in J^v} \left(f(u) \overset{T}{*} g(v) \right) / u \overset{S}{*} v. \qquad (5.50)$$

The introduction of extended triangular norms operating on type-1 sets allows to simplify considerably the notation of complicate operations on type-2 fuzzy sets.

Remark 5.4
The extended function ϕ may also be a function of many variables. Then the operations of the extended t-norm and t-conorm take the following forms:

$$\overset{\tilde{T}}{\underset{i=1}{T}} F_i = \int_{u_1 \in J_1} \cdots \int_{u_n \in J_n} \overset{n}{\underset{i=1}{T}}^* f_i(u_i) / \overset{n}{\underset{i=1}{T}} u_i, \qquad (5.51)$$

$$\overset{\tilde{S}}{\underset{i=1}{S}} F_i = \int_{u_1 \in J_1} \cdots \int_{u_n \in J_n} \overset{n}{\underset{i=1}{T}} f_i(u_i) / \overset{n}{\underset{i=1}{S}} u_i, \qquad (5.52)$$

where $F_i = \int_{u_i \in J_i} f_i(u_i)/u_i, i = 1, \ldots, n.$

Remark 5.5
The operations of extended t-norm and t-conorm are easier to be made with specified assumptions concerning the membership function of particular fuzzy sets. We are going to discuss n convex, normal type-1 fuzzy sets $F_1, \ldots, F_n$ with membership functions $f_1, \ldots, f_n$. Let us assume that $f_1(v_1) = f_2(v_2) = \cdots = f_n(v_n) = 1$, where $v_1, v_2, \ldots, v_n$ are real numbers such that $v_1 \leq v_2 \leq \cdots \leq v_n$ Then the extended minimum type t-norm, known as the meet operation, is specified as follows ([97, 134]):

$$\mu_{\cap_{i=1}^n F_i}(\theta) = \begin{cases} \vee_{i=1}^n f_i(\theta), & \theta < v_1, \\ \wedge_{i=1}^k f_i(\theta), & v_k \leq \theta < v_{k+1}, 1 \leq k \leq n-1, \\ \wedge_{i=1}^n f_i(\theta), & \theta \geq v_n, \end{cases} \qquad (5.53)$$

whereas the extended maximum type t-conorm takes the form

$$\mu_{\cup_{i=1}^n F_i}(\theta) = \begin{cases} \wedge_{i=1}^n f_i(\theta), & \theta < v_1, \\ \wedge_{i=k+1}^n f_i(\theta), & v_k \leq \theta < v_{k+1}, 1 \leq k \leq n-1, \\ \vee_{i=1}^n f_i(\theta), & \theta \geq v_n. \end{cases} \qquad (5.54)$$

Remark 5.6
Let us discuss n Gaussian fuzzy sets $F_1, F_2, ..., F_n$ with means $m_1, m_2, ..., m_n$ and with standard deviations $\sigma_1, \sigma_2, ..., \sigma_n$. Then, as a result of an approximate extended operation of the algebraic t-norm we have [97]

$$\mu_{F_1 \cap F_2 \cap \cdots \cap F_n}(\theta) \approx e^{(-1/2)((\theta - m_1 m_2 \cdots m_n)/\overline{\sigma})^2}, \qquad (5.55)$$

while

$$\overline{\sigma} = \sqrt{\sigma_1^2 \prod_{i; i \neq 1} m_i^2 + \cdots + \sigma_j^2 \prod_{i; i \neq j} m_i^2 + \cdots + \sigma_n^2 \prod_{i; i \neq n} m_i^2}, \qquad (5.56)$$

where $i = 1, ..., n$.

5.6 Type-2 fuzzy relations

At first, we are going to define the Cartesian product of type-2 fuzzy sets.

Definition 5.8
The Cartesian product of n type-2 fuzzy sets $\tilde{A}_1 \subseteq X_1, \tilde{A}_2 \subseteq X_2, \ldots, \tilde{A}_n \subseteq X_n$ is the fuzzy set $\tilde{A} = \tilde{A}_1 \times \tilde{A}_2 \times \ldots \times \tilde{A}_n$ defined on set $X_1 \times X_2 \times \ldots \times X_n$, while the membership function of set $\tilde{A}$ is given by the formula

$$\mu_{\tilde{A}}(\mathbf{x}) = \mu_{\tilde{A}_1 \times \tilde{A}_2 \times \ldots \times \tilde{A}_n}(x_1, x_2, \ldots, x_n) = \overset{n}{\underset{i=1}{\tilde{T}}} \mu_{\tilde{A}_i}(x_n), \qquad (5.57)$$

where $x_1 \in X_1, ..., x_n \in X_i$, and the operation of the extended t-norm is described by dependency (5.51).

Definition 5.9
The binary type-2 fuzzy relation $\tilde{R}$ between two non-empty non-fuzzy sets X and Y is the type-2 fuzzy set determined on the Cartesian product $X \times Y$, i.e.

$$\tilde{R}(X, Y) = \int_{X \times Y} \mu_{\tilde{R}}(x, y) / (x, y), \qquad (5.58)$$

while $x \in X, y \in Y$, and the membership grade of the pair (x, y) to the fuzzy set $\tilde{R}$ is a type-1 fuzzy set defined in the interval $J_{x,y}^v \subset [0, 1]$, i.e.

$$\mu_{\tilde{R}}(x, y) = \int_{v \in J_{x,y}^v} r_{x,y}(v) / v, \qquad (5.59)$$

where $r_{x,y}(v)$ is the secondary grade.

Example 5.13

Let $X = \{3, 4\}$ and $Y = \{4, 5\}$. We are going to formalize an imprecise statement "y is more or less equal to x". At first, we are going to determine the type-1 relation R in the following way:

$$R = \frac{0.8}{(3.4)} + \frac{0.6}{(3.5)} + \frac{1}{(4.4)} + \frac{0.8}{(4.5)}. \tag{5.60}$$

An analogic type-2 fuzzy relation may take on the form

$$\begin{aligned}
\widetilde{R} = &\ (0.6/0.7 + 1/0.8 + 0.5/0.6) / (3, 4) \\
&+ (0.3/0.5 + 1/0.6 + 0.4/0.3) / (3, 5) \\
&+ (1/1 + 1/1 + 1/1) / (4, 4) \\
&+ (0.6/0.7 + 1/0.8 + 0.5/0.6) / (4.5).
\end{aligned} \tag{5.61}$$

Example 5.14

We are going to formalize an imprecise statement "number x slightly differs from number y". This problem may be solved with the type-1 fuzzy relation described by the membership function

$$\mu_R (x, y) = \max \left\{ (4 - |x - y|) / 4.0 \right\}. \tag{5.62}$$

An analogic type-2 fuzzy relation may take on the form

$$\mu_{\widetilde{R}} (x, y) = \int_{v \in [0,1]} \exp \left[- \left(\frac{v - m (x, y)}{\sigma} \right)^2 \right] / v, \tag{5.63}$$

where $\sigma > 0$ and

$$m (x, y) = \max \left\{ (4 - |x - y|) / 4.0 \right\}. \tag{5.64}$$

Alternatively, the secondary membership function of Gaussian type may be substituted by a fuzzy triangular number. Figure 5.7a depicts the illustration of the type-1 fuzzy relation given by formula (5.62). Figure 5.7b depicts the possibility of uncertainty in the specification of the statement "number x slightly differs from number y". The figure depicts the footprint of uncertainty while the level of shading corresponds to the value of the secondary grade. Figure 5.7c presents the triangular secondary membership function defined in the interval $J_x = [0.2, 0.4]$.

It is worth mentioning that fuzzy relations may be made with the use of extended norms. We are going to discuss the membership function of the type-2 fuzzy set defined on set $X, \widetilde{A} \subseteq X$, i.e.

$$\mu_{\widetilde{A}} (x) = \int_{u \in J_x^u} f_x (u) / u \tag{5.65}$$

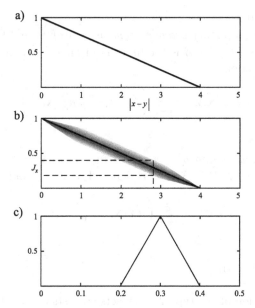

FIGURE 5.7. Illustration of type-1 and type-2 fuzzy relations

and the membership function of fuzzy set $\widetilde{B}$ defined on another set Y, $\widetilde{B} \subset Y$, i.e.

$$\mu_{\widetilde{B}}(y) = \int_{v \in J_y^v} g_y(v) / v, \qquad (5.66)$$

where $J_x^u, J_y^v \subset [0, 1]$. The extended t-conorm of the type-2 fuzzy sets defined on different spaces forms a certain fuzzy relation $\widetilde{R}$, determined as follows:

$$\mu_{\widetilde{R}}(x, y) = \mu_{\widetilde{A}}(x) \overset{\widetilde{S}}{*} \mu_{\widetilde{B}}(y) = \int_{u \in J_x^u} \int_{v \in J_y^v} f_x(u) \overset{T}{*} g_y(v) / u \overset{S}{*} v \qquad (5.67)$$

$$= \int_{w \in J_{x,y}^w} r_{x,y}(w) / w.$$

Similarly, the extended t-norm creates a fuzzy relation in the form of

$$\mu_{\widetilde{R}}(x, y) = \mu_{\widetilde{A}}(x) \overset{\widetilde{T}}{*} \mu_{\widetilde{B}}(y) = \int_{u \in J_x^u} \int_{v \in J_y^v} f_x(u) \overset{T^*}{*} g_y(v) / u \overset{T}{*} v \qquad (5.68)$$

$$= \int_{w \in J_{x,y}^w} r_{x,y}(w) / w.$$

In the application of the theory of fuzzy sets to the construction of inference systems it is necessary to use the concept of the composition of fuzzy relations, which, in the context of type-2 fuzzy sets, are defined as follows:

Definition 5.10
The sup-T type (sup-star) *extended composition of* type-2 fuzzy relations $\widetilde{R} \subseteq X \times Y$ and $\widetilde{S} \subseteq Y \times Z$ is the fuzzy relation $\widetilde{R} \circ \widetilde{S} \subseteq X \times Z$ with the membership function

$$\mu_{\widetilde{R} \circ \widetilde{S}} (\mathbf{x}, \mathbf{z}) = \underset{y \in Y}{\widetilde{S}} \left(\mu_{\widetilde{R}} (\mathbf{x}, \mathbf{y}) \overset{\widetilde{T}}{*} \mu_{\widetilde{S}} (\mathbf{x}, \mathbf{z}) \right). \tag{5.69}$$

Definition 5.11
The extended composition of the type-2 fuzzy set $\widetilde{A}$, $\widetilde{A} \subset X$, and of the type-2 fuzzy relation $\widetilde{R} \subseteq X \times Y$ is denoted as $\widetilde{A} \circ \widetilde{R}$ and determined as follows:

$$\mu_{\widetilde{B}} (y) = \underset{x \in X}{\widetilde{S}} \left(\mu_{\widetilde{A}} (\mathbf{x}) \overset{\widetilde{T}}{*} \mu_{\widetilde{R}} (\mathbf{x}, y) \right). \tag{5.70}$$

5.7   Type reduction

The defuzzification of the type-2 fuzzy sets consists of two stages: At first, a so-called *type reduction* should be made, which is the transformation of the type-2 fuzzy set into the type-1 fuzzy set. This way we are going to obtain the type-1 fuzzy set called a *centroid*, which may be defuzzified to a non-fuzzy value. We are going to show the method for the determination of the centroid of the type-2 fuzzy set.

Let us discuss a fuzzy set A (type-1) defined on set X. Let us assume that set X has been discretized and takes R values $x_1, ..., x_R$. The centroid of fuzzy set A is determined as follows:

$$C_A = \frac{\sum\limits_{k=1}^{R} x_k \mu_A (x_k)}{\sum\limits_{k=1}^{R} \mu_A (x_k)}. \tag{5.71}$$

We are going to determine the centroid of the type-2 fuzzy set, $\widetilde{A} = \left\{ (x, \mu_{\widetilde{A}} (x)) \,|x \in X \right\}$, which, as a result of an analogic discretizaton is notated as follows:

$$\widetilde{A} = \sum_{k=1}^{R} \left[\int_{u \in J_{x_k}} f_{x_k} (u) \,/u \right] /x_k. \tag{5.72}$$

Applying the extension principle to formula (5.71) we get

$$C_{\widetilde{A}} = \int_{\theta_1 \in J_{x_1}} \cdots \int_{\theta_R \in J_{x_R}} [f_{x_1} (\theta_1) * \cdots * f_{x_R} (\theta_R)] \, / \frac{\sum\limits_{k=1}^{R} x_k \theta_k}{\sum\limits_{k=1}^{R} \theta_k}. \tag{5.73}$$

Of course, the centroid $C_{\widetilde{A}}$ is a type-1 fuzzy set. Let us note that any selection of elements $\theta_1 \in J_{x_1}, ..., \theta_R \in J_{x_R}$ along with corresponding secondary grades $f_{x_1}(\theta_1), ..., f_{x_R}(\theta_R)$, creates an embedded fuzzy set $\widetilde{A}_o$ (type-2).

Example 5.15
Let $\mathbf{X} = \{2, 5\}$. We are going to perform the type reduction of the following type-2 fuzzy set:

$$\widetilde{A} = (0.6/0.4 + 1/0.8)/2 + (0.3/0.7 + 1/0.6)/5. \tag{5.74}$$

The centroid of the type-2 fuzzy set given by formula (5.74) is a type-1 fuzzy set taking the form

$$C_{\widetilde{A}} = \frac{0.6 \times 0.3}{a_1} + \frac{0.6 \times 1}{a_2} + \frac{1 \times 0.3}{a_3} + \frac{1 \times 1}{a_4} \tag{5.75}$$

$$= \frac{0.18}{a_1} + \frac{0.6}{a_2} + \frac{0.3}{a_3} + \frac{1}{a_4},$$

while

$$a_1 = \frac{2 \times 0.4 + 5 \times 0.7}{0.4 + 0.7} = \frac{43}{11},$$

$$a_2 = \frac{2 \times 0.4 + 5 \times 0.6}{0.4 + 0.6} = 3.8,$$

$$a_3 = \frac{2 \times 0.8 + 5 \times 0.7}{0.8 + 0.7} = 3.4,$$

$$a_4 = \frac{2 \times 0.8 + 5 \times 0.6}{0.8 + 0.6} = \frac{23}{7}.$$

In a continuous case, the determination of the centroid of the type-2 fuzzy set is a much more complicated task from the computational point of view. The problem becomes easier to solve, if the secondary membership functions are interval ones. Then formula (5.73) takes the form

$$C_{\widetilde{A}} = \int_{\theta_1 \in J_{x_1}} \cdots \int_{\theta_R \in J_{x_R}} 1 \Big/ \frac{\sum\limits_{k=1}^{R} x_k \theta_k}{\sum\limits_{k=1}^{R} \theta_k}. \tag{5.76}$$

We are going to show the method of the determination of the centroid of the type-2 fuzzy set having an interval secondary membership function. With reference to formula (5.76), let us define

$$s(\theta_1, ..., \theta_R) = \frac{\sum\limits_{k=1}^{R} x_k \theta_k}{\sum\limits_{k=1}^{R} \theta_k}. \tag{5.77}$$

It is obvious that centroid (5.76) will be an interval type-1 fuzzy set, i.e.

$$C_{\tilde{A}} = \int_{x \in [x_l, x_p]} 1/x \equiv [x_l, x_p]. \tag{5.78}$$

From the observations shown above, it may be concluded that the determination of centroid (5.76) comes down to the optimization (maximization and minimization) with respect to θ_k of function given by formula (5.77), taking account of constraints

$$\theta_k \in \left[\underline{\theta}^k, \overline{\theta}^k \right], \tag{5.79}$$

where $k = 1, ..., R$ and

$$\underline{\theta}^k = \underline{J_x}, \overline{\theta}^k = \overline{J}_x. \tag{5.80}$$

Differentiating expression (5.77) with respect to θ_j, we get

$$\frac{\partial}{\partial \theta_j} s(\theta_1, ..., \theta_R) = \frac{\partial}{\partial \theta_j} \left[\frac{\sum_{k=1}^{R} x_k \theta_k}{\sum_{k=1}^{R} \theta_k} \right] = \frac{\partial}{\partial \theta_j} \left[\frac{x_j \theta_j + \sum_{k \neq j} x_k \theta_k}{\theta_j + \sum_{k \neq j} \theta_k} \right] \tag{5.81}$$

$$= \left[\frac{1}{\theta_j + \sum_{k \neq j} \theta_k} \right] (x_j) \left(x_j \theta_j + \sum_{k \neq j} x_k \theta_k \right) \left[\frac{-1}{\left(\theta_j + \sum_{k \neq j} \theta_k \right)^2} \right]$$

$$= \frac{x_j}{\sum_{k=1}^{R} \theta_k} - \frac{\sum_{k=1}^{R} x_k \theta_k}{\left(\sum_{k=1}^{R} \theta_k \right)^2} = \frac{x_j}{\sum_{k=1}^{R} \theta_k} - \frac{\sum_{k=1}^{R} x_k \theta_k}{\sum_{k=1}^{R} \theta_k} \frac{1}{\sum_{k=1}^{R} \theta_k}$$

$$= \frac{x_j - s(\theta_1, ..., \theta_R)}{\sum_{k=1}^{R} \theta_k}.$$

Of course $\sum_{k=1}^{R} \theta_k > 0$. Hence, from the last equality we have

$$\frac{\partial}{\partial \theta_j} s(\theta_1, ..., \theta_R) \geq 0, \quad \text{if} \quad x_j \geq s(\theta_1, ..., \theta_R) \tag{5.82}$$

and

$$\frac{\partial}{\partial \theta_j} s(\theta_1, ..., \theta_R) \leq 0, \quad \text{if} \quad x_j \leq s(\theta_1, ..., \theta_R). \tag{5.83}$$

When equating the right side of expression (5.81) to zero we get

$$\frac{\sum_{k=1}^{R} x_k \theta_k}{\sum_{k=1}^{R} \theta_k} = x_j. \tag{5.84}$$

Therefore

$$\sum_{k=1}^{R} x_k \theta_k = x_j \sum_{k=1}^{R} \theta_k \tag{5.85}$$

and

$$x_j \theta_j + \sum_{\substack{k=1 \\ k \neq j}}^{R} x_k \theta_k = x_j \theta_j + x_j \sum_{\substack{k=1 \\ k \neq j}}^{R} \theta_k. \tag{5.86}$$

In consequence

$$\frac{\sum_{k \neq j} x_k \theta_k}{\sum_{k \neq j} \theta_k} = x_j. \tag{5.87}$$

We find out that the necessary condition for the extremum s to exist does not depend in any way on parameter θ_k with respect to which the derivative was calculated. However, inequalities (5.82) and (5.83) show in which direction we should go in order to increase or decrease the value of expression $s(\theta_1, ..., \theta_R)$. Upon the basis of these inequalities we conclude that

i) if $x_j > s(\theta_1, ..., \theta_R)$, then $s(\theta_1, ..., \theta_R)$ is increasing along with the decrease of parameter θ_j,

ii) if $x_j < s(\theta_1, ..., \theta_R)$, then $s(\theta_1, ..., \theta_R)$ is increasing along with the increase of parameter θ_j.

Let us remind that $\underline{\theta}_k \leq \theta_k \leq \overline{\theta}_k$. Hence, function s reaches the maximum if

a) $\theta_k = \overline{\theta}_k$ for these values k, for which $x_k > s$,

b) $\theta_k = \underline{\theta}_k$ for these values k, for which $x_k < s$,

Upon this basis we are going to present an iterative algorithm (known as Karnik - Mendel type reduction algorithm) for the search of the maximum of function s:

1) Determine $\theta_k = \frac{\underline{\theta}_k + \overline{\theta}_k}{2}$, $k = 1, ..., R$, calculate $s' = s(\theta_1, ..., \theta_R)$.

2) Find j $(1 \leq j \leq R - 1)$ so that $x_j \leq s' < x_{j+1}$.

3) Substitute $\theta_k = \underline{\theta}_k$ for $k \leq j$ and $\theta_k = \overline{\theta}_k$ for $k > j$.
Calculate $s'' = s(\underline{\theta}_1, ..., \underline{\theta}_j, \overline{\theta}_{j+1}, ..., \overline{\theta}_R)$.

4) If $s'' = s'$ then s'' is the maximum value of function s.
If $s'' \neq s'$ then pass on to step 5.

5) Substitute $s' = s''$ and pass on to step 2.

In an analogic way, we may determine the minimum of function s. This function reaches the minimum if

a) $\theta_k = \overline{\theta}_k$ for these values k, for which $x_k < s$,

b) $\theta_k = \underline{\theta}_k$ for these values k, for which $x_k > s$,

The iterative algorithm for the search of the minimum of function s is given as follows:

1) Determine $\theta_k = \frac{\underline{\theta}_k + \overline{\theta}_k}{2}, k = 1, ..., R$, calculate $s' = s\left(\theta_1, ..., \theta_R\right)$.

2) Find j $(1 \le j \le R - 1)$ so that $x_j < s' \le x_{j+1}$.

3) Substitute $\theta_k = \overline{\theta}_k$ for $k < j$ and $\theta_k = \underline{\theta}_k$ for $k \ge j$, calculate $s'' = s\left(\overline{\theta}_1, ..., \overline{\theta}_j, \underline{\theta}_{j+1}, ..., \underline{\theta}_R\right)$.

4) If $s'' = s'$ then, s'' is the minimum value of function s. If $s'' \ne s'$ then pass on to step 5.

5) Substitute $s' = s''$ and pass on to step 2.

Example 5.16

Figures 5.8 – 5.10 depict the method of working of the iterative algorithm for the search of the centroid of the type-2 fuzzy set with the interval secondary membership function. In Fig. 5.8, the footprint of uncertainty of the type-2 fuzzy set, which will be subject to type reduction, is marked. The thick line in this picture corresponds with point 1 of the iterative algorithm, which starts with the determination of the centre of particular intervals $J_x, x \in X$ and the value of expression (5.77).

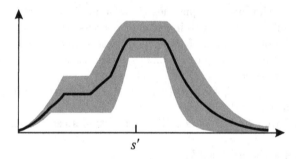

FIGURE 5.8. The footprint of uncertainty of type-2 fuzzy set; the thick line corresponds with point 1 of the K-M iterative type-reduction algorithm

The centroid is a type-1 fuzzy set given by formula (5.78). By iteration, we search for point x_p, determining the centroid of an embedded fuzzy set (Fig. 5.9) consisting first of a piece of the lower membership function, and then of a piece of the upper membership function. Similarly, we search for point x_l, determining the centroid of an embedded fuzzy set (Fig. 5.10) consisting first of a piece of the upper membership function, and then of a piece of the lower membership function.

The obtained fuzzy set $C_{\tilde{A}} = [x_l, x_p]$ may be deffuzified (Fig. 5.11) in the following way:

$$\widehat{x}_w = \frac{x_l + x_p}{2}. \tag{5.88}$$

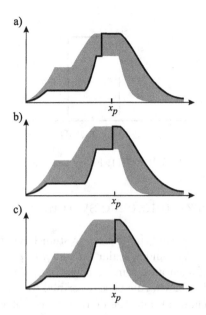

FIGURE 5.9. Iterative search for point x_p determining the centroid of an embedded fuzzy set

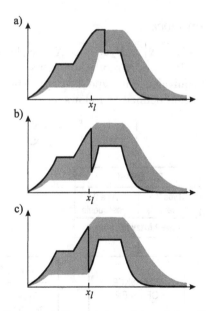

FIGURE 5.10. Iterative search for point x_l determining the centroid of an embedded fuzzy set

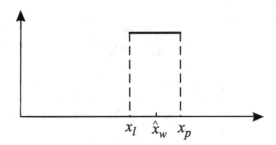

FIGURE 5.11. Fuzzy set $C_{\widetilde{A}}$

5.8 Type-2 fuzzy inference systems

We are going to discuss the type-2 fuzzy system having n input variables $x_i \in X_i \subset R, i = 1, \ldots, n$, and a scalar output $y \in Y$. Figure 5.12 depicts the block diagram of such a system. It consists of the following elements: the type-2 fuzzification block, rule base described by type-2 fuzzy relations, type-2 inference mechanism, and the deffuzification block.

The deffuzification has two stages: at first, the type reduction is performed (Subchapter 5.7) and then the classic defuzzification is applied (Subchapter 4.9).

5.8.1 Fuzzification block

Let $\bar{x} = (x_1, \ldots, x_n)^T \in \mathbf{X} = X_1' \times X_2 \times \cdots \times X_n$ be the input signal of the fuzzy inference system. In type-1 fuzzy systems, the singleton type fuzzification is applied. Its equivalence in type-2 fuzzy systems is the singleton-singleton type fuzzification defined as follows:

$$\widetilde{\mu}_{A'}(\mathbf{x}) = \begin{cases} 1/1, & \text{if } \mathbf{x} = \bar{\mathbf{x}}, \\ 1/0, & \text{if } \mathbf{x} \neq \bar{\mathbf{x}}. \end{cases} \tag{5.89}$$

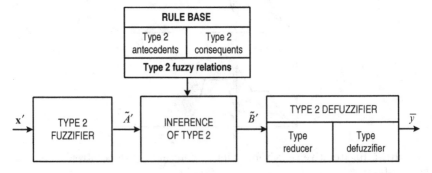

FIGURE 5.12. Block diagram of a type-2 fuzzy inference system

In case of the independence of particular input variables, the above-mentioned operation takes the form of:

$$\widetilde{\mu}_{A'}(x_i) = \begin{cases} 1/1, & \text{if} \quad x_i = \overline{x}_i, \\ 1/0, & \text{if} \quad x_i \neq \overline{x}_i. \end{cases} \tag{5.90}$$

for $i = 1, ..., n$. As a result of the fuzzification, we obtain n input type-2 fuzzy sets described by:

$$\widetilde{A}'_i = (1/1)/\overline{x}_i, \quad i = 1, \ldots, n, \tag{5.91}$$

where $\overline{x}_i$ is a specific value of i-th input variable.

It is worth mentioning that other methods for the fuzzification of the input signal are also possible. Figure 5.13 depicts a graphic illustration of these methods. For instance, the fuzzification of singleton-interval type (Fig. 5.13b) means that the secondary membership function is an interval fuzzy set. The non-singleton-singleton fuzzification (Fig. 5.13c) means that the secondary membership function is a singleton type fuzzy set, and in this case the fuzzification is identical to the non-singleton type fuzzification for type-1 fuzzy sets. The non-singleton-triangular fuzzification (Fig. 5.13e) means that the secondary membership function is triangular fuzzy set, while the level of shading on Fig. 5.13e reflects the value of the secondary membership function (triangular) for given element $u \in J_x$.

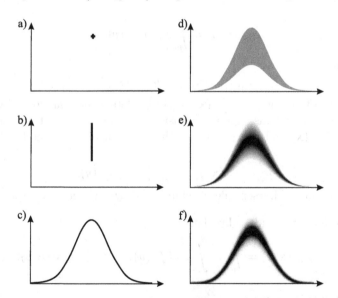

FIGURE 5.13. Illustration of different fuzzification methods: a) singleton-singleton, b) singleton-interval, c) nonsingleton-singleton, d) nonsigleton-interval, e) nonsingleton-triangular, f) nonsingleton-gaussoidal

5.8.2 Rules base

The linguistic model consists of N rules in the form of:

$$\widetilde{R}^k : \textbf{IF } x_1 \text{ is } \widetilde{A}_1^k \textbf{ AND } x_2 \text{ is } \widetilde{A}_2^k \textbf{ AND} \ldots \textbf{AND } x_n \text{ is } \widetilde{A}_n^k$$
$$\textbf{THEN } y \text{ is } \widetilde{B}^k, k = 1, \ldots, N. \tag{5.92}$$

Denote

$$\widetilde{A} = \widetilde{A}^k = \widetilde{A}_1^k \times \widetilde{A}_2^k \times \ldots \times \widetilde{A}_n^k. \tag{5.93}$$

Of course

$$\mu_{\widetilde{A}^k}(\mathbf{x}) = \overset{n}{\underset{i=1}{\mathbf{T}}} \mu_{\widetilde{A}_i^k}(\overline{x}_i). \tag{5.94}$$

It is easily seen that rule (5.92) may be presented in the form of implication

$$\widetilde{A}^k \to \widetilde{B}^k, \quad k = 1, ..., n. \tag{5.95}$$

5.8.3 Inference block

At first, we are going to determine membership function $\mu_{\widetilde{A}^k \to \widetilde{B}^k}(\mathbf{x}, y)$. Each k-th rule is represented in a fuzzy system by a certain type-2 fuzzy relation.

$$\widetilde{R}^k(\mathbf{x}, y) = \int_{\mathbf{X} \times Y} \mu_{\widetilde{R}^k}(\mathbf{x}, y) / (\mathbf{x}, y), \tag{5.96}$$

where

$$\mu_{\widetilde{R}^k}(\mathbf{x}, y) = \int_{v \in V_{\mathbf{x}, y}} r_{\mathbf{x}, y}^k(v) / v. \tag{5.97}$$

Therefore

$$\mu_{\widetilde{A}^k \to \widetilde{B}^k}(\mathbf{x}, y) = \mu_{\widetilde{R}(k)}(\mathbf{x}, y). \tag{5.98}$$

Membership function $\mu_{\widetilde{A}^k \to \widetilde{B}^k}(\mathbf{x}, y)$ will be determined, analogically as in case of type-1 systems, upon the basis of the knowledge of membership function $\mu_{\widetilde{A}^k}(\mathbf{x})$ and $\mu_{\widetilde{B}^k}(y)$. Using the operator of the extended t-norm we have

$$\mu_{\widetilde{A}^k \to \widetilde{B}^k}(\mathbf{x}, y) = \mu_{\widetilde{A}^k}(\mathbf{x}) \overset{\widetilde{T}}{*} \mu_{\widetilde{B}^k}(y). \tag{5.99}$$

The Mamdani and Larsen rules used in type-1 systems now take the form of

- extended min rule (Mamdami)

$$\mu_{\widetilde{A}^k \to \widetilde{B}^k}(\mathbf{x}, y) = \int_{u \in J_{\mathbf{x}}^u} \int_{v \in J_y^v} \left(f_{\mathbf{x}}(u) \overset{T}{*} g_y(v) \right) / \min(u, v), \tag{5.100}$$

- extended product rule (Larsen)

$$\mu_{\widetilde{A}^k \to \widetilde{B}^k}(\mathbf{x}, y) = \int_{u \in J_{\mathbf{x}}^u} \int_{v \in J_y^v} \left(f_{\mathbf{x}}(u) \overset{T}{*} g_y(v) \right) / uv. \tag{5.101}$$

At the output of the inference block, we obtain a type-2 fuzzy set $\widetilde{B}'^k$. This set is determined by the composition of the input fuzzy set $\widetilde{A}'$ and the fuzzy relation $\widetilde{R}^k$, i.e.

$$\widetilde{B}'^k = \widetilde{A}' \circ \widetilde{R}^k = \widetilde{A}' \circ \left(\widetilde{A}^k \rightarrow \widetilde{B}^k \right). \tag{5.102}$$

Using Definition 5.11, we determine the membership function of the fuzzy set $\widetilde{B}'^k$

$$\mu_{\widetilde{B}'^k}(y) = \mu_{\widetilde{A}' \circ \widetilde{R}^k}(y) = \widetilde{\underset{\mathbf{x} \in \mathbf{X}}{S}} \left(\mu_{\widetilde{A}'}(\mathbf{x}) \overset{\widetilde{T}}{*} \mu_{\widetilde{B}^k}(\mathbf{x}, y) \right) \tag{5.103}$$

$$= \widetilde{\underset{\mathbf{x} \in \mathbf{X}}{S}} \left(\mu_{\widetilde{A}'}(\mathbf{x}) \overset{\widetilde{T}}{*} \mu_{\widetilde{A}' \rightarrow \widetilde{B}^k}(\mathbf{x}, y) \right).$$

In case of singleton-singleton type fuzzification (5.84) the formula above takes the form

$$\mu_{\widetilde{B}'^k}(y) = \mu_{\widetilde{A}^k \rightarrow \widetilde{B}^k}(\overline{\mathbf{x}}, y). \tag{5.104}$$

Using formulae (5.99) and (5.94), we obtain

$$\mu_{\widetilde{B}'^k}(y) = \mu_{\widetilde{A}_1^k \times \ldots \times \widetilde{A}_n^k}(\overline{\mathbf{x}}) \overset{\widetilde{T}}{*} \mu_{\widetilde{B}^k}(y) = \left(\overset{n}{\underset{i=1}{\widetilde{T}}} \mu_{\widetilde{A}_i^k}(\overline{x}_i) \right) \overset{\widetilde{T}}{*} \mu_{\widetilde{B}^k}(y). \tag{5.105}$$

Let us denote the firing strength of k-th rule in the following way:

$$\tau_k = \overset{n}{\underset{i=1}{\widetilde{T}}} \mu_{\widetilde{A}_i^k}(\overline{x}_i). \tag{5.106}$$

Then dependency (5.105) takes the form

$$\mu_{\widetilde{B}'^k}(y) = \tau_k \overset{\widetilde{T}}{*} \mu_{\widetilde{B}^k}(y). \tag{5.107}$$

Remark 5.7
In case of type-1 fuzzy sets the firing strength of τ_k rule is a real number while $\tau_k \in [0, 1]$. In case of type-2 fuzzy sets the firing strength of τ_k rule is a type-1 fuzzy set defined in $[0, 1]$.

Having inference results $\widetilde{B}'^k$ for all N rules, we make an aggregation using the operator of the extended t-conorm

$$\mu_{\widetilde{B}'}(y) = \overset{N}{\underset{k=1}{\widetilde{S}}} \mu_{\widetilde{B}'^k}(y). \tag{5.108}$$

We are going to show the inference process in interval systems. In such systems the secondary membership functions of fuzzy sets $\widetilde{A}_i^k$ and $\widetilde{B}^k$, $i = 1, \ldots, n$, $k = 1, \ldots, N$, are constant functions taking value 1 in all intervals $J_x, x \in X$. Within further discussion we are going to apply two properties ([97, 134]) of interval type-1 fuzzy sets $F_1, ..., F_n$, defined on intervals $[l_1, p_1], ..., [l_n, p_n]$, where $l_i \geq 0$ and $p_i \geq 0$, $i = 1, ..., n$.

1) Extended t-norm $\widetilde{T}_{i=1}^n F_i$ is an interval type-1 fuzzy set defined on interval $\left[\left(l_1 \overset{T}{*} l_2 \overset{T}{*} ... \overset{T}{*} l_n \right), \left(p_1 \overset{T}{*} p_2 \overset{T}{*} ... \overset{T}{*} p_n \right) \right]$, where $\overset{T}{*}$ denotes t-norm of minimum type or product.

2) Extended t-conorm $\widetilde{S}_{i=1}^n F_i$ is an interval type-1 fuzzy set defined in the interval $[(l_1 \vee l_2 \vee ... \vee l_n), (p_1 \vee p_2 \vee ... \vee p_n)]$, where $\vee$ means a maximum operation.

We are going to introduce a symbolic notation, according to which the interval fuzzy set A will be denoted as

$$A = \int_{x \in [a,b]} 1/x \equiv [a, b]. \tag{5.109}$$

Using property 1, we are going to express the firing strength of rule τ_k, being now an interval type-1 fuzzy set, through the values of the lower and upper membership functions of fuzzy sets $\widetilde{A}_i^k$. Based on property 1, we may denote

$$\tau_k = [\underline{\tau}_k, \overline{\tau}_k], \tag{5.110}$$

where

$$\underline{\tau}_k(\mathbf{\overline{x}}) = \underline{\mu}_{\widetilde{A}_1^k}(\overline{x}_1) * ... * \underline{\mu}_{\widetilde{A}_n^k}(\overline{x}_n) \tag{5.111}$$

and

$$\overline{\tau}_k(\mathbf{\overline{x}}) = \overline{\mu}_{\widetilde{A}_1^k}(\overline{x}_1) * \cdots * \overline{\mu}_{\widetilde{A}_n^k}(\overline{x}_n). \tag{5.112}$$

Using formulas (5.107), (5.110), and property 1, we obtain

$$\mu_{\widetilde{B}'^k}(y) = \mu_{\widetilde{B}^k}(y) \overset{\widetilde{T}}{*} [\underline{\tau}^k, \overline{\tau}^k] \equiv \left[\underline{b}^k(y), \overline{b}^k(y) \right], \quad y \in Y, \tag{5.113}$$

where

$$\underline{b}^k(y) = \underline{\tau}^k \overset{T}{*} \underline{\mu}_{\widetilde{B}^k}(y) \tag{5.114}$$

and

$$\overline{b}^k(y) = \overline{\tau}^k \overset{T}{*} \overline{\mu}_{\widetilde{B}^k}(y). \tag{5.115}$$

Using formulas (5.113), (5.108), and property 2, we may determine

$$\mu_{\widetilde{B}'}(y) = \overset{N}{\underset{k=1}{S}} \mu_{\widetilde{B}'^k}(y) = \overset{N}{\underset{k=1}{S}} \left[\underline{b}^k(y), \overline{b}^k(y) \right] = [\underline{b}(y), \overline{b}(y)], \tag{5.116}$$

where

$$\underline{b}(y) = \underline{b}^1(y) \vee \underline{b}^2(y) \vee ... \vee \underline{b}^N(y) \tag{5.117}$$

and

$$\overline{b}(y) = \overline{b}^{1}(y) \vee \overline{b}^{2}(y) \vee \ldots \vee \overline{b}^{N}(y). \sum_{k \neq j}. \tag{5.118}$$

Example 5.17

Figure 5.14 shows the method of determining the firing strength of a type-2 system with two rules. As the t-norm the minimum operation was chosen. Therefore,

$$\underline{\tau}_{k} = \min\left[\underline{\mu}_{\widetilde{A}_{1}^{k}}(\overline{x}_{1}), \underline{\mu}_{\widetilde{A}_{2}^{k}}(\overline{x}_{2})\right] \tag{5.119}$$

and

$$\overline{\tau}_{k} = \min\left[\overline{\mu}_{\widetilde{A}_{1}^{k}}(\overline{x}_{1}), \overline{\mu}_{\widetilde{A}_{2}^{k}}(\overline{x}_{2})\right] \tag{5.120}$$

for $k = 1, 2$. As we have emphasized earlier, the firing strengths are interval type-1 fuzzy sets.

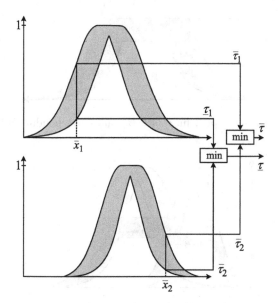

FIGURE 5.14. The method of determining the firing strength of a type-2 fuzzy system with singleton-singleton fuzzification

Example 5.18

Figures 5.15 and 5.16 show output type-2 fuzzy sets $\widetilde{B}^{1}$ and $\widetilde{B}^{2}$, as well as fuzzy sets (shaded ones) $\widetilde{B}'^{1}$ and $\widetilde{B}'^{2}$ resulting from inference given by formula (5.113).

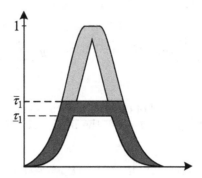

FIGURE 5.15. Output type-2 fuzzy set $\widetilde{B}^1$ and corresponding inferred type-2 fuzzy set $\widetilde{B}'^1$

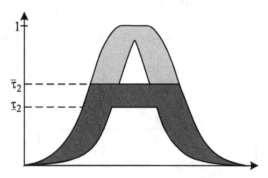

FIGURE 5.16. Output type-2 fuzzy set $\widetilde{B}^2$ and corresponding inferred type-2 fuzzy set $\widetilde{B}'^2$

FIGURE 5.17. Type-2 fuzzy set $\widetilde{B}'$ resulting from the aggregation of fuzzy sets $\widetilde{B}'^1$ and $\widetilde{B}'^2$

Figure 5.17 presents fuzzy set (shaded one) $\widetilde{B}'$ given by formula (5.116) and resulting from the aggregation of fuzzy sets $\widetilde{B}'^1$ and $\widetilde{B}'^2$. In order to determine this set we have used the operation

$$\max\left(\min\overline{\tau}_1,\overline{\mu}_{\widetilde{B}^1}\left(y\right),\min\overline{\tau}_2,\overline{\mu}_{\widetilde{B}^2}\left(y\right)\right) \qquad (5.121)$$

and

$$\max \left(\min \underline{\tau}_1, \underline{\mu}_{\widetilde{B}^1}(y), \min \underline{\tau}_2, \underline{\mu}_{\widetilde{B}^2}(y) \right). \tag{5.122}$$

Example 5.19

Examples 5.17 and 5.18 present the results obtained for interval type-2 fuzzy systems with singleton fuzzification given by formula (5.89). These results can be generalized for the case where the input signal is a type-1 fuzzy set (nonsingleton-singleton fuzzification) or an interval type-2 fuzzy

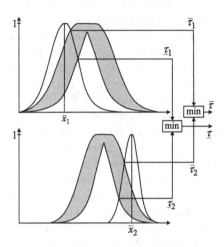

FIGURE 5.18. The method of determining the firing strength of a type-2 fuzzy system with nonsigleton-singleton fuzzification

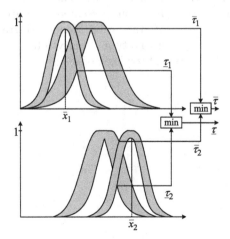

FIGURE 5.19. The method of determining the firing strength of a type-2 fuzzy system with nonsigleton (type-2) – interval fuzzification

set (type-2 nonsingleton fuzzification – interval). Figures 5.18 and 5.19 show the method of determining the firing strength illustrating both cases.

5.9 Notes

The notion of the type-2 fuzzy set has been introduced by Lotfi Zadeh [266]. In his article the author also defines the amount and the intersection of the type-2 fuzzy sets using the extension principle for that purpose. The basic notions characterizing type-2 fuzzy sets, i.e. the secondary membership functions and grades, the upper and lower membership functions, as well as the notions of embedded fuzzy sets and the footprint of uncertainty, have been successively introduced to the global literature by Mendel, and their review is contained in his monography [134]. The method of inference with the use of interval type-2 fuzzy sets was first described by Gorzałczany [64]. Basic operations on type-2 fuzzy sets have been provided by Dubois and Prade [42], and Karnik and Mendel [97,100]. The interval fuzzy sets of higher levels have been examined by Hisdal [80]. The iterative algorithm of type reduction for the interval type-2 fuzzy sets has been introduced by Karnik and Mendel [97, 101]. This has allowed to construct the interval type-2 fuzzy logic systems. The first such constructions have been presented by Karnik, Mendel and Liang [99]. The analysis of the differences between the interval inference systems and type-1 systems is presented in an article by Starczewski [240]. An interesting method of type reduction has been presented by Wu and Mendel in article [261]. The interval type-2 systems have been used for the prediction of chaotic series [98]. A novelty is the construction of the type-2 fuzzy inference system with the triangular secondary membership function, presented by Starczewski [238]. On the webpage http://ieee-cis.org/standards/ Mendel, Hagras and John have presented basic information on the type-2 fuzzy sets. This subject is also discussed on http://www.type2fuzzylogic.org/.

6
Neural networks and their learning algorithms

6.1 Introduction

For many years, scientists have tried to learn the structure of the brain and discover how it works. Unfortunately, it still remains a fascinating riddle not solved completely. Based on observation of people crippled during different wars or injured in accidents, the scientists could assess the specialization of particular fragments of the brain. It was found, for example, that the left hemisphere is responsible for controlling the right hand, whereas the right hemisphere – for the left hand. The scientists still do not have any detailed information on higher mental functions. We can assume hypothetically that the left hemisphere controls speech function and scientific thinking, whereas the right hemisphere is its opposite as it manages artistic capabilities, spatial imagination etc. The nervous system is made of cells called neurons. There are about 100 billion of them in the human brain. The functioning of a single neuron consists in the flow of so-called nerve impulses. The impulse induced by a specific stimulus encountering a neuron causes its spreading along all its dendrones. As a result, a muscle contraction can occur or another neuron can be stimulated. Why, then, appropriately connected artificial neurons could not, instead of controlling muscles, manage, for example, the work of a device or solve various problems requiring intelligence? This chapter discusses artificial neural networks. We will present a mathematical model of a single neuron, various structures of artificial neural networks and their learning algorithms.

6.2   Neuron and its models

6.2.1   Structure and functioning of a single neuron

The basic element of the nervous system is a nervous cell called *neuron*. Figure 6.1 presents its simplified scheme. In a neuron, we can distinguish *nerve cell body* (called soma) and two types of surrounding dendrones: dendrones introducing information to neuron, so called *dendrites* and a dendrone leading the information out of the neuron, so-called *axon*. Each neuron has exactly one dendrone leading information out through which it can send impulses to many other neurons. A single neuron receives stimulation from an enormous number of neurons reaching as much as a thousand. As mentioned before, in a human brain there are about 100 billion of neurons which interact with one another through an enormous number of connections. One neuron stimulates other neurons through neuron junctions called *synapses*, while signals are transmitted through complex chemical and electric processes. The synapses function as information transmitters and as a result of their functioning the stimulation can be strengthened or weakened. As a result, the neuron receives signals and some of them are stimulating whereas others are suppressing. The neuron sums stimulating and suppressing impulses. If their algebraic sum exceeds a certain threshold value, the signal at neuron output is transmitted – via axon – to other neurons.

We will present now a model of neuron referring to the first attempts to formalize the description of the nerve cell functioning. Let us introduce the following notations: n – a number of inputs in a neuron, $x_1, .., x_n$ – input signals, $\mathbf{x} = [x_1, \ldots, x_n]^T$, $w_0, \ldots, w_n$ – synaptic weights, $\mathbf{w} = [w_0, \ldots, w_n]^T$, y – neuron output value, w_0 – threshold value, f – activation function.

Formula describing neuron functioning is expressed by the dependency

$$y = f(s),\tag{6.1}$$

where

$$s = \sum_{i=0}^{n} x_i w_i.\tag{6.2}$$

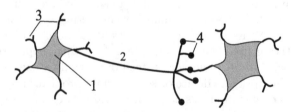

FIGURE 6.1. Simplified scheme of the neuron: 1- soma, 2 - axon, 3 - dendrites, 4 - synapses

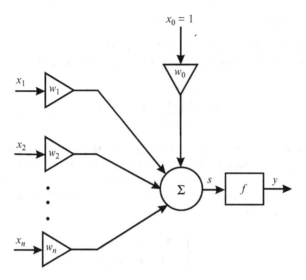

FIGURE 6.2. Model of a neuron

Formulas (6.1) and (6.2) describe a neuron presented in Fig. 6.2. Activation function f can take various forms depending on a specific model of the neuron.

As it can be inferred from the above formulas, neuron functioning is very simple. First, input signals $x_0, x_1, \ldots, x_n$ are multiplied by corresponding weights $w_0, w_1, \ldots, w_n$. The values obtained in this way, should be then summed up. The result is a s signal reflecting the functioning of linear part of the neuron. This signal is subject to the operation of activation function, which is most frequently not a linear one. We assume that the value of the signal x_0 equals 1, whereas the weight w_0 is called a *bias*. Where then is the knowledge hidden in the neuron described in this way? The knowledge is encrypted precisely in weights. And the biggest phenomenon is that neurons may be easily (using algorithms described in the following part of this chapter) trained, by changing weights appropriately. In Fig. 6.2, we presented a general scheme of neuron, however in networks various models are used. Some of them will be discussed in following points. It should be mentioned that similarly to brain where the nerve cells join one with another, also in case of mathematical models artificial neurons presented in Fig. 6.2 are connected to one another creating multilayer neural networks. Method of neuron connection as well as learning methods for structures created in this manner will be described further in this chapter.

6.2.2 Perceptron

Figure 6.3, illustrates the scheme of perceptron.

Operation of perceptron can be described by the formula

$$y = f\left(\sum_{i=1}^{n} w_i x_i + \theta\right). \tag{6.3}$$

Let us notice that formula (6.3) corresponds to the general notation (6.1) if $\theta = w_0$. Function f can be a discontinuous step function – bipolar function (takes the value -1 or 1) or unipolar function (takes the value 0 or 1). For the purpose of further discussion we will assume that the activation function is bipolar, i.e.

$$f(s) = \begin{cases} 1, & \text{if} \quad s > 0, \\ -1, & \text{if} \quad s \leq 0. \end{cases} \tag{6.4}$$

The perceptron, due to its activation function, takes only two different output values, so it may classify signals applied at its input in the form of vectors $\mathbf{x} = [x_1, ..., x_n]^T$ to one of two classes. For example, perceptron with one input can evaluate if the input signal is positive or negative. In case of two inputs x_1 and x_2 perceptron divides the plane into two parts. The partition is determined by a line of equation

$$w_1 x_1 + w_2 x_2 + \theta = 0. \tag{6.5}$$

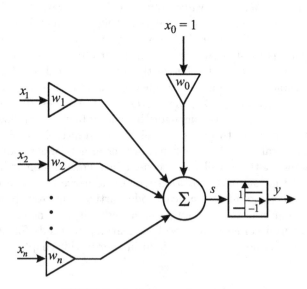

FIGURE 6.3. Scheme of perceptron

Therefore, equation (6.5) can be written as follows

$$x_2 = -\frac{w_1}{w_2} \cdot x_1 - \frac{\theta}{w_2}. \tag{6.6}$$

In general, when perceptron has n inputs, it divides n-dimensional space of input vectors $\mathbf{x}$ into two half spaces. They are separated by $n -$ 1-dimensional hyperplane, called *decision boundary* given by the formula

$$\sum_{i=1}^{n} w_i x_i + \theta = 0. \tag{6.7}$$

Figure 6.4, illustrates the decision boundary for $n = 2$. It should be noted that the line determining the partition of space is always perpendicular to the vector of weights $\mathbf{w} = [w_1, w_2]^T$.

According to our introduction, the perceptron can learn. During this process its weights are modified. The perceptron learning method belongs to the group of algorithms called *learning with teacher* or *supervised learning*. Learning of this type consists in applying signals $\mathbf{x}(t) = [x_0(t), x_1(t), \ldots, x_n(t)]^T, t = 1, 2, \ldots$, at perceptron input for which we know correct values of output signals $d(t), t = 1, 2, \ldots$, called output desired signals. A set of such input samples together with corresponding values of output desired signals is called a *learning sequence*. In these methods, after input values are applied, the output signal of neuron is computed. Then the weights are modified to minimize the error between output desired signal and perceptron output. As the teacher determines the desired value, this method is called "learning with teacher". Of course, we can presume that there are algorithms for learning networks without teacher, but they will be presented in the following points of this chapter. The algorithm for perceptron learning is presented below:

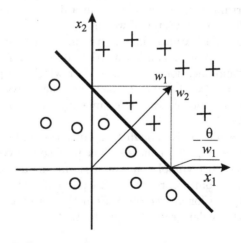

FIGURE 6.4. Decision boundary for $n = 2$

1. We select initial weights of perceptron at random.

2. At neuron inputs, we present a learning vector $\mathbf{x}$, while $\mathbf{x} = \mathbf{x}(t) = [x_0(t), x_1(t), \ldots, x_n(t)]^T, t = 1, 2, \ldots$.

3. We compute the output value of the perceptron y, according to formula (6.3).

4. We compare the output value $y(t)$ with the desired output value $d = d(\mathbf{x}(t))$ occurring in the learning sequence.

5. We modify weights according to dependencies:

   a) if $y(\mathbf{x}(t)) \neq d(\mathbf{x}(t))$, then $w_i(t+1) = w_i(t) + d(\mathbf{x}(t))x_i(t)$;

   b) if $y(\mathbf{x}(t)) = d(\mathbf{x}(t))$, then $w_i(t+1) = w_i(t)$, i.e. weights remain unchanged.

6. We go back to point 2.

The algorithm is repeated until for all input vectors included in the learning sequence the error at the output will be smaller than the assumed tolerance. Figure 6.5, illustrates the flowchart of perceptron learning. Operation of internal loop in this figure refers to so-called one *epoch*, which consists of data creating the learning sequence. The operation of the external loop reflects the possibility of multiple use of the same learning sequence until the algorithm stopping criterion is satisfied.

We shall demonstrate that the algorithm for perceptron learning is convergent. The theorem on the convergence of algorithm for perceptron learning is formulated as follows:

If a set of weights $\mathbf{w}^* = [w_1^*, \ldots, w_n^*]^T$ exists which correctly classifies learning signals $\mathbf{x} = [x_1, \ldots, x_n]^T$ i.e. determines mapping $y = d(\mathbf{x})$, then the learning algorithm will find a solution in a finite number of iterations for any initial values of the vector of weights $\mathbf{w}$.

We assume that the learning data represent linearly separable classes because only then perceptron can learn. We will demonstrate that a finite number of weight modification steps exist after which the perceptron will realize mapping $y = d(\mathbf{x})$. Due to the fact that the activation function is of sgn type in the perceptron, we can assume any length of vector $\mathbf{w}^*$, e.g. equal 1, i.e. $\|\mathbf{w}^*\| = 1$.

Thus, during learning it is enough to modify the vector $\mathbf{w}$ so that the angle α presented in Fig. 6.6 equals 0. Then, of course $\cos(\alpha) = 1$. The fact that $|\mathbf{w}^* \circ \mathbf{x}| > 0$ (symbol $\circ$ in this case means a scalar product of vectors) and $\mathbf{w}^*$ is a solution, results in existence of such a constant $\delta > 0$ for which $|\mathbf{w}^* \circ \mathbf{x}| > \delta$ for all vectors $\mathbf{x}$ from the learning sequence. From the definition of the scalar product it results that

$$\cos(\alpha) = \frac{\mathbf{w}^* \circ \mathbf{w}}{\sqrt{\|\mathbf{w}^*\|^2 \|\mathbf{w}\|^2}}. \tag{6.8}$$

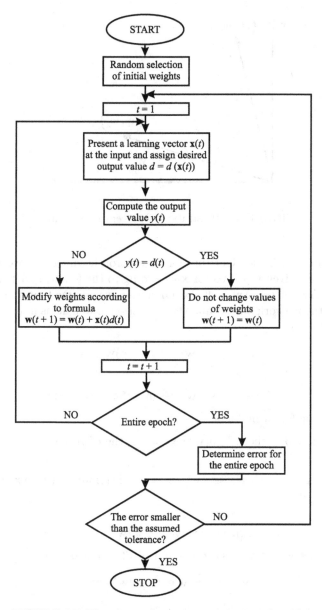

FIGURE 6.5. Flowchart of perceptron learning algorithm

Since

$$\sqrt{\|\mathbf{w}^*\|^2} = \|\mathbf{w}^*\| = 1, \tag{6.9}$$

therefore

$$\cos(\alpha) = \frac{\mathbf{w}^* \circ \mathbf{w}}{\|\mathbf{w}\|}. \tag{6.10}$$

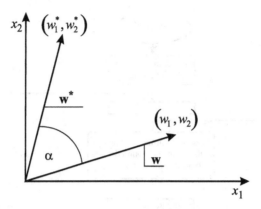

FIGURE 6.6. Illustration of perceptron learning

In accordance with the algorithm for perceptron learning, the weights are modified for a given input vector $\mathbf{x}$ according to the following dependency: $\mathbf{w}' = \mathbf{w} + \Delta\mathbf{w}$, where $\Delta\mathbf{w} = d(\mathbf{x})\mathbf{x}$. Of course, we assume that an error occurred at the network output and the correction of weights is indispensable. Let us notice that

$$\mathbf{w}' \circ \mathbf{w}^* = \mathbf{w} \circ \mathbf{w}^* + d(\mathbf{x})\mathbf{w}^* \circ \mathbf{x} \qquad (6.11)$$

therefore

$$\mathbf{w}' \circ \mathbf{w}^* = \mathbf{w} \circ \mathbf{w}^* + \mathrm{sgn}\,(\mathbf{w}^* \circ \mathbf{x})\,\mathbf{w}^* \circ \mathbf{x}. \qquad (6.12)$$

There are the following facts:

(i) If $\mathbf{w}^* \circ \mathbf{x} < 0$,   then $\mathrm{sgn}(\mathbf{w}^* \circ \mathbf{x}) = -1$, therefore $\mathrm{sgn}(\mathbf{w}^* \circ \mathbf{x})\mathbf{w}^* \circ \mathbf{x} = -1(\mathbf{w}^* \circ \mathbf{x}) > 0$,

(ii) If $\mathbf{w}^* \circ \mathbf{x} > 0$,   then $\mathrm{sgn}(\mathbf{w}^* \circ \mathbf{x}) = 1$, therefore $\mathrm{sgn}(\mathbf{w}^* \circ \mathbf{x})\mathbf{w}^* \circ \mathbf{x} = 1(\mathbf{w}^* \circ \mathbf{x}) > 0$.

Therefore

$$\mathrm{sgn}(\mathbf{w}^* \circ \mathbf{x})\mathbf{w}^* \circ \mathbf{x} = |\mathbf{w}^* \circ \mathbf{x}|. \qquad (6.13)$$

In accordance with formulas (6.12) and (6.13), we can write

$$\mathbf{w}' \circ \mathbf{w}^* = \mathbf{w} \circ \mathbf{w}^* + |\mathbf{w}^* \circ \mathbf{x}|. \qquad (6.14)$$

We also know that $|\mathbf{w}^* \circ \mathbf{x}| > \delta$, hence

$$\mathbf{w}' \circ \mathbf{w}^* > \mathbf{w} \circ \mathbf{w}^* + \delta. \qquad (6.15)$$

Let us now estimate the value $\|\mathbf{w}'\|^2$, bearing in mind that we are analyzing a case where, after applying a learning vector $\mathbf{x}$ at input, an error occurs at the network output, i.e.

$$d(\mathbf{x}) = -\mathrm{sgn}(\mathbf{w} \circ \mathbf{x}). \qquad (6.16)$$

Of course,

$$\|\mathbf{w}'\|^2 = \|\mathbf{w} + d(\mathbf{x})\mathbf{x}\|^2 = \|\mathbf{w}\|^2 + 2d(\mathbf{x})\mathbf{w} \circ \mathbf{x} + \|\mathbf{x}\|^2. \tag{6.17}$$

Using the dependencies (6.16) and (6.17) and assuming the input signals are bounded, we have

$$\|\mathbf{w}'\|^2 < \|\mathbf{w}\|^2 + \|\mathbf{x}\|^2 = \|\mathbf{w}\|^2 + C. \tag{6.18}$$

After t steps of network weights modification, dependencies (6.15) and (6.18) take the form

$$\mathbf{w}(t) \circ \mathbf{w}^* > \mathbf{w} \circ \mathbf{w}^* + t\delta \tag{6.19}$$

and

$$\|\mathbf{w}(t)\|^2 < \|\mathbf{w}\|^2 + tC. \tag{6.20}$$

Using formulas (6.10), (6.19) and (6.20), we get

$$\cos \alpha(t) = \frac{\mathbf{w}^* \circ \mathbf{w}(t)}{\|\mathbf{w}(t)\|} > \frac{\mathbf{w}^* \circ \mathbf{w} + t\delta}{\sqrt{\|\mathbf{w}\|^2 + tC}}. \tag{6.21}$$

Therefore a $t = t_{\max}$, must exist for which $\cos(\alpha) = 1$. Hence, there is a finite number of weights modification steps, after which the vector of initial weights will satisfy the mapping $y = d(\mathbf{x})$. If we assume that the initial values of weights equal 0, then

$$t_{\max} = \frac{C}{\delta^2}. \tag{6.22}$$

Example 6.1

Now, we will present an example of perceptron learning. When discussing its operation, we have stated that this two-input neuron model divides the plane into two parts (cf. 6.5). Therefore, if we place on the plane two classes of samples, which may be separated by means of a line, then the perceptron in the learning process will be able to find this division line. In our experience, we shall draw an output desired line, denoted by the letter L in Fig. 6.7. Let us assume that all the points of the plane located over this line represent the class 1 samples whereas the points located under line L represent class 2. There are infinitely many such points on both half-planes, and that is why we have to select a few samples of each class. We want that after training, the perceptron for samples from class one gives an output signal equal to 1, and for class two samples a signal equal to -1. So we have built a learning sequence presented in Table 6.1.

We shall assume the following initial values of perceptron weights: $w_1 = 2$, $w_2 = 2$, $\theta = -4$. Based on these parameters as well as earlier information, we shall draw a line K, which shows the division of space (decision boundary), defined by the perceptron before starting the learning process. After

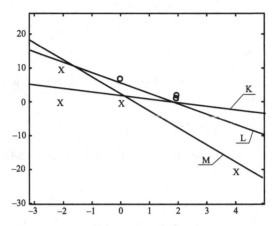

FIGURE 6.7. Decision boundaries in Example 6.1

ten epochs of the learning algorithm (we applied at neuron input 10 times all the elements of the learning sequence), the perceptron started to classify the learning sequence vectors correctly. Its weights took the following values: $w_1 = 4$, $w_2 = 1$, $\theta = -1$, which is reflected by line M being the decision boundary. In Fig. 6.7, we can see that after learning, the perceptron classifies the learning samples correctly, although line M is not identical to the desired output line L.

TABLE 6.1. A learning sequence from Example 6.1

x_1	x_2	$d(\mathbf{x})$
2	1	1
2	2	1
0	6	1
−2	8	−1
−2	0	−1
0	0	−1
4	−20	−1

6.2.3 Adaline model

Figure 6.8 illustrates the scheme of Adaline (*Adaptive Linear Neuron*) neuron. The construction of this neuron is very similar to the perceptron model, and the only difference relates to the learning algorithm. The determination method of output signal is identical to the one presented in previous point concerning the perceptron. However, in case of the Adaline neuron,

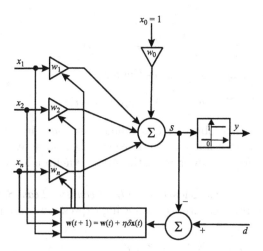

FIGURE 6.8. The scheme of Adaline neuron

the output desired signal d is compared to signal s at the output of the linear part of the neuron (adder). Thus, the name of this type of neurons. Therefore, we get an error given by the formula

$$\varepsilon = d - s. \tag{6.23}$$

Neuron learning, or the selection of weights, comes down to minimization of the function defined as follows:

$$Q\left(\mathbf{w}\right) = \frac{1}{2}\varepsilon^2 = \frac{1}{2}\left[d - \left(\sum_{i=0}^{n} w_i x_i\right)\right]^2. \tag{6.24}$$

The error measure (6.24) is called *the mean squared error*. By taking into account only the linear part of the neuron, we may use the gradient algorithms to modify the weights, as the objective function defined by dependency (6.24) is differentiable. We shall use the steepest descent method to minimize this function. This method will be discussed in more detail when we will be describing the backpropagation algorithm. The weights in an Adaline neuron are modified according to the formula

$$w_i(t+1) = w_i(t) - \eta\frac{\partial Q(w_i)}{\partial w_i}, \tag{6.25}$$

where η is the learning coefficient. Let us notice that

$$\frac{\partial Q(w_i)}{\partial w_i} = \frac{\partial Q(w_i)}{\partial s} \cdot \frac{\partial s}{\partial w_i}. \tag{6.26}$$

As s is a linear function with relation to the weights vector, we have

$$\frac{\partial s}{\partial w_i} = x_i. \tag{6.27}$$

Moreover,

$$\frac{\partial Q(w_i)}{\partial s} = -(d - s). \tag{6.28}$$

Therefore, dependency (6.25) takes the form

$$w_i(t + 1) = w_i(t) + \eta \delta x_i, \tag{6.29}$$

where $\delta = d - s$. The above rule is called *delta rule* (it is a special form of this rule, as it does not take into account neuron activation function). Figure 6.9 presents the flowchart of the Adaline neuron learning algorithm using this rule.

Adaline neurons may also learn using *the Recursive Least Squares* method (RLS). As the error measure, the following expression is adopted:

$$Q(t) = \sum_{k=1}^{t} \lambda^{t-k} \varepsilon^2 (k) \tag{6.30}$$

$$= \sum_{k=1}^{t} \lambda^{t-k} \left[d(k) - \mathbf{x}^T(k) \, \mathbf{w}(t) \right]^2,$$

in which λ is *the forgetting factor* selected from the interval $[0, 1]$. Let us notice that the previous errors have a lesser influence on the value of expression (6.30). When computing the error measure gradient, we get the following dependency:

$$\frac{\partial Q(t)}{\partial \mathbf{w}(t)} = \frac{\partial \sum_{k=1}^{t} \lambda^{t-k} \varepsilon^2}{\partial \mathbf{w}(t)}$$

$$= \frac{\partial \sum_{k=1}^{t} \lambda^{t-k} \left[d(k) - \mathbf{x}^T(k) \, \mathbf{w}(t) \right]^2}{\partial \mathbf{w}(t)} \tag{6.31}$$

$$= -2 \sum_{k=1}^{t} \lambda^{t-k} \left[d(k) - \mathbf{x}^T(k) \, \mathbf{w}(t) \right] \mathbf{x}(k).$$

The optimum values of weights should satisfy the so-called *normal equation*

$$\sum_{k=1}^{t} \lambda^{t-k} \left[d(k) - \mathbf{x}^T(k) \, \mathbf{w}(t) \right] \mathbf{x}(k) = \mathbf{0}. \tag{6.32}$$

Equation (6.32) may be presented in the form

$$\mathbf{r}(t) = \mathbf{R}(t) \, \mathbf{w}(t), \tag{6.33}$$

where

$$\mathbf{R}(t) = \sum_{k=1}^{t} \lambda^{t-k} \mathbf{x}(k) \, \mathbf{x}^T(k) \tag{6.34}$$

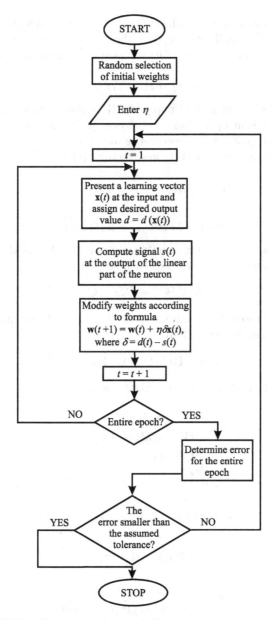

FIGURE 6.9. Flowchart of the Adaline neuron learning algorithm

is $n \times n$-dimensional autocorrelation matrix, and

$$\mathbf{r}(t) = \sum_{k=1}^{t} \lambda^{t-k} d(k) \mathbf{x}(k) \tag{6.35}$$

is $n \times 1$-dimensional cross-correlation vector of the input signal and output desired signal. We shall assume that these signals are the realization of stationary stochastic processes. The solution of the normal equation (6.33) takes the form

$$\mathbf{w}(t) = \mathbf{R}^{-1}(t)\mathbf{r}(t), \tag{6.36}$$

if det $\mathbf{R}(t) \neq 0$. We now shall apply the RLS algorithm in order to avoid the matrix inversion operation in equation (6.36) and we will solve the normal equation (6.33) using recurrence method.

Let us notice that matrix $\mathbf{R}(t)$ and vector $\mathbf{r}(t)$ may be presented in the form

$$\mathbf{R}(t) = \lambda\mathbf{R}(t-1) + \mathbf{x}(t)\mathbf{x}^T(t) \tag{6.37}$$

and

$$\mathbf{r}(t) = \lambda\mathbf{r}(t-1) + \mathbf{x}(t)d(t). \tag{6.38}$$

Now, we shall apply the inverse matrix lemma. Let $\mathbf{A}$ and $\mathbf{B}$ be positively defined $n \times n$-dimensional matrices such as

$$\mathbf{A} = \mathbf{B}^{-1} + \mathbf{C}\mathbf{D}^{-1}\mathbf{C}^T \tag{6.39}$$

where $\mathbf{D}$ is a positively definite $m \times m$-dimensional matrix, while $\mathbf{C}$ is $n \times m$-dimensional matrix. Then

$$\mathbf{A}^{-1} = \mathbf{B} - \mathbf{B}\mathbf{C}\left(\mathbf{D} + \mathbf{C}^T\mathbf{B}\mathbf{C}\right)^{-1}\mathbf{C}^T\mathbf{B}. \tag{6.40}$$

By comparing formulas (6.40) and (6.37), we receive

$$\mathbf{A} = \mathbf{R}(t), \tag{6.41}$$
$$\mathbf{B}^{-1} = \lambda\mathbf{R}(t-1),$$
$$\mathbf{C} = \mathbf{x}(t),$$
$$\mathbf{D} = 1.$$

Therefore,

$$\mathbf{P}(t) = \lambda^{-1}\left[\mathbf{I} - \mathbf{g}(t)\mathbf{x}^T(t)\right]\mathbf{P}(t-1), \tag{6.42}$$

where

$$\mathbf{P}(t) = \mathbf{R}^{-1}(t) \tag{6.43}$$

and

$$\mathbf{g}(t) = \frac{\mathbf{P}(t-1)\mathbf{x}(t)}{\lambda + \mathbf{x}^T(t)\mathbf{P}(t-1)\mathbf{x}(t)}. \tag{6.44}$$

We shall demonstrate the truth of the following equation:

$$\mathbf{g}(t) = \mathbf{P}(t)\mathbf{x}(t). \tag{6.45}$$

As a result of some simple algebraic operations, we get

$$\mathbf{g}(t) = \frac{\mathbf{P}(t-1)\mathbf{x}(t)}{\lambda + \mathbf{x}^T(t)\mathbf{P}(t-1)\mathbf{x}(t)}$$

$$= \frac{\lambda^{-1} \left[\lambda \mathbf{P}\left(t-1\right) \mathbf{x}\left(t\right) + \mathbf{P}\left(t-1\right) \mathbf{x}\left(t\right) \mathbf{x}^T\left(t\right) \mathbf{P}\left(t-1\right) \mathbf{x}\left(t\right) \right]}{\lambda + \mathbf{x}^T\left(t\right) \mathbf{P}\left(t-1\right) \mathbf{x}\left(t\right)}$$

$$- \frac{\lambda^{-1} \left[\mathbf{P}\left(t-1\right) \mathbf{x}\left(t\right) \mathbf{x}^T\left(t\right) + \mathbf{P}\left(t-1\right) \mathbf{x}\left(t\right) \right]}{\lambda + \mathbf{x}^T\left(t\right) \mathbf{P}\left(t-1\right) \mathbf{x}\left(t\right)}$$

$$= \frac{\lambda^{-1} \left[\left(\lambda + \mathbf{x}^T\left(t\right) \mathbf{P}\left(t-1\right) \mathbf{x}\left(t\right) \right) \mathbf{I} \right] \mathbf{P}\left(t-1\right) \mathbf{x}\left(t\right)}{\lambda + \mathbf{x}^T\left(t\right) \mathbf{P}\left(t-1\right) \mathbf{x}\left(t\right)} \tag{6.46}$$

$$- \frac{\lambda^{-1} \left[\mathbf{P}\left(t-1\right) \mathbf{x}\left(t\right) \mathbf{x}^T\left(t\right) \right] \mathbf{P}\left(t-1\right) \mathbf{x}\left(t\right)}{\lambda + \mathbf{x}^T\left(t\right) \mathbf{P}\left(t-1\right) \mathbf{x}\left(t\right)}$$

$$= \lambda^{-1} \left[\mathbf{I} - \frac{\mathbf{P}\left(t-1\right) \mathbf{x}\left(t\right) \mathbf{x}^T\left(t\right)}{\lambda + \mathbf{x}^T\left(t\right) \mathbf{P}\left(t-1\right) \mathbf{x}\left(t\right)} \right] \mathbf{P}\left(t-1\right) \mathbf{x}\left(t\right)$$

$$= \lambda^{-1} \left[\mathbf{I} - \mathbf{g}\left(t\right) \mathbf{x}^T\left(t\right) \right] \mathbf{P}\left(t-1\right) \mathbf{x}\left(t\right) = \mathbf{P}\left(t\right) \mathbf{x}\left(t\right).$$

It results from the dependencies (6.38) and (6.36) that

$$\mathbf{w}\left(t\right) = \mathbf{R}^{-1}\left(t\right) \mathbf{r}\left(t\right) = \lambda \mathbf{P}\left(t\right) \mathbf{r}\left(t-1\right) + \mathbf{P}\left(t\right) \mathbf{x}\left(t\right) d\left(t\right). \tag{6.47}$$

From equation (6.42) and (6.47) we get

$$\mathbf{w}\left(t\right) = \left[\mathbf{I} - \mathbf{g}\left(t\right) \mathbf{x}^T\left(t\right) \right] \mathbf{P}\left(t-1\right) \mathbf{r}\left(t-1\right) + \mathbf{P}\left(t\right) \mathbf{x}\left(t\right) d\left(t\right). \tag{6.48}$$

The consequence of the dependency (6.38) and (6.36) is the following relation:

$$\mathbf{w}\left(t\right) = \mathbf{w}(t-1) - \mathbf{g}\left(t\right) \mathbf{x}^T\left(t\right) \mathbf{w}\left(t-1\right) + \mathbf{P}\left(t\right) \mathbf{x}\left(t\right) d\left(t\right). \tag{6.49}$$

Taking into consideration relation (6.45) in dependency (6.49), we get the following recursion:

$$\mathbf{w}\left(t\right) = \mathbf{w}\left(t-1\right) + \mathbf{g}\left(t\right) \left[d\left(t\right) - \mathbf{x}^T\left(t\right) \mathbf{w}\left(t-1\right) \right]. \tag{6.50}$$

In consequence, the RLS algorithm used for learning of Adaline neuron takes the following form:

$$\varepsilon\left(t\right) = d\left(t\right) - \mathbf{x}^T\left(t\right) \mathbf{w}\left(t-1\right) = d\left(t\right) - y\left(t\right), \tag{6.51}$$

$$\mathbf{g}\left(t\right) = \frac{\mathbf{P}\left(t-1\right) \mathbf{x}\left(t\right)}{\lambda + \mathbf{x}^T\left(t\right) \mathbf{P}\left(t-1\right) \mathbf{x}\left(t\right)}, \tag{6.52}$$

$$\mathbf{P}\left(t\right) = \lambda^{-1} \left[\mathbf{I} - \mathbf{g}\left(t\right) \mathbf{x}^T\left(t\right) \right] \mathbf{P}\left(t-1\right), \tag{6.53}$$

$$\mathbf{w}\left(t\right) = \mathbf{w}\left(t-1\right) + \mathbf{g}\left(t\right) \varepsilon\left(t\right). \tag{6.54}$$

As initial values, it is usually assumed

$$\mathbf{P}\left(0\right) = \gamma \mathbf{I}, \quad \gamma > 0, \tag{6.55}$$

where γ is a constant, while $\mathbf{I}$ is an identity matrix.

6.2.4 Sigmoidal neuron model

Construction of a sigmoidal neuron is analogical to the two models previously discussed, i.e. the perceptron and Adaline neuron. The name derives from an activation function which takes the form of a unipolar or bipolar sigmoidal function. These are continuous functions and are expressed by the following dependencies:

$$f(x) = \frac{1}{1 + e^{-\beta x}} \quad - \text{ unipolar function}$$

and

$$f(x) = \tanh(\beta x) = \frac{1 - e^{\beta x}}{1 + e^{-\beta x}} \quad - \text{ bipolar function}$$

Figure 6.10 presents characteristics of a unipolar function for different values of parameter β. The Reader may notice that with a low value of the coefficient β, the function has a gentle shape, when the coefficient value increases, the graph becomes increasingly steeper, and finally the function presents threshold characteristics. The feature which is undeniably a great advantage of sigmoidal neurons, is the differentiability of the activation function.

Moreover, the derivatives of these functions may be easily calculated, as they take the following form:

$$f'(x) = \beta f(x)(1 - f(x)) \quad \text{for unipolar function,} \tag{6.56}$$

$$f'(x) = \beta \left(1 - f^2(x)\right) \quad \text{for bipolar function.} \tag{6.57}$$

Figure 6.11 illustrates the scheme of a sigmoidal neuron.

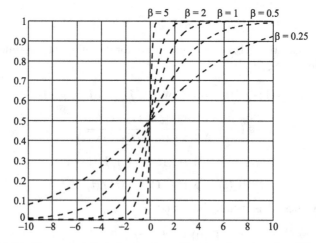

FIGURE 6.10. Characteristics of unipolar activation functions for different values of parameter β

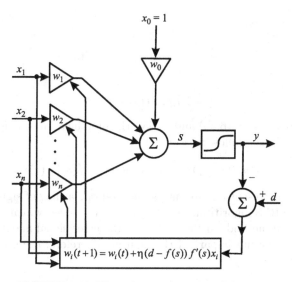

FIGURE 6.11. The scheme of a sigmoidal neuron

The output signal is given by the following formula

$$y(t) = f\left(\sum_{i=0}^{n} w_i(t) x_i(t)\right). \tag{6.58}$$

The error measure Q is defined as a square of difference of the desired output value and value obtained at the neuron output, i.e.

$$Q(\mathbf{w}) = \frac{1}{2}\left[d - f\left(\sum_{i=0}^{n} w_i x_i\right)\right]^2. \tag{6.59}$$

Like in the case of Adaline type neuron, the steepest descent rule is used for learning purposes; however, now in derivations we will take into account the activation function as well. The neuron weights are updated according to the formula

$$w_i(t+1) = w_i(t) - \eta\frac{\partial Q(w_i)}{\partial w_i}. \tag{6.60}$$

We will determine the derivative of the error measure with respect to the weights. Of course,

$$\frac{\partial Q(w_i)}{\partial w_i} = \frac{\partial Q(w_i)}{\partial s} \cdot \frac{\partial s}{\partial w_i} \tag{6.61}$$

and

$$\frac{\partial s}{\partial w_i} = x_i. \tag{6.62}$$

Hence,

$$\frac{\partial Q(w_i)}{\partial w_i} = \frac{\partial Q(w_i)}{\partial s} \cdot x_i. \tag{6.63}$$

It is easy to see that

$$\frac{\partial Q(w_i)}{\partial s} = -(d - f(s)) \cdot f'(s).$$
(6.64)

Let us denote

$$\delta = -(d - f(s)) \cdot f'(s).$$
(6.65)

According to formulas (6.60) and (6.65), the modification of weights in step $t + 1$ is made in the following way:

$$w_i(t + 1) = w_i(t) - \eta \delta x_i = w_i(t) + \eta(d - f(s))f'(s)x_i.$$
(6.66)

Now, we will present an alternative method of learning of sigmoidal neuron using the RLS algorithm. Let us consider two cases which differ in the error definition method. In the first case, the error signal is determined at the output of the linear part of the neuron. Therefore, the error measure has the form

$$Q(t) = \sum_{k=1}^{t} \lambda^{t-k} e^2(k)$$
(6.67)

$$= \sum_{k=1}^{t} \lambda^{t-k} \left[b(k) - \mathbf{x}^T(k)\mathbf{w}(t) \right]^2,$$

where

$$b(k) = f^{-1}(d(k)) = \begin{cases} \ln \dfrac{d(k)}{1 - d(k)} & \text{in case of a unipolar function} \\ \\ \dfrac{1}{2} \ln \dfrac{1 + d(k)}{1 - d(k)} & \text{in case of a bipolar function} \end{cases}$$
(6.68)

has the interpretation of a desired signal at the output of the linear part of neuron. At present, the normal equation takes the form

$$\frac{Q(t)}{\partial \mathbf{w}(t)} = -2 \sum_{k=1}^{t} \lambda^{t-k} \left[b(k) - \mathbf{x}^T(k)\mathbf{w}(t) \right] \mathbf{x}^T(k)$$
(6.69)

or in vector form

$$\mathbf{r}(t) = \mathbf{R}(t)\mathbf{w}(t),$$
(6.70)

where

$$\mathbf{R}(t) = \sum_{k=1}^{t} \lambda^{t-k} \mathbf{x}(k)\mathbf{x}^T(k)$$
(6.71)

and

$$\mathbf{r}(t) = \sum_{k=1}^{t} \lambda^{t-k} b(k)\mathbf{x}(k).$$
(6.72)

Let us notice that equation (6.71) and (6.72) are similar to equation (6.34) and (6.35). Thus, the RLS algorithm takes the following form:

$$e(t) = b(t) - \mathbf{x}^T(t)\mathbf{w}(t-1) = b(t) - s(t), \tag{6.73}$$

$$\mathbf{g}(t) = \frac{\mathbf{P}(t-1)\mathbf{x}(t)}{\lambda + \mathbf{x}^T(t)\mathbf{P}(t-1)\mathbf{x}(t)}, \tag{6.74}$$

$$\mathbf{P}(t) = \lambda^{-1}\left[\mathbf{I} - \mathbf{g}(t)\mathbf{x}^T(t)\right]\mathbf{P}(t-1), \tag{6.75}$$

$$\mathbf{w}(t) = \mathbf{w}(t-1) + \mathbf{g}(t)e(t), \tag{6.76}$$

while the initial conditions are defined by formula (6.55).

In the second case, the error is determined at the output of the nonlinear part of neuron. The error measure takes the form

$$Q(t) = \sum_{k=1}^{t} \lambda^{t-k}\varepsilon^2(k) \tag{6.77}$$

$$= \sum_{k=1}^{t} \lambda^{t-k}\left[d(k) - f\left(\mathbf{x}^T(k)\mathbf{w}(t)\right)\right]^2.$$

By determining a partial derivative of measure (6.77) with respect to vector $\mathbf{w}(t)$ and equating the result to $\mathbf{0}$, we get

$$\frac{\partial Q(t)}{\partial \mathbf{w}(t)} = 2\sum_{k=1}^{t} \lambda^{t-k}\frac{\partial \varepsilon(k)}{\partial \mathbf{w}(t)}\varepsilon(k) \tag{6.78}$$

$$= -2\sum_{k=1}^{t} \lambda^{t-k}\frac{\partial y(k)}{\partial s(k)}\frac{\partial s(k)}{\partial \mathbf{w}(t)}\varepsilon(k) = \mathbf{0}.$$

As a result of some further computations, we get

$$\sum_{k=1}^{t} \lambda^{t-k}\frac{\partial y(k)}{\partial s(k)}\frac{\partial s(k)}{\partial \mathbf{w}(t)}[d(k) - y(k)]$$

$$= \sum_{k=1}^{t} \lambda^{t-k}\frac{\partial y(k)}{\partial s(k)}\mathbf{x}^T(k)[d(k) - y(k)] \tag{6.79}$$

$$= \sum_{k=1}^{t} \lambda^{t-k}\frac{\partial y(k)}{\partial s(k)}\mathbf{x}^T(k)[f(b(k)) - f(s(k)))] = \mathbf{0}.$$

As a result of applying the Taylor expansion to the expression in square bracket of formula (6.79), we get

$$f(b(k)) \approx f(s(k)) + f'(s(k))(b(k) - s(k)), \tag{6.80}$$

where
$$b(t) = f^{-1}(d(t)).\qquad(6.81)$$

As a consequence of formulas (6.79) and (6.80), we get the equation

$$\sum_{k=1}^{t}\lambda^{t-k}f'^2\left(s\left(k\right)\right)\left[b\left(k\right)-\mathbf{x}^T\left(k\right)\mathbf{w}\left(t\right)\right]\mathbf{x}\left(k\right)=\mathbf{0}.\qquad(6.82)$$

Equation (6.82) in the vector form takes the form

$$\mathbf{r}(t) = \mathbf{R}(t)\mathbf{w}(t),\qquad(6.83)$$

where

$$\mathbf{R}(t) = \sum_{k=1}^{t}\lambda^{t-k}f'^2\left(s\left(k\right)\right)\mathbf{x}\left(k\right)\mathbf{x}^T\left(k\right)\qquad(6.84)$$

and

$$\mathbf{r}(t) = \sum_{k=1}^{t}\lambda^{t-k}f'^2\left(s\left(k\right)\right)b\left(k\right)\mathbf{x}\left(k\right).\qquad(6.85)$$

By applying the following substitutions in formulas (6.73) – (6.76):

$$\mathbf{x}(k) \rightarrow f'\left(s\left(k\right)\right)\mathbf{x}\left(k\right),\qquad(6.86)$$

$$b(k) \rightarrow f'\left(s\left(k\right)\right)b\left(k\right),\qquad(6.87)$$

we get the following form of the RLS algorithm applied to learning sigmoidal neuron:

$$\varepsilon(t) = f'\left(s\left(t\right)\right)\left[b\left(t\right)-\mathbf{x}^T\left(t\right)\mathbf{w}\left(t-1\right)\right] \approx d(t)-y(t),\qquad(6.88)$$

$$\mathbf{g}(t) = \frac{f'\left(s\left(t\right)\right)\mathbf{P}\left(t-1\right)\mathbf{x}\left(t\right)}{\lambda+f'^2\left(s\left(t\right)\right)\mathbf{x}^T\left(t\right)\mathbf{P}\left(t-1\right)\mathbf{x}\left(t\right)},\qquad(6.89)$$

$$\mathbf{P}(t) = \lambda^{-1}\left[\mathbf{I}-f'\left(s\left(t\right)\right)\mathbf{g}\left(t\right)\mathbf{x}^T\left(t\right)\right]\mathbf{P}\left(t-1\right),\qquad(6.90)$$

$$\mathbf{w}(t) = \mathbf{w}(t-1)+\mathbf{g}(t)\varepsilon(t).\qquad(6.91)$$

The initial conditions are defined by the dependency (6.55).

6.2.5 Hebb neuron model

Figure 6.12 illustrates the Hebb neuron model. It is a structure identical to Adaline neuron model and sigmoidal neuron model, but is characterized by a specific learning method, known as *Hebb rule*. This rule occurs in the version without the teacher and with the teacher. Hebb [74] studied the functioning of neural cells. During his research, he noticed that the connection between two cells was strengthened if the two cells became active at the same time.

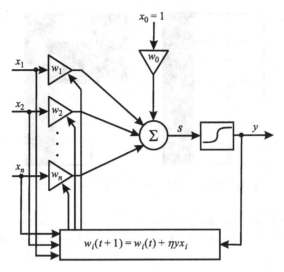

FIGURE 6.12. Scheme of the Hebb neuron model

In an analogical way, he proposed an algorithm pursuant to which the modification of weights is made as follows:

$$w_i(t+1) = w_i(t) + \Delta w_i, \tag{6.92}$$

whereas

$$\Delta w_i = \eta x_i y. \tag{6.93}$$

In case of a single neuron, during learning we will modify the value of weight w_i proportionally both to the value of the signal input to the i-th input, and the output signal y, taking into account the learning coefficient η. Let us notice that in this case we do not present a desired output value, therefore, we apply here the learning without teacher. A slight modification of dependency (6.93) leads to the second learning method of Hebb neuron – learning with teacher

$$\Delta w_i = \eta x_i d, \tag{6.94}$$

where d is the output desired signal. A certain disadvantage of the algorithm discussed is the fact that the values of weights may increase to any high values. That is why different modifications of the Hebb rule are introduced in the literature.

Example 6.2

Now, we will present an example of neuron learning using Hebb rule in the version with teacher. Our task will be to modify the neuron weights in such a way as to recognize digits 1 and 4, schematically presented in Fig. 6.13.

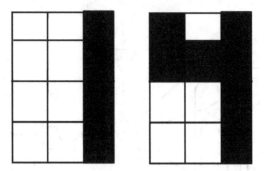

FIGURE 6.13. Illustration to Example 6.2

By assigning digit 1 to the white fields and digit -1 to the black fields in the figure, we will get two vectors included in a learning sequence:

$$[-1 \quad -1 \quad 1 \quad -1 \quad -1 \quad 1 \quad -1 \quad -1 \quad 1 \quad -1 \quad -1 \quad 1] \; - \; \text{for digit 1,}$$
$$[1 \quad -1 \quad 1 \quad 1 \quad 1 \quad 1 \quad -1 \quad -1 \quad 1 \quad -1 \quad -1 \quad 1] \; - \text{for digit 4.}$$

For the first pattern (digit 1), we will demand that at the neuron output the signal $d = -1$ appeared, while for the second (digit 4) we have determined the desired output value as $d = 1$. As we know the input and output patterns, the neuron weights in subsequent iterations of the algorithm will be subject to modification according to dependency (6.94). Their initial values are equal to 0. The neuron with the signum type activation function learnt during 100 epochs, we have assumed the learning coefficient equal to 0.2. After having the neuron learnt and giving the first learning vector at its input, the signal $s = -120$ appeared at the output, while in case of the second learning vector the signal $s = 120$ appeared at the output. We may justly assume that if the number of epochs increases, these values will also increase. The vector of weights after being learnt took the following form: $\mathbf{w} = [40 \quad 0 \quad 0 \quad 40 \quad 40 \quad 0 \quad 0 \quad 0 \quad 0 \quad 0 \quad 0]$. We can see that only the components of the vector of weights that corresponded to the differences between individual components of the learning vectors have changed.

6.3 Multilayer feed-forward networks

6.3.1 Structure and functioning of the network

In the previous chapter, we have discussed different neuron models. We have shown that these neurons may learn, i.e. adjust the values of their weights to the values of a learning sequence. Moreover, when describing the perceptron, we have demonstrated that if it had n inputs, it divided n-dimensional space into two half spaces. They are divided by $n - 1$-dimensional hyperplane. The scope of possible problems which may be solved using a single

perceptron is rather limited. As an example, we present the problem of
the logical XOR function. The learning sequence for this problem has been
presented in Table 6.2.

Values of the learning sequence from Table 6.2 are marked in Fig. 6.14.
As we may see in this figure there is no line which divides the points with
XOR function values equal to -1 from points with values equal to 1. In this
case, a decision boundary is described by an ellipse, therefore the algorithm
presented in Fig. 6.5 would not be convergent. We are unable to find the
weights of a single perceptron as to solve the XOR problem. Fortunately,
we are supported by multilayer networks. We will return to XOR problem
when we present the learning methods of these networks. What actually are
multilayer networks? Nothing more than appropriately connected neurons
arranged in two or more layers. Generally, these are sigmoidal neurons,
but also neurons with other activation functions are used, e.g. linear –
most commonly used in the last layers of the neural network structure.
In multilayer neural networks, at least two layers must exist: input and
output one. Between them, however, there may be some hidden layers.
If the network contains only two layers, then the input layer is identified

TABLE 6.2. Learning sequence for the XOR problem

x_1	x_2	$d = \mathrm{XOR}(x_1, x_2)$
$+1$	$+1$	-1
$+1$	-1	1
-1	$+1$	1
-1	-1	-1

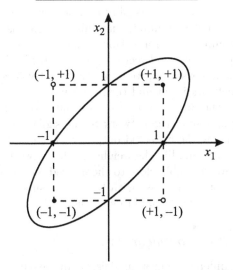

FIGURE 6.14. Illustration of the XOR problem

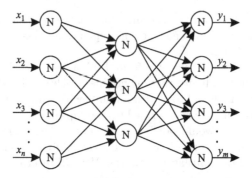

FIGURE 6.15. A typical scheme of a three-layer feed-forward neural network

with a hidden layer. In some studies, the input layer is understood as the vector of input signals applied to the neural network. In the discussed structures the neurons transmit the signals only between different layers. Within the same layer, the neurons can not connect with each other. The signals are transmitted from the input to the output layer (hence the name: feed-forward), and there is no feedback to the previous layers. Figure 6.15 illustrates a typical scheme of a three-layer feed-forward neural network.

In the following points, we will present different algorithms for learning of multilayer neural networks with teacher. At the beginning, we will present the operation of *the backpropagation* algorithm and its several modifications. Another algorithm to be discussed in this chapter will be the RLS algorithm. Then we will demonstrate, what is the influence of selecting an appropriate network structure on network learning and operation. As we have mentioned before, the algorithms presented in this chapter belong to the group of algorithms for learning networks with teacher, also called *supervised learning*. Let us remind what this term means. In case of these algorithms, we assume that in a learning sequence there are the following pairs: vector of input values and vector of desired output signals. Hence, the learning will go as follows: first, we apply the input values from a learning sequence at the network input and then we compute the output values for each neuron from the input layer to the output layer, one by one. Thus, we will get the response of the network to the signal (pattern) entered at its input. As we know the expected output value (it is in the learning sequence), we will try to modify the weights in the network in such a way so the output value was possibly close to the desired output value. Hence the name of this type of algorithms (with teacher or supervised learning), as the "teacher" indicates what should be the network's response.

6.3.2 Backpropagation algorithm

When presenting different neuron models, we discussed some of their basic learning techniques. Most often, it was as follows: we computed the sum of values of input signal products and their corresponding weights. Then we subjected the value thus obtained to the operation of appropriately defined activation function and we received the neuron output value. As we knew the expected output value (desired output value obtained from the learning sequence), we could easily define the error at the neuron output. This error was the difference between the value obtained at its output and the desired output value. In the same way we may define the error for the last layer in case of multilayer networks. However, the problem of defining error for the hidden layers appears, as the teacher does not know what should be the output value of particular hidden layer neurons. The most common technique of learning of multilayer neural networks called *error backpropagation algorithm* or *backpropagation algorithm* is very helpful in solving the problem.

In order to derive this algorithm, we must appropriately define the error measure. It will be a function in which all weights of a multilayer neural network play the role of variables. Let us denote this function as $Q(\mathbf{w})$, where $\mathbf{w}$ is the vector of all network weights. During the network learning, we will aim to find a minimum of function Q with respect to vector $\mathbf{w}$. Let us therefore expand the function considered into a Taylor series in the closest neighborhood of the known current solution $\mathbf{w}$. This expansion will be presented along the direction $\mathbf{p}$ as follows:

$$Q(\mathbf{w} + \mathbf{p}) = Q(\mathbf{w}) + [\mathbf{g}(\mathbf{w})]^T \mathbf{p} + 0.5\mathbf{p}^T \mathbf{H}(\mathbf{w})\mathbf{p} + \ldots, \tag{6.95}$$

where $\mathbf{g}(\mathbf{w})$ means the gradient vector, i.e.

$$\mathbf{g}(\mathbf{w}) = \left[\frac{\partial Q}{\partial w_1}, \frac{\partial Q}{\partial w_2}, \frac{\partial Q}{\partial w_3}, \ldots, \frac{\partial Q}{\partial w_n} \right]^T, \tag{6.96}$$

while $\mathbf{H}(\mathbf{w})$ is the Hessian, i.e. the matrix of the second-order derivatives

$$\mathbf{H}(\mathbf{w}) = \begin{bmatrix} \dfrac{\partial^2 Q}{\partial w_1 \partial w_1} & \cdots & \dfrac{\partial^2 Q}{\partial w_1 \partial w_n} \\ \vdots & & \vdots \\ \dfrac{\partial^2 Q}{\partial w_n \partial w_1} & \cdots & \dfrac{\partial^2 Q}{\partial w_n \partial w_n} \end{bmatrix}. \tag{6.97}$$

The modification of these weights is made as follows:

$$\mathbf{w}(t + 1) = \mathbf{w}(t) + \eta(t)\mathbf{p}(t), \tag{6.98}$$

where η is the learning coefficient (the method of selecting this parameter will be described further in this chapter). The modification of weights may be repeated as many times as function Q reaches the minimum or its value drops below the assumed threshold. So, the task comes down to determining such a direction vector $\mathbf{p}$ that in further steps of the algorithm the error at network output would decrease. This means that we require the satisfaction of the inequality $Q(\mathbf{w}(t+1)) < Q(\mathbf{w}(t))$ in subsequent steps of iterations. Let us limit the Taylor series approximating the Q error function to a linear expansion, i.e.

$$Q(\mathbf{w} + \mathbf{p}) = Q(\mathbf{w}) + [g(\mathbf{w})]^T \mathbf{p}. \tag{6.99}$$

As function $Q(\mathbf{w})$ depends on the weights determined in step t, and $Q(\mathbf{w} + \mathbf{p})$ depends on the weights determined in step $t + 1$, then in order to obtain the dependency $Q(\mathbf{w}(t+1)) < Q(\mathbf{w}(t))$, we should select the vector $\mathbf{p}(t)$, so that $\mathbf{g}(\mathbf{w}(t))^T \mathbf{p}(t) < 0$. It is easy to notice that this condition is satisfied if we assume

$$\mathbf{p}(t) = -\mathbf{g}(\mathbf{w}(t)). \tag{6.100}$$

By substituting dependency (6.100) to formula (6.98), we get the following formula, which defines the method of changing the weights of a multilayer neural network

$$\mathbf{w}(t+1) = \mathbf{w}(t) - \eta \mathbf{g}(\mathbf{w}(t)). \tag{6.101}$$

Dependency (6.101) is known in the literature as *the steepest descent rule*. In order to use effectively dependency (6.101) to derive the backpropagation algorithm, we must formally describe the scheme of a multilayer neural network and introduce appropriate notations.

Such scheme is presented in Fig. 6.16. In each layer, there are N_k elements, $k = 1, ..., L$, denoted as N_i^k, $i = 1, ..., N_k$. Elements N_i^k will be called neurons, and each of them may be a sigmoidal neuron. The discussed neural network has N_0 inputs, to which signals $x_1(t), ..., x_{N_0}(t)$ are applied, notated in the form of a vector

$$\mathbf{x} = [x_1(t), ..., x_{N_0}(t)]^T \quad t = 1, 2, ... \tag{6.102}$$

The output signal of i-th neuron in k-th layer is denoted as $y_i^{(k)}(t)$, $i = 1, ..., N_k$, $k = 1, ..., L$.

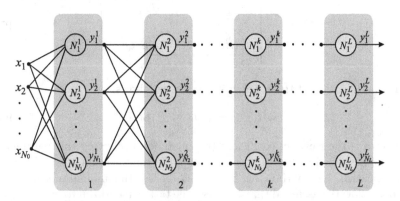

FIGURE 6.16. Scheme of a multilayer neural network

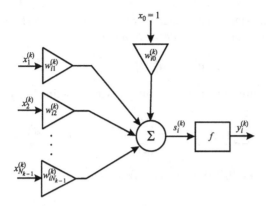

FIGURE 6.17. A detailed scheme of i-th neuron in k-th layer

Figure 6.17 presents a detailed scheme of i-th neuron in k-th layer. Neuron N_i^k has N_k inputs making up the vector

$$\mathbf{x}^{(k)}(t) = \left[x_0^{(k)}(t), ..., x_{N_{k-1}}^{(k)}(t) \right]^T , \qquad (6.103)$$

while $x_i^{(k)}(t) = +1$ for $i = 0$ and $k = 1, ..., L$. Let us notice that the input signal of neuron N_i^k is related to the output signal of layer $k-1$ in the following way

$$x_i^{(k)}(t) = \begin{cases} x_i(t) & \text{for} \quad k = 1, \\ y_i^{(k-1)}(t) & \text{for} \quad k = 2, ..., L, \\ +1 & \text{for} \quad i = 0, \ k = 1, ..., L. \end{cases} \qquad (6.104)$$

In Fig. 6.17 $w_{ij}^{(k)}(t)$ is the weight of i-th neuron, $i = 1, ..., N_k$, of layer k, connecting this neuron to the j-th input signal $x_j^{(k)}(t)$, $j = 0, 1, ..., N_k$. The vector of weights of neuron N_i^k shall be denoted as

$$\mathbf{w}_i^{(k)}(t) = \left[w_{i,0}^{(k)}(t), ..., w_{i,N_{k-1}}^{(k)}(t) \right]^T , \quad k = 1, ..., L, \ i = 1, ..., N_k. \quad (6.105)$$

The output signal of neuron N_i^k at the instant t, $t = 1, 2, ...$, is defined as

$$y_i^{(k)}(t) = f \left(s_i^{(k)}(t) \right) , \qquad (6.106)$$

while

$$s_i^{(k)}(t) = \sum_{j=0}^{N_{k-1}} w_{ij}^{(k)}(t) x_j^{(k)}(t) . \qquad (6.107)$$

Let us notice that output signals of neurons in L-th layer

$$y_1^L(t), y_2^L(t), ..., y_{N_L}^L(t) \qquad (6.108)$$

are at the same time output signals of the whole network. They are compared to the so-called *output desired signals*

$$d_1^L(t), d_2^L(t), ..., d_{N_L}^L(t). \tag{6.109}$$

The error measure Q at network output will be defined as follows:

$$Q(t) = \sum_{i=1}^{N_L} \left(\varepsilon_i^{(L)}\right)^2 (t) = \sum_{i=1}^{N_L} \left(d_i^{(L)}(t) - y_i^{(L)}(t)\right)^2. \tag{6.110}$$

Using dependency (6.101) and (6.110), we get

$$w_{ij}^{(k)}(t+1) = w_{ij}^{(k)}(t) - \eta \frac{\partial Q(t)}{\partial w_{ij}^{(k)}(t)}. \tag{6.111}$$

Let us notice that

$$\frac{\partial Q(t)}{\partial w_{ij}^{(k)}(t)} = \frac{\partial Q(t)}{\partial s_i^{(k)}(t)} \frac{\partial s_i^{(k)}(t)}{\partial w_{ij}^{(k)}(t)} = \frac{\partial Q(t)}{\partial s_i^{(k)}(t)} x_j^{(k)}(t). \tag{6.112}$$

By determining

$$\delta_i^{(k)}(t) = -\frac{1}{2} \frac{\partial Q(t)}{\partial s_i^{(k)}(t)}, \tag{6.113}$$

we get the equality

$$\frac{\partial Q(t)}{\partial w_{ij}^{(k)}(t)} = -2\delta_i^{(k)}(t) x_j^{(k)}(t), \tag{6.114}$$

and hence algorithm (6.111) takes the form

$$w_{ij}^{(k)}(t+1) = w_{ij}^{(k)}(t) + 2\eta\delta_i^{(k)}(t) x_j^{(k)}(t). \tag{6.115}$$

The method of determining the value $\delta_i^{(k)}$ depends on the network layer. For the last layer we get

$$\delta_i^{(L)}(t) = -\frac{1}{2} \frac{\partial Q(t)}{\partial s_i^{(L)}(t)} = -\frac{1}{2} \frac{\partial \sum_{m=1}^{N_L} Q_m^{(L)^2}(t)}{\partial s_i^{(L)}(t)}$$

$$= -\frac{1}{2} \frac{\partial Q_i^{(L)^2}(t)}{\partial s_i^{(L)}(t)} = -\frac{1}{2} \frac{\partial \left(d_i^{(L)}(t) - y_i^{(L)}(t)\right)^2}{\partial s_i^{(L)}(t)} \tag{6.116}$$

$$= Q_i^{(L)}(t) \frac{\partial y_i^{(L)}(t)}{\partial s_i^{(L)}(t)} = Q_i^{(L)}(t) f'\left(s_i^{(L)}(t)\right).$$

For any layer $k \neq L$ we have

$$\delta_i^{(k)}(t) = -\frac{1}{2}\frac{\partial Q(t)}{\partial s_i^{(L)}(t)} = -\frac{1}{2}\sum_{m=1}^{N_{k+1}}\frac{\partial Q(t)}{\partial s_m^{(k+1)}(t)}\frac{\partial s_m^{(k+1)}(t)}{\partial s_i^{(k)}(t)}$$

$$= \sum_{m=1}^{N_{k+1}}\delta_m^{(k+1)}(t)\,w_{mi}^{(k+1)}(t)\,f'\left(s_i^{(k)}(t)\right) \qquad (6.117)$$

$$= f'\left(s_i^{(k)}(t)\right)\sum_{m=1}^{N_{k+1}}\delta_m^{(k+1)}(t)\,w_{mi}^{(k+1)}(t)\,.$$

We shall define the error in the k-th layer (except for the last one) for the i-th neuron

$$\varepsilon_i^{(k)}(t) = \sum_{m=1}^{N_{k+1}}\delta_m^{(k+1)}(t)\,w_{mi}^{(k+1)}(t)\,, \quad k = 1, ..., L-1. \qquad (6.118)$$

By substituting expression (6.118) to formula (6.117), we get

$$\delta_i^{(k)}(t) = \varepsilon_i^{(k)}(t)\,f'\left(s_i^{(k)}(t)\right)\,. \qquad (6.119)$$

In consequence, the backpropagation algorithm takes the form:

$$y_i^{(k)}(t) = f\left(s_i^{(k)}(t)\right), s_i^{(k)}(t) = \sum_{j=0}^{N_{k-1}}w_{ij}^{(k)}(t)\,x_j^{(k)}(t)\,, \qquad (6.120)$$

$$Q_i^{(k)}(t) = d_i^{(L)}(t) - y_1^{(L)}(t) \quad \text{for} \quad k = L, \qquad (6.121)$$

$$Q_i^{(k)}(t) = \sum_{m=1}^{N_{k+1}}\delta_m^{(k+1)}(t)\,w_{mi}^{(k+1)}(t) \quad \text{for} \quad k = 1, ..., L-1, \qquad (6.122)$$

$$\delta_i^{(k)}(t) = \varepsilon_i^{(k)}(t)\,f'\left(s_i^{(k)}(t)\right)\,, \qquad (6.123)$$

$$w_{ij}^{(k)}(t+1) = w_{ij}^{(k)}(t) + 2\eta\delta_i^{(k)}(t)\,x_j^{(k)}(t)\,. \qquad (6.124)$$

Above, we have presented a series of mathematical dependencies, describing the learning method of multilayer neural network. The operation of the algorithm begins with presenting a learning signal at the network input. At first, it is processed by neurons of the first layer. By processing, here we shall understand determining of output signal (formulas (6.106) and (6.107)) for each neuron in a given layer. The signals thus obtained become the inputs for the neurons of the next layer. This cycle is repeated, i.e. we determine the values of signals at neurons outputs of the next layer and transfer them further, finally to the last layer. Knowing the output signal of the last layer and the output desired signal from the learning sequence, we may compute the error at network output according to formula (6.121). Using the delta rule, like in case of a single sigmoidal neuron, we may modify the weights of neurons of the last layer, using formulas (6.121), (6.123) and (6.124). However, in such a way, we will not modify the weights in neurons of the hidden layers (we do not know the value $\delta_i^{(k)}$ for neurons of these layers), and after all the objective function defined by formula (6.110) is a function in which the variables are all the weights of the network. That is why the output error is propagated from the back (from the output layer to the input layer) according to neuron connections between layers and taking into consideration their activation functions (see formulas: (6.122), (6.123) and (6.124)). The name of the algorithm derives from the method of its realization, i.e. the error is "backed" from the output to the input layer.

In the discussion concerning the backpropagation method, we have stated that network learning (modification of weights) is made each time after applying a learning vector at the input. This operation is called an *on-line updating of weights* or *instantaneous training of weights*. However, there is another method of operation. At the input of network we may in turn apply learning vectors, determine their corresponding signals at the network output, and then, having them compared to their desired output values, sum up the errors obtained in subsequent iterations. When we complete applying of samples of the entire epoch at the network input, we perform a correction of values of all weights using the cumulated error value. This algorithm is called an *accumulative updating of weights* or *batch procedure*.

A very important problem, often discussed in articles concerning neural networks, is the initiation of weights of networks. It is obvious that the more the initial values of weights are close to the optimum values (minimizing error measure (6.110)), the less time the learning will take. As mentioned before, network learning consists in finding the minimum of an error function. It is the function of many variables, which are all the weights. Such function may have many local minima. Figure 6.18 presents a hypothetical graph of the error function for one variable (one weight).

Even in such a simple case, there may be many local minima, which occurs very often. Unfortunately, the steepest descent method is not a method resistant (robust) to the occurrence of local minima and the algorithm, looking for solution, may get stuck in them. That is why the modifications of backpropagation algorithm are often used, which will be discussed further in this chapter. The simplest method that may eliminate the problem of local minima is to start learning the neural network for different

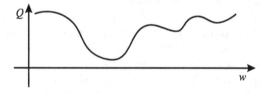

FIGURE 6.18. A hypothetical graph of the error function for one variable (one weight)

values of the initial weights. We proceed as follows. We select the weights at random, assuming a uniform distribution in a defined interval. Then, by applying the backpropagation algorithm, we train the neural network. When the learning error stops decreasing or starts to increase, we again select at random the weights, only now from a different interval of their values. We repeat the operation of network learning and observe whether the final value of error in this cycle is smaller than in the previous one. If it is so, then in the first cycle we could get stuck in a local minimum. Of course, the procedure described prolongs the network learning process. When selecting high values of initial weights the mean squared error at the network output is basically constant. This may be caused by the fact that the output signal of the linear part of the neuron is very large, and therefore we are dealing with a saturation of the activation function. The change of values of weights in the error backpropagation algorithm is proportional to the derivative of this function. As shown in Fig. 6.19, this derivative has the shape of a bell function with the midpoint in zero point.

Therefore, the higher the absolute value of the signal at output of the linear part of neuron is, the smaller the correction of weights, and hence the learning process will be very slow. In general, the initial weights selected at random should give a signal close to one at the output of the linear part of the neuron. The element that has a large influence on the convergence of the error backpropagation algorithm is the learning coefficient η. Unfortunately, there is no general method of selecting its value. In principle, the correction step is taken from the interval $(0, 1)$. If we are dealing with a flat objective function, then the values of the gradient are low, and with higher values of the learning coefficient the algorithm will faster find the solution. On the other hand, if the objective function is steep, then adopting a high value of the coefficient will cause an oscillation around the solution and, thus, will lengthen the learning process. The learning coefficient is selected depending

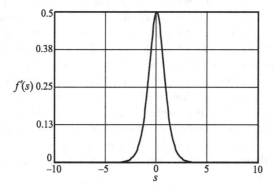

FIGURE 6.19. The derivative of sigmoidal activation function

on the problem which is to be solved and to a large extent depends on the experience of the person performing the learning.

An important issue is the selection of the stopping criterion of the error backpropagation algorithm. It is obvious that in case the minimum (local or global) is reached, the gradient vector (6.96) takes the value **0**. Therefore, the operation of the algorithm may be stopped if the value of the Euclidean norm of the gradient vector drops below a fixed threshold. Alternatively, we may check whether the mean squared error determined within one epoch dropped below another fixed threshold. Sometimes, a combination of both mentioned methods is applied, i.e. the algorithm is stopped if one of the listed values drops below the assumed fixed threshold. Another option is to test continuously the neural network during learning (after each iteration). The learning process may be stopped when the network has good generalization properties.

6.3.3 Backpropagation algorithm with momentum term

As we have mentioned before, during learning of a neural network the error backpropagation algorithm is not doing its best: it may get stuck in a local minimum or oscillate around the final solution. That is why the standard form of this method is introduced with an additional coefficient α called *momentum*, according to which formula (6.124) is modified as follows:

$$w_{ij}^{(k)}(t+1) = w_{ij}^{(k)}(t) + 2\eta \delta_i^{(k)} x_j^{(k)}(t) + \alpha \left[w_{ij}^{(k)}(t) - w_{ij}^{(k)}(t-1) \right]. \quad (6.125)$$

This coefficient makes the value of weight in the next step $(t+1)$ dependent not only on its value in the current step (as in the classic backpropagation method) but also on the previous step $(t-1)$. If in subsequent iterations the direction of the weight modification was the same, then the term containing the momentum coefficient causes an increase in the weight increment value and its shifting with increased strength towards the minimum. In opposite case this term causes slowing down of sharp changes of weights. In general, the operation of the momentum term is activated on flat sections of the objective function, as well as near the local minimum. In particular on flat sections of the objective function a significant acceleration of the learning process takes place, owing to that term. But, near a local minimum the value of the objective function gradient is near zero and the momentum term becomes dominant in formula (6.125), which allows to leave the area of the local minimum. Coefficient α takes the values in the interval $(0, 1)$, most often $\alpha = 0.9$ is assumed. Defining its specific value depends on the problem considered and on the experience of the person conducting the learning process.

Example 6.3

In Fig. 6.20, we present the graph of the algorithm of searching for the minimum of function $F(w_1, w_2) = w_1^2 + w_2^{0.08}$ using the steepest descent method. The learning coefficient η is equal to 0.9 and the starting point $[w_1(0), w_2(0)] = [-2.90]$. For such value of coefficient η the subsequent solutions slowly converge to the optimum point $\mathbf{w} = [0, 0]^T$, and the whole course of the algorithm is characterized by the occurrence of oscillations. Figure 6.21 illustrates the solution of the same problem using the algorithm (6.125), where $\eta = 0.2$, and $\alpha = 0.9$. As we may see in the figure, the momentum coefficient causes the algorithm to be convergent faster, and there are no oscillations around the minimum.

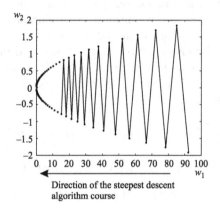

Direction of the steepest descent
algorithm course

FIGURE 6.20. Illustration to Example 6.3 - the steepest descent method

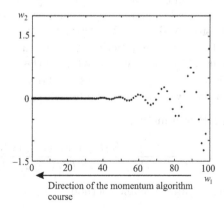

Direction of the momentum algorithm
course

FIGURE 6.21. Illustration to Example 6.3 - the momentum algorithm

6.3.4 Variable-metric algorithm

Another modification of the error backpropagation algorithm is the variable-metric algorithm. In order to derive this algorithm, we will consider the three initial terms in the expansion of the objective function into a Taylor series. The objective functions will now be defined as follows:

$$Q(\mathbf{w}(t) + \mathbf{p}(t)) = Q(\mathbf{w}(t)) + (\mathbf{g}(\mathbf{w}(t)))^T \mathbf{p}(t) + 0.5\mathbf{p}^T \mathbf{H}(\mathbf{w}(t)) \mathbf{p}(t). \tag{6.126}$$

Let us now remind the necessary and sufficient conditions for the existence of the minimum of the function. The necessary condition is zeroing of the first derivative in this point. Unfortunately, it is not a sufficient condition, which may be easily checked on the example of function $f(x) = x^3$ (in point $x = 0$, the first derivative is equal to zero but this point is not a minimum). A sufficient condition for the existence of the function minimum in point x_0 is the satisfaction of two conditions: 1) zeroing of the first derivative in this point and 2) value of the second derivative in this point should be bigger than 0. Analogous conditions must be satisfied in case of function of many variables. Let us now derive the variable-matrix algorithm. We shall minimize the criterion Q given by formula (6.126) with respect to the vector $\mathbf{p}$. Of course,

$$\frac{\partial Q(\mathbf{w}(t) + \mathbf{p}(t))}{\partial \mathbf{p}(t)} = (\mathbf{g}(\mathbf{w}(t)))^T + \mathbf{H}(\mathbf{w}(t)) \mathbf{p}(t). \tag{6.127}$$

Vector $\mathbf{p}(t)$ minimizing the criterion Q must satisfy the equation

$$\mathbf{g}(\mathbf{w}(t)) + \mathbf{H}(\mathbf{w}(t)) \mathbf{p}(t) = \mathbf{0}. \tag{6.128}$$

In consequence

$$\mathbf{p}(t) = -\left[\mathbf{H}(\mathbf{w}(t))\right]^{-1} \mathbf{g}(\mathbf{w}(t)). \tag{6.129}$$

In order to reach the minimum in a given point by the objective function $Q(\mathbf{w}(t)+\mathbf{p}(t))$, the Hessian (matrix of the second-order derivatives) should be in this point positively defined. Unfortunately, this condition is very difficult to satisfy. That is why in practice, its approximated value $\mathbf{G}(\mathbf{w}(t))$ is determined. Let us assume that $\mathbf{c}(t) = \mathbf{w}(t) - \mathbf{w}(t-1)$, $\mathbf{r}(t) = \mathbf{g}(\mathbf{w}(t)) - \mathbf{g}(\mathbf{w}(t-1))$, $\mathbf{V}(t) = [\mathbf{G}(\mathbf{w}(t))]^{-1}$ and $\mathbf{V}(t-1) = [\mathbf{G}(\mathbf{w}(t-1))]^{-1}$. To update the values of matrix $\mathbf{V}$, two known methods are applied:

i) Davidon-Fletcher-Powell method

$$\mathbf{V}(t) = \mathbf{V}(t-1) + \frac{\mathbf{c}(t)\mathbf{c}^T(t)}{\mathbf{c}^T(t)\mathbf{r}(t)} - \frac{\mathbf{V}(t-1)\mathbf{r}(t)\mathbf{r}^T(t)\mathbf{V}(t-1)}{\mathbf{r}^T(t)\mathbf{V}(t-1)\mathbf{r}(t)}; \tag{6.130}$$

ii) Broyden-Fletcher-Goldfarb-Shano method

$$\mathbf{V}(t) = \mathbf{V}(t-1) + \left[1 + \frac{\mathbf{r}^T(t)\mathbf{V}(t-1)\mathbf{r}(t)}{\mathbf{c}^T(t)\mathbf{r}(t)}\right] \frac{\mathbf{c}(t)\mathbf{c}^T(t)}{\mathbf{c}^T(t)\mathbf{r}(t)} \tag{6.131}$$

$$-\frac{\mathbf{c}(t)\mathbf{r}^T(t)\mathbf{V}(t-1) + \mathbf{V}(t-1)\mathbf{r}(t)\mathbf{c}^T(t)}{\mathbf{c}^T(t)\mathbf{r}(t)},$$

where $\mathbf{V}(0) = \mathbf{I}$. By combining formula (6.129) with formula (6.130) or (6.131), we get

$$\mathbf{p}(t) = -\mathbf{V}(t)\mathbf{g}(\mathbf{w}(t)). \tag{6.132}$$

The variable-matrix algorithm is characterized by a quite fast convergence. Its disadvantage is a relatively high computational complexity resultant from the need to determine, in every step, all the Hessian elements. That is why this algorithm is applied rather to small networks.

6.3.5 Levenberg-Marquardt algorithm

Like in previous points, we shall assume that $Q(\mathbf{w})$ is an error function and we will attempt to find its minimum. The Levenberg-Marquardt algorithm uses the expansion of the function $Q(\mathbf{w})$, expressed by formula (6.95), to the third term. The search for the minimum of the expression obtained in this way has been already described in the previous point. The minimization direction is identical as in the case of the variable-matrix algorithm, i.e. $\mathbf{p}(t) = -[\mathbf{H}(\mathbf{w}(t))]^{-1}\mathbf{g}(\mathbf{w}(t))$.

Let us consider a neural network with N_L outputs. Let us assume the error measure in the form

$$Q(\mathbf{w}) = \frac{1}{2}\sum_{i=1}^{N_L} e_i^2(\mathbf{w}), \tag{6.133}$$

where

$$e_i = d_i - y_i. \tag{6.134}$$

In literature the following formulas defining the gradient vector

$$\mathbf{g}(\mathbf{w}) = [J(\mathbf{w})]^T e(\mathbf{w}) \tag{6.135}$$

and approximated Hessian matrix

$$\mathbf{H}(\mathbf{w}) = [J(\mathbf{w})]^T J(\mathbf{w}) + S(\mathbf{w}), \tag{6.136}$$

are presented where

$$J(\mathbf{w}) = \begin{bmatrix} \dfrac{\partial e_1}{\partial w_1} & \dfrac{\partial e_1}{\partial w_2} & \cdots & \dfrac{\partial e_1}{\partial w_n} \\ \dfrac{\partial e_2}{\partial w_1} & \dfrac{\partial e_2}{\partial w_2} & \cdots & \dfrac{\partial e_2}{\partial w_n} \\ \cdots & \cdots & \cdots & \cdots \\ \dfrac{\partial e_{N_L}}{\partial w_1} & \dfrac{\partial e_{N_L}}{\partial w_2} & \cdots & \dfrac{\partial e_{N_L}}{\partial w_n} \end{bmatrix} \tag{6.137}$$

and

$$e(\mathbf{w}) = [e_1(w), ..., e_{N_L}(w)]^T. \qquad (6.138)$$

The part $S(\mathbf{w})$ in formula (6.136) corresponds to the terms of the Hessian expansion, containing the higher derivatives with respect to vector $\mathbf{w}$. This part may be approximated as $S(\mathbf{w}) = \mu \mathbf{I}$, where μ is the so-called Levenberg-Marquardt parameter. Using this approximation and substituting dependencies (6.135) and (6.136) to formula (6.129), we get

$$\mathbf{p}(t) = -[J^T(\mathbf{w}(t))J(\mathbf{w}(t)) + \mu(t)\mathbf{I}]^{-1}[J^T(\mathbf{w}(t))]^T e(\mathbf{w}(t)). \qquad (6.139)$$

Parameter μ is selected depending on the error at network output during its learning process. This parameter takes on high values at the algorithm start, and as it is closer to optimum solution, its value decreases to zero.

Example 6.4

We shall compare the algorithms discussed so far using the example of the XOR problem. The experiments were repeated ten times for each algorithm (we changed the learning parameters, e.g. learning coefficient) and the best results were selected. All the experiences were carried out using the Matlab package. The learning was stopped when the mean squared error dropped below 0.012. The results are illustrated in Table 6.3. As it may be noticed, the fastest method is the Levenberg – Marquardt method. Unfortunately, this algorithm may not always be used, as it requires large sizes of computers' RAM .

TABLE 6.3. Comparison of operation of neural networks learning algorithms

Name of learning algorithm	Number of epochs
Steepest descent	415
Momentum	250
Variable-matrix	8
Levenberg-Marquardt	3

6.3.6   Recursive least squares method

In Sections 6.2.3 and 6.2.4, we have presented models of neurons trained using the RLS method. Currently, we will apply this method to learning of multilayer neural networks. Let as define the following error measure

$$Q(t) = \sum_{l=1}^{t} \lambda^{t-l} \sum_{j=1}^{N_L} \varepsilon_j^{(L)^2}(l) \qquad (6.140)$$

$$= \sum_{l=1}^{t} \lambda^{t-l} \sum_{j=1}^{N_L} \left[d_j^{(L)}(l) - f\left(\mathbf{x}^{(L)^T}(l)\,\mathbf{w}_j^{(L)}(t)\right)\right]^2,$$

where λ is the forgetting factor selected from the interval $(0, 1]$. Let us notice that the previous errors have a lesser influence on the value of expression (6.140). Further in our discussion, we will apply notations introduced in Section 6.3.2. Moreover, let us denote

$$\varepsilon_i^{(k)}(l) = d_i^{(k)}(l) - y_i^{(k)}(l) \tag{6.141}$$

and

$$b_i^{(k)}(l) = f^{-1}\left(d_i^{(k)}(l)\right), \tag{6.142}$$

assuming that f is an invertible function, $l = 1, ..., t$, $i = 1, ..., N_k$, $k = 1, ..., L$.

When calculating the error measure gradient and equating it to zero, we get the equation

$$\frac{\partial Q(t)}{\partial \mathbf{w}_i^{(k)}(t)} = 2 \sum_{l=1}^{t} \lambda^{t-l} \sum_{j=1}^{N_L} \frac{\partial \varepsilon_j^{(L)}(l)}{\partial \mathbf{w}_i^{(k)}(t)} \varepsilon_j^{(L)}(l) \tag{6.143}$$

$$= -2 \sum_{l=1}^{t} \lambda^{t-l} \sum_{j=1}^{N_L} \frac{\partial y_j^{(L)}(l)}{\partial \mathbf{w}_i^{(k)}(t)} \varepsilon_j^{(L)}(l) = \mathbf{0}.$$

Using dependency (6.106) and (6.107), the equation (6.143) will be converted as follows:

$$\sum_{l=1}^{t} \lambda^{t-l} \sum_{j=1}^{N_L} \frac{\partial y_j^{(L)}(l)}{\partial s_j^{(L)}(l)} \sum_{p=1}^{N_{L-1}} \frac{\partial s_t^{(L)}(l)}{\partial y_p^{(L-1)}(l)} \frac{\partial y_p^{(L-1)}(l)}{\partial \mathbf{w}_i^{(k)}(t)} \varepsilon_j^{(L)}(l)$$

$$= \sum_{l=1}^{t} \lambda^{t-l} \sum_{p=1}^{N_{L-1}} \frac{\partial y_p^{(L-1)}(l)}{\partial \mathbf{w}_i^{(k)}(t)} \sum_{j=1}^{N_L} \frac{\partial y_j^{(L-1)}(l)}{\partial s_j^{(L)}(l)} w_{jp}^{(L)} \varepsilon_j^{(L)}(l) \tag{6.144}$$

$$= \sum_{l=1}^{t} \lambda^{t-l} \sum_{p=1}^{N_{L-1}} \frac{\partial y_p^{(L-1)}(l)}{\partial \mathbf{w}_i^{(k)}(t)} \varepsilon_p^{(L-1)}(l) = \sum_{l=1}^{t} \lambda^{t-l} \sum_{q=1}^{N_k} \frac{\partial y_p^{(k)}(l)}{\partial \mathbf{w}_i^{(k)}(t)} \varepsilon_q^{(k)}(l) = \mathbf{0},$$

where

$$\varepsilon_p^{(k)}(l) = \sum_{j=1}^{N_{k+1}} \frac{\partial y_j^{(k+1)}(l)}{\partial s_j^{(k+1)}(l)} w_{jp}^{(k+1)}(t) \varepsilon_j^{(k+1)}(l). \tag{6.145}$$

Expression (6.145) defines the error determination method in subsequent layers, starting from the last one. By further converting expression (6.144), we get a sequence of equalities

$$\sum_{l=1}^{t} \lambda^{t-l} \sum_{q=1}^{N_k} \frac{\partial y_q^{(k)}(l)}{\partial \mathbf{w}_i^{(k)}(t)} \varepsilon_q^{(k)}(l) = \sum_{j=1}^{t} \lambda^{t-l} \frac{\partial y_i^{(k)}(l)}{\partial s_i^{(k)}(t)} \mathbf{y}^{(k-1)^T}(l) \varepsilon_i^{(k)}(l) \tag{6.146}$$

$$= \sum_{l=1}^{t} \lambda^{t-l} \frac{\partial y_i^{(k)}(l)}{\partial s_i^{(k)}(t)} \mathbf{y}^{(k-1)^T}(l) \left[d_i^{(k)}(l) - y_i^{(k)}(l)\right] = \mathbf{0}.$$

where

$$\mathbf{y}^{(k)} = \left[y_1^{(k)}, ..., y_{N_k}^{(k)} \right]^T.$$

By applying the approximation

$$f\left(b_i^{(k)}(l)\right) \approx f\left(s_i^{(k)}(l)\right) + f'\left(s_i^{(k)}(l)\right)\left(b_i^{(k)}(l) - s_i^{(k)}(l)\right), \quad (6.147)$$

we get the normal equality

$$\sum_{l=1}^{t} \lambda^{t-l} f'^2\left(s_i^{(k)}(l)\right)\left[b_i^{(k)}(l) - \mathbf{x}^{(k)^T}(l)\,\mathbf{w}_i^{(k)}(t)\right]\mathbf{x}^{(k)^T}(l) = \mathbf{0}, \quad (6.148)$$

the vector notation of which is

$$\mathbf{r}_i^{(k)}(t) = \mathbf{R}_i^{(k)}(t)\,\mathbf{w}_i^{(k)}(t), \quad (6.149)$$

where

$$\mathbf{R}_i^{(k)}(t) = \sum_{l=1}^{t} \lambda^{t-l} f'^2\left(s_i^{(k)}(l)\right)\mathbf{x}^{(k)}(l)\,\mathbf{x}^{(k)^T}(l), \quad (6.150)$$

$$\mathbf{r}_i^{(k)}(t) = \sum_{l=1}^{t} \lambda^{t-l} f'^2\left(s_i^{(k)}(l)\right)b_i^{(k)}(l)\,\mathbf{x}^{(k)}(l). \quad (6.151)$$

Equation (6.149) may be solved using recurrence method, without the need to inverse the matrix $\mathbf{R}_i^{(k)}(t)$. This calls for the application of the RLS algorithm, as it was done in points 6.2.3 and 6.2.4 in case of a single neuron model. As a result, the adaptive correction of all weights $\mathbf{w}_i^{(k)}$ is made as follows

$$\varepsilon_i^{(k)}(t) = \begin{cases} d_i^{(L)}(t) - y_i^{(L)}(t) & \text{for } k = L, \\ \sum_{j=1} f'\left(s_j^{(k+1)}(t)\right)w_{ji}^{(k+1)}(t)\,\varepsilon_j^{(k+1)}(t) & \text{for } k = 1, ..., L-1, \end{cases} \quad (6.152)$$

$$\mathbf{g}_i^{(k)}(t) = \frac{f'\left(s_i^{(k)}(t)\right)\mathbf{P}_i^{(k)}(t-1)\,\mathbf{x}^{(k)}(t)}{\lambda + f'^2\left(s_i^{(k)}(t)\right)\mathbf{x}^{(k)^T}(t)\,\mathbf{P}_i^{(k)}(t-1)\,\mathbf{x}^{(k)}(t)}, \quad (6.153)$$

$$\mathbf{P}_i^{(k)}(t) = \lambda^{-1}\left[\mathbf{I} - f'\left(s_i^{(k)}(t)\right)\mathbf{g}_i^{(k)}(t)\,\mathbf{x}^{(k)^T}(t)\right]\mathbf{P}_i^{(k)}(t-1), \quad (6.154)$$

$$\mathbf{w}_i^{(k)}(t) = \mathbf{w}_i^{(k)}(t-1) + \mathbf{g}_i^{(k)}(t)\varepsilon_i^{(k)}(t) \quad (6.155)$$

where $i = 1, ..., N_k, k = 1, ..., L$.
Initial values of the RLS algorithm are usually assumed as follows:

$$\mathbf{P}^{(k)}(0) = \delta\mathbf{I}, \quad \delta \gg 0, \quad (6.156)$$

$$\mathbf{w}_i^{(k)}(0) = 0. \quad (6.157)$$

Initial weights $\mathbf{w}_i^{(k)}(0)$ of the neural network may also be selected randomly in a given interval.

6.3.7 Selection of network architecture

The term of designing neural networks architecture the Reader should understand as the selection of the number of network layers and selection of the number of neurons in each layer. As we may suppose, these values depend on the problem we are trying to solve. Let us think how the number of network layers impacts its operation. According to what was said earlier, a single neuron divides a plane into two parts. But two layers may map simplexes, i.e. convex areas bounded with hyperplanes. Using three layers, we may define any area. Therefore, a three-layer network is able to solve a wide range of classification and approximation problems. Here, it is well worth to present the Kolmogorov theorem, pursuant to which any given continuous real function $f(x_1, .., x_n)$, defined on $[0,1]^n$, $n \geq 2$, may be approximated using the function F given by formula

$$F(\mathbf{x}) = \sum_{j=1}^{2n+1} g_j \left(\sum_{i=1}^{n} \phi_{ij}(x_i) \right), \qquad (6.158)$$

where $\mathbf{x} = [x_1, .., x_n]^T$, g_j, $j = 1, ..., 2n+1$ are appropriately selected continuous functions of one variable and ϕ_{ij}, $i = 1, ..., n$, $j = 1, ..., 2n+1$, are continuous and monotonically increasing functions which are independent of f. Let us consider for a moment a dependency (6.158) and let us analyze how it may be related to the neural network structure. We may easily notice that the structure corresponding to the dependency (6.158) is created by the two-layer neural network with n inputs, $2n+1$ neurons in the hidden layer and one neuron with linear activation function in the output layer. Based on the above considerations, we may state that the structure of network presented in Fig. 6.22 is able to approximate any continuous function of n variables defined on $[0,1]^n$. The Kolmogorov theorem is theoretical in nature, as it does not present any type of non-linear functions and network learning methods.

In literature we may also find another theorem, which directly relates to a multilayer neural network.

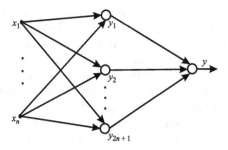

FIGURE 6.22. Structure of neural network which is able to approxamate any continuous function of n variables

Let us assume that ϕ is any continuous sigmoidal function. Then for each continuous function f defined on $[0,1]^n$, $n \geq 2$, and for any $\varepsilon > 0$, there is an integer N and a set of constants α_i, θ_i and $w_{ij}, i = 1, \ldots, N$, $j = 1, \ldots, n$, so that the function

$$F(x_1, .., x_n) = \sum_{i=1}^{N} \alpha_i \phi \left(\sum_{j=1}^{n} w_{ij} x_j - \theta_i \right) \tag{6.159}$$

approximates the function f, i.e.

$$|F(x_1, .., x_n) - f(x_1, .., x_n)| < \varepsilon$$

for all $\{x_1, ..., x_n\} \in [0,1]^n$. Based on the above theorem, we may state that the neural network with a linear output neuron and one hidden layer with neurons having sigmoidal activation function may approximate any real continuous function defined on $[0,1]^n$. In practice, it appears that are very few problems which require more than two hidden layers to be solved by feed-forward neural networks.

The number of neurons in particular layers has a very large impact on network operation. It should be stressed that an extensive number of neurons lengthens the learning process. Moreover, if the number of learning samples is small compared to the size of the network, we may "over-train" the structure and it will lose its capability of generalizing knowledge. The network will learn the learning sequence "by heart" and will correctly map only the samples included in it.

Example 6.5

In order to familiarize the Reader with this problem, we will try to perform an experiment, in which we will learn the neural network so that it approximates the function $f(x) = \sin(x)$ in a closed interval $[0, 2\pi]$. Before we start building an appropriate structure of the neural network, we must create a learning sequence. As it can be easily guessed, different arguments x of function $f(x) = \sin(x)$ create the network inputs and the corresponding output values y are the desired output values. It follows that the learning sequence will be composed of pairs $\{x, \sin(x)\}$. Let us notice that the function we approximate is continuous. Hence, we cannot build the learning sequence of all x from the interval $[0, 2\pi]$ and the corresponding values y. We have to select a certain number of characteristic pairs $\{x, f(x)\}$ and train the neural network using these pairs. At this moment, the Reader should notice a very important fact: the network will learn based on a certain subset of samples from interval $[0, 2\pi]$, but we want the network to operate correctly in the entire interval $[0, 2\pi]$. This is exactly the phenomenon of generalization of knowledge included in the weights of the network. Despite learning only some selected examples, the network is able to generalize the knowledge and answer correctly to signals applied at its input which were not in the learning sequence. That is why after having network learnt, we should check its operation on a testing sequence, consisting of samples which did not participate in the learning process. After a successful completion of this test, we may state that the structure (i.e. weights) has been learnt and operates correctly. Let us assume that the learning sequence consists of samples presented in Table 6.4.

TABLE 6.4. Learning sequence in Example 6.5

Sample No.	1	2	3	4	5	6	7	8
Input x	0	$\dfrac{\pi}{6}$	$\dfrac{\pi}{3}$	$\dfrac{\pi}{4}$	π	2π	$\dfrac{7\pi}{6}$	$\dfrac{4\pi}{3}$
Desired output $d = f(x)$	0	0.5	$\dfrac{\sqrt{3}}{2}$	$\dfrac{\sqrt{2}}{2}$	0	0	-0.5	$-\dfrac{\sqrt{3}}{2}$

Sample No.	9	10	11	12	13	14	15
Input x	$\dfrac{5\pi}{4}$	$\dfrac{5\pi}{6}$	$\dfrac{2\pi}{3}$	$\dfrac{3\pi}{4}$	$\dfrac{5\pi}{3}$	$\dfrac{11\pi}{6}$	$\dfrac{7\pi}{4}$
Desired output $d = f(x)$	$-\dfrac{\sqrt{2}}{2}$	$\dfrac{1}{2}$	$\dfrac{\sqrt{3}}{2}$	$\dfrac{\sqrt{2}}{2}$	$-\dfrac{\sqrt{3}}{2}$	$-\dfrac{1}{2}$	$-\dfrac{\sqrt{2}}{2}$

Network structure used in the simulation is made of one hidden layer with sigmoidal neurons and one linear neuron in the output layer. The simulations will be performed three times with a different number of neurons in the hidden layer: 2, 3 and 15. The learning sequence is created by discretization of the interval $[0, 2\pi]$ with step 0.1. Figures 6.23, 6.24 and 6.25 illustrate the result of network operation, i.e. the abilities of generalization when presenting a testing sequence at its input.

In each of the below figures, a graph of function $f(x) = \sin(x)$ has been presented, as well as the points of the learning sequence given in Table 6.4 and points reflecting the network output signals after presenting at its input the testing sequence. As it can be observed in Fig. 6.23, a too small number of neurons as compared to the number of the learning samples will cause the network be unable to approximate the function. On the other hand, Fig. 6.25 illustrates the fact that a too large number of neurons will cause an excessive adjustment of the network to the learning sequence.

Error at network output decreases to 0 during learning, and increases dramatically in case of a testing sequence. Therefore, it is natural to raise the question: how to define the number of neurons hidden in the network? To a certain extent, we may be helped by the so-called *Vapnik-Chervonenkis*

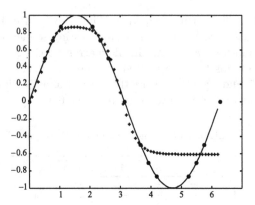

FIGURE 6.23. Illustration of the generalization effect of the neural network - 2 neurons in the hidden layer

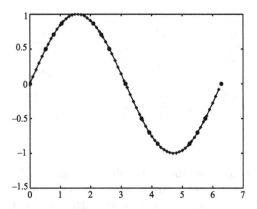

FIGURE 6.24. Illustration of the generalization effect of the neural network - 3 neurons in the hidden layer

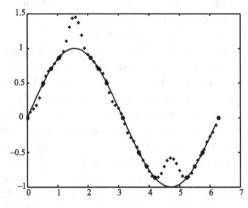

FIGURE 6.25. Illustration of the generalization effect of the neural network - 15 neurons in the hidden layer

dimension (VC). The VC dimension for the set of functions is denoted by letter h and defined as the maximum number of vectors that may be separated (shattered) in all possible ways using the functions from this set. Let us notice that in case of a binary classification l vectors may be divided into two classes in all 2^l possible ways. If we find a set of functions which will be able to make such classification (in all 2^l ways), then the VC dimension for this set of functions will be $h = l$. Let $i_f(\mathbf{x}, \mathbf{w})$ be a function taking only two values, i.e.

$$i_f(\mathbf{x}, \mathbf{w}) \in \{-1, 1\}. \tag{6.160}$$

The above condition is met by the function describing the perceptron

$$i_f(\mathbf{x}, \mathbf{w}) = \text{sign}(\mathbf{x}^T, \mathbf{w}). \tag{6.161}$$

In the two-dimensional case, we have

$$i_f(\mathbf{x}, \mathbf{w}) = \text{sign}\left(\sum_{i=1}^{2} w_i x_i + w_0\right), \tag{6.162}$$

where $\mathbf{x} = [x_1, x_2]^T$ and $\mathbf{w} = [w_0, w_1, w_2]^T$. As mentioned before, the operation of the perceptron with two inputs is illustrated by a line which divides the plane into two parts. This means that depending on the value of the weights, it creates a set of functions which may assign to input vectors the values $+1$ or -1, depending on which side of the line their coordinates are located. Figure 6.26 presents all the possible partitions of the plane determined by the perceptron with appropriately selected weights for the three input vectors. In this figure, the arrows mark the half spaces corresponding to positive values of function i_f. In this case, the VC dimension is 3.

If VC dimension is h, this means that there is at least one set h of vectors which may be divided into all possible ways. This does not mean that this

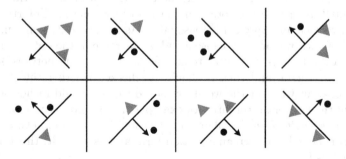

FIGURE 6.26. The partition determined by the perceptron with appropriately selected weights for the three input vectors

property must concern all the vectors of the given set. In the case of a function defined by dependency (6.161) in an n-dimensional space, the VC dimension is $h = n + 1$.

In particular, it equals 3 for a plane. Form the above statement, it may be concluded that the VC dimension increases together with the increase of number of weights in the neuron (when the dimension of the space increases, the number of weights in the neuron increases as well). However, it has been demonstrated that the increase in the number of weights does not always impact the increase in the VC [108] dimension. In fact, it is very difficult to define the value of the VC dimension. However, this number should be treated as an indication when defining the generalization possibilities of multilayer neural networks, which is illustrated by dependency [73]

$$Q_G(\mathbf{w}) \leq Q_U(\mathbf{w}) + \Phi\left(\frac{M}{h}, Q_U(\mathbf{w})\right), \qquad (6.163)$$

where M is the number of learning samples, Q_U means the learning error, Q_G is a generalization error, while Φ is a certain function dependent on M, Q_U and VC dimension. When building and training a neural network, we should attempt to get the generalization error as small as possible. We should also remember that together with the increase in the number of neurons in the network, its time of operation also increases. That is why many methods were created which are used to extend and reduce the network structure.

In the algorithms that reduce the network structure, the neurons or connections (weights) which have no or little significance are removed from the network. An intervention in the network structure is always made after it has been learnt. The network error for the entire testing sequence is defined before and then after removing the weight. If the error does not increase, this means that the appropriate weights were removed. The simplest method is the removal of those weights which have lower values than the fixed threshold. One might expect that the low value of the weight has a small impact on the total stimulation of the neuron. Unfortunately, not always such an approach is reasonable. As it turns out, sometimes the removal of such weight may cause a sudden increase of the output error of the network. That is why after removing the weight, the network should always learn again, and then it should be checked whether the structure works better or whether the weight removed earlier should rather be reinserted in the network structure. An example of such procedure may be *the weight decay method*. In this algorithm, the standard form of network error function is added with a regularization term and as a result, this function is defined as follows:

$$Q = \frac{1}{2}\sum_{i=1}^{N_L}(y_i - f(\mathbf{x}_i))^2 + \lambda\sum_{i=1}^{I}w_i^2, \qquad (6.164)$$

where I is the number of all weights in the network. After having the network learnt, the weights the values of which are lower that the fixed threshold may be removed from the structure. However, the above change causes that as a result of learning, a large number of weights with low values is created in the structure. That is why the literature (e.g. [12]) proposes many modifications of formula (6.164). Another method of network structure reduction is the OBD algorithm (*Optimal Brain Damage*). As we want to remove certain weights in the learnt structure, we should estimate their impact on the network operation. Let us denote the network error after removing the weights by $Q(\mathbf{w})$, and before removal but after network learning – by $Q(\mathbf{w}^*)$. As we remember, deriving the backpropagation algorithm, we approximated the error function with the Taylor series, using dependency (6.95). As the reduction of weights is made when the neural network has already learnt, we may assume that the error function has reached a minimum. Accordingly, the components of the gradient vector of function Q would take zero values. If we want to estimate the value of the difference of errors before and after reduction of the structure, omitting the gradient vector and cutting the expansion of the series to the third term, we get

$$Q(\mathbf{w}) - Q(\mathbf{w}^*) \approx 0.5(\mathbf{w} - \mathbf{w}^*)^T \mathbf{H}(\mathbf{w}^*)(\mathbf{w} - \mathbf{w}^*). \qquad (6.165)$$

Determination of the hessian in formula (6.165) is problematic due to a large number of neuron weights in the network. That is why the OBD method authors assumed that the value of the hessian is most impacted by its diagonal elements. Therefore, the sensitivity measure for the j-th weight in i-th neuron has been defined as

$$S_{ij} = \frac{1}{2} \frac{\partial^2 Q}{\partial w_{ij}^2} w_{ij}^2. \qquad (6.166)$$

The OBD algorithm consists in the following stages:

1. Initial selection of the network structure and its learning.

2. Calculation of the sensitivity coefficient for all the weights according to formula (6.166) and removing the ones, for which the value S_{ij} is rather low.

3. Repeated learning of the network.

The hessian matrix of weights is a non-diagonal matrix and the OBD algorithm may cause the removal of significant weights from the network. That is why the so-called OBS method (*Optimal Brain Surgeon*), in which all the hessian components are taken into account has been proposed. At first, we compute the coefficients

$$s_j = \frac{1}{2} \frac{w_j^2}{\mathbf{H}_{jj}^{-1}}. \qquad (6.167)$$

Then the weights with the lowest values of coefficient (6.167) are removed. The corrections, with which the weights should be modified, are defined according to the dependency

$$\Delta \mathbf{w} = -\frac{w_j}{\mathbf{H}_{jj}^{-1}} \mathbf{H}^{-1} \cdot \mathbf{i}_j, \qquad (6.168)$$

where $\mathbf{i}j$ is the vector composed of only zeros and 1 on the j-th position.

6.4   Recurrent neural networks

In all networks discussed so far, we did not consider the case in which the signal received at the output was sent again to the network input. Such circulation of the signal is called *feedback*. The structures in which this phenomenon occurs are called *recurrent neural networks*. A single stimulation of the structure with feedback may generate the sequence of many new phenomena and signals, as the signals from the network output are resent again to its inputs, generating new signals until the output signals are stabilized. Suppressions, oscillations, sudden rises or falls of the signals often occur during such circulation. We shall present the architectures and will briefly discuss the operation of the best known recurrent neural networks, namely the Hopfield neural network, Hamming neural network, RTRN (*Real Time Recurrent Network*), Elman neural network and BAM (*Bidirectional Associative Memory*). It should be mentioned that the recurrent neural networks are applied as associative memories. For instance, the Hopfield neural networks may serve as autoassociative memories, whereas the Hamming neural networks and BAM networks are examples of heteroassociative memories (association of two different vectors).

6.4.1   Hopfield neural network

Figure 6.27 presents the Hopfield neural network. It is a one-layer network with a regular structure, made of many neurons connected one to the other. There are no feedbacks in the same neuron. This means that the output signal of a given neuron does not reach its input, and thus the values of weights w_{ii} equal 0. The weights in this network are symmetrical, i.e. the weight w_{kj} connecting the neuron k to the neuron j is equal to the weight w_{jk} connecting the neuron j to the neuron k. The Hopfield neural network during learning modifies its weights w_{kj} depending on the value of learning vector $\mathbf{x}$. In retrieval mode, the weights are not subject to modifications, but the input signal stimulates the network which, through the feedback, repeatedly receives the output signal at its input, until the answer is stabilized. If we assume that the neuron activation function is of the signum type, then the operation of the network in step t may be described as follows

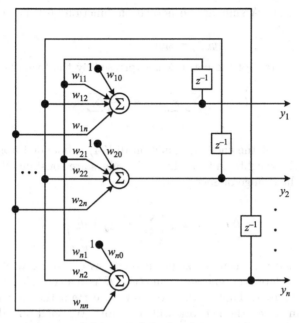

FIGURE 6.27. Scheme of the Hopfield neural network

$$y_k(t) = \text{sgn}\left(\sum_{j=1, j\neq k}^{n} w_{kj} y_j(t-1) + \theta_k \right), \quad k = 1, ..., N, \qquad (6.169)$$

while $y_j(0) = x_j$. The signal at the output will change until in the step $t-1$ will be equal to signal in step t, and so $y_k(t) = y_k(t-1)$ for all N neurons making up the network. Let us demonstrate that if the weights in the Hopfield neural network are symmetrical, then this network always gets stabilized. Let us denote the output of the linear part of k-th neuron in the moment $t+1$ by $s_k(t+1)$, i.e.

$$s_k(t+1) = \sum_{j=1, j\neq k}^{n} w_{kj} y_j(t) + \theta_k, \qquad (6.170)$$

where θk is the threshold value of the neuron. The activation function may be defined as follows:

$$y_k(t+1) = \text{sgn}\left(s_k(t+1) \right). \qquad (6.171)$$

The network is stabilized, if

$$y_k(t) = y_k(t-1) \qquad (6.172)$$

for each neuron. Assuming the activation function (6.171), dependency (6.172) becomes

$$y_k(t) = \mathrm{sgn}\,(s_k(t-1)).$$
(6.173)

The energetic state of the network is expressed by the Lyapunov function in the form

$$E = -\frac{1}{2}\sum_{\substack{j \\ j \neq k}}\sum_k y_j y_k w_{kj} - \sum_k \theta_k y_k.$$
(6.174)

Let us notice that the energetic function (6.174) is bounded from the bottom, while the weights and the threshold values are constant. It is easy to check that the change of energy

$$\Delta E = -\Delta y_k \left(\sum_{j \neq k} y_j w_{kj} + \theta_k\right)$$
(6.175)

is always negative, when the signal y_k changes according to dependency (6.171). Therefore, the energetic function E is a decreasing function in subsequent steps t. That is why we may be certain that it will reach a minimum, in which the network will be stable. We shall now discuss the method of selecting the weights in Hopfield neural networks. One of the learning methods for Hopfield neural networks in *the generalized Hebb rule*. According to this rule, the weights are modified using the dependency

$$w_{kj} = \frac{1}{N}\sum_{i=1}^{M} x_k^i x_j^i,$$
(6.176)

in which $\mathbf{x}^i = [x_1^i, ..., x_n^i]$, $i = 1, \ldots, M$. Unfortunately, the Hebb rule does not guarantee the best results. It can be proved that the network capacity (maximum number of patterns which the network is able to memorize) trained using this rule is only 13.8% of the number of neurons. That is why in practice, pseudoinverse method is often applied. Let $\mathbf{X}$ be the matrix of M learning vectors, i.e. $\mathbf{X} = \left[\mathbf{x}^1, \mathbf{x}^2...\mathbf{x}^M\right]$. It is assumed that the objective of network learning is such a selection of weights, so that after providing signal $\mathbf{x}$ at its input the same signal was created at the output, i.e.

$$\mathbf{WX} = \mathbf{X},$$
(6.177)

where $\mathbf{W}$ denotes the matrix of weights with the dimension $n \times n$. The solution of the system of equations (6.177) is as follows:

$$\mathbf{W} = \mathbf{XX}^+,$$
(6.178)

where symbol $+$ means the pseudoinverse. If we assume that the learning vectors are linearly independent, then the equation (6.178) takes the form

$$\mathbf{W} = \mathbf{X}(\mathbf{X}^T\mathbf{X})^{-1}\mathbf{X}^T.$$
(6.179)

Example 6.6

Now, we will attempt to train the Hopfield neural network to recognize three digits: 1, 4, 7. This experiment will be carried out in the Matlab environment. Based on the patterns presented in Fig. 6.28, we create a learning sequence (white fields are denoted as -1, black as 1)

$$\mathbf{x}(1) = [-1 - 1\, 1 - 1 - 1\, 1 - 1 - 1\, 1 - 1 - 1\, 1]$$
$$\mathbf{x}(2) = [1 - 1\, 1\, 1\, 1\, 1 - 1 - 1\, 1\, 1 - 1 - 1\, 1]$$
$$\mathbf{x}(3) = [1\, 1\, 1 - 1 - 1\, 1\, 1 - 1 - 1\, 1\, 1 - 1 - 1\, 1]$$

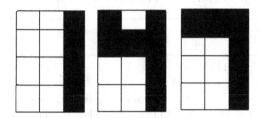

FIGURE 6.28. Patterns in Example 6.6

Next, we train the Hopfield neural network using the above patterns. In order to check the correctness of operation of a trained network, we shall apply subsequent learning signals at its input. As it results from Table 6.5, the network perfectly solved the association problem for noise-free learning vectors.

TABLE 6.5. Result of operation of Hopfield neural network for noise-free patterns

Input	Output
$-1 - 11 - 1 - 11 - 1 - 11 - 1 - 11$	$-1 - 11 - 1 - 11 - 1 - 11 - 1 - 11$
$1 - 11111 - 1 - 11 - 1 - 11$	$1 - 11111 - 1 - 11 - 1 - 11$
$111 - 1 - 11 - 1 - 11 - 1 - 11$	$111 - 1 - 11 - 1 - 11 - 1 - 11$

Now we will distort the signals (Fig. 6.29) and will check out the network answer. In case of noisy signals, the input signals $\mathbf{x}_z$ have the form

$$\mathbf{x}_z(1) = [-1 - 11 - 1 - 11 - 111 - 1 - 11]$$
$$\mathbf{x}_z(2) = [1 - 111 - 11 - 1 - 11 - 1 - 11]$$
$$\mathbf{x}_z(3) = [1 - 11 - 1 - 11 - 1 - 11 - 1 - 11]$$

The effect of noisy signals is presented in Table 6.6.

For the third noisy vector the network output did not stabilize even after 12 iterations, which means that the network could not recognize the distorted sample.

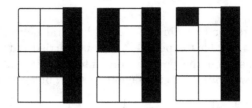

FIGURE 6.29. Distorted patterns in Example 6.6

TABLE 6.6. Result of operation of Hopfield neural network for noise patterns

Input	Output	Number of iterations
$-1 - 11 - 1 - 11 - 111$ $-1 - 11$	$-1 - 11 - 1 - 11 - 1 - 11 - 1$ -11 (the network recognized the digit 1)	2
$1 - 111 - 11 - 1 - 11$ $-1 - 11$	$1 - 11111 - 1 - 11 - 1 - 11$ (the network recognized the digit 4)	2
$1 - 11 - 1 - 11 - 1$ $-11 - 1 - 11$	$-0.0571 \ 0.0571111 - 1 - 11$ $-1 - 11 - 1 - 11$	12

6.4.2 Hamming neural network

The structure of the Hamming neural network presented in Fig. 6.30 is a three-layer network used to classify images.

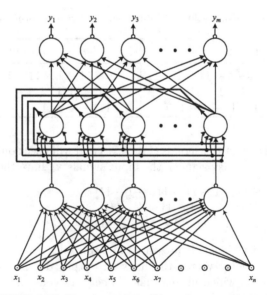

FIGURE 6.30. Structure of the Hamming neural network

In the first layer, there are p neurons, which determine the Hamming distance between the input vector and each of the p desired vectors coded in the weights of this layer. The second layer is called MAXNET. It is a layer corresponding to the Hopfield network the operation of which has been discussed earlier. However, in this layer feedbacks covering the same neuron are added. The weights in these feedbacks are equal to 1. The values of weights of other neurons of this layer are selected so that they inhibit the process, e.g. taking negative values. Thus, in the MAXNET layer there is the extinction of all outputs except the one which was the strongest in the first layer. The neuron of this layer, which is identified with the winner, through the weights of output neurons with a linear activation function, will retrieve the output vector associated with the vector coded in the first layer.

Example 6.7
In order to present the operation of the Hamming neural network, we have created a learning sequence in the form of a binary representation of the subsequent digits: 1, 2, 3 and 4 (Fig. 6.31a).

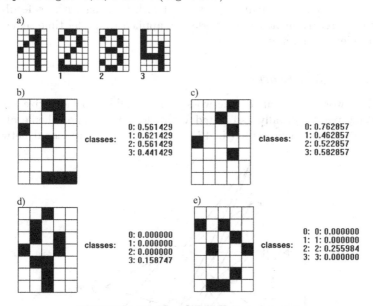

FIGURE 6.31. Illustration to Example 6.7

In the first phase the network learnt, using each of these patterns. Next, the vectors, which represented noisy output desired signals were applied at the input. The results have been presented in Fig. 6.31b, 6.31c, 6.31d, 6.31e. In the simulations, the NetLab [271] program was used, which numbers classes corresponding to digits 1 to 4 from 0 to 3. The numbers on the right side of the Fig. 6.31 mean the states of output neurons for each of the desired classes. As it may be noted, in all cases the network classified the noisy images correctly.

6.4.3 Multilayer neural networks with feedback

The literature describes different structures of multilayer neural networks with feedback. Most often, Elman neural networks and RTRN (*Real Time Recurrent Network*) networks are used. Figure 6.32 presents a scheme of *the Elman neural network*, which was named after its originator. In this structure, we may differentiate neurons with feedback which are contained in the hidden layers. Each of the neurons of the hidden layer processes the external input signals and as well as signals from feedback. The signals from the output layer are not subjected to the feedback operation.

Another well-known structure is *the RTRN network* presented in Fig. 6.33. Contrary to Elman neural network, the feedback connects both network output signals and the hidden neurons. The Elman neural network and RTRN network may learn using gradient algorithms, which take a more complex form than in case of network learning without feedback. Multilayer recurrent neural networks are applied to model time sequences and to identify dynamic objects.

6.4.4 BAM network

BAM network is a recurrent network, which enables to memorize the set of vector pairs mutually associated with each other. The scheme of this structure has been presented in Fig. 6.34. It is a bi-directional network.

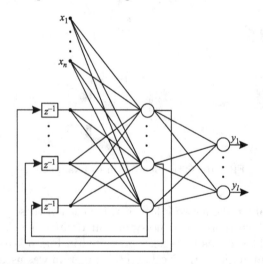

FIGURE 6.32. Scheme of the Elman neural network

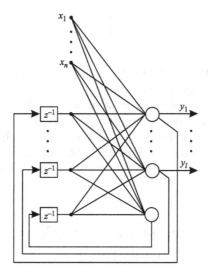

FIGURE 6.33. Scheme of the RTRN neural network

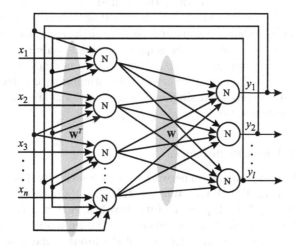

FIGURE 6.34. Scheme of the BAM neural network

Signals received at the neurons output of the first layer are transmitted through the weights making up the matrix $\mathbf{W}$ to the neurons inputs in the second layer. The neuron output signals of this layer are directed through the matrix $\mathbf{W}^T$ again to the inputs of the neurons in the first layer. This cycle is repeated until the equilibrium state of the network is obtained. The matrix of weights $\mathbf{W}$ is determined based on the following formula:

$$\mathbf{W} = \sum_{i=1}^{M} \mathbf{x}_i^T \mathbf{y}_i, \tag{6.180}$$

where M means the number of learning vectors, $\mathbf{x} = [x_i, ..., x_n]^T$ is the vector of the input signals, while $\mathbf{y} = [y_i, ..., y_l]^T$ is the vector of output signals. The activation functions in neurons are unipolar or bipolar. In the learning process, the learning unipolar vectors are changed into bipolar vectors, as the network which learns using bipolar signals shows better properties in retrieval mode. Moreover, different modifications of formula (6.180) are used in order to improve the operation of the BAM network in the retrieval phase.

6.5 Self-organizing neural networks with competitive learning

The networks presented in Subchapter 6.3 learn using algorithms with teacher. This means that we knew the desired answer of the structure to the input signal. Now, we will demonstrate self-organizing neural networks. Their learning is called *unsupervised learning* or *learning without teacher*. As we might expect, the learning sequence is made only of input values, without the desired output signal. Self-organizing neural networks have a simple structure, and no great knowledge of mathematics is sufficient to understand them. Despite their simplicity, these networks are widely applied, e.g. in data clustering tasks.

6.5.1 WTA neural networks

The first self-organizing neural network we will describe, will be the network which learns using the WTA (*Winner Takes All*) algorithm. Figure 6.35 presents a scheme of this type of neural network. Signal $\mathbf{x} = [x_1, x_2, ..., x_n]^T$ applied at the network input is directed to the inputs of all N neurons. Here, the similarity measure (distance) of the input signal $\mathbf{x}$ to all vectors of weights $\mathbf{w}_i = [w_{i1}, w_{i2}, ..., w_{in}]^T$, $i = 1, ..., N$, is determined. Most often the Euclidean measure is applied

$$d(\mathbf{x}, \mathbf{w}_i) = \|\mathbf{x} - \mathbf{w}_i\| = \sqrt{\sum_{j=1}^{n} (x_j - w_{ij})^2} \tag{6.181}$$

or the scalar product

$$d(\mathbf{x}, \mathbf{w}_i) = 1 - \mathbf{x} \circ \mathbf{w}_i = 1 - \|\mathbf{x}\| \, \|\mathbf{w}_i\| \cos(\mathbf{x}, \mathbf{w}_i). \tag{6.182}$$

The neuron characterized by the smallest distance of the vector of weights from the input signal takes the value 1 at its output, and all the remaining

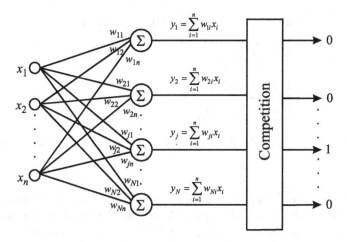

FIGURE 6.35. Scheme of the WTA neural network

neurons take the value 0 (hence the name: "winner takes all"). Therefore for the j-th neuron – the winner, we have

$$d(\mathbf{x}, \mathbf{w}_j) = \min_{1 \leq i \leq N} d(\mathbf{x}, \mathbf{w}_i), \qquad (6.183)$$

where d is one of the above similarity measures. In the learning process only the weights of the winner are modified, pursuant to the rule

$$\mathbf{w}_j(t+1) = \mathbf{w}_j(t) + \eta(t)\left[\mathbf{x}(t) - \mathbf{w}_j(t)\right], \qquad (6.184)$$

where η is the parameter of the weight modification step. In self-organizing networks, it is recommended to perform the normalization of input signals (norm of vector $\mathbf{x}$ is equal to 1), as it will ensure a coherent partition of the data space. The normalization may be made by redefining the components of vector $\mathbf{x}$ as follows:

$$x_i' = \frac{x_i}{\sqrt{\sum_{i=1}^{n} x_i^2}}. \qquad (6.185)$$

The normalization of the learning vector $\mathbf{x}$ causes the normalization of the vector of weights during the learning process. Moreover, the operation of the neuron will not be impacted by the length of the vector of weights, but only by the cosine of the angle between the vector of weights and the input vector $\mathbf{x}$ (Euclidean measure and the scalar product are equal in this case).

Figure 6.36a presents the vectors of weights and normalized inputs while Fig. 6.36b presents vectors without normalization with the same directions. It is easy to note that if the vectors are standardized, then the scalar product $\mathbf{x} \circ \mathbf{w}_1 = \|\mathbf{x}\| \, \|\mathbf{w}_1\| \cos \alpha$ is bigger than the scalar product of vectors $\mathbf{x}$ and $\mathbf{w}_2$ (because $\cos \alpha$ increases when the angle α decreases). In Fig. 6.36b also the scalar product $\mathbf{x}$ and $\mathbf{w}_1$ is bigger than $\mathbf{x}$ and $\mathbf{w}_2$, even though when watching the scheme, we have the impression that vectors $\mathbf{x}$ and $\mathbf{w}_2$ are more "similar to each other".

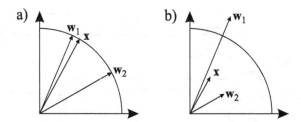

FIGURE 6.36. a) Normalized vectors of weights and inputs; b) vectors without normalization

The operation of a self-organizing neural network comes down to dividing M learning signals into N classes (N means the number of neurons in the network). Following learning, each neuron represents a different class, the vectors of weights $\mathbf{w}$ become centers of classes discriminated by the structure we trained. We may attempt to define the error measure of the classification for self-organizing neural network. We may state that the task of the network is such a selection of values of weights so that the distance of the input vector $\mathbf{x}$, which belongs to i-th class (i-th neuron should be the winner), from the i-th vector of weights would be as small as possible. Generally speaking: we should modify the weights in such a way that for all M learning signals $\mathbf{x}$, their distances from appropriate mid-points of classes were as small as possible. Our task is to minimize the quantization error given by the formula

$$Q = \frac{1}{M} \sum_{t=1}^{M} \|\mathbf{w}^*(t) - \mathbf{x}(t)\|, \qquad (6.186)$$

where $\mathbf{w}^*(t)$ is the weight of the winner neuron when applying the vector $\mathbf{x}(t)$.

Example 6.8
We will apply the WTA algorithm to solve the problem of Iris flower. Based on its four features: length and width of the petal and length and width of the leaf, we will define to which of the three species it belongs. As we might guess, each input learning vector is made of four components ($n = 4$). It may be inferred that each neuron must have four inputs so also four weights. The number of differentiated species suggests that we need three neurons in the network ($N = 3$). Therefore, in the learning process we will select the values of 12 weights in total. We have 120 learning vectors in total.

Figure 6.37 presents the distribution of values of weights before learning and after learning for different combinations of features. Before learning, the initial values of weights in particular figures are equal. After completion of the learning process, the values of weights may be presented in the form of the following matrix:

$$\mathbf{W} = \begin{bmatrix} 58.3060 & 27.2904 & 43.2095 & 13.969 \\ 50.0810 & 34.3020 & 14.6303 & 2.4686 \\ 67.7765 & 30.3668 & 56.0270 & 20.0741 \end{bmatrix}.$$

Based on the distribution of values of features (Fig. 6.37), we may infer which values determine a given class. To test the neural network, we decided to apply 150 testing vectors (including 120 learning vectors). Each testing vector consisted of 4 input signals and the output desired signal informing to which species a given pattern belongs. The simulation results have been presented in Table 6.7.

TABLE 6.7. The testing results in Example 6.8

Sample number	1 ... 50	51 , 53	52, 54 ... 75	77 , 78	79 ... 100	101	102	103 ... 106	107	118, 114, 113	115	119	120, 122 ... 127, 128	121, 123 ... 126, 129 ... 138	139, 143	140, 142 ... 144 ... 150
Number of winner - the neuron	2	3	1	3	1	3	1	3	1	3	1	3	1	3	1	3
The value of the desired signal	1	2	2	2	2	3	3	3	3	3	3	3	3	3	3	3
Error Yes/No	N	Y	N	Y	N	N	Y	N	Y	N	Y	N	Y	N	Y	N

In the first row of the table, there are numbers of subsequent samples. The second row contains information which neuron won after presenting samples with specific numbers at the input of the self-organizing neural network. In the third row, we entered the information, included in the testing sequence, stating to which class a given sample belongs. The Reader, when analyzing data in the table, may find that the structure learnt does not operate correctly, as the numbers of the winning neurons and the numbers of classes to which a given sample should belong are not the same. In fact, it is not true. While teaching the network, we did not "tell" it which neuron corresponds to a given species (class). That is why neuron 2 should be treated as a representative of species 1, neuron 1 as representative of species 2, neuron three should be identified with species 3. Only now we can come to correct conclusions. Namely, the first species of the iris is recognized by the neural network correctly (for the first 50 samples the network

Matching	Distribution of neuron weights values before learning and values of learning data depending on matching of features weights are marked with crosses	Distribution of neuron weights values after learning (300 epochs) and values of learning data depending on matching of features weights are marked with crosses
(X axis) values of the first feature in relation to (Y axis) values of the second feature		
(X axis) values of the second feature in relation to (Y axis) values of the third feature		
(X axis) values of the third feature in relation to (Y axis) values of the fourth feature		
(X axis) values of the fourth feature in relation to (Y axis) values of the first feature		

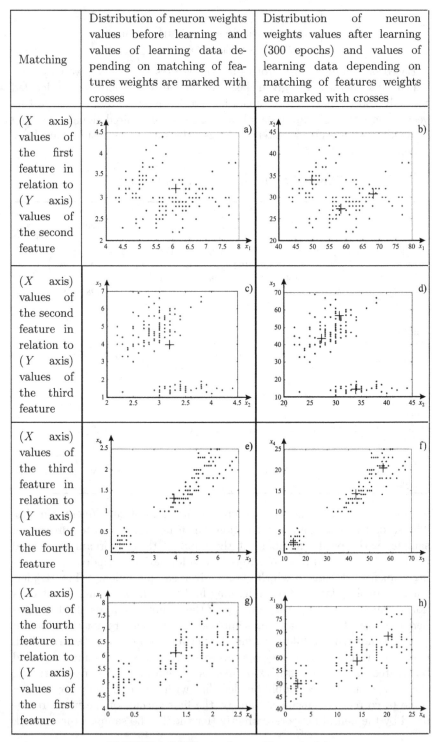

FIGURE 6.37. Illustration to Example 6.8

indicates the same class – neuron 2 wins), whereas errors occur in case of species two and three.

Example 6.9

Another problem is the task of recognizing a circle and a square. At first, we defined a circle on a plane and then we set up initial values of the weights of ten neurons in its center. The learning sequence has been generated based on the selected points located on the circle. During 250 epochs of operating the WTA algorithm ($n = 2$, $N = 10$), every 25 epochs we drew the relocation of weights. As a result the values of weights of neurons moved to the boundary of the circle, creating "paths" shown in Fig. 6.38.

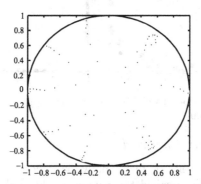

FIGURE 6.38. Relocation of weights during learning process (experiment with a circle) in Example 6.9

The experiment for a square was planned analogically. In Fig. 6.39, we may observe changes of the values of weights which occurred in the learning process every 40 epochs.

When initial values of weights of networks are selected (initialized) at random, it may happen that some neurons will never win during competition. The distance of their weights from learning samples and other neurons becomes too large, that is why these weights will not be subject to adaptation. Such units are called *dead neurons*. In order to increase their chances in the competition, the notion of potential p_i is introduced as follows:

$$p_i(t+1) = \begin{cases} p_i(t) + \frac{1}{N} & \text{for } i \neq j, \\ p_i(t) - p_{\min} & \text{for } i = j, \end{cases} \tag{6.187}$$

where j denotes the number of the winning neuron. If the value of the potential drops for the i-th neuron below $p_{\min}$, then this neuron does not participate in the competition and the winner is one neuron from the group with potential $p_i \geq p_{\min}$. It is assumed that the maximum value of the potential may amount to 1. If $p_{\min} = 0$ all the neurons take part in the competition. If $p_{\min} = 1$, only one neuron takes part in the competition, as it has the potential allowing it to win.

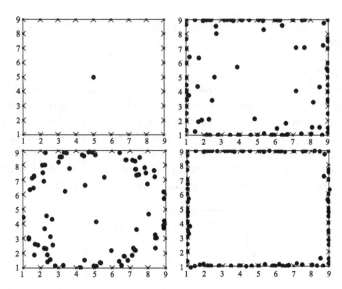

FIGURE 6.39. Relocation of weights during learning process (experiment with a square) in Example 6.9

6.5.2 WTM neural networks

So far, we considered self-organizing networks, where in the learning process only one neuron, the so-called the winner, could modify its weights. Now we will deal with algorithms, operation of which causes not only correction of the weight of the winner, but also the weights of the neurons in its neighborhood. These algorithms are called WTM (*Winner Takes Most*) *algorithms*. The first of them will be *the algorithm of neuronal gas*. The correction of weights in this case is very similar to the WTA method. At first, we calculate the distance of the input vector from the vectors of all weights in the network. In case of the WTA method, we modified the weights of the neuron which was the closest to the given input according to dependency (6.184). In the algorithm of "neuronal gas", we sort in ascending order all the vectors of neuron weights depending on their distance from the input signal. The smaller the distance, the bigger change of the value of neuron weights. Let $m(i)$ means the order of i-th neuron obtained in the process of sorting; for the j neuron (winner) we have $m(j) = 0$, and for the most distant neuron $m(i) = N - 1$. Let us therefore introduce the neighborhood function for i-th neuron as follows:

$$G(i, \mathbf{x}) = \exp\left(-\frac{m(i)}{R}\right),\qquad(6.188)$$

where R is the neighborhood radius. The weights of neurons are modified according to the dependency

$$\mathbf{w}_i(t + 1) = \mathbf{w}_i(t) + \eta G(i, \mathbf{x})[\mathbf{x} - \mathbf{w}_i],\qquad(6.189)$$

where $\mathbf{w}_i$ means the vector of weights in i-th neuron, t-number of iteration step, η-learning coefficient. In formula (6.188), in case where $R \to 0$, only the winner modifies its weights (we then get the WTA algorithm). Whereas for $R > 0$ the weights of all neurons are updated, but the learning coefficient of distant neurons quickly goes to zero. To get good results, at the beginning of the algorithm a high R value must be assumed and then it should be decreased as the number of its iterations increases.

Another WTM algorithm is the classic *Kohonen algorithm*. In this method, the neighborhood is introduced for good, i.e. it is defined which neurons are connected with each other, creating a net. Figure 6.40 presents some examples which may be defined in Matlab environment.

NAME OF THE NET AND ITS SIZE	SCHEMA
GRIDTOP 12 × 12	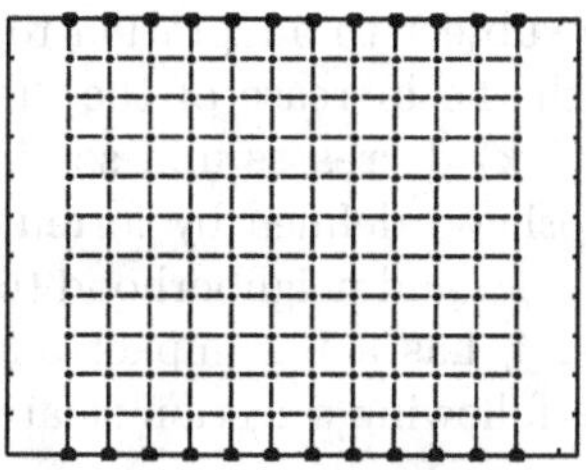
NEXTOP 12 × 12	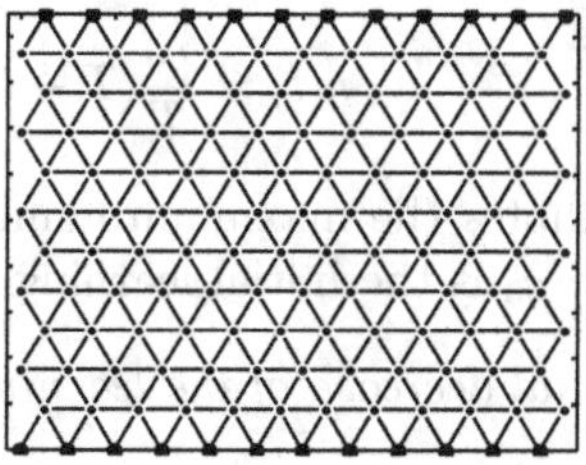
RANDTOP 12 × 12	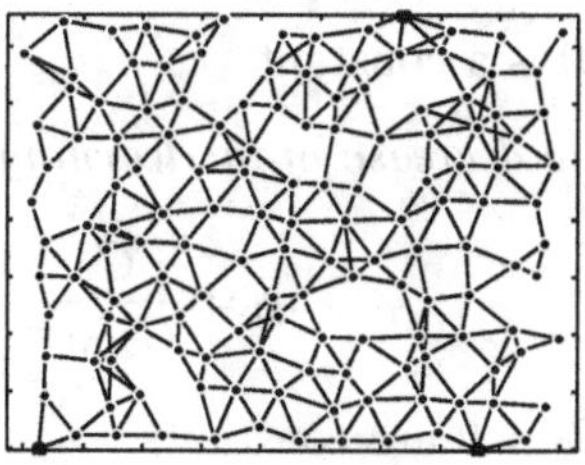

FIGURE 6.40. Examples of different neighborhood topologies (dots denote location of neurons)

In this algorithm, distances of all neurons from the input signal are calculated. Next, the winner is selected (the smallest distance) and its weights and the weights of its neighbors are modified according to dependency (6.189), where

$$G(i, \mathbf{x}) = \begin{cases} 1 & \text{for } d(i, j) \leq R, \\ 0 & \text{for others.} \end{cases} \tag{6.190}$$

In dependency (6.190) $d(i, j)$ means the Euclidean distance between the j neuron (winner) and i-th neuron from neighborhood G or the distance calculated as the number of neurons. The neighborhood radius R should decrease as the learning time increases. The size of correction often depends on the distance from the winner and also on the size of radius R, i.e.

$$G(i, \mathbf{x}) = \exp(-\frac{d^2(i, j)}{2R^2}), \tag{6.191}$$

where j is the number of the winning neuron. In such case, particular neurons are subject to adaptation to various degrees. As it can be noted, together with the increase of the distance from the winner, the value of function $G(i, \mathbf{x})$ decreases and so the value of the correction of weights. The neighborhood defined by formula (6.190) is called *neighborhood of a rectangular type*, and neighborhood (6.191) *of the Gaussian type*. The value of parameter η has a big impact on the learning process. The literature contains the following selection strategies of this parameter:

a) *Linear decrease of the learning coefficient*

$$\eta(t) = \frac{\eta_0}{T}(T - t), \quad t = 1, 2, ...,$$

where T denotes the maximum number of iterations of the learning algorithm, while η_0 is the initial learning coefficient.

b) *Exponential decrease of the learning coefficient*

$$\eta(t) = \eta_0 e^{-Ct}, \quad t = 1, 2, ...,$$

where $C > 0$ is a constant.

c) *Hyperbolic decrease of the learning coefficient*

$$\eta_0 = \frac{C_1}{C_2 + t}, \quad t = 1, 2, ...,$$

where $C_1, C_2 > 0$ are constants.

Example 6.10

We will now present an example of application of the Kohonen network which is used to recognize a circle with different topologies of networks. On a plane we have defined a circle. Our task is to modify the weights which are located in the middle of the circle when the algorithm starts, in such a way that they would be on its boundary after completion of the learning process. Figure 6.41 presents the location of the neuron weights after the learning process for different network topologies, while $n = 2$, $N = 25$.

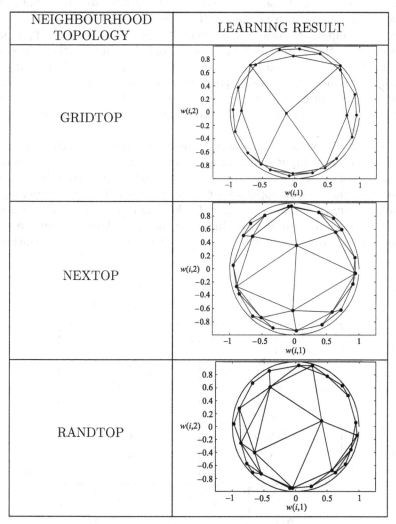

NEIGHBOURHOOD TOPOLOGY	LEARNING RESULT
GRIDTOP	
NEXTOP	
RANDTOP	

FIGURE 6.41. Distribution of neuron weights in Example 6.10 after the learning process for different network topologies

6.6 ART neural networks

Adaptive Resonance Theory, in short ART, relates to learning of neural networks without a teacher. The objective of network operation is the competitive recognition of binary images memorized earlier. Moreover, this network should be able to learn to differentiate the unknown shapes, if there is such a need. The new patterns are memorized in the network when the degree of similarity to the currently stored images is too small. It is determined by the so-called *vigilance parameter* τ. There is a network equivalent for continuous images, the so-called ART2 modified later to ART3, both are suitable to classify binary images. ART network consists of two layers. The first layer, input, consists of n neurons. Its task is to compare the input images with the ones stored in memory and determining the degree of similarity. It is also called *the comparising layer*. The second layer, output, consists of m neurons and is supposed to recognize an input shape, in other words, to recognize the class to which the image belongs. Both layers interact and work out the final decision on recognition of the image or learning a new one.

Figure 6.42 presents a schematic connection structure of ART network. Let us denote the input signal (shape) vector as $\mathbf{x} = [x_1, \ldots, x_n]^T$ and upper layer output signals vector as $\mathbf{y} = [y_1, \ldots, y_l]^T$. We denote the weights of connections between upper layer neuron outputs and inputs of lower layer neurons by $v_{ij}, i = 1, .., n, j = 1, \ldots, l$. The weights of connections between lower layer neuron outputs and inputs of the upper layer neurons are denoted by w_{ij}. It is therefore characteristic that two types of inter-neuron

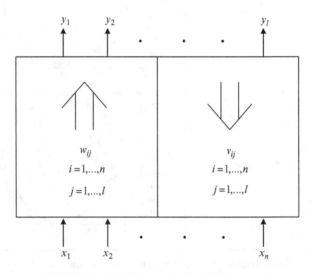

FIGURE 6.42. Scheme of the ART network

connections of both layers exist, "from bottom up" (weights w_{ij}) and "from top down" (weights v_{ij}). Now, we will present the learning algorithm of ART1 network. In this algorithm two following steps may be distinguished:

Step 1. The operation of the algorithm is started from initialization of the values of weights in both sets of connections. Usually for the initial values of weights of top-down connections we have $v_{ij}(0) = 1$, and for bottom-up connections we take $w_{ij}(0) = \frac{1}{1+n}$, where n is the number of neurons of the input layer. As the vigilance parameter τ, a small positive number from the interval $(0, 1)$ is assumed.

Step 2. In this step after applying an input vector $\mathbf{x} = [x_1, \ldots, x_n]^T$, where $x_i \in \{0, 1\}, i = 1, 2, \ldots n$, the sum $\sum_{i=1}^{n} w_{ij} x_i$, $j = 1, \ldots, l$ is calculated. This sum is treated as the measure of image adjustment to the currently memorized patterns. Its maximum value decides which neuron of the output layer (let us denote its number as j^*) will take the value 1 or will become the winner of the competition. At the same time it indicates the class to which the recognized shape is classified. The remaining neurons take the value 0 at their outputs. It is a normal phase of network operation recognizing a previously memorized image.

Step 3. After the winner neuron indicates the class to which the investigated image fits best, it is necessary to determine how similar it is to the vectors earlier assigned to this class. It is checked whether

$$D = \frac{\sum_{i=1}^{n} v_{ij^*} x_i}{\sum_{i=1}^{n} x_i} > \tau. \tag{6.192}$$

If not, then the output of the current winner neuron is zeroed and the most stimulated neuron from among other neurons is considered to be the winner. We are coming next to step 3. If as a result of this operation (which may be repeated many times) a positive result is not achieved, then the studied image must be categorized as belonging to a new, yet unknown class, assigning it the first, free so far neuron of the output layer. It the test of degree of similarity is positive, we say that the network and the input signal are in "resonance", which justifies the name ART. The selection of a low value of the vigilance parameter τ (i.e. close to zero) causes that the meeting of condition (6.192) is possible even with a relatively small number of the same pixels of the pattern and of the image recognized. In this case, the network classifies the images, sometimes not very similar, to a small number of possible classes. One could say that the network then classifies the images only "roughly". If the vigilance parameter is closer to 1, then many categories are created with quite subtle differences between them and quite a precise recognition of images by the network takes place.

Step 4. After selecting the winning neuron, the correction of related weights is made, both for input weights w_{ij^*} and v_{ij^*} according to formulas

$$v_{ij*}(t+1) = v_{ij*}(t)\, x_i, \tag{6.193}$$

$$w_{ij*}(t+1) = \frac{v_{ij*}(t+1)}{0.5 + \sum_{j=1}^{n} v_{ij*}(t+1)\, x_i}. \tag{6.194}$$

In this step the network is "tuned", which should be understood as adaptation of the stored pattern to the currently recognized shape. It may be achieved by a specific superposition of both shapes and aims to achieve their binary compliance. The weights of connections for the same bits are reinforced and for others weakened or left unchanged.

Step 5. Processing of an input signal by the ART network is stopped when the values of weights are not changed or for neither of neurons condition (6.192) is met and there is no "free" neuron to which a new class could be assigned in the first layer.

ART network uses two sets of weights which in fact decide on the network memory, i.e. its possibility to recognize. The set of weights "from bottom up" is responsible for the so-called *long-term memory* and the set of weights "from top down" for the so-called *short-term memory*. It should be noted that due to the method of "tuning" (modifications of weights in step 4 of the algorithm) and a relatively uncomplicated test of similarity (6.192), the network does not well recognize images distorted by interferences or the noisy ones. In these conditions, only a rough recognition is possible, i.e. the selection of a small vigilance parameter τ.

Example 6.11

In order to illustrate the operation of ART1 networks, we will consider the problem of classification of subsequent digits from 1 to 4. Their binary representation has been entered at the network input for different values of the vigilance parameter. Figure 6.43 presents the results of simulation, in which the sequence of shapes applied at the input was the following: 1, 3,

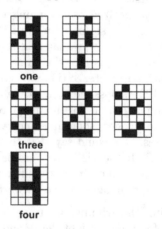

FIGURE 6.43. Results of simulations in Example 6.11: $\tau = 0.6$

2, 4, distorted digit 1 and distorted digit 3. The vigilance parameter was equal to 0.6, or we should say "average". As it may be seen, the network did not differentiate the shapes 3 and 2 and classified them as identical (identified with pattern 3). On the other hand the distorted digit 1 was correctly classified as class of pattern 1, and the distorted digit 3 also correctly, i.e. to the class of pattern 3.

Figure 6.44 illustrates a situation where the vigilance parameter of the network was increased to $\tau = 0.75$. The sequence of applied images was as before. In this case the distorted digit 1 was deemed to be a new (fourth) pattern.

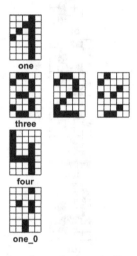

FIGURE 6.44. Results of simulations in Example 6.11: $\tau = 0.75$

Figure 6.45 illustrates the result of simulation where the vigilance parameter is 0.4 (is "small"). Then the digits 1 and 4 belong to the same class (which means that they are not differentiated by the network), patterns 2,

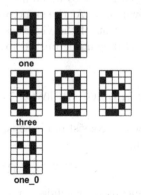

FIGURE 6.45. Results of simulations in Example 6.11: $\tau = 0.4$

3 and the distorted number 3, like previously, were qualified to class two, and the distorted digit 1 is a pattern of a new class. Figure 6.46 illustrates the case for vigilance parameter equal to 0.9 ("very big"). Currently each shape of the sequence studied is a pattern of a separate class, but providing the distorted digit 3 triggers a message on the impossibility to create a new class, which means that this shape does not "match" any of the earlier patterns.

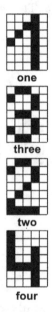

one

three

two

four

FIGURE 6.46. Results of simulations in Example 6.11: $\tau = 0.9$

The above example is only an illustration of the ART1 network properties, showing its very interesting possibilities of shapes recognition, learning without supervision, i.e. possibility to make independent decisions on creating of a new and unknown class. At the same, we may see how a difficult issue is the appropriate selection of the vigilance parameter which should be determined by an expert depending on the specific task.

6.7   Radial-basis function networks

Radial-basis function networks consist of neurons, of which activation functions perform the mapping

$$\mathbf{x} \to \varphi(\|\mathbf{x} - \mathbf{c}\|), \quad \mathbf{x} \in \mathbf{R}^n, \tag{6.195}$$

where $(\| \cdot \|)$ most often means an Euclidean norm. Functions $\varphi(\|\mathbf{x} - \mathbf{c}\|)$ are called *radial-basis functions*. Their values change radially around the

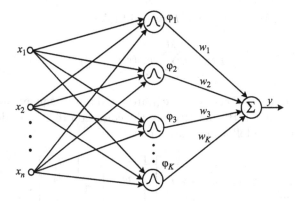

FIGURE 6.47. Scheme of the radial-basis neural network

center $\mathbf{c}$. Figure 6.47 presents a scheme of radial-basis neural network called an RBF (*Radial Basis Functions*) network. Input signals $x_1, x_2, \ldots, x_n$ making up the vector $\mathbf{x}$ are applied to each neuron in the hidden layer (identical to input layer). The neurons in the hidden layer satisfy mapping (6.195). The number of these neurons is equal to the number of vectors $\mathbf{x}$ in a learning set or it is smaller. The neuron in the output layer executes the operation of weighted sum of output signals of neurons in the hidden layer, which may be expressed using the following formula:

$$y = \sum_i w_i \varphi_i = \sum_i w_i \varphi(\|\mathbf{x} - \mathbf{c}_i\|).$$ (6.196)

Below, we will present some typical examples of radial basis functions:

$$\varphi(\|\mathbf{x} - \mathbf{c}_i\|) = \exp(-\|\mathbf{x} - \mathbf{c}\|^2 / r^2),$$ (6.197)

$$\varphi(\|\mathbf{x} - \mathbf{c}_i\|) = \|\mathbf{x} - \mathbf{c}\| / r^2,$$ (6.198)

$$\varphi(\|\mathbf{x} - \mathbf{c}_i\|) = (r^2 + \|\mathbf{x} - \mathbf{c}\|^2)^{-\alpha}, \quad \alpha > 0,$$ (6.199)

$$\varphi(\|\mathbf{x} - \mathbf{c}_i\|) = (r^2 + \|\mathbf{x} - \mathbf{c}\|^2)^\beta, \quad 0 < \beta < 1,$$ (6.200)

$$\varphi(\|\mathbf{x} - \mathbf{c}_i\|) = (r \|\mathbf{x} - \mathbf{c}\|)^2 \ln(r \|\mathbf{x} - \mathbf{c}\|),$$ (6.201)

while $r > 0$. In order to simplify the notation, we shall temporarily assume that the scalar parameters in formulas (6.197) – (6.201) are the same for all neurons in the network. The analysis of operation of a radial network includes two cases depending on the number of neurons in the hidden layer:

Case a. Let us assume that at the input of a network which solves interpolation problems M different input vectors $\mathbf{x}(1), \ldots, \mathbf{x}(M)$, were entered; the vectors are to be mapped into a set of real numbers $d(1), \ldots, d(M)$. The problem consists in finding such function F satisfying the mapping $\mathbf{R}^n \to \mathbf{R}$, so that the following equality occurs

$$F(\mathbf{x}_i) = d_i$$ (6.202)

for $i = 1, 2, \ldots, M$. As function F, we will take

$$F(\mathbf{x}) = \sum_{i=1}^{M} w_i \varphi(\|\mathbf{x} - \mathbf{c}_i\|). \tag{6.203}$$

Let us assume that centers $\mathbf{c}_1, \ldots, \mathbf{c}_M$ are equal to subsequent values of vectors $\mathbf{x}(1), \ldots, \mathbf{x}(M)$. By substituting the condition (6.202) to dependency (6.203), we get the following matrix equation:

$$\begin{bmatrix} \varphi_{11} & \varphi_{12} & \cdots & \varphi_{1M} \\ \varphi_{21} & \varphi_{22} & \cdots & \varphi_{2M} \\ \cdots & \cdots & \cdots & \cdots \\ \varphi_{M1} & \varphi_{M2} & \cdots & \varphi_{MM} \end{bmatrix} \begin{bmatrix} w_1 \\ w_2 \\ \cdots \\ w_M \end{bmatrix} = \begin{bmatrix} d_1 \\ d_2 \\ \cdots \\ d_M \end{bmatrix}, \tag{6.204}$$

where $\varphi_{ji} = \varphi(\|\mathbf{x}_j - \mathbf{x}_i\|)$. With notations

$$\mathbf{d} = [d_1, d_2, ..., d_M]^T, \tag{6.205}$$

$$\mathbf{w} = [w_1, w_2, ..., w_M]^T, \tag{6.206}$$

$$\Phi = \{\varphi_{ji} \mid j, i = 1, 2, ..., M\} \tag{6.207}$$

dependency (6.204) takes the form

$$\Phi \mathbf{w} = \mathbf{d}. \tag{6.208}$$

For a given class of radial basis functions $\varphi(\|\mathbf{x} - \mathbf{c}\|)$, e.g. for function (6.197) and (6.199), the matrix Φ is positive definite if the input vectors satisfy the condition

$$\mathbf{x}(1) \neq \mathbf{x}(2) \neq \cdots \mathbf{x}(M). \tag{6.209}$$

Then the solution of equation (6.208) takes the form

$$\mathbf{w} = \Phi^{-1}\mathbf{d}. \tag{6.210}$$

Case b. The assumption of a number of radial neurons equal to the number of learning signals causes the network to lose its generalization abilities. Moreover, with a large number of output desired signals, the network structure would have to grow to a huge size. Therefore we now assume that the radial structure satisfies the following mapping

$$F(\mathbf{x}) = \sum_{i=1}^{K} w_i \varphi \|\mathbf{x} - \mathbf{c}_i\|, \tag{6.211}$$

where $K < M$. In this way, we are searching for an approximated solution. We must appropriately select not only the weights $\mathbf{w}_i$, but also to find centres $\mathbf{c}_i$ for the radial neurons. The proper minimization criterion takes the form

$$E = \sum_{i=1}^{M} \left[\sum_{j=1}^{K} w_j \varphi(\|\mathbf{x}_i - \mathbf{c}_j\|) - d_i \right]^2. \tag{6.212}$$

Let us denote:

$$\mathbf{G} = \begin{bmatrix} \varphi(\|\mathbf{x}_1 - \mathbf{c}_1\|) & \varphi(\|\mathbf{x}_1 - \mathbf{c}_2\|) & \dots & \varphi(\|\mathbf{x}_1 - \mathbf{c}_K\|) \\ \varphi(\|\mathbf{x}_2 - \mathbf{c}_1\|) & \varphi(\|\mathbf{x}_2 - \mathbf{c}_2\|) & \dots & \varphi(\|\mathbf{x}_2 - \mathbf{c}_K\|) \\ \dots & \dots & \dots & \dots \\ \varphi(\|\mathbf{x}_M - \mathbf{c}_1\|) & \varphi(\|\mathbf{x}_M - \mathbf{c}_2\|) & \dots & \varphi(\|\mathbf{x}_M - \mathbf{c}_K\|) \end{bmatrix}, \tag{6.213}$$

$$\mathbf{d} = [d_1, d_2, ..., d_M]^T, \tag{6.214}$$

$$\mathbf{w} = [w_1, w_2, ..., w_M]^T. \tag{6.215}$$

The matrix $\mathbf{G}$ defined by (6.213) is the so-called *Green matrix*. By minimizing criterion (6.212) and taking into account formulas (6.213) – (6.215), we get

$$\mathbf{Gw} = d, \tag{6.216}$$

$$\mathbf{w} = \mathbf{G}^+ d, \tag{6.217}$$

where $\mathbf{G}^+$ denotes the pseudoinverse of the rectangular matrix $\mathbf{G}$, i.e.

$$\mathbf{G}^+ = (\mathbf{G}^T \mathbf{G})^{-1} \mathbf{G}^T. \tag{6.218}$$

Criterion (6.212) may be supplemented by the so-called *regularization term* which results from the Tichonov regularization method

$$E(f) = \sum_{i=1}^{M} \left[\sum_{j=1}^{K} w_j \varphi(\|\mathbf{x}_i - \mathbf{c}_j\|)^2 - d_i \right] + \lambda \|\mathbf{P}f\|^2 \tag{6.219}$$

$$= \|\mathbf{Gw} - \mathbf{d}\|^2 + \lambda \|\mathbf{P}f\|^2,$$

where $\mathbf{P}$ is a certain linear differential operator. In approximation tasks, the introduction of this operator is related to the assumption on the smoothness of the function which approximates the unknown solution. The right side of expression (6.219) may be presented in the form [73]

$$\|\mathbf{P}f\|^2 = \mathbf{w}^T \mathbf{G}_0 \mathbf{w}, \tag{6.220}$$

where matrix $\mathbf{G}_0$ is a square matrix, of the dimension $K \times K$, defined as follows:

$$\mathbf{G}_0 = \begin{bmatrix} \varphi(\|\mathbf{c}_1 - \mathbf{c}_1\|) & \varphi(\|\mathbf{c}_1 - \mathbf{c}_2\|) & \dots & \varphi(\|\mathbf{c}_1 - \mathbf{c}_K\|) \\ \varphi(\|\mathbf{c}_2 - \mathbf{c}_1\|) & \varphi(\|\mathbf{c}_2 - \mathbf{c}_2\|) & \dots & \varphi(\|\mathbf{c}_2 - \mathbf{c}_K\|) \\ \dots & \dots & \dots & \dots \\ \varphi(\|\mathbf{c}_K - \mathbf{c}_1\|) & \varphi(\|\mathbf{c}_K - \mathbf{c}_2\|) & \dots & \varphi(\|\mathbf{c}_K - \mathbf{c}_K\|) \end{bmatrix}. \tag{6.221}$$

By minimizing dependency (6.219) and using equations (6.220) and (6.221), we obtain the vector of weights **w** given by

$$\mathbf{w} = (\mathbf{G}^T\mathbf{G} + \lambda\mathbf{G}_0)^{-1}\mathbf{G}^T\mathbf{d}. \qquad (6.222)$$

It has been proved that the neural network with radial-basis functions based on Green functions is a universal approximator. The radial-basis function network learning algorithm consists of two stages:

1. At first, the location and the shape of base functions is selected by means of the following methods:
– random selection,
– selection with application of the self-organization process,
– selection using the error backpropagation method.

2. In the second phase, the matrix of weights of the output layer is selected, and this problem is much simpler to solve than the selection of the radial basis functions parameters. The matrix of weights **w** is determined in one step by pseudoinverse of Green matrix **G**, i.e. $\mathbf{w} = \mathbf{G}^+\mathbf{d}$.

Now we will present one of the selected methods of parameters selection using the self-organization process. The self-organization process divides the space of input signals into the so-called *Voronoi diagrams*. Each such diagram is called a *group* or *cluster*. It contains a central point which is the average of all elements in the group. At the same time, it is the centre of the radial-basis function. There are two versions of the algorithm:

1. direct – updating of centers is made after each presentation of vector **x** from the learning sequence;

2. cumulated – the update is made after all the learning vectors have been presented.

Also a variation of the direct version of the algorithm is used, in which the adaptation involves the centers from the nearest neighborhood of the winner. They are modified according to the WTM rule.

Another method of selecting parameters for the RBF network is applied in the Matlab software. In this environment, there are two methods used to create such a structure. In case of the first one, by means of NEWRBE instruction, for each learning vector $\mathbf{x}(1), \ldots, \mathbf{x}(M)$ we create a separate radial-basis neuron in the first layer. Then, weights for the neuron of the second layer are selected. It is a situation identical to the case discussed earlier and concerning the analysis of operation of a radial-basis function network. The second method, resulting from the operation of another instruction called NEWRB, triggers the following algorithm:

1. A two-layer network is created, with no neuron in the first layer.

2. At the network input, subsequent learning vectors are applied and for them the output error is calculated (for each learning sample the error is recorded separately).

3. The learning vector with the biggest error is chosen and the radial-basis neuron with weights equal to the components of this learning vector is added to the first layer.

4. Next, the weights of the linear neuron in the second layer are selected so as to minimize the network error.

5. Points 2 – 4 are repeated until the network error drops below the threshold set by the user as the NEWRB function parameter. Another parameter of the NEWRB method is the number of neurons above which the operation of the algorithm should be stopped. By default, this number is equal to the size of the learning sequence.

Example 6.12
Using the NEWRB instruction from Matlab package, we made an experiment consisting in the RBF network learning to map the function $f(x) = \sin(x)$, where $x \in [-10, 10]$. The learning sequence consists of points x generated by discretization of the interval $[-10, 10]$ with step 0.1 and the corresponding values of function $f(x)$. The error threshold, below which the network is deemed to be learnt, amounts to 0.000001. In Fig. 6.48, we present the operation of the structure created for a different number of neurons. The solid line represents the graph of the function $y = \sin(x)$. The symbol + denotes the network answers to the testing signals applied.

Radial-basis neural networks are applied in classification and approximation problems, as well as in prediction tasks. These are tasks in which sigmoidal neural networks have been applied for many years. However, the radial-basis structures utilize a different method of data processing, resulting in shortening of the learning process. When solving classification tasks, the radial-basis function network does not only provide information about the pattern class, but it also indicates the possibility of creation of a new class. A significant advantage is a much simplified network learning algorithm. The starting point may be selected so as to locate it much closer to the optimum solution than in the case of sigmoidal neural networks. The basic differences between the radial-basis and the sigmoidal neural networks are:

1. Radial-basis function neural networks have a pre-defined architecture consisting of two layers while sigmoidal neural networks may consist of any number of layers.

NUMBER OF NEURONS	ERROR	GRAPH
2	65.3489	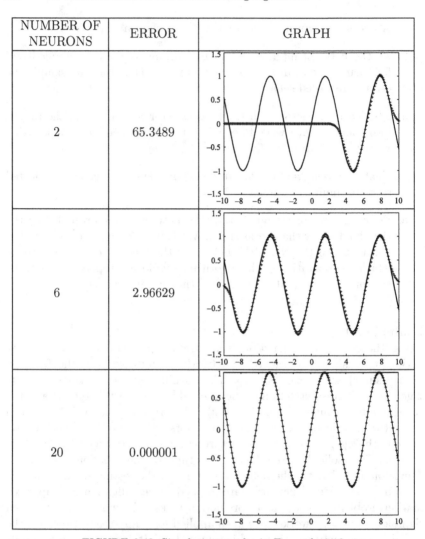
6	2.96629	
20	0.000001	

FIGURE 6.48. Simulation results in Example 6.12

2. Radial-basis function networks may apply any base functions in the hidden layer, while a multi-layer network most often applies sigmoidal functions.

3. In radial-basis function neural networks, different learning techniques may be used for both layers, e.g. the first (hidden) layer may learn using the gradient method or self-organization method, and the second layer the pseudoinverse method. In case of multilayer sigmoidal neural networks, neurons of all layers most often learn using the backpropagation method.

6.8   Probabilistic neural networks

In many identification, classification and prediction problems, there is a
need to estimate the probability density function. In order to estimate this
function, we may apply the estimator

$$\widehat{f}_M\left(\mathbf{x}\right) = \frac{1}{M} \sum_{i=1}^{M} K_M\left(\mathbf{x}, \mathbf{X}_i\right), \tag{6.223}$$

where $\mathbf{X}_1, ..., \mathbf{X}_M$ is a sequence of observation of an n-dimensional random
variable $\mathbf{X}$ with probability density f, while K_M is an appropriately selected kernel. Figure 6.49 presents a network realization of the estimator
(6.223).
It should be stressed that the proposed network does not require the learning process (optimum selection of the connections weights), as the role of
weights is played by subsequent components of observation vectors $\mathbf{X}_i$. As
function K_M the so-called *Parzen kernel* may be assumed, in the following
form

$$K_M\left(\mathbf{x}, \mathbf{u}\right) = h_M^{-n}\, K\left(\frac{\mathbf{x} - \mathbf{u}}{h_M}\right), \tag{6.224}$$

while the sequence h_M is a function of the length of the learning sequence
M and should meet the conditions

$$\lim_{M \to \infty} h_M = 0 \quad \text{and} \quad \lim_{M \to \infty} M h_M^n = \infty. \tag{6.225}$$

It may be demonstrated (Cacoullos [20]) that

$$E\left[\widehat{f}_M\left(\mathbf{x}\right) - f_M\left(\mathbf{x}\right)\right]^2 \xrightarrow{M} 0 \tag{6.226}$$

in continuity points of density f. Function K in formula (6.224) may be
presented in the form

$$K\left(\mathbf{x}\right) = \prod_{i=1}^{n} H\left(x^{(i)}\right). \tag{6.227}$$

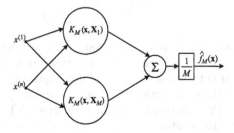

FIGURE 6.49. Probabilistic neural network for density estimation

Assuming that the function H is of Gaussian type, we have

$$\widehat{f}_M(\mathbf{x}) = \frac{1}{(2\pi)^{\frac{n}{2}} nh_M^n} \sum_{i=1}^{M} \exp\left(-\frac{(\mathbf{x} - \mathbf{X}_i)^T(\mathbf{x} - \mathbf{X}_i)}{2h_M^2}\right). \tag{6.228}$$

In analogical way, we may construct a probabilistic neural network in order to estimate the regression function. Let $(\mathbf{X}, Y)$ be a pair of random variables. Let us assume that $\mathbf{X}$ takes the values in set R^n, while Y in set R. Let f be the probability density function of the random variable $\mathbf{X}$. Based on M independent observations $(\mathbf{X}_1, Y_1), ..., (\mathbf{X}_M, Y_M)$ of variables $(\mathbf{X}, Y)$, we should estimate the regression function R of the random variable Y with respect to $\mathbf{X}$, i.e.

$$\phi(\mathbf{x}) = E[Y \mid \mathbf{X} = \mathbf{x}]. \tag{6.229}$$

Let us define the function

$$R(\mathbf{x}) = \phi(\mathbf{x}) \cdot f(\mathbf{x}). \tag{6.230}$$

To estimate function (6.230), we may apply an estimator similar to procedure (6.223), i.e.

$$\widehat{R}_M(\mathbf{x}) = \frac{1}{M} \sum_{i=1}^{M} Y_i K_M(\mathbf{x}, \mathbf{X}_i). \tag{6.231}$$

Therefore the regression function shall be estimated using

$$\widehat{\phi}_M(\mathbf{x}) = \frac{\widehat{R}_M(\mathbf{x})}{\widehat{f}_M(\mathbf{x})}. \tag{6.232}$$

In consequence, we get the following estimator:

$$\widehat{\phi}_M(\mathbf{x}) = \frac{\sum_{i=1}^{M} Y_i K\left(\frac{\mathbf{x} - \mathbf{X}_i}{h_M}\right)}{\sum_{i=1}^{M} K\left(\frac{\mathbf{x} - \mathbf{X}_i}{h_M}\right)}. \tag{6.233}$$

Figure 6.50 presents the neural realization of the regression function estimator referring in its structure to the structure given in Fig. 6.49. The probabilistic neural networks structures do not require learning. Moreover, the application of these networks with an appropriately selected h_M sequence guarantees the convergence of the estimators. These are, however, asymptotic results. In practice, we must ensure that we have learning sequences of a significant length. Both structures may be used to solve the classification problem and then we achieve a convergence to the Bayes rule.

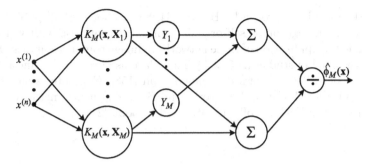

FIGURE 6.50. Probabilistic neural network for regression estimation

6.9 Notes

Knowledge concerning biological neuron contributed to creating its mathematical model. This domain was pioneered by American scientists
McCulloch and Pitts, who in 1943 created the first neuron model [133].
Their idea was developed until 1969, when Minsky and Papert published
book [138] stressing the limited abilities of artificial neural networks with
one-layer and at the same time demonstrating the lack of algorithms for
learning of multilayer neural networks. This book caused a suspension of
research on neural networks. By this time however, many publications in
this scope saw light worldwide. Among others, Donald Hebb developed a
network learning rule called today the "Hebb rule" [74], Rosenblatt built
the perceptron [181], while Widrow constructed the model of neuron called
Adaline [257 – 259]. The research on neural networks was restarted as a
result of publication in 1986 of work [183]. This study presented a description of learning method of multilayer neural network which was called the
"error backpropagation" method. This fact caused a return to neural networks. Still today, they are studied by dozens of thousands of researchers
worldwide. Probabilistic neural networks have been proposed by Specht
[236, 237]. Their concept is derived from non-parametrical methods of estimation of density and regression functions [20, 69, 157, 159, 188 – 203].
Probabilistic neural networks are also applied in case of non-stationary
distributions of probabilities [221 – 224]. The RLS algorithm of neural networks learning has been explained in study [11]. Among many monographs
and handbooks on neural network, several may be referenced [12, 26, 47,
51, 72, 73, 76, 86, 91, 93, 96, 115, 117, 121, 123, 131, 155, 156, 178, 204,
241, 242, 244, 270, 271]. The issue of neural networks has been the subject of many conferences organized by Polish Neural Network Society e.g.:
[206, 209, 219, 226, 243]. The Levenberg-Marquardt algorithm has been
explained in study [68]. Gradient optimization methods are discussed in
detail in monographs [24, 53, 108]. Elman and RTRN neural networks have
been presented in works [49] and [52]. Approximation properties of neural

networks have been proven by Hornik [84, 85]. Kohonen [113] published a well-known monograph on self-organizing networks with competitive learning, while the application of these networks in image compression tasks have been presented in studies [205, 207]. The relations of rough sets and neural networks have been discussed in monograph [158]. Computer simulations concerning the ART neural networks and Hamming neural networks were made using the NetLab software attached to book [271].

7
Evolutionary algorithms

7.1 Introduction

The beginning of research into evolutionary algorithms was inspired by the imitation of nature. All the living organisms live in certain environment. They have a specific genetic material containing information about them and allowing them to transfer their features to new generations. During reproduction, a new organism is created, which takes certain features after its parents. These features are coded in genes, and these are stored in chromosomes, which in turn constitute genetic material – genotype. During the transfer of features, genes become modified. Then the crossover of different paternal and maternal chromosomes occurs. Mutation often occurs additionally, which is the exchange of single genes in a chromosome. An organism is created which differs from that of its parents and contains genes of its predecessors but also has certain features specific to itself. This organism starts to live in a given environment. If it turns out that it is well fit to the environment, in other words – if the combination of genes turns out to be advantageous – it will transfer its genetic material to its offspring. The individual that is poorly fit to the environment will find it difficult to live in this environment and transfer its genes to subsequent generations.

The presented idea has been applied to solve optimization problems. It turns out that an analogous approach to numerical calculations can be proposed – using so-called *evolutionary algorithms*. The environment is defined upon the basis of the solved problem. A population of individuals constituting potential solutions of a given problem lives in this environment.

With the use of appropriately defined fitness function, we check to what extent they are adapted to the environment. Individuals exchange genetic material with each other, crossover and mutation operators are introduced in order to generate new solutions. Among potential solutions, only the best fit ones "survive".

This chapter will discuss the family of evolutionary algorithms, i.e. the classical genetic algorithm, evolution strategies, evolutionary programming, and genetic programming. We are also going to present advanced techniques used in evolutionary algorithms. The second part of the chapter will discuss connections between evolutionary techniques and neural networks and fuzzy systems.

7.2   Optimization problems and evolutionary algorithms

Literature [63] lists three types of methods of search of optimum solutions. They include analytical methods, enumerative methods, and random methods. *The analytical methods* comprise two classes: indirect methods and direct methods. In indirect methods, we search for local function extrema, solving the system of equations (usually nonlinear). We obtain this system as a result of equating the objective function gradient to zero. Direct methods search for a local optimum through "jumping" on the function graph towards the direction specified by the gradient. Both methods are not free from disadvantages. First of all, they have a local scope, because they search for optimum solutions in the neighborhood of a given point. Their application depends on the existence of derivatives. In practice, many of the solved problems have discontinued functions in the complicated space of solutions. Thus, analytical methods have a limited scope of application.

Enumerative methods exist in many forms. Let us assume that we have a finite search space. The simplest method would consist in calculating the objective function value and reviewing all points of space one after another. In spite of its simplicity and similarity to human reasoning, this method has one serious disadvantage – ineffectiveness. Many problems have such a large search space that it is impossible to search all the points within a reasonable time-limit.

The last one of the search method is *the random method*. It became popular at the moment when scientists became aware of the weaknesses of analytical and enumerative methods. Random search algorithms, which consisted in random space searching and remembering the best solution, also turned out to be ineffective. However, they should be distinguished from techniques based on the pseudorandom numbers and evolutionary search of the space of solutions. Evolutionary algorithms are an example of the approach where the random selection is only a tool for supporting search in the coded space of solutions.

The evolutionary algorithm is a method of solving problems – mainly optimization problems – that is based on natural evolution. Evolutionary algorithms are search procedures based on the natural selection and inheritance mechanisms. They apply the evolutionary principle of the survival of the individuals that are the best fit. They are different from traditional optimization methods by the following elements:

1. Evolutionary algorithms do not directly process the task parameters, but their coded form.

2. Evolutionary algorithms make a search starting from a population of points instead of a single point.

3. Evolutionary algorithms use only the objective function and not its derivatives or other supplementary information.

4. Evolutionary algorithms apply probabilistic selection rules instead of deterministic rules.

These four features, i.e. parameters coding, operation on populations, using the minimum information about the task, and randomized operations, provide for the robustness of the evolutionary algorithm and for its consequent advantage over the other above-mentioned techniques.

7.3 Type of algorithms classified as evolutionary algorithms

Evolutionary algorithms apply terms borrowed from genetics. For instance, we speak about *a population of individuals*, and basic terms are *gene, chromosome, genotype, phenotype*, and *allel*. The terms corresponding to them and coming from the technical vocabulary are also used, such as *chain, binary sequence*, and *structure*.

- **Population** is a set of *individuals* of a specified size.

- **Individuals** of a population in genetic algorithms are sets of *task parameters* coded in the form of chromosomes, which means solutions otherwise called *search space points. Individuals* are sometimes called *organisms*.

- **Chromosomes** – otherwise *chains* or *code sequences* – are ordered sequences of *genes*.

- **Gene** – also called *a feature, sign, or detector* – constitutes a single element of *the genotype*, of the chromosome in particular.

- **Genotype** – otherwise *structure* – is a set of chromosomes of a given individual. Thus, individuals of a population may be *genotypes* or single chromosomes (if a genotype consists of only one chromosome, and such is often the assumption).

- **Phenotype** is a set of values corresponding to a given genotype, which is a *decoded structure*, and thus, *a set of task parameters (a solution, search space point)*.

- **Allel** is the value of a given *gene*, also specified as *the feature value* or *the feature variant*.

- **Locus** is a *position* indicating the place of the location of a given gene in *the chain*, that is in the chromosome (its plural form, that means "positions", is *loci*).

A very important notion in genetic algorithms is *the fitness function* otherwise called the *adaptation function* or *evaluation function*. It constitutes the measure of fitness (adaptation) of a given individual in the population. This function is extremely important, because it allows to evaluate the degree of fitness of particular individuals in a population, and based on this degree select the individuals that are the best fit (that is, having the highest fitness function), in accordance with the evolutionary principle of the survival of "the strongest" (the best fit) ones. The fitness function also takes its name directly from genetics. It has a strong impact on the operation of evolutionary algorithms and must be appropriately defined. In optimization related questions, the fitness function is usually an optimized function (strictly speaking, a maximized function) called *the objective function*. In minimization issues the objective function is transformed and the problem is reduced to the maximization issue. In the control theory the fitness function can be *the error function*, and in the theory of games – *the cost function*. In an evolutionary algorithm, in each of its iterations the fitness of each individual of a given population is assessed by the fitness function, and upon this basis a new population of individuals is created, which individuals constitute a set of potential solutions of the problems, e.g. optimization tasks.

Another iteration in the evolutionary algorithm is called *generation,* and a newly created population of individuals is also called *the new generation* or *the offspring generation*.

7.3.1 Classical genetic algorithm

The basic (classical) genetic algorithm, also called the elementary or simple genetic algorithm comprises the following steps:

1) initiation, which is the selection of the initial population of chromosomes,

2) evaluation of the fitness of chromosomes in the population,

3) checking the stopping criterion,

4) selection of chromosomes,

5) using genetic operators,

6) creating a new population,

7) presentation of the "best" chromosome.

The flowchart for the basic genetic algorithm is depicted in Fig. 7.1. Let us present particular components of this algorithm in more details.

Initiation, which is the creation of an initial population, consists of the random selection of the demanded number of chromosomes (individuals) represented by binary sequences having a determined length.

The evaluation of the fitness of chromosomes in a population consists in calculating the value of the fitness function for each chromosome of this population. The higher the value of this function, the better the "quality" of the chromosome. The form of the fitness function depends on the type

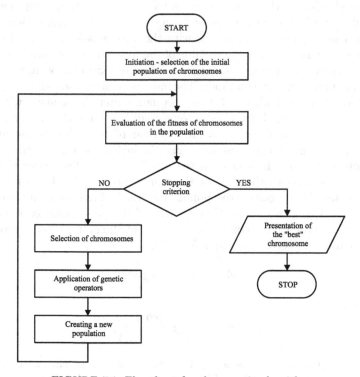

FIGURE 7.1. Flowchart for the genetic algorithm

of the solved problem. It is assumed that the fitness function always takes nonnegative values and, furthermore, the solved optimization problem is a problem of searching for the maximum of this function. If the initial form of the fitness function does not meet these assumptions then an appropriate transformation is made (e.g. the problem of searching for the minimum of the function may easily be reduced to the problem of searching for the maximum).

Checking the stopping criterion. The determination of the criterion for stopping the genetic algorithm depends on a specific application of this algorithm. In optimization issues, if the maximum (or minimum) value of the fitness function is known the stopping of the algorithm may occur after obtaining the desired optimum value, possibly with a specified accuracy. The stopping of the algorithm may also occur if its further operation no longer improves the best obtained value. The algorithm may also be stopped after the lapse of a determined period of time or after a determined number of generations. If the stopping criterion is met then the last step is taken, that is the presentation of the "best" chromosome. Otherwise the next step is selection.

The selection of chromosomes consists in selecting, based on the calculated values of the fitness function (step 2), these chromosomes which will take part in the creation of offspring until the next generation. This selection takes place in accordance with the natural selection rule, i.e. the chromosomes having the highest value of the fitness function have the most of the chances for the participation in the creation of new individuals. There are many selection methods. The most popular is a so-called roulette-wheel selection method, which takes its name after the analogy with the roulette wheel game. Each chromosome may be assigned a sector of the roulette wheel of a size that is proportional to the value of the fitness function of the given chromosome. Thus, the higher the value of the fitness function, the larger the sector on the roulette wheel. The entire roulette wheel corresponds to the sum of the fitness functions' values of all the chromosomes in the considered population. Each chromosome denoted by ch_i for $i = 1, 2, ..., K$, where K is the size of the population, corresponds to a wheel sector $\nu (ch_i)$ constituting a part of the entire wheel expressed in percentage, in accordance with formula:

$$\nu (ch_i) = p_s(ch_i) \cdot 100\%, \tag{7.1}$$

in which

$$p_s(ch_i) = \frac{F (ch_i)}{\sum_{j=1}^{K} F (ch_j)}, \tag{7.2}$$

while $F (ch_i)$ means the value of the fitness function of chromosome ch_i, and $p_s(ch_i)$ is the *probability of selecting* chromosome ch_i. The selection of a chromosome may be perceived as a turn of a roulette wheel, as a result of which the chromosome belonging to the roulette wheel sector drawn this way "wins" (is selected). Certainly, the larger the wheel sector, the higher the probability of the "victory" of a given chromosome. Thus, the probability of selecting a given chromosome grows in proportion to the growth of its fitness function. If we treat the entire circle of the roulette wheel as a numeral interval $[0, 100]$ then the drawing of a chromosome may be treated as drawing a number of range $[a, b]$ where a and b mean, respectively, the beginning and the end of the circle fragment corresponding to that wheel sector, certainly $0 \leq a < b \leq 100$. Then, selection with the use of the roulette wheel is reduced to drawing a number of range $[0, 100]$, which corresponds to a specific point on the circle of the roulette wheel. Other methods will be presented in point 7.4.2.

As a result of the selection process *the parents population* is created, also called as *the mating pool*, with the size equal to K, i.e. the same as the size of the current population.

Applying genetic operators to chromosomes selected with the selection method leads to the creation of a new population constituting the offspring population derived from the parents population.

In the classical genetic algorithm two basic genetic operators are used: *crossover operator* and *mutation operator*. It should, however, be emphasized that the mutation operator has a definitely secondary role in comparison to the crossover operator. This means that in the classical genetic algorithm, crossover is almost always present while mutation occurs quite rarely. The probability of crossover is usually assumed to be high (generally $0.5 \leq p_c \leq 1$), and, in turn, a very small probability of the occurrence of mutation is assumed (often $0 \leq p_m \leq 0.1$). This also results from the analogy to the world of living organisms, where mutations rarely occur.

In the genetic algorithm the mutation of a chromosome may be made on a parents population before the crossover operation, or on a population of offspring created as a result of the crossover.

Crossover operator. The first stage of crossover is the selection of pairs of chromosomes of the parents population (mating pool). This is a temporary population consisting of chromosomes selected with the selection method and intended for further processing with the crossover and mutation operators in order to create a new offspring population. At this stage the chromosomes of the parents population are mated in pairs. This is made randomly, in accordance with the probability of crossover p_k. Next, for each pair of the parents selected this way, the gene position in the chromosome (locus) is drawn, which specifies the so-called *crossover point*. If a chromosome of each of the parents consists of L genes then, certainly, the crossover point l_k is a natural number less than L. Therefore, the selection of the crossover point is reduced to drawing a number from interval $[1, L - 1]$. As a result of the crossover of a pair of parent chromosomes the following pair of offspring is created:

1) offspring whose chromosome consists of genes on positions from 1 to l_k coming from the first parent, and next genes, on positions from $l_k + 1$ to L coming from the second parent;

2) offspring whose chromosome consists of genes on positions from 1 to l_k coming from the second parent, and next genes, on positions from $l_k + 1$ to L coming from the first parent.

Example 7.1

Let us discuss two chromosomes $ch_1 = [1001001110]$ and $ch_2 = [1001111110]$ which undergo the crossover operation . In this case the chromosomes consist of 10 genes ($L = 10$), we therefore draw an integer number from interval $[1, 9]$. Let us assume that number 5 was drawn. The course of the crossover operation is as follows:

$$
\begin{array}{lcl}
\text{Pair of parents:} & & \text{Pair of offspring:} \\
ch_1 = [10010 \mid 01110] & \xrightarrow{\text{crossover}} & [10010 \mid \mathbf{11110}] \\
ch_2 = [10011 \mid 11110] & & [10011 \mid \mathbf{01110}]
\end{array}
$$

where symbol $\mid$ denominates the crossover point and the replaced genes are in bold.

The mutation operator, in accordance with the probability of mutation p_m replaces the gene value in the chromosome to the opposite value (i.e. from 0 into 1, or from 1 into 0). As it has already been mentioned, the probability of the occurrence of mutation is usually very small and certainly, whether or not a given gene in the chromosome will undergo mutation depends on this probability. Making mutation in accordance with probability p_m consists in, for instance, drawing a number from interval $[0, 1]$ for each gene and selecting those genes for mutation, for which the drawn number is equal to probability p_m or less.

Example 7.2

We perform mutation on chromosome $[1001101010]$. The value of p_m is 0.02. We draw the following numbers from interval $[0, 1]$:

$$0.23 \quad 0.76 \quad 0.54 \quad 0.10 \quad 0.28 \quad 0.68 \quad 0.01 \quad 0.30 \quad 0.95 \quad 0.12.$$

The gene located on position 7, undergoes mutation since the drawn random number 0.01 is less than the value of the probability of mutation p_m. Therefore, its value is changed from 1 to 0, and we obtain chromosome $[1001100010]$.

Creation of a new population. The chromosomes obtained as a result of the operation of genetic operators belong to a new population. This population becomes the so-called *current population* for a given generation of the genetic algorithm. In each subsequent generation, the value of the fitness function of each of this population's chromosomes is calculated. Next, the algorithm stopping criterion is checked and either a result in the form of chromosome with the highest value of the fitness function is presented, or the next step of the genetic algorithm, i.e. selection, is taken. In the classical genetic algorithm, the entire previous population of chromosomes is replaced by an equally large new offspring population.

Presentation of the "best" chromosome. If the stopping criterion for the genetic algorithm is satisfied then the result of the algorithm operation should be presented, that means the solution of the problem. The best solution is a chromosome with the highest value of the fitness function.

Example 7.3

We are going to show, on a simple example, how the genetic algorithm works in practice. We will find the maximum of function

$$y = 2x + 1,$$

assuming that x takes integer numbers in interval $[0, 31]$. In this case *the task parameter* is x. The set $\{0, 1, ..., 31\}$ constitutes the search space. This is at the same time a set of potential *solutions of the task*. Each of 32 numbers belonging to this set is called *the search point, solution, parameter's value, phenotype*. The solution optimizing the function is called the best solution or optimum solution. The solution of the task undergoes binary coding with the use of five bits. The created *code sequences* are also called *chains* or *chromosomes*. In our example they are also *genotypes*. The value of a gene on a specified position is called *allel*, these are certainly values 0 or 1. Thus, the optimization task consists in searching a space composed of 32 points and finding the point for which the function takes on the highest value. We can guess that the solution is 31, that is the chromosome containing only numbers 1.

In accordance with the algorithm, we begin with drawing the initial population. We are going to operate on a small population comprising eight individuals. As a result of drawing, we obtain

$$\text{ch}_1 = [00110], \qquad \text{ch}_2 = [00101],$$
$$\text{ch}_3 = [01101], \qquad \text{ch}_4 = [10101],$$
$$\text{ch}_5 = [11010], \qquad \text{ch}_6 = [10010],$$
$$\text{ch}_7 = [01000], \qquad \text{ch}_8 = [00101].$$

We start the main loop of the genetic algorithm, which means that we calculate the fitness of particular individuals. We decode information from chromosomes and obtain the following phenotypes:

$$\text{ch}_1^* = 6, \qquad \text{ch}_2^* = 5,$$
$$\text{ch}_3^* = 13, \qquad \text{ch}_4^* = 21,$$
$$\text{ch}_5^* = 26, \qquad \text{ch}_6^* = 18,$$
$$\text{ch}_7^* = 8, \qquad \text{ch}_8^* = 5.$$

We calculate fitness with the use of a function which is the same as the function we optimize. Since we search for the maximum, the individuals with the highest value of the fitness function will be considered the best fit. In place of parameter x we substitute the value of the phenotype. For instance, for the first two individuals we obtain

$$F\left(\mathrm{ch}_1\right) = 2 \cdot \mathrm{ch}_1^* + 1 = 13, \qquad F\left(\mathrm{ch}_2\right) = 2 \cdot \mathrm{ch}_2^* + 1 = 11.$$

By analogy, we determine

$$\begin{array}{ll} F\left(\mathrm{ch}_3\right) = 27, & F\left(\mathrm{ch}_4\right) = 43, \\ F\left(\mathrm{ch}_5\right) = 53, & F\left(\mathrm{ch}_6\right) = 37, \\ F\left(\mathrm{ch}_7\right) = 17, & F\left(\mathrm{ch}_8\right) = 11. \end{array}$$

We may now distinguish between the best and the worst fit individuals. As it is shown, chromosome ch_5 has the highest value of the fitness function, as opposed to chromosomes ch_2 and ch_8 which have identical the lowest fitness. The next step is the selection of chromosomes. We are going to apply the roulette wheel method. Upon the basis of formulae (7.1) and (7.2) we obtain sectors of the roulette wheel expressed in percentage (Fig. 7.2)

$$\begin{array}{ll} \nu\left(\mathrm{ch}_1\right) = 6.13, & \nu\left(\mathrm{ch}_2\right) = 5.19, \\ \nu\left(\mathrm{ch}_3\right) = 12.74, & \nu\left(\mathrm{ch}_4\right) = 20.28, \\ \nu\left(\mathrm{ch}_5\right) = 25, & \nu\left(\mathrm{ch}_6\right) = 17.45 \\ \nu\left(\mathrm{ch}_7\right) = 8.02, & \nu\left(\mathrm{ch}_8\right) = 5.19. \end{array}$$

Drawing with the use of the roulette wheel is reduced to a random selection of a number from interval $[0, 100]$ indicating the appropriate sector on the wheel, a consequently a specific chromosome. Let us assume that 8 following numbers were drawn:

$$79 \quad 44 \quad 9 \quad 74 \quad 45 \quad 86 \quad 48 \quad 23.$$

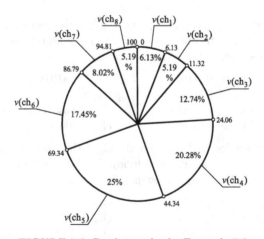

FIGURE 7.2. Roulette wheel - Example 7.3

This means that the following chromosomes were selected:

$$ch_6, \quad ch_4, \quad ch_2, \quad ch_6, \quad ch_5, \quad ch_6, \quad ch_5, \quad ch_3.$$

As it is shown, chromosome ch_5 was drawn twice. It should be noted that this is a chromosome with the highest value of the fitness function. Furthermore, chromosome ch_6 with a quite high value of the fitness function was drawn. However, chromosome ch_2 with the lowest value of the fitness function was also drawn. All the chromosomes selected in this way belong to the so-called *mating pool*.

Let us assume that none of the chromosomes selected during the selection undergoes mutation, i.e. probability $p_m = 0$. We only make crossover, assuming the crossover probability as $p_c = 0.75$. We mate the individuals in pairs as they are set in the mating pool. We draw a number from interval $[0, 1]$ for each of them

$$0.12 \quad 0.73 \quad 0.65 \quad 0.33.$$

All the drawn numbers are lower than the crossover probability p_c. Thus, crossover occurs for each of the pairs. Next, for each pair we find crossover points through drawing integer numbers from interval $[1, 4]$. As a result, we obtain

First pair of parents:
$ch_6 = [10010]$ $\xrightarrow{\text{crossover}}$ First pair of offspring:
$[10001]$
$ch_4 = [10101]$ $[10110]$
$l_k = 3$

Second pair of parents:
$ch_2 = [00101]$ $\xrightarrow{\text{crossover}}$ Second pair of offspring:
$[00100]$
$ch_6 = [10010]$ $[10011]$
$l_k = 4$

Third pair of parents:
$ch_5 = [11010]$ $\xrightarrow{\text{crossover}}$ Third pair of offspring:
$[11010]$
$ch_6 = [10010]$ $[10010]$
$l_k = 3$

Fourth pair of parents:
$ch_5 = [11010]$ $\xrightarrow{\text{crossover}}$ Fourth pair of offspring:
$[11101]$
$ch_3 = [01101]$ $[01010]$
$l_k = 2$

As a result of the crossover operation, we obtain the following offspring population:

$$Ch_1 = [10001], \qquad Ch_2 = [10110],$$
$$Ch_3 = [00100], \qquad Ch_4 = [10011],$$
$$Ch_5 = [11010], \qquad Ch_6 = [10010],$$
$$Ch_7 = [11101], \qquad Ch_8 = [01010].$$

Chromosomes of the new population are marked by a capital letter. Now, we pass on again to step 2 of the algorithm, that is to the evaluation of the fitness function of the new population chromosomes, which population will become the current population. When decoding the information from the new population of chromosomes, we obtain the phenotype values

$$Ch_1^* = 17, \qquad Ch_2^* = 22,$$
$$Ch_3^* = 4, \qquad Ch_4^* = 19,$$
$$Ch_5^* = 26, \qquad Ch_6^* = 18,$$
$$Ch_7^* = 29, \qquad Ch_8^* = 10,$$

and next, the values of the fitness function

$$F(Ch_1) = 35, \qquad F(Ch_2) = 45,$$
$$F(Ch_3) = 9, \qquad F(Ch_4) = 39,$$
$$F(Ch_5) = 53, \qquad F(Ch_6) = 37,$$
$$F(Ch_7) = 59, \qquad F(Ch_8) = 21.$$

As it is shown, the offspring population is characterized by a much higher average value of the fitness function than the parents population. Let us notice that as a result of crossover, chromosome Ch_7 was obtained, which has the highest value of the fitness function that none of the parent chromosome has had. The opposite could happen, that is – after the first generation, as a result of the crossover operation, a chromosome which had the highest value of the fitness function in the parents population could be "lost". In spite of this, the average "fitness" of the new population would be better than the one of the previous population, and chromosomes having a higher value of the fitness function would have the chance to appear in subsequent generations.

Example 7.4
Next example illustrating the operation of the classical genetic algorithm will consist in finding a chromosome having the highest number of numbers one. Let us assume that chromosomes consist of 9 genes, and the size of the population is 8 chromosomes. We presume that the probability of crossover is $p_c = 0.75$ and the probability of mutation is $p_m = 0.02$. The determination of the fitness function in this example is very simple. This function will reflect the number of "ones" in a chromosome. The individuals

that are better fit will have a higher number of "ones", which means that the value of the fitness function will be higher accordingly.

As it happened in the previous example, we start with drawing the initial population of individuals.

$$ch_1 = [100111001], \qquad ch_2 = [001011000],$$
$$ch_3 = [010100110], \qquad ch_4 = [101011110],$$
$$ch_5 = [110100010], \qquad ch_6 = [100101001],$$
$$ch_7 = [010011011], \qquad ch_8 = [001010011].$$

Let as make a simulation of one generation. We start with determining the value of the fitness function for particular chromosomes

$$F(ch_1) = 5, \qquad F(ch_2) = 3,$$
$$F(ch_3) = 4, \qquad F(ch_4) = 6,$$
$$F(ch_5) = 4, \qquad F(ch_6) = 4,$$
$$F(ch_7) = 5, \qquad F(ch_8) = 4.$$

As it is shown, the number of "ones" and "zeros" in the chromosomes is rather balanced. Only chromosome ch_4 is distinguished, which has as many as 6 "ones". We pass on to the next step, that is selection. We are going to apply the roulette wheel method, which is already known. Upon the basis of formulae (7.1) and (7.2) we obtain the following sectors of the roulette wheel (expressed in the percentage of the probability of particular chromosomes selection):

$$\nu(ch_1) = 14.29, \qquad \nu(ch_2) = 8.57,$$
$$\nu(ch_3) = 11.43, \qquad \nu(ch_4) = 17.14,$$
$$\nu(ch_5) = 11.43, \qquad \nu(ch_6) = 11.43,$$
$$\nu(ch_7) = 14.29, \qquad \nu(ch_8) = 11.43.$$

The partition of the roulette wheel is presented in Fig. 7.3. We draw 8 numbers from interval $[0, 100]$ in order to select appropriate chromosomes. Let us assume that the following numbers were drawn:

$$67 \quad 7 \quad 84 \quad 50 \quad 68 \quad 38 \quad 83 \quad 11.$$

This means that the following chromosomes were selected:

$$ch_6, \quad ch_1, \quad ch_7, \quad ch_4, \quad ch_6, \quad ch_4, \quad ch_7, \quad ch_1.$$

It turned out that the best fit chromosome ch_4 was selected as many times as twice, and the worst of the individuals, ch_2, was not selected at all. Next step of this generation is the modification of chromosomes with the use of genetic operators. We start with the crossover operator. Let us assume that chromosomes are combined in the following pairs:

$$ch_6 \text{ and } ch_1, \quad ch_7 \text{ and } ch_4, \quad ch_6 \text{ and } ch_4, \quad ch_7 \text{ and } ch_1.$$

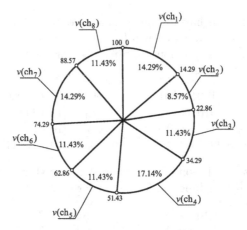

FIGURE 7.3. Roulette wheel - Example 7.4

We draw the following numbers from interval $[0, 1]$:

$$0.42 \quad 0.30 \quad 0.18 \quad 0.19.$$

and next, we compare them with the probability of crossover $p_c = 0.75$. All the drawn numbers are less than the probability of crossover and thus, each pair undergoes the operation of crossover. Thus, let us make a simulation of this process, similarly to the previous example.

First pair of parents: First pair of offspring:
$ch_6 = [100101001]$ crossover $[100\mathbf{11001}]$
$ch_1 = [100111001]$ $[100\mathbf{01001}]$

    $l_k = 4$

As the place of crossover, position No. 4 in the chromosome was drawn. As it is shown, this crossover did not principally impact the improvement of the fitness of chromosomes ch_6 and ch_1. The number of "ones" remains the same, i.e. 4 and 5. Let's cross the next pair

Second pair of parents: Second pair of offspring:
$ch_7 = [010011011]$ crossover $[010011\mathbf{010}]$
$ch_4 = [101011110]$ $[101011\mathbf{111}]$

    $l_k = 7$

This time, the place of crossover turned out to be advantageous. One of the offspring chromosomes has already 7 "ones". Let us have a look at subsequent crossovers:

Third pair of parents: Third pair of offspring:

$ch_6 = [100101001]$ $\xrightarrow{\text{crossover}}$ $[100\mathbf{011110}]$

$ch_4 = [101011110]$ $[101\mathbf{101001}]$

$l_k = 3$

Fourth pair of parents: Fourth pair of offspring:

$ch_7 = [010011011]$ $\xrightarrow{\text{crossover}}$ $[01001\mathbf{1001}]$

$ch_1 = [100111001]$ $[10011\mathbf{1011}]$

$l_k = 5$

Let us see how the new population looks like after the crossover. Individuals of the new population are marked by a capital letter.

$$Ch_1 = [100111001], \qquad Ch_2 = [100101001],$$
$$Ch_3 = [010011010], \qquad Ch_4 = [101011111],$$
$$Ch_5 = [100011110], \qquad Ch_6 = [101101001],$$
$$Ch_7 = [010011001], \qquad Ch_8 = [100111011].$$

The mutation is made on the new population. As we remember, the probability of mutation is $p_m = 0.02$. For each of the genes, we draw a number from interval $[0, 1]$ and check whether this number is less than probability p_m. The mutation consists in replacing the gene's value with the opposite value. Let us see the course of this process:

$$Ch_1 = [100111001], \qquad Ch_2 = [100101001],$$
$$Ch_3 = [010011010], \qquad Ch_4 = [111011111],$$
$$Ch_5 = [100011110], \qquad Ch_6 = [101101101],$$
$$Ch_7 = [010011001], \qquad Ch_8 = [100111011].$$

The probability of mutation is so small that it occurred only in two places, i.e. in individuals Ch_4 and Ch_6. Let us check the value of the fitness function of particular individuals after one generation:

$$F(Ch_1) = 5, \qquad F(Ch_2) = 4,$$
$$F(Ch_3) = 4, \qquad F(Ch_4) = 8,$$
$$F(Ch_5) = 5, \qquad F(Ch_6) = 6,$$
$$F(Ch_7) = 4, \qquad F(Ch_8) = 6.$$

As it is shown, after one generation we managed to find an individual having a higher value of the fitness function. Besides, the average value of the fitness function of the entire population improved. Individuals that are better fit will have the chance to appear more often in subsequent generations and give rise to new ones, which are better fit.

Example 7.5

Let us apply the classical genetic algorithm in a so-called "knapsack problem". Let us imagine a bag which has a certain capacity W. It is possible to put n items having the weight w_i and the importance p_i, $i = 1, \ldots, n$ in there. The task consists in loading the bag in a manner so that we have items that will bring us the most of the profits possible (that is, so that we get the highest possible sum p_i of the importance of the items). Of course, we cannot surpass capacity W, i.e.

$$\sum_{i=1}^{n} w_i x_i \leq W. \tag{7.3}$$

Without taking condition (7.3) into consideration, the chromosomes corresponding to the occurrence of all the items in the bag would turn out to be the best ones. The solution to the problem is imposing a penalty on such chromosomes. In this example, the method of punishment of individuals that do not meet specified requirements will be the assumption of zero value of the fitness function. Such individuals have no chance for reproduction.

Let us try to code the solution using the notation of the binary chromosome. Values 0 and 1 of subsequent gene x_i would correspond to the lack or the occurrence of the i-th item in the bag. The fitness function takes the form

$$f(\mathbf{x}) = \sum_{i=1}^{n} p_i x_i. \tag{7.4}$$

In our task, we take $n = 10$ items into consideration, each of importance $p_i = 1$ and weight from 1 to 10, which makes $w_i = i$, $i = 1, \ldots, 10$. Since the importance of all the items is the same, the solution comes down to placing as much items as possible in the bag. We assume that the capacity is 27, which means that it is equal to almost the half sum of the weights of all the items. In order to find an optimum filling of the bag, we are going to use the classical genetic algorithm with the proportional selection (roulette wheel) and with the probability of crossover $p_c = 0.7$, and the probability of mutation $p_m = 0.01$. The number of individuals in the population is 10. Let us follow the changes in the value of the fitness function for individuals during 100 subsequent generations of the algorithm (Fig. 7.4).

The graph reflects the change in the average value of the fitness of individuals in subsequent generations (dotted curve). These values were not significantly different from the results of the best chromosome (continuous curve). Therefore, it can also be concluded that the diversity of the population was not very high. In the 43rd generation the best solution was found and the fitness of the entire population was improved. The solution of the problem turned out to be six items with weight 1, 2, 4, 5, 7 and 8 placed in the bag.

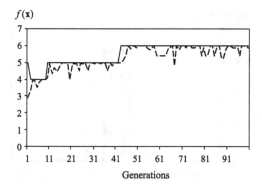

FIGURE 7.4. Fitness function for the best chromosome (continous curve) and average value (dotted curve) of the fitness of individuals in 100 subsequent generations in Example 7.5

The above-mentioned algorithm was started 100 times. The following were received:
- 2 solutions with 4 items;
- 24 solutions with 5 items;
- 74 solutions with 6 items.

Example 7.6
We are going to solve again the bag problem presented in Example 7.5. This time, we are going to assume that $n = 50$, and the bag capacity W is 318, which makes almost half of the sum of all the items. We assume $p_i = 1$ and $w_i = i$, identical as in Example 7.5. Due to the complicated task, the size of the population will be set as 100 individuals. We use the same genetic algorithm, but the number of generations is 200.

Figure 7.5 depicts the operation of the algorithm, and Fig. 7.6 depicts the frequency of the occurrence of various solutions in 100 starts of the genetic

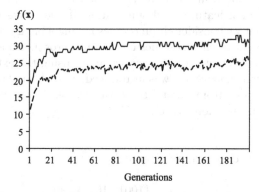

FIGURE 7.5. Fitness function for the best chromosome (continous curve) and average value (dotted curve) of the fitness of individuals in 200 subsequent generations in Example 7.6

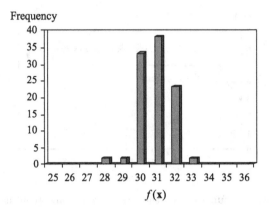

FIGURE 7.6. Histogram of various solutions in 100 runs of the genetic algorithm in Example 7.6

algorithm. From the analysis of the histogram (Fig. 7.6) it results that 38 solutions guaranteeing the placement of 31 items in the bag were found, and only 2 solutions with 33 items in the bag. As it is shown, the problem is more complex than the one described in Example 7.5. The chromosome has now a much larger length and it is worth thinking about a method for improving the operation of the algorithm, which is going to be presented in Example 7.16.

7.3.1.1. Theoretical bases for the operation of genetic algorithms

In order to better understand the genetic algorithm, we are going to discuss the topic based on the notion of *scheme,* and present basic theorem concerning the genetic algorithms, a so-called *schemata theorem.* The notion of schema was introduced in order to determine the set of chromosomes having certain common features and similarities. If allels take values 0 or 1, which means that we consider chromosomes having a binary alphabet, then the schema is a set of chromosomes containing zeros and ones on selected positions. It is convenient to consider schemata using the extended alphabet $\{0, 1, *\}$ in which symbol $*$ was introduced, apart from 0 and 1, in order to specify any value from among these values, i.e. 0 or 1, symbol $*$ on a given position means "don't care".

Example 7.7
Two examples of schemata are given below

$$10*1 = \{1001, \ 1011\},$$

$$*01*10 = \{001010, \ 001110, \ 101010, \ 101110\}.$$

We say that *a chromosome belongs to a given schema* if for each position (locus) $j = 1, 2, ..., L$, where L is the length of the chromosome, and the symbol on j-th position of the chromosome corresponds to the symbol on j-th position of the schema, while both 0 and 1 correspond to symbol $*$. The same can be expressed by saying that *a chromosome fits to the schema* or *is a representative of the schema*. Let us note that if there are m symbols $*$ in a schema then this schema contains 2^m chromosomes. Furthermore, each chromosome (chain) with length L belongs to 2^L schemata.

Example 7.8
Chain 01 fits to 4 schemata: $**$, $*1$, $0*$, 01, and chain 100 fits to 8 schemata: $***$, $**0$, $*0*$, $1**$, $*00$, $1*0$, $10*$, 100.

The genetic algorithm is based on the rule of processing the best fit individuals (chromosomes). Let $\mathbf{P}(0)$ mean the initial population of individuals, and $\mathbf{P}(t)$ the current population in generation t of the algorithm operation. The chromosomes with the best fitness are selected, using the selection method, out of each population $\mathbf{P}(t)$, $t = 0, 1, ...$ to the mating pool denoted by $\mathbf{M}(t)$. Next, when mating parent individuals from $\mathbf{M}(t)$ population in pairs and performing the crossover operation with the probability of crossover p_c and the operation of mutation with the probability of mutation p_m, we obtain a new population $\mathbf{P}(t+1)$, containing the offspring of individuals of population $\mathbf{M}(t)$.

We would like that the number of chromosomes fitting the schema representing a good solution in population $\mathbf{P}(t)$ increase along with an increase in the number of generations t.

Three factors impact the appropriate processing of schemata in the genetic algorithm: the selection, crossover and mutation of chromosomes. We are going to analyze the operation of each of them, examining their impact on the expected number of representatives of a given schema.

Let S be a given schema, and $c(S,t)$ – the number of chromosomes in population $\mathbf{P}(t)$ fitting to schema S. Therefore, $c(S,t)$ is the number of elements of set $\mathbf{P}(t) \cap S$.

Let us begin with examining the impact of selection. During selection, chromosomes of population $\mathbf{P}(t)$ are copied to the mating pool $\mathbf{M}(t)$ with the probability given by formula (7.2). Let $F(S,t)$ means the average value of the fitness function of chromosomes in population $\mathbf{P}(t)$ fitting to schema S. If

$$\mathbf{P}(t) \cap S = \left\{ \mathrm{ch}_i, ..., \mathrm{ch}_{c(S,t)} \right\},$$

then

$$F(S,t) = \frac{\sum_{i=1}^{c(S,t)} F(\mathrm{ch}_i)}{c(S,t)}. \tag{7.5}$$

$F(S,t)$ is also called *the fitness of schema S in generation t*.

Let $\Im(t)$ means the sum of the values of the fitness function of chromosomes in population $\mathbf{P}(t)$ with size K, i.e.

$$\Im(t) = \sum_{i=1}^{K} F\left(\mathrm{ch}_i^{(t)}\right). \qquad (7.6)$$

Let $\overline{F}(t)$ denotes the average value of the fitness function of the chromosomes in this population, that means:

$$\overline{F}(t) = \frac{1}{K}\Im(t). \qquad (7.7)$$

Let $\mathrm{chr}^{(t)}$ be an element of the mating pool $\mathbf{M}(t)$. For each $\mathrm{chr}^{(t)} \in \mathbf{M}(t)$ and for each $i = 1, .., c(S, t)$ the probability that $\mathrm{chr}^{(t)} = \mathrm{ch}_i$ is given by formula $F\left(\mathrm{ch}_i\right)/\Im(t)$. Therefore, the expected number of chromosomes in population $\mathbf{M}(t)$ equal to ch_i is

$$K\frac{F\left(\mathrm{ch}_i\right)}{\Im(t)} = \frac{F\left(\mathrm{ch}_i\right)}{\overline{F}(t)}.$$

The expected number of chromosomes in set $\mathbf{P}(t) \cap S$ selected to mating pool $\mathbf{M}(t)$ is, therefore, equal to

$$\sum_{i=1}^{c(S,t)} \frac{F\left(\mathrm{ch}_i\right)}{\overline{F}(t)} = c(S,t)\frac{F\left(S,t\right)}{\overline{F}(t)},$$

which results from formula (7.5). Since each chromosome from mating pool $\mathbf{M}(t)$ is at the same time a chromosome belonging to population $\mathbf{P}(t)$, then chromosomes from set $\mathbf{M}(t) \cap S$ are simply the same chromosomes which were selected to population $\mathbf{M}(t)$ from set $\mathbf{P}(t) \cap S$. If $b(S,t)$ denotes the number of chromosomes from mating pool $\mathbf{M}(t)$ that fit to schema S, that is the number of elements of set $\mathbf{M}(t) \cap S$, then the following is concluded from the above-mentioned discussion:

Corollary 7.1 *(impact of selection)*
The expected value $b(S,t)$, that is the expected number of chromosomes in mating pool $\mathbf{M}(t)$ that fit to schema S, is given by formula:

$$E\left[b\left(S,t\right)\right] = c\left(S,t\right)\frac{F\left(S,t\right)}{\overline{F}(t)}. \qquad (7.8)$$

The conclusion is that if schema S contains chromosomes with the value of the fitness function that is above the average, that means that the fitness of schema S in generation t is higher than the average value of the fitness function of chromosomes in population $\mathbf{P}(t)$, which means that $F\left(S,t\right)/\overline{F}(t) > 1$, then the expected number of chromosomes that fit to schema S in mating pool $\mathbf{M}(t)$ is higher than the number of chromosomes that fit to schema S of population $\mathbf{P}(t)$. Therefore, it can be concluded that selection causes the dissemination of schemata with fitness that is "better" than the average one and the vanishing of schemata with a "worse" fitness.

Before we start the analysis of the impact of the operation of genetic operators, i.e. crossover and mutation, on chromosomes from the mating pool, we are going to define the notion of *order* and *length* of schema, which are necessary for further discussions. Let L means the length of the chromosomes belonging to schema S.

Definition 7.1
The order of schema S, otherwise called *the size* of a schema and denoted by $o(S)$, is the number of fixed positions in a schema, i.e. the number of zeros and ones in case of alphabet $\{0, 1, *\}$.

Example 7.9
Orders for 4 schemata are determined below:

$$o(10*1) = 3, \quad o(*01*10) = 4, \quad o(**0*1*) = 2, \quad o(*101**) = 3.$$

Order $o(S)$ of the schema is equal to length L minus the number of symbols $*$, which can be easily verified on the above-mentioned examples (for $L = 4$ with one symbol $*$ and for $L = 6$ with two, four and three symbols $*$). It is easy to notice that the order of a schema having nothing but symbols $*$ is equal to zero, i.e. $o(****) = 0$, and the order of a schema with no symbol $*$ is equal to L, e.g. $o(10011010) = 8$. The order of schema $o(S)$ is an integer number from interval $[0, L]$.

Definition 7.2
The defining length of schema S, also referred to as *the length* of a schema (which should not be confused with length L), denoted by $d(S)$, is a distance between the first and the last fixed symbol.

Example 7.10
The lengths of schemata indicated in Example 7.9 are as follows:

$$d(10*1) = 4 - 1 = 3, \quad d(*01*10) = 6 - 2 = 4,$$
$$d(**0*1*) = 5 - 3 = 2, \quad d(*101**) = 4 - 2 = 2.$$

The length of schema $d(S)$ is an integer number from interval $[0, L - 1]$. Let us note that the length of a schema with fixed symbols on the first and the last position is equal to $L - 1$ (as in the first of the above-mentioned examples). The length of a schema with one fixed position is equal to zero, e.g. $d(**1*) = 0$. The length of the schema specifies the content of the information contained in the schema.

Let us move on to discuss the impact of the crossover operation on the processing of schemata in the genetic algorithm. Let us note first that some schemata are more susceptible to damage during crossover than other ones. For instance, let us have a look at schemata $S_1 = 1{*}{*}{*}{*}0{*}$ and $S_2 = {*}{*}01{*}{*}{*}$ and chromosome ch $= [1001101]$ that fits both schemata. It is shown that schema S_2 has a higher chance to survive the crossover operation than schema S_1, which is more prone to a "split" in crossover point 1, 2, 3, 4 or 5. Schema S_2 may be splitted only when the crossover point is equal to 3. Let us note the length of both schemata, which – as it is shown – is important in the crossover process.

When analyzing the impact of the crossover operation on mating pool $\mathbf{M}(t)$, let us discuss a given chromosome from set $\mathbf{M}(t) \cap S$, that is a chromosome from the mating pool, which fits schema S. The probability that this chromosome will be selected for crossover is p_c. If none of the offspring of this chromosome belongs to schema S this means that the crossover point must be placed between the first and the last fixed symbol in schema S. The probability of this is $d(S)/(L-1)$. Further corollaries are drawn below.

Corollary 7.2 *(impact of crossover)*
For a given chromosome in $\mathbf{M}(t) \cap S$, the probability that this chromosome is selected for crossover and none of its offspring belongs to schema S has an upper bound given by

$$p_c \frac{d(S)}{L-1}$$

which is called *the probability of destruction of schema S*.

Corollary 7.3
For a given chromosome in $\mathbf{M}(t) \cap S$, the probability that this chromosome will not be selected for crossover or at least one of its offspring will belong to schema S after the crossover has a lower bound given by

$$1 - p_c \frac{d(S)}{L-1}$$

which is called *the probability of survival of schema S*.

It is easy to demonstrate that when a given chromosome belongs to schema S and is selected for crossover, and the other parent chromosome also belongs to schema S, then both chromosomes being their offspring also belong to schema S. Corollaries 7.2 and 7.3 confirm the significant role of the length of schema $d(S)$ in the probability of destruction or survival of a schema.

Let us discuss the impact of mutation on mating pool $\mathbf{M}(t)$. The mutation operator changes at random, with probability p_m, the value on a fixed position from 0 to 1 or in the opposite direction. It is evident that if the schema is to survive the mutation then all the fixed positions in the schema must remain unchanged. A given chromosome from the mating pool belonging to schema S, that is the chromosome from set $\mathbf{M}(t) \cap S$, remains in schema S only when none of the symbols in this chromosome, corresponding to the fixed symbols in schema S, is changed during mutation. The probability of such an event is $(1 - p_m)^{o(S)}$. This result is presented as yet another corollary.

Corollary 7.4 *(impact of mutation)*
For a given chromosome in $\mathbf{M}(t) \cap S$, the probability that this chromosome will belong to schema S after the operation of mutation is given by

$$(1 - p_m)^{o(S)}. \tag{7.9}$$

This notion is called *the probability of mutation survival* by schema S.

Corollary 7.5
If the probability of mutation p_m is low $(p_m << 1)$, then it may be assumed that the probability of survival of mutation of schema S specified in Corollary 7.4 is roughly equal to

$$1 - p_m o(S). \tag{7.10}$$

The effect of combining the impact of selection, crossover and mutation (Corollaries 7.1 – 7.4), and taking into consideration the fact that if a chromosome from set $\mathbf{M}(t) \cap S$ gives an offspring that fits to schema S, then it belongs to $\mathbf{P}(t+1) \cap S$, leads to the following reproduction schema [24]:

$$E\left[c\left(S, t+1\right)\right] \geq c\left(S, t\right) \frac{F\left(S, t\right)}{\overline{F}\left(t\right)} \left(1 - p_c \frac{d\left(S\right)}{L-1}\right) \left(1 - p_m\right)^{o(S)}. \tag{7.11}$$

Dependency (7.11) shows how the number of chromosomes that fit to a given schema changes from one population to another. Three factors reflected on the right side of expression (7.11) have an impact on this change, that is: $F(S, t) / \overline{F}(t)$ indicating the role of the average value of the fitness function, $1 - p_c d(S) / (L-1)$ indicating the impact of crossover, and $(1 - p_m)^{o(S)}$ indicating the impact of mutation. The higher the value of each of these factors, the higher the expected number of adaptations to schema S in the subsequent population. Corollary 7.5 allows to present dependency (7.11) in the form of

$$E\left[c\left(S, t+1\right)\right] \geq c\left(S, t\right) \frac{F\left(S, t\right)}{\overline{F}\left(t\right)} \left(1 - p_c \frac{d\left(S\right)}{L-1} - p_m o\left(S\right)\right). \tag{7.12}$$

For large populations formula (7.12) may be approximated by

$$c\left(S, t+1\right) \geq c\left(S, t\right) \frac{F\left(S, t\right)}{\overline{F}\left(t\right)} \left(1 - p_c \frac{d\left(S\right)}{L-1} - p_m o\left(S\right)\right). \tag{7.13}$$

Dependencies (7.11) and (7.12) imply that the expected number of chromosomes that fit to schema S in the subsequent generation is the function of the current number of chromosomes belonging to this schema, relative fitness of the schema and the order and length of the schema. It is shown that schemata with a fitness above the average and with a low order and low length are characterized by a growing number of their representatives in subsequent populations. This growth is exponential, which results from dependency (7.8), which, for large populations, can be replaced by an approximated recurrence dependency in the form of [136]

$$c(S, t+1) = c(S, t) \frac{F(S, t)}{\overline{F}(t)}. \qquad (7.14)$$

If we assume that schema S has a fitness of ϵ % above the average, i.e.

$$F(S, t) = \overline{F}(t) + \epsilon \overline{F}(t), \qquad (7.15)$$

then substituting dependency (7.13) to inequality (7.12) and assuming that ε does not change with time, and starting from $t = 0$, we obtain

$$c(S, t) = c(S, 0)(1 + \epsilon)^t$$

and

$$\epsilon = \left(F(S, t) - \overline{F}(t)\right) / \overline{F}(t), \qquad (7.16)$$

i.e. $\varepsilon > 0$ for a schema with the fitness above the average and $\varepsilon < 0$ for a schema with the fitness below the average.

Equation (7.16) describes a geometrical sequence. It results that in the reproduction process schemata that are better (worse) than the average are selected in an ascending (descending) number exponentially in subsequent generations of the genetic algorithm. Let us notice that dependencies (7.8) – (7.14) are based on the assumption that fitness function F takes only positive values. In case where genetic algorithms are applied to optimization problems in which the function to be optimized may take negative values, some supplementary mapping between the function to be optimized and the fitness function is required. The final effect resulting from dependencies (7.11) – (7.13) may be formulated in the form of theorem. This is the basic theorem for genetic algorithms, that is *the schemata theorem* [82].

Theorem 7.1
Schemata with low order and a low length and with the fitness above the average obtain the exponentially ascending number of their representatives in subsequent generations of the genetic algorithm.

Given the above-mentioned theorem, an important issue is coding, which should bring good schemata of a low range, low length and fitness above the average. The following hypothesis may be considered as an intermediary result of Theorem 7.1. This is a so-called *bricks hypothesis* (or *building blocks hypothesis*) [63, 136].

Hypothesis 7.1
The genetic algorithm aims at reaching a result that is close to the optimum by listing good schemata (with fitness above the average), of a low order and low length. These schemata are called *bricks* (or building blocks).

The building blocks hypothesis was made upon the basis of the schemata theorem, by inferring that genetic algorithms explore search space via schemata having a low order and a low length, which schemata will next take part in the exchange of information during crossover.

Although some research was made towards proving this hypothesis, still in case of the majority of non-trivial applications we base mainly on empirical results. During the last dozen of years, many papers have been produced on the application of genetic algorithms supporting this hypothesis. When assuming that the hypothesis is correct the coding problem seems to be critical for the operation of the genetic algorithm; the coding should fulfill the concept of small building blocks. Undoubtedly, the strength of the genetic algorithms is based on the processing of a large number of schemata, owing to which these algorithms have advantage over other traditional methods.

7.3.2   Evolution strategies

The best way to get familiarized with evolution strategies is to begin with comparing them to the operation of genetic algorithms. Certainly, the main similarity is that both methods operate on the populations of potential solutions and use the selection principle and the principle of processing the best fit individuals. However, there are many differences between these two algorithms. The first difference concerns the method of representing the individuals. Evolution strategies operate on vectors of floating-point numbers while classical genetic algorithms operate on binary vectors.

The second difference between evolution strategies and genetic algorithms is hidden in the selection process. In the genetic algorithm, certain number of individuals corresponding to the size of the parents population is selected to the new population. This is made by means of sampling during which the chance to select an individual depends on the value of its fitness function. This results in a fact that the worst chromosomes may also be drawn. In turn, in evolution strategies, a temporary population is created during the selection procedure, and its size differs from the size of the parents population. The subsequent generation of individuals is created through the selection of the best ones. Irrespectively of the selection method applied in the genetic algorithm (the roulette wheel method or e.g. the ranking method), the individuals that are better fit may be selected several times. In evolution strategies, individuals are selected without recurrence. The selection is deterministic.

The third difference between evolution strategies and genetic algorithms concerns the order of processes of selection and recombination (that is the change of genes through the operation of genetic operators). In evolution strategies, the recombination process occurs first, and next – the selection process. An offspring is a result of the crossover of two parent individuals and mutation, while in some versions of evolution strategies only mutation is applied. The above mentioned temporary population created in this way undergoes the selection process, which reduces the size of this population down to the size of the parents population. Genetic algorithms work the opposite way. First, the selection of individuals occurs on which genetic operators (crossover and mutation) are applied later in accordance with the determined probability. Next difference between evolution strategies and genetic algorithms consists in the fact that genetic algorithms' parameters, such as the probability of crossover and the probability of mutation, remain constant during the evolution process, while in evolution strategies these parameters undergo a permanent change (self-adaptation of parameters).

Evolution strategies may be divided into several basic types. We are going present below three strategies.

7.3.2.1. Evolution strategy $(1 + 1)$

The first algorithm that gave rise to the entire family of evolution strategies is strategy $(1 + 1)$. The general schema of strategy $(1 + 1)$ is presented in Fig. 7.7.

In the $(1 + 1)$ evolution strategy one base chromosome $\mathbf{x}$ is processed. The algorithm begins with a random choosing of the values of particular components of vector $\mathbf{x}$.

In each generation, a new individual $\mathbf{y}$ is created as a result of mutation. When comparing the values of the fitness function of both individuals, i.e. $F(\mathbf{x})$ and $F(\mathbf{y})$, the one having higher value of the fitness function is selected (in case when the maximum is searched). It is this chromosome that becomes the new base chromosome $\mathbf{x}$ in the next generation. In this algorithm, the crossover operator does not occur. Chromosome $\mathbf{y}$ is created through adding, to each gene of chromosome $\mathbf{x}$, certain random number generated in accordance with normal distribution, i.e.

$$y_i = x_i + \sigma N_i (0, 1), \tag{7.17}$$

where y_i means i-th gene of chromosome $\mathbf{y}$; x_i – i-th gene of chromosome $\mathbf{x}$; σ – a parameter determining the range of mutation; $N_i (0, 1)$ – a random number generated in accordance with normal distribution for the i-th gene.

The adaptation of parameters in this strategy is made through a change of the mutation range, i.e. parameter σ. The $1/5$ *success rule* is applied in this case. In terms of this rule, the best results in the search of the optimum solution are obtained where the relation between the successful mutations and all the mutations is precisely $1/5$. If during subsequent k generations the relation between successful mutations and all the mutations is higher than value $1/5$, then we increase the value of parameter σ. If this relation is

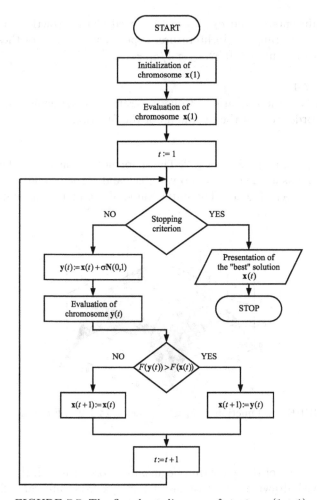

FIGURE 7.7. The flowchart diagram of strategy $(1 + 1)$

lower than $1/5$ then we reduce the range of mutation. Parameter σ remains unchanged when the relation between the successful mutations and all the mutations is precisely $1/5$. The $1/5$ success rule appeared in literature as a result of the process of the optimization of the speed of convergence of some multidimensional test functions, the so-called corridor and spherical model. Let $\varphi(k)$ be the coefficient of the mutation operator's success in previous k generations. The $1/5$ success rule may be formally denoted as follows:

$$\sigma' = \begin{cases} c_1 \cdot \sigma & \text{for } \varphi(k) < 1/5, \\ c_2 \cdot \sigma & \text{for } \varphi(k) > 1/5, \\ \sigma & \text{for } \varphi(k) = 1/5, \end{cases}$$

where coefficients c_1 and c_2 govern the speed of the growth or reduction of the mutation range σ. In literature, experimental values of these coefficients were chosen: $c_1 = 0.82$, $c_2 = 1/c_1 = 1.2$.

Example 7.11

Let us analyze the operation of one generation of the evolution strategy $(1+1)$ in order to find the maximum of the function

$$f(x_1, x_2) = -x_1^2 - x_2^2 + 2 \qquad (7.18)$$

for $-2 \leq x_1 \leq 2$ and $-2 \leq x_2 \leq 2$. The shape of this function is depicted in Fig. 7.8. This is a function of two variables and chromosome $\mathbf{x}$ consists of the same number of genes. The initial value of the mutation range $\sigma = 1$.

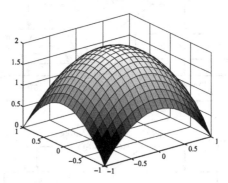

FIGURE 7.8. Function $f(x_1, x_2) = -x_1^2 - x_2^2 + 2$ in Example 7.11

The operation of the algorithm begins, in accordance with the schema of Fig. 7.7, with drawing initial values of vector $\mathbf{x}$ from interval $[-2, 2]$. The range of our search will be limited to this interval. The vector of numbers $[-1.45, -1.95]$ was drawn. We evaluate the fitness of the individual with the use of the fitness function, which in our case will be function (7.18). We obtain $F(\mathbf{x}) = -3.91$. The main loop of the algorithm begins. In order to carry out the operation of mutation, we draw two numbers from the normal distribution $N_1(0,1) = -0.90$ and $N_2(0,1) = -0.08$. Applying formula (7.17), we create a new individual $\mathbf{y} = [-2.35, -2.03]$. We check whether its fitness $F(\mathbf{y}) = -7.64$ is higher than the fitness of individual $\mathbf{x}$. We notice that in the future generation $\mathbf{x}$ will again be the base chromosome. We are going to determine the parameters of the $(1+1)$ strategy. We assume that if the $1/5$ success rule requires an increase in the mutation range then we increase this range applying coefficient $c_2 = 1.2$. In case the mutation range needs to be reduced this coefficient is $c_1 = 0.82$.

A change of parameter σ will be made every $k = 5$ generations. Changes in the value of the fitness of the base chromosome during 100 generations

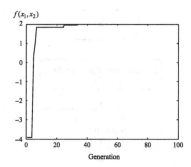

FIGURE 7.9. Fitness function in Example 7.11

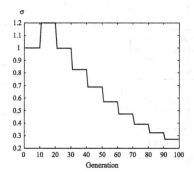

FIGURE 7.10. Mutation range changes in Example 7.11

are presented in Fig. 7.9, and changes in the value of the mutation range are depicted in Fig. 7.10.

7.3.2.2. Evolution strategy $(\mu + \lambda)$

The extension of the $(1 + 1)$ evolution strategy is the $(\mu + \lambda)$ strategy, of which flowchart is depicted in Fig. 7.11. This algorithm, owing to a larger number of individuals and, in consequence, a larger diversity of genotypes, allows to avoid the final solution obtained in the form of the local minimum, which is often found in the previously described $(1 + 1)$ strategy.

We start the algorithm with a random generation of the initial parents population $\mathbf{P}$ containing μ individuals. Next, a temporary population $\mathbf{T}$ is created by means of reproduction, which population contains λ individuals, while $\lambda \geq \mu$. Reproduction consists in a multiple random selection of λ individuals out of population $\mathbf{P}$ (multiple sampling) and placing the selected ones in temporary population $\mathbf{T}$. Individuals of population $\mathbf{T}$ undergo crossover and mutation operations as a result of which an offspring population $\mathbf{O}$ is created, which also has size λ. The last step is the selection

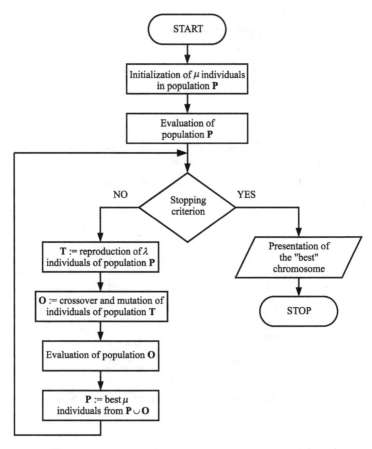

FIGURE 7.11. Flowchart of the evolution strategy $(\mu+\lambda)$

of μ best offspring from both populations $\mathbf{P} \cup \mathbf{O}$, which will constitute a new parent population $\mathbf{P}$.

Strategy $(\mu + \lambda)$ is extended with the self-adaptation of the mutation range, which replaces the previously described 1/5 success method. For the needs of self-adaptation, chromosome σ was added to the description of each individual, which chromosome contains the values of standard deviations applied during the mutation of particular genes of chromosome $\mathbf{x}$. Additionally, the crossover operator was introduced. It is important that both chromosomes undergo genetic operations, both vector $\mathbf{x}$ and standard deviation vector σ.

The operation of the crossover operator consists in drawing two individuals and exchanging or averaging the value of their genes. Two new individuals created this way replace their parents. We are going to describe a crossover consisting in exchanging the value of genes. Let us select two individuals

$$(\mathbf{x}^1, \sigma^1) = \left(\left[x_1^1, ..., x_n^1 \right]^T, \left[\sigma_1^1, ..., \sigma_n^1 \right]^T \right)$$

and

$$(\mathbf{x}^2, \sigma^2) = \left(\left[x_1^2, ..., x_n^2 \right]^T, \left[\sigma_1^2, ..., \sigma_n^2 \right]^T \right).$$

In case where the value of the genes is replaced, the offspring takes on the form

$$(\mathbf{x}', \sigma') = \left(\left[x_1^{q_1}, ..., x_n^{q_n} \right]^T, \left[\sigma_1^{q_1}, ..., \sigma_n^{q_n} \right]^T \right), \tag{7.19}$$

where $q_i = 1$ or $q_i = 2$ (each gene comes from the first or the second selected parent). In case of a crossover consisting in averaging the value of the genes of two individuals, the new offspring is created following the formula

$$(\mathbf{x}', \sigma') = \left(\left[\left(x_1^1 + x_1^2 \right) / 2, ..., \left(x_n^1 + x_n^2 \right) / 2 \right]^T, \right. \\ \left. \left[\left(\sigma_1^1 + \sigma_1^2 \right) / 2, ..., \left(\sigma_n^1 + \sigma_n^2 \right) / 2 \right]^T \right). \tag{7.20}$$

Averaging may also be made with coefficient α drawn from the uniform distribution $U(0,1)$, and then the crossover is following:

$$\begin{aligned} x_i'^1 &= \alpha x_i^1 + (1 - \alpha) x_i^2, \\ x_i'^2 &= \alpha x_i^2 + (1 - \alpha) x_i^1, \\ \sigma_i'^1 &= \alpha \sigma_i^1 + (1 - \alpha) \sigma_i^2, \\ \sigma_i'^2 &= \alpha \sigma_i^2 + (1 - \alpha) \sigma_i^1. \end{aligned} \tag{7.21}$$

The mutation operation is made on a single individual. The first one to undergo this operation is chromosome $\sigma = [\sigma_1, ..., \sigma_n]^T$ according to the formula

$$\sigma_i' = \sigma_i \exp \left(\tau' N(0,1) + \tau N_i(0,1) \right), \tag{7.22}$$

in which $i = 1, ..., n$, n is the length of the chromosome, $N(0,1)$ – a random number from the normal distribution (drawn once for the entire chromosome), $N_i(0,1)$ – a random number from the normal distribution (drawn separately for each gene), τ' and τ are parameters of the evolution strategy that impact the convergence of the algorithm. The following formula can be found in literature:

$$\tau' = \frac{C}{\sqrt{2n}}, \qquad \tau = \frac{C}{\sqrt{2n}}, \tag{7.23}$$

where C takes value 1 the most frequently. New ranges of mutation σ_i' impact the change of value x_i in the following manner:

$$x_i' = x_i + \sigma_i' N_i(0,1), \tag{7.24}$$

where $N_i(0,1)$ is a random number from the normal distribution, $i = 1, ..., n$.

As it is easy to notice, the introduction of parameter σ_i' given by formula (7.22) allows the self-adaptation of the mutation process.

Example 7.12

We are going to present the operation of the evolution strategy $(\mu + \lambda)$ and follow some of its steps. The task on the basis of which we are going to learn about this algorithm is the minimization of function

$$f(x_1, x_2) = x_1^2 + x_2^2 \qquad (7.25)$$

with constraints $-1 \leq x_1 \leq 1$ and $-1 \leq x_2 \leq 1$. The shape of this function is depicted in Fig. 7.12.

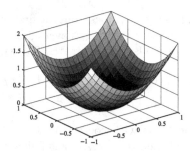

FIGURE 7.12. Function $f(x_1, x_2) = x_1^2 + x_2^2$ in Example 7.12

The global minimum is in point $(0,0)$, while $f(0,0) = 0$. We assume the following parameters of the algorithm: $\mu = 4$ and $\lambda = 4$. We are going to apply only the mutation operator within genetic operations. The fitness of particular individuals is specified by function (7.25), and the individuals that we are going to be considered as the best fit will be the ones for which this function's value is the lowest.

The parents population $\mathbf{P}$ consists of four individuals ($\mu = 4$), and each of them contains two-element vectors $\mathbf{x} = [x_1, x_2]^T$ and $\sigma = [\sigma_1, \sigma_2]^T$. The initial parents population will be generated at random. Values x_1 and x_2 are drawn from interval $[-1, 1]$ and we assume that $\sigma_1 = \sigma_2 = 1$. Table 7.1 presents chromosomes of the initial population $\mathbf{P}$ and the values of their fitness function.

TABLE 7.1. Parents population P

Individual's number	x_1	x_2	σ_1	σ_2	$f(x_1, x_2)$
1	0.63	0.41	1	1	0.57
2	0.57	-0.91	1	1	1.15
3	-0.67	-0.62	1	1	0.83
4	0.38	0.65	1	1	0.57

It is easy to notice that chromosomes number 1 and 4 are characterized by the lowest value of the fitness function. Now we create a temporary population $\mathbf{T}$ having a size of $\lambda = 4$. With multiple drawing out of the parents

TABLE 7.2. Temporary population T

Individual's number	x_1	x_2	σ_1	σ_2	$f(x_1, x_2)$
1	0.38	0.65	1	1	0.57
2	0.57	−0.91	1	1	1.15
3	−0.67	−0.62	1	1	0.83
4	0.38	0.65	1	1	0.57

population **P**, we select individuals holding numbers 4, 2, 3 and 4. The temporary population **T** is presented in Table 7.2. Let us note that only one best chromosome was selected from population **P** to population **T**.

Next step are genetic operations made on individuals of the temporary population, the result of which will be offspring population **O**. The mutation of chromosome σ needs the determination of parameters τ' and τ according to formula (7.23). We assume that $C = 1$. Then, for $n = 2$, parameters τ' and τ are equal, respectively, 0.5 and 0.5946. The course of the mutation of elements σ_i is presented in Table 7.3, and of elements x_i – in Table 7.4.

TABLE 7.3. Course of mutation of chromosome σ of particular individuals of population T

Individual's number	$N(0,1)$	Gen 1			
		σ_1	$N_1(0,1)$	$\exp(\tau'N(0,1) + \tau N_1(0,1))$	σ_1'
1	1.27	1	0.47	2.50	2.50
2	−0.58	1	0.05	0.77	0.77
3	0.47	1	−0.82	0.78	0.78
4	−2.38	1	0.31	0.37	0.37
Individual's number	$N(0,1)$	Gen 2			
		σ_2	$N_2(0,1)$	$\exp(\tau'N(0,1) + \tau N_2(0,1))$	σ_2'
1	1.27	1	−0.38	1.51	1.51
2	−0.58	1	−0.46	0.57	0.57
3	0.47	1	0.44	1.64	1.64
4	−2.38	1	−1.05	0.16	0.16

After carrying out genetic operations, we obtain eventually population O presented in Table 7.5.

According to the mechanism of the operation of strategy $(\mu + \lambda)$, the new mating pool **P** is created by μ best chromosomes of the old population **P** (Table 7.1) and population **O** (Table 7.5). Thus, we select the individual holding number 4 out of population **O** and individuals holding numbers 1, 4 and 3 out of population **P**. The new population is presented in Table 7.6.

TABLE 7.4. Course of mutation of chromosome **x** of particular individuals of population T

Individual's number	x_1	Gen 1		
		$N_1(0,1)$	$\sigma_1 N_1(0,1)$	x_1'
1	0.38	−0.27	−0.67	−0.29
2	0.57	0.20	0.15	0.72
3	−0.67	−1.14	−0.89	−1.56
4	0.38	−0.27	−0.10	0.28
Individual's number	x_2	Gen 2		
		$N_2(0,1)$	$\sigma_2 N_2(0,1)$	x_2'
1	0.65	−1.03	−1.55	−0.90
2	−0.91	−0.30	−0.17	−1.08
3	−0.62	−1.32	−2.17	−2.79
4	0.65	−1.71	−0.28	0.37

TABLE 7.5. Offspring population O

Individual's number	x_1	x_2	σ_1	σ_2	$f(x_1, x_2)$
1	−0.29	−0.90	2.50	1.51	0.89
2	0.72	−1.08	0.77	0.57	1.68
3	−1.56	−2.79	0.78	1.64	10.22
4	0.28	0.37	0.37	0.16	0.22

TABLE 7.6. New parents population P

Individual's number	x_1	x_2	σ_1	σ_2	$f(x_1, x_2)$
1	0.28	0.37	0.37	0.16	0.22
2	0.63	0.41	1	1	0.57
3	0.38	0.65	1	1	0.57
4	−0.67	−0.62	1	1	0.83

Let us note the best individual in the new population. It was created as a result of the operation of genetic operations and is characterized by values x_1 and x_2 that are close to the optimum solution (the minimum of the function). Let us note that the values of elements σ_1 and σ_2 corresponding to the best individual are clearly lower that the ones that were initially assumed. Such a small range of mutation allowed to obtain a more accurate solution and, which is more important, will be transferred to the new population, thus enabling to narrow the searched space. Individual number 1 from the old population **P** noticed earlier is still one of the best individuals and found its place in the new mating pool.

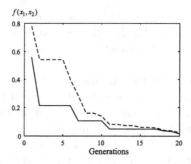

FIGURE 7.13. The best (continuous curve) and average (dotted curve) values of the fitness function of individuals in the 20 subsequent generations of the evolution strategy $(\mu + \lambda)$ in Example 7.12

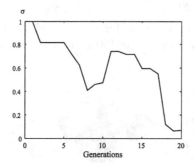

FIGURE 7.14. Average values of the mutation range in 20 subsequent generations in Example 7.12

Figure 7.13 presents the best (continuous curve) and average (dotted curve) values of the fitness function of individuals in the 20 subsequent generations of the evolution strategy $(\mu + \lambda)$. Convergence to the minimum of function (7.25) is shown clearly.

Figure 7.14 shows average values of the mutation range in 20 subsequent generations. These values, in spite of certain fluctuations, have a descending tendency.

Example 7.13

Next example of applying the evolution strategy $(\mu + \lambda)$ will be the search of the minimum of the Ackley test function. The test functions are used to check the correctness and effectiveness of the operation of evolutionary algorithms. *The Ackley function* is given by formula

$$f(\mathbf{x}) = -20\exp\left(-0.2\sqrt{\frac{1}{n}\sum_{i=1}^{n}x_i^2}\right) - \exp\left(\frac{1}{n}\sum_{1}^{n}\cos\left(2\pi x_i\right)\right) + 20 + e \quad (7.26)$$

with constraints $-30 \leq x_i \leq 30$, while n is the number of variables. There is one global minimum of this function $\mathbf{x} = \mathbf{0}$ and then $f(\mathbf{x}) = 0$. For two variables x_1 and x_2, a fragment of the function graph is presented in Fig. 7.15.

In our task, we search for the minimum of the 10 – dimensional Ackley function. A single individual consists of 10 – element vectors $\mathbf{x}$ and σ. We are going to adopt the following parameters of strategy $(\mu + \lambda)$ for experiment: $\mu = 50$ and $\lambda = 200$. The number of generations is 100. The components of vectors $\mathbf{x}$ of particular chromosomes will be initiated by random values out of interval $[-30, 30]$, but all the mutation ranges, i.e. components of vectors σ initially take the value that is equal to 1. In accordance with formula (7.23) parameters τ' and τ for $C = 1$ and $n = 10$ will be determined in the following manner:

$$\tau' = \frac{1}{\sqrt{20}} = 0.2236 \qquad \tau = \frac{1}{\sqrt{2\sqrt{10}}} = 0.3976. \qquad (7.27)$$

The crossover operator is not used in our example. In Fig. 7.16 we may observe the course of the operation of 100 generations of the evolution strategy

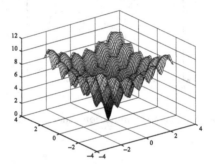

FIGURE 7.15. Function (7.26) for $n = 2$

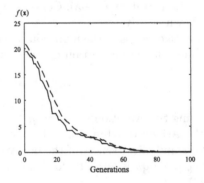

FIGURE 7.16. The best (continuous curve) and average (dotted curve) values of the fitness function of individuals in the 100 subsequent generations of the evolution strategy $(\mu + \lambda)$ in Example 7.13

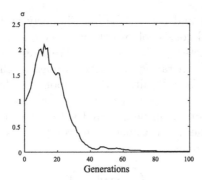

FIGURE 7.17. Average values of the mutation range in 20 subsequent generations in Example 7.13

$(\mu + \lambda)$. In the vicinity of the 70th generation, the best values of the fitness function are getting close to the global minimum. Let us compare this with a change of the average value of the mutation range (Fig. 7.17). Its initial growth may be interpreted as an increase in the variety in the population and thus, an extension of the search range. Just after that moment a strong decrease occurs and the differences in chromosomes caused by the genetic operator are getting reduced. Individuals begin to oscilate around the final solution.

7.3.2.3. Evolution strategy (μ, λ)

A strategy used more frequently than $(\mu + \lambda)$ is the strategy (μ, λ). The operation of both algorithms is almost identical. The difference is that the new population $\mathbf{P}$ containing μ individuals is selected only out of the best λ individuals of population $\mathbf{O}$. Condition $\mu > \lambda$ must be met in order to make it possible. The flowchart presenting the operation of the algorithm is shown in Fig. 7.18.

This method has an advantage over strategy $(\mu + \lambda)$ in one point that is quite important: so far the population could be dominated by one individual with a high value of the fitness function, but too high or too low values of standard deviations. This would hamper the determination of better solutions. Strategy (μ, λ) does not have this disadvantage since old individuals are not transferred to the new mating pool. The genetic operators that can be applied are not different from crossover (7.19) – (7.21) that is already known and mutation (7.22) and (7.24).

Example 7.14

Let us follow some steps of the operation of evolution strategy (μ, λ), by solving a simple task of finding the maximum of function

$$f(x_1, x_2) = -x_1^2 - x_2^2, \tag{7.28}$$

while $-1 \le x_1 \le 1$ and $-1 \le x_2 \le 1$. Similarly as in Example 7.12 a single individual will consist of two-element vectors $\mathbf{x}$ and σ. We assume the following parameters: $\mu = 4$, $\lambda = 8$. We begin the evolution strategy algorithm with generating the initial values of chromosomes of the parents population. We will initiate the components of vector $\mathbf{x}$ with random numbers from the range of the searched solution, that is $[-1, 1]$, and the elements of vector σ will be initiated with values that are equal to 1. Table 7.7 presents the initial values of the components of vectors $\mathbf{x}$ and σ of four drawn individuals.

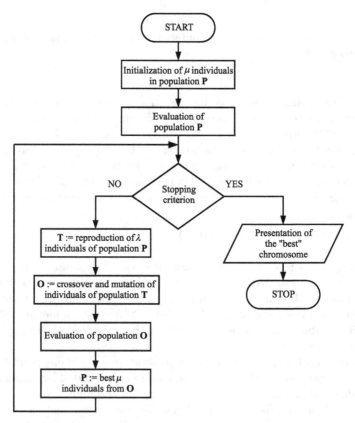

FIGURE 7.18. Flowchart for the evolution strategy (μ, λ)

TABLE 7.7. Initial parents population P

Individual's number	x_1	x_2	σ_1	σ_2	$f(x_1, x_2)$
1	−0.67	−0.68	1	1	−0.91
2	0.36	−1.00	1	1	−1.13
3	−0.97	−0.83	1	1	−1.63
4	−0.19	0.98	1	1	−1.00

The chromosomes in the population that are considered to be the best fit will be those which have the highest value of function (7.28). We select now, by means of multiple drawing, 8 individuals to the temporary population **T**. Let us assume that the chromosomes holding the following numbers were drawn: $2, 2, 2, 3, 3, 1, 3$, and 4. Population **T** is shown in Table 7.8.

Similarly as in Example 7.12 for $C = 1$ and $n = 2$, $\tau' = 0.5$ and $\tau = 0.5946$. The course of the mutation operation for particular components of vectors σ are shown in Table 7.9.

The second column of Table 7.9 contains the values of numbers (random variables with the normal distribution) drawn for this chromosome. However, the remaining columns present the values of parameter σ_i before

TABLE 7.8. Temporary population T

Individual's number	x_1	x_2	σ_1	σ_2	$f(x_1, x_2)$
1	0.36	−1.00	1	1	−1.13
2	0.36	−1.00	1	1	−1.13
3	0.36	−1.00	1	1	−1.13
4	−0.97	−0.83	1	1	−1.63
5	−0.97	−0.83	1	1	−1.63
6	−0.67	−0.68	1	1	−0.91
7	−0.97	−0.83	1	1	−1.63
8	−0.19	0.98	1	1	−1.00

TABLE 7.9. Course of the mutation of genes of chromosome σ

Individual's number	$N(0,1)$	Gen 1			
		σ_1	$N_1(0,1)$	$\exp(\tau' N(0,1) + \tau N_1(0,1))$	σ_1'
1	−0.94	1	1.16	1.25	1.25
2	0.15	1	−0.67	0.72	0.72
3	0.80	1	−1.74	0.53	0.53
4	−1.44	1	−1.01	0.27	0.27
5	−0.77	1	0.05	0.70	0.70
6	−1.67	1	0.27	0.51	0.51
7	−0.11	1	1.31	2.06	2.06
8	−2.36	1	0.71	0.47	0.47

Individual's number	$N(0,1)$	Gen 2			
		σ_2	$N_2(0,1)$	$\exp(\tau' N(0,1) + \tau N_2(0,1))$	σ_2'
1	−0.94	1	−0.04	0.61	0.61
2	0.15	1	−1.00	0.59	0.59
3	0.80	1	−0.42	1.16	1.16
4	−1.44	1	−1.37	0.22	0.22
5	−0.77	1	−0.36	0.55	0.55
6	−1.67	1	0.80	0.70	0.70
7	−0.11	1	−0.03	0.93	0.93
8	−2.36	1	0.49	0.41	0.41

mutation, drawn values $N_i(0,1)$ for a given gene, coefficient $\exp(\tau'N(0,1)+\tau N_i(0,1))$ by which the previous value σ_i will be multiplied, and the new value of the mutation range for both genes of chromosome $\sigma = [\sigma_1, \sigma_2]$. Table 7.10 shows the course of the mutation of chromosomes $\mathbf{x}$ with the use of the new values of the range of mutation of σ'.

TABLE 7.10. Course of the mutation of genes of chromosome $\mathbf{x}$

Individual's number	Gen 1			
	x_1	$N_1(0,1)$	$\sigma_1 N_1(0,1)$	x_1'
1	0.36	−0.50	−0.62	−0.26
2	0.36	1.04	0.75	1.11
3	0.36	0.78	0.41	0.77
4	−0.97	1.23	0.33	−0.64
5	−0.97	−0.48	−0.34	−1.31
6	−0.67	0.84	0.43	−0.24
7	−0.97	−0.42	−0.87	−1.84
8	−0.19	0.81	0.38	0.19

Individual's number	Gen 2			
	x_2	$N_2(0,1)$	$\sigma_2 N_2(0,1)$	x_2'
1	−1.00	−0.05	−0.03	−1.03
2	−1.00	−0.70	−0.42	−1.42
3	−1.00	−1.90	2.21	1.21
4	−0.83	−1.54	−0.33	−1.16
5	−0.83	−0.70	−0.38	−1.21
6	−0.68	0.44	0.31	−0.37
7	−0.83	0.51	0.47	−0.36
8	0.98	−0.12	−0.05	0.93

Subsequent columns of Table 7.10 present the values of genes x_i before mutation, the drawn numbers $N_i(0,1)$, coefficients by which genes x_i will be changed and the new values of genes x_i'. Offspring population $\mathbf{O}$ is presented in Table 7.11.

TABLE 7.11. Offspring population O

Individual's number	x_1	x_2	σ_1	σ_2	$f(x_1, x_2)$
1	−0.26	−1.03	1.25	0.61	−1.13
2	1.11	−1.42	0.72	0.59	−3.25
3	0.77	1.21	0.53	1.16	−2.06
4	−0.64	−1.16	0.27	0.22	−1.76
5	−1.31	−1.21	0.70	0.55	−3.18
6	−0.24	−0.37	0.51	0.70	−0.19
7	−1.84	−0.36	2.06	0.93	−3.52
8	0.19	0.93	0.47	0.41	−0.90

The new parents population is created by four best individuals from population **O**. In our case they hold numbers 6, 8, 1 and 4. Individuals of the new population **P** are shown in Table 7.12.

When comparing the Tables: 7.7 and 7.12, we can see an improvement in the results. The new population **P** is close to the optimum solution. It is also worth noting that the mutation ranges are reduced. Let us follow the operation of the algorithm and analyze the course of subsequent generations. Figure 7.19 shows the values of the best fitness function (continuous curve) and the values of the average fitness function (dotted curve)

TABLE 7.12. New parents population P

Individual's number	x_1	x_2	σ_1	σ_2	$f(x_1, x_2)$
1	−0.24	−0.37	0.51	0.70	−0.19
2	0.19	0.93	0.47	0.41	−0.90
3	−0.26	−1.03	1.25	0.61	−1.13
4	−0.64	−1.16	0.27	0.22	−1.76

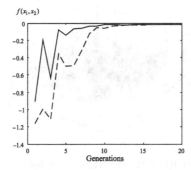

FIGURE 7.19. The best (continuous curve) and average (dotted curve) values of the fitness function of individuals in the 20 subsequent generations of the evolution strategy $(\mu + \lambda)$ in Example 7.14

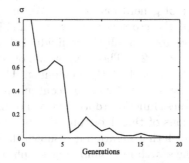

FIGURE 7.20. Average values of the mutation range in 20 subsequent generations in Example 7.14

of the entire population. Figure 7.20 presents the course of changes in the mutation range in this example

It is worth reminding that the best fit individuals from the previous population are not transferred to the newly created population, as in strategy $(\mu + \lambda)$. It is clearly shown in the graph that in the 2nd generation the result of the operation of the algorithm was improved, but we observe worse results in the next step.

Example 7.15

We are going to apply strategy (μ, λ) to search for the minimum of quite complicated *Rastingir test function,* which is described by formula

$$f(\mathbf{x}) = An + \sum_{i=1}^{n} x_i^2 - A \cos(2\pi x_i). \tag{7.29}$$

We assume that $A = 10$ and $n = 10$, $-5.21 \le x_i \le 5.21$, $i = 1, .., 10$. This is a function (Fig. 7.21) with a grid of local minima and one global minimum in point $\mathbf{x} = \mathbf{0}$ for which $f(\mathbf{x}) = 0$.

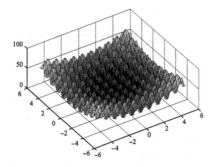

FIGURE 7.21. Function (7.29) for $n = 2$

We determine the following parameters of the algorithm: $\mu = 100$ and $\lambda = 400$. The number of generations is 100. We are going to apply the mutation operator and the crossover operator (7.21).Initially, the drawn values of elements of vector $\mathbf{x}$ of all the individuals from the mating pool come from interval $[-5.21, 5.21]$. The components of vector σ, similarly to previous examples, take value 1 at the beginning. The course of changes in the value of the best fitness function (continuous curve) and in the average value of the fitness function (dotted curve) of the entire population in subsequent generations of the algorithm may be observed in Fig. 7.22. As early as after around 40th generation it is shown that the solution comes close to the value of the search minimum. The graph presenting the values of the fitness of the best individuals is clearly "jagged". This results from the fact that information on the previous best solution is lost in subsequent steps. As we have mentioned before, this is important in order to avoid local minima. Figure 7.23 presents the course of changes in the mutation range. As we are getting closer to the optimum solution, the average value of the mutation range is definitely decreasing.

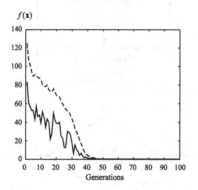

FIGURE 7.22. The best (continuous curve) and average (dotted curve) values of the fitness function of individuals in the 100 subsequent generations of the evolution strategy $(\mu + \lambda)$ in Example 7.15

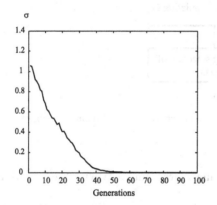

FIGURE 7.23. Average values of the mutation range in 20 subsequent generations in Example 7.15

7.3.3 Evolutionary programming

Initially, evolutionary programming was developed within the context of discovering the grammar of an unknown language. Grammar used to be modelled with the use of a finite automat which was subject to evolution. The results turned out to be promising, however, evolutionary programming gained its popularity when it developed towards numerical optimization. Figure 7.24 presents a flowchart of the evolutionary programming algorithm.

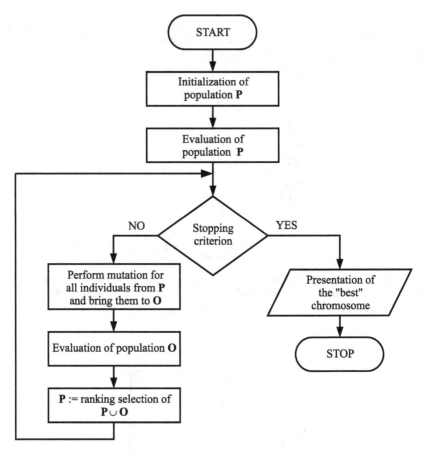

FIGURE 7.24. Flowchart for the evolutionary programming algorithm

Looking at this schema a large similarity with the evolution strategy $(\mu + \lambda)$ may be observed.

However, there is quite a significant difference. During each generation of the evolutionary programming algorithm, the new population **O** is created by means of mutation of each of the individuals of the parents population **P**. In turn, in the evolution strategy $(\mu + \lambda)$ each of the individuals has the same chance to appear in the temporary population **T** on which genetic operations are carried out, while $\lambda \geq \mu$. In the evolutionary programming, populations **P** and **O** are of the same size, i.e. $\mu = \lambda$. Finally, the new parents population **P** is created with the use of ranking selection which operates on both individuals from the old population **P** and mutated individuals from population **O**. It should be noted that mutations of individuals in evolutionary programming consist of a random perturbation of the value of particular genes.

7.3.4 Genetic programming

Genetic programming is an extension of the classical genetic algorithm and may be used in the automatic generation of computer programs. Genetic programming uses programming language LISP in which the program is represented in the same way as data, i.e. in the form of a tree. Therefore, in genetic programming, binary coding was replaced by tree coding. Non-binary values appear in the nodes of the tree, e.g. numerical values, variables, functions or symbols of an alphabet. The introduction of such coding entailed the introduction of new genetic operators. Chromosomes are of a specific construction and thus, traditional crossover and mutation methods that are known from the classical genetic algorithm are of no use here. In genetic programming, chromosomes are coded as trees consisting of nodes and edges. The proper information is contained in nodes. And edges specify mutual relations between nodes. We distinguish between terminal nodes (not having subordinated nodes) and intermediate nodes (opposition to terminal nodes – have connections with subsequent nodes). In tree coding we have one node constituting the root of a tree. This node has no superordinated node. Figure 7.25 shows an example of tree coding. Coded information is a function $x + b \cdot b - (8 + \cos(x))$.

As we have already mentioned, in the case of tree coding there are specific crossover and mutation operators. Crossover consists in using two parents chromosomes and creating two new offspring individuals upon their basis. We select one node at random in each of the chromosomes intended for crossover. It may be both a terminal node and an intermediate node. These nodes, along with corresponding sub-trees, replace each other. As a result of this operation we obtain two individuals that are different from their parents (Fig 7.26).

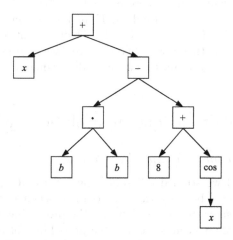

FIGURE 7.25. Example of tree coding of function $x + b \cdot b - (8 + \cos(x))$

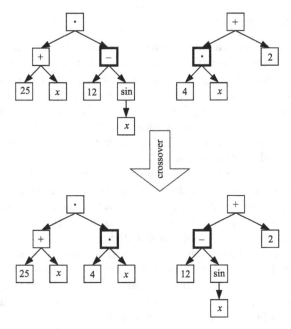

FIGURE 7.26. Example of crossover in tree coding

Mutation is an operation that is a little bit more complicated. As in the previous case, one node is selected at random in the chromosome on which the operation is to be carried out. One of several operations may be made on the content of this node. If it is a terminal node its content may be replaced with another one, and this node still remains a terminal node. Alternatively, a terminal node may be replaced by an intermediate node with a sub-tree generated at random. If an intermediate node is drawn it may be replaced with a terminal node. Then the entire sub-tree of the node is removed. Another possibility is to replace the intermediate node with another node with a sub-tree generated at random.

7.4   Advanced techniques in evolutionary algorithms

7.4.1   Exploration and exploitation

In Subchapter 7.3 we studied different types of evolutionary algorithms. In spite of differences between them, a common schema of operation may be drawn, which generally comes down to the processing of certain population of individuals. During selection, the probability of multiplication of individuals with a better fitness to the new population is higher than in the case of individuals whose fitness is worse. Owing to this operation, the evolutionary algorithm has a tendency to go towards better solutions. This capacity to improve the average value of the fitness of the parents population is called *selective pressure*. We say that the algorithm has a high selective pressure if the expected value of the number of copies of a better individual is higher than the expected value of the number of copies of a worse individual.

Selective pressure is strictly connected with the relation between *the exploration* and *the exploitation* of search space. We speak of two extreme cases where, on one hand, the evolutionary algorithm performs exploration, that is the search of the entire space of solutions in order to approach the global point being the solution to the problem. Another extreme case is exploitation, that is moving within a fragment of space close to the global solution. Exploration is reached by reducing selective pressure. Individuals that are selected are not the ones which are the best fit but the ones which can bring us closer to the optimum solution in the future. Exploitation is reached by increasing selective pressure, because individuals that are transferred to the next population have values which are each time closer to the expected albeit not necessarily global solution. Other parameters of the evolutionary algorithm also have impact on exploration and exploitation. Increasing the probability of mutation and crossover causes a higher diversity of individuals, which, as we can guess, leads to the extension of search space and prevents premature convergence. In consequence, decreasing these probabilities contributes to the creation of individuals that become each time more similar to one another. The evolutionary algorithm may balance between these two extreme cases, and the instrument intended for steering are its parameters. If we balance exploration and exploitation, the average value of fitness of individuals in the population should increase, which may significantly influence the effectiveness of the operation of the evolutionary algorithm.

7.4.2 Selection methods

We are going to present several selection methods that are the most frequently used in genetic algorithms. These methods consist in selecting individuals from the parents population for further processing with the use of genetic operators. One of them, the roulette wheel method, has already been described on the occasion of discussing the classical genetic algorithm (Section 7.3.1). The remaining methods to be presented are the ranking method and the tournament method. There are, of course, other selection methods apart from these ones.

7.4.2.1. Roulette wheel method

A selection method using the roulette wheel mechanism, although it is a random procedure, allows the selection of parent individuals proportionally to the value of their fitness function, i.e. in accordance with the probability of selection given by formula (7.2). Each of the individuals obtains such a number of its copies in the mating pool as it results from the formula

$$e\left(\text{ch}_i\right) = p_s\left(\text{ch}_i\right) \cdot K, \tag{7.30}$$

where K is the number of chromosomes $\text{ch}_i, i = 1, 2, ..., K$, in the population, and $p_s s\left(\text{ch}_i\right)$ means the probability of selecting chromosome ch_i given by formula (7.2). Precisely, the number of copies of a given individual in the mating pool equals the number representing the integer part of $e\left(\text{ch}_i\right)$. Let us note when applying formulae (7.2) and (7.30) that $e\left(\text{ch}_i\right) = F\left(\text{ch}_i\right)/\overline{F}$, where $\overline{F}$ is the average value of the fitness function in the population. It is obvious that the roulette wheel method may be applied if the values of the fitness function are positive. This method may only be applied in tasks of function maximization (and not minimization).

The minimization problem may, of course, be easily converted to the issue of function maximization and the opposite. However, the possibility to apply the roulette wheel method to one class of tasks only, i.e. to maximization only (or only to minimization) is undoubtedly a disadvantage. Another weak point of the roulette wheel method is the fact that individuals with a very low value of fitness function are eliminated from the population too early, which may cause a premature convergence of the genetic algorithm. In order to prevent this, the scaling of fitness function is applied (Section 7.4.3).

7.4.2.2. Ranking selection

In *the ranking selection* otherwise called *the rank selection*, the individuals of a population are set in an order proportionally to the value of their fitness function. It can be figured out as a ranking list of individuals ranked from the best fit one to the worst (or in the opposite direction), where a number specifying the order of an individual on the list and called *rank* is assigned to each of the individuals. The number of copies of each individual entered to the mating pool $\mathbf{M}\left(t\right)$ is determined according to the previously defined function depending on the rank of the individual. An example of such a function is shown in Fig. 7.27.

The advantage of the ranking method is the possibility to apply this method both to the function maximization and minimization. There is no need of scaling due to the problem of premature convergence, which may occur with the roulette method.

7.4.2.3. Tournament selection

In *the tournament selection*, individuals of a population are divided into sub-groups and next the individual with the best fitness is selected out of each of the sub-groups. There are two types of such selection: *deterministic*

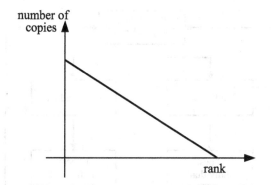

FIGURE 7.27. Ranking selection

tournament selection and *stochastic tournament selection*. In the deterministic case, the selection is made with the probability equal to 1, and in the case of the stochastic selection – with the probability less than 1. The subgroups may be of any size, most frequently a population is divided into sub-groups composed of 2 or 3 individuals. The tournament method can be applied both to maximization and minimization problems. Besides, it can be easily extended to tasks concerning multi-criterion optimization, that is the optimization of several functions at a time. In *the tournament method,* the size of the sub-groups constituting a population may be changed. Research shows that the tournament method works better than the roulette method.

Figure 7.28 presents a block diagram illustrating the tournament selection method for sub-groups composed of 2 individuals. It is easy to generalize it to a larger size of sub-groups.

7.4.2.4. Other selection methods

There are many different types of selection algorithms. The methods previously presented (the roulette, tournament and ranking methods) are the most frequently used. Other methods constitute a modification or combination of them.

Threshold selection is a particular case of ranking selection where the function determining the probability of transferring an individual to the mating pool has a form of threshold. This allows the determination of an appropriate selective pressure by steering the threshold value, on which the selection of fit individuals depends.

An interesting selection method is *crowding selection* where the newly created individuals replace parent individuals that are the most similar to them, irrespectively of the value of their fitness function. The purpose of such a procedure is to keep the biggest possible diversity of the population. The introduced parameter *CF (crowding factor)* determines the number of

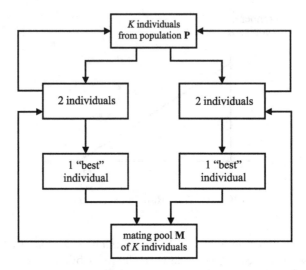

FIGURE 7.28. Block diagram illustrating the tournament selection method for sub-groups composed of 2 individuals

parents that are similar to the newly created individual and out of which the individual to be removed will be drawn. The similarity between the new and the old individuals is indicated by the Hamming distance.

7.4.3   Scaling the fitness function

The scaling of the fitness function is made mainly due to two reasons. Firstly, in order to prevent a premature convergence of the genetic algorithm. Secondly, in the final phase of the algorithm, in case where the population keeps a significant diversity but the difference between the average fitness values and the maximum value is minor. The scaling of the fitness function may then prevent such a situation in which average and best individuals obtain almost the same number of offspring in next generations, which is an unwanted phenomenon. But a premature convergence of the algorithm means that the best but not yet optimal chromosomes dominate in the population. This phenomenon may occur in an algorithm with the roulette wheel selection method. During several generations, with a selection that is proportional to the value of the fitness function, the population will contain only the copies of the best chromosome from the initial population. It is not probable that such a chromosome represent an optimum solution given the fact that the initial population is only a small sample of the entire search space. Scaling the fitness function protects the population against the domination of a chromosome that is not optimal and thus, prevents a premature convergence of the evolutionary algorithm.

Scaling consists of an appropriate transformation of the fitness function. We can distinguish between 3 basic scaling methods: linear scaling, sigma truncation, and power law scaling.

7.4.3.1. Linear scaling

Linear scaling consists in transforming the fitness function F to form F' through the following linear dependency:

$$F' = a \cdot F + b, \tag{7.31}$$

where a and b are constants selected in a way allowing the average value of the fitness function after scaling to be equal to the average value of the fitness function before scaling, and the maximum value of the fitness function after scaling to be a multiple of the average value of the fitness function. The multiplication coefficient is often assumed to be between 1.2 and 2. It should be ensured that function F' does not take negative values.

7.4.3.2. Sigma truncation

Sigma truncation is a scaling method consisting in transforming the fitness function F to F' according to the following dependency:

$$F' = F + \left(\overline{F} - c \cdot \sigma\right), \tag{7.32}$$

where $\overline{F}$ is an average value of the fitness function in the population, c is a small natural number (usually from 1 to 5), and σ is the standard deviation in the population. In case where negative values of F' occur then they are assumed to be equal to zero.

7.4.3.3. Power law scaling

Power law scaling is a scaling method transforming the fitness function F according to the dependency:

$$F' = F^k, \tag{7.33}$$

where k is a number close to 1. The selection of k usually depends on the problem. For instance, it can be assumed that $k = 1.005$.

In paper [75] other types of scaling are also described, for example, logarithmic scaling, "window scaling", Boltzmann scaling and exponential ranking scaling.

7.4.4   Specific reproduction procedures

The specific reproduction procedures are the elitist strategy and the genetic algorithm with a partial replacement of the population.

7.4.4.1. Elitist strategy

The *elitist strategy* consists in protecting the best chromosomes in subsequent generations. In the classical genetic algorithm the best fit individuals are not always transferred to the next generation. It does not always happen that the new population $\mathbf{P}(t+1)$ contains a chromosome from population $\mathbf{P}(t)$, having the highest value of the fitness function. The elitist strategy is applied in order to protect populations against the loss of such an individual. It is always included in the new population.

7.4.4.2. Genetic algorithm with a partial replacement of the population

The genetic algorithm with a partial replacement of population, also called the *steady-state algorithm*, is characterized by the fact that a part of the population is transferred to the next generation without any changes. This means that this part of the population does not undergo crossover and mutation operations. It often happens in a specific implementation of this algorithm that only one or two individuals are replaced in a given moment instead of crossover and mutation within the entire population.

Example 7.16

We are going to solve again the knapsack problem presented in Example 7.6. This time we are going to use the roulette wheel method and the elitist strategy. The elitist strategy will consist of an automatic transfer of 2 best fit individuals to the next population.

Figure 7.29 shows graphs representing the value of the fitness function of the best individual (continuous curve) and the average value of all the individuals of the population (dotted curve) in subsequent generations of the genetic algorithm. It is easy to notice that changes in the value of the fitness function of the best individual occur in a less chaotic way than in Example 7.6. In subsequent generations the best solutions are not lost and thus, in Fig. 7.29 the maximum value of the fitness function grows or does not change. It is worth to emphasize one more advantage of applying the elitist strategy. New solutions have been found, which allow packing 34 or 35 items into the bag. Let us analyse the histogram (Fig. 7.30) of solutions in 100 subsequent runs of the genetic algorithm and compare with a similar graph (Fig. 7.6) from Example 7.6. We can see a visible improvement in the results. Only one solution with 33 items was found (which was the best solution last time), 57 solutions gave 34 items, and in 42 cases the chromosome was the solution allowing to pack 35 items to the bag.

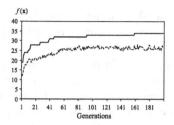

FIGURE 7.29. Fitness function for the best chromosome (continous curve) and average value (dotted curve) of the fitness of individuals in 200 subsequent generations in Example 7.16

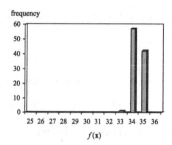

FIGURE 7.30. Histogram of various solutions in 100 runs of the genetic algorithm in Example 7.16

7.4.5 Coding methods

In the classical genetic algorithm the binary coding of chromosomes is used. We apply the known method of representing decimal numbers in the binary system where each bit of the binary code corresponds to the subsequent power of number 2. For instance, the binary sequence $[10011]$ is a code of number 19, because $1 \cdot 2^4 + 0 \cdot 2^3 + 0 \cdot 2^2 + 1 \cdot 2^1 + 1 \cdot 2^0 = 19$. In case of coding real numbers the value of independent variable $x_i \in [a_i, b_i] \in R$ coded with the use of ni bits is determined upon the basis of the value of genes of chromosome $\mathbf{x}$ having subsequent numbers from $s(i)$ to $s(i) + n_i - 1$:

$$x_i = a_i + \frac{b_i - a_i}{2^{n_i} - 1} \sum_{j=0}^{n_i - 1} 2^j x_{s(i)+j}. \tag{7.34}$$

The formula presented above results from a simple mapping of the linear interval $[a_i, b_i]$ into interval $[0, 2^{n_i} - 1]$ where $2^{n_i} - 1$ is a decimal number coded in the form of binary sequence with length n_i composed of nothing but numbers one, and 0 is, of course, a decimal value of a binary sequence with length n_i composed of nothing but zeros.

In genetic algorithms, the Gray code may be applied for instance, which code is characterized by the fact that binary sequences corresponding to two subsequent integer numbers differ by one bit only. Such a manner of coding the chromosomes may turn out to be justified due to the operation of mutation.

Logarithmic coding is applied in order to reduce the length of chromosomes in the genetic algorithm. This coding is used mainly in optimization problems with many parameters and having large search spaces.

In the logarithmic coding the first bit (α) of a code sequence is a bit of the exponential function's sign, the second bit (β) is a bit of the sign of the exponential function's exponent, and the remaining bits (bin) are a representation of the exponential function's exponenty, i.e.

$$[\alpha\beta bin] = (-1)^\beta e^{(-1)^\alpha [bin]_{10}}, \qquad (7.35)$$

while $[bin]_{10}$ means a decimal value of a number coded in the form of binary code bin.

Example 7.17

It is easy to verify that $[10110]$ is a code sequence of number

$$x_1 = (-1)^0 e^{(-1)[110]_{10}} = e^{-6} = 0.002478752.$$

Similarly, $[01010]$ is a code sequence of number

$$x_2 = (-1)^0 e^{(-1)[010]_{10}} = -e^2 = -7.389056099.$$

Let us note that this way, using 5 bits we can code numbers from interval $\left[-e^7, e^7\right]$. This is a much larger range than $[0, 31]$ used in binary coding.

The advantage of the binary coding is the simplicity in usage and the possibility to use simple genetic operators; however, there are also disadvantages. The binary alphabet is not natural for the majority of optimization problems. This causes the appearance of huge solution spaces. Even the coding of several independent variables in a chromosome, with a quite high accuracy (a significant number of bits intended for coding the value of the variables), causes a rapid growth of the chromosome length. Coding with floating-point numbers seems to be a much better method, especially in problems of numerical optimization.

Example 7.18

We are going to use the classical genetic algorithm in order to find the maximum of function

$$f(x_1, x_2) = 2(1 - x_1)e^{-x_1^2 - x_2^2} + 3e^{-x_1^2 - (x_2-2)^2} + 4e^{-(x_1-2)^2 - x_2^2} \qquad (7.36)$$

with constraints $-3 < x_1 < 3$ and $-3 < x_2 < 3$. The graph of this function is depicted in Fig. 7.31. Variables x_1 and x_2 are coded with the use of linear mapping (7.34), using 20 bits for each variable. For instance, for chromosome

$$[00010111101101011101 \quad 00011001100101101000]$$

we may determine the phenotype

$$[1.3797, \ -2.4703],$$

where $\alpha_i = -3$, $b_i = 3$, $i = 1, 2$. We assume the probability of mutation $p_m = 0.01$ and the probability of crossover $p_c = 0.7$, and the population comprises 20 individuals. The algorithm is going to be stopped after 100 generations. Changes in the maximum (continuous curve) and average (dotted curve) value of the fitness function may be observed in Fig. 7.32. We can see that the solution was found quite quickly, that is as early as in the 37th generation.

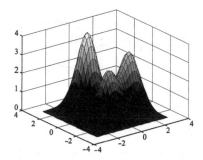

FIGURE 7.31. Function (7.36) for $n = 2$

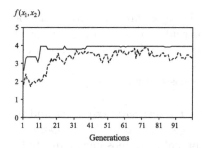

FIGURE 7.32. Fitness function for the best chromosome (continous curve) and average value (dotted curve) of the fitness of individuals in 100 subsequent generations in Example 7.18

Let us have a look at Figs. 7.33 and 7.34. Figure 7.33 shows the solutions generated initially. Points which are solutions coded in particular chromosomes are marked by "x". Let us note that at the end of the algorithm operation (Fig. 7.34) the majority of individuals are found in the neighborhood of the searched maximum.

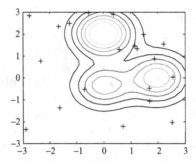

FIGURE 7.33. Distribution of solutions after generation of initial population in Example 7.18

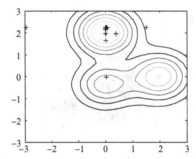

FIGURE 7.34. Distribution of solutions after 100 generations in Example 7.18

7.4.6   Types of crossover

In the classical genetic algorithm, the crossover operation is a so-called one-point crossover presented in section 7.3.1. Other types of crossover are also applied: two-point, multiple-point or uniform crossover.

7.4.6.1. Two-point crossover

The difference between *the two-point crossover* and the one-point crossover is, as the very name suggests, that in *the two-point crossover* the offspring inherit fragments of parents chromosomes indicated by 2 drawn crossover points.

Example 7.19

Let us discuss two chromosomes $ch_1 = [1101001110]$ and $ch_2 = [1011111100]$ which undergo two-point crossover. Two crossover points were drawn: 2 and 7. The crossover process is as follows:

Pair of parents: Pair of offspring:

$ch_1 = [11 \mid 01001 \mid 110]$ $\xrightarrow{\text{crossover}}$ $[11 \mid \mathbf{11111} \mid 110]$

$ch_2 = [10 \mid 11111 \mid 100]$ $[10 \mid \mathbf{01001} \mid 100]$

where symbol | denotes the crossover point and the bold font denotes the replaced fragments of chromosomes.

7.4.6.2. Multiple-point crossover

The multiple-point crossover is a generalization of previous operations and is characterized by an appropriately larger number of crossover points.

Example 7.20
We are going to perform multiple-point crossover using four crossover points, on chromosomes ch_1 and ch_2, which are presented in Example 7.19. The following crossover points were drawn: 1, 4, 6, and 9. The crossover process is as follows:

Pair of parents:
$ch_1 = [1 \mid 101 \mid 00 \mid 111 \mid 0]$ $\xrightarrow{\text{crossover}}$
$ch_2 = [1 \mid 011 \mid 11 \mid 110 \mid 0]$

Pair of offspring:
$[1 \mid \mathbf{011} \mid \mathbf{00} \mid \mathbf{110} \mid 0]$
$[1 \mid \mathbf{101} \mid \mathbf{11} \mid \mathbf{111} \mid 0]$

Example 7.21
For chromosomes ch_1 and ch_2, presented in Example 7.19 we are going to perform multiple-point crossover taking into consideration three crossover points. The following crossover points were drawn: 4, 6 and 8. The crossover process is as follows:

Pair of parents:
$ch_1 = [1101 \mid 00 \mid 11 \mid 10]$ $\xrightarrow{\text{crossover}}$
$ch_2 = [1011 \mid 11 \mid 11 \mid 00]$

Pair of offspring:
$[1101 \mid \mathbf{11} \mid 11 \mid \mathbf{00}]$
$[1011 \mid \mathbf{00} \mid 11 \mid \mathbf{10}]$

Crossover for 5 or a higher odd number of crossover points is performed analogically. One-point crossover is of course a particular case of such a crossover.

7.4.6.3. Uniform crossover

The uniform crossover also called *steady crossover* is performed according to a drawn pattern indicating which genes are inherited from the first one of the parents (the remaining genes come from the other parent). This type of crossover may be applied to different types of the coding of the chromosome. Only one condition must be met – the chromosomes must be of the same length.

Example 7.22
Let us assume that for the same pair of parents as in Example 7.19 the following pattern was drawn: 0101101110 in which 1 means the takeover of

the gene on an appropriate position (locus) from parent 1, and 0 – from parent 2. Then the crossover is performed in the following way:

Pair of parents: Pair of offspring:
$ch_1 = [1101001110]$ $\xrightarrow{\text{crossover}}$ [**100**1**101**1**00**]
$ch_2 = [1011111100]$ [**111**1**011**1**10**]

The drawn pattern

$$0101101110$$

The replaced genes are in bold.

7.4.7 Types of mutation

The classical genetic algorithm is equipped with a mutation operator. Its task is to introduce certain diversity in the population; however its role is rather small. In other types of evolutionary algorithms, it is the dominant operator. We are going to describe several methods of mutation.

7.4.7.1. Mutation in case of binary coding

We met one of the types of mutation intended for binary coding in Section 7.3.1, on the occasion of discussing the classical genetic algorithm. The mutation operation was applicable to those genes in the chromosome for which the drawn number from interval $[0,1]$ was less than the probability of mutation p_m. Mutation may consist of the negation of the bit value or of the replacement of the bit with the value drawn from set $\{0,1\}$.

7.4.7.2. Mutation in case of coding with floating-point numbers

If a chromosome is coded with the use of real numbers a simple negation can not be performed. Certain generalization of the binary mutation should be made. We assume that the value of the i-th gene x_i is bound by interval $[a_i, b_i]$. For each gene, we draw a number from interval $[0,1]$, and if it is lower than the probability of mutation p_m, then we perform the mutation according to formula

$$y_i = a_i + (b_i - a_i)\, U_i\,(0,1)\,, \tag{7.37}$$

where y_i is a new value of the gene, and $U_i\,(0,1)$ – random variable generated from the uniform distribution in interval $(0,1)$. Let us note that the value of y_i does not depend on x_i thus, the higher the value of probability p_m, the more similar this operator's operation is to the random search of the space.

A more frequently used method is a mutation consisting in adding certain random variable Z_i to each value of gene x_i, i.e.

$$y_i = x_i + Z_i. \tag{7.38}$$

The most frequently used distributions are the normal distribution or Cauchy distribution.

7.4.8 Inversion

Holland [82] presents three techniques allowing to obtain offspring different than parent chromosomes. These are: crossover, mutation and inversion. Inversion operates on a single chromosome, changing the order of allels between two randomly selected positions (locus) of the chromosome. Although this operator was defined by the analogy to the biological process of chromosome inversion, still it is not frequently used in genetic algorithms.

Example 7.23
Let us discuss chromosome [001100111010] as an example of inversion. Let us assume that the following positions were drawn: 4 and 10. After the inversion we will obtain chromosome [001**101110**010], where the replaced genes are in bold.

7.5 Evolutionary algorithms in the designing of neural networks

The use of neural networks for solving any task requires that these networks be designed earlier. An appropriate networks architecture should be determined, and the number of layers, number of neurons in each of them, and the manner of their connection should be specified. Next, learning process is carried out with the use, for instance, of the error backpropagation algorithm. These actions sometimes need quite a lot of time and experience, but evolutionary algorithms can be helpful here.

Evolutionary methods may be applied to solving the following tasks:

- to the learning of weights of neural networks,

- to the determination of optimal architecture,

- to the simultaneous determination of structure and weights value.

The idea that neural networks may be learned with the use of the evolutionary algorithms has appeared in papers of many researchers. First papers on this subject concerned the application of the genetic algorithm as a method for learning small feed-forward neural networks, and a successful use of this algorithm in case of relatively large networks was noted next. The two of the most important arguments for the use of evolutionary algorithms to the problems of optimization of neural networks' weights are, first of all, a global search of the space of weights and avoiding local minima.

Besides, evolutionary algorithms may be used in problems where obtaining information about gradients is difficult or expensive. The evolutionary learning of neural networks' weights is discussed further in Subsection 7.5.1. Optimal designing of the architecture of neural network can be treated as the search of structure which is the best for a specific task. This means the search of the space of architectures and the selection of the best element of this space using a specific criterion of optimum. A typical cycle of the evolution of architectures is presented in Section 7.5.2.

Methods of simultaneous evolutionary learning of the neural network's weights and of searching the optimal architecture can be combined within one evolutionary algorithm. This subject is further discussed in Section 7.5.3.

7.5.1 Evolutionary algorithms applied to the learning of weights of neural networks

The learning of the neural network consists in determining the values of the weights of the network, the topology of which has been determined earlier. Weights are coded in a chromosome in the form of binary sequence or vector of real numbers. Each individual of the population is determined by a total set of the neural network's weights. An example of such a chromosome is presented in Fig. 7.35. In this chromosome, the weights of a network with two inputs, two hidden layers, two neurons in each of these layers and one neuron in the output layer have been mapped. Along with polarization weights, the chromosome contains information about 15 weights. The order of placing the weights in the chromosome is arbitrary, but cannot be changed from the beginning of learning.

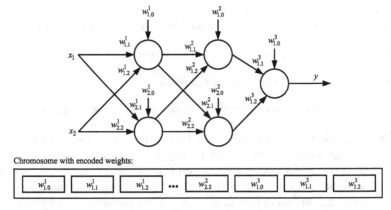

FIGURE 7.35. Coding of the neural network weights in a chromosome

The fitness of individuals will be evaluated upon the basis of the fitness function defined as the sum of squares of errors, being the differences between the network desired signal and network output signal for different input data. Let us now discuss the selection of the schema of weights representation. It should be decided whether binary representation or the presentation of weights in form of real numbers is to be applied. Apart from the natural binary code, the Gray code, logarithmic coding (Section 7.4.5) or other more complex coding methods can also be applied to the binary coding of weights. In some cases we have problems with the accuracy of weights representation. If too few bits were used for the presentation of each weight then learning may take too long and even bring no result, since some combinations of real weights cannot be approximated by discrete values within certain range of tolerance. On the other hand, if we use too many bits then the binary sequences representing large neural networks will be very long, which will considerably prolong the evolution process and make the evolutionary approach to learning impractical. In order to avoid disadvantages of the binary representation schema, the representation of weights with real number was proposed, i.e. one real number to represent one weight.

After selecting the schema of the chromosomes representation, for instance in the way it is depicted in Fig. 7.35, the evolutionary algorithm operates on the population of individuals (chromosomes representing neural networks with the same architecture but with different weights' values) according to the typical evolution cycle comprising the following steps:

1) Decoding each individual of the current population to the set of weights and constructing the corresponding neural network with this set of weights, while the network architecture and the learning rule are specified earlier.

2) Calculating the total mean squared error of the difference between the desired signals and output signals for all the networks. This error determines the fitness of the individual (constructed network); the fitness function can be defined in a different way depending on the type of network.

3) The reproduction of individuals with the probability that is appropriate to their fitness or rank, depending on the selection method applied.

4) Using genetic operators, such as crossover, mutation and/or inversion and obtaining a new generation.

Example 7.24

Let us consider the neural network presented in Fig. 7.36. This network will be used for the realization of XOR system (see Chapter 6). We are going to determine weights $w_{1.1}^1$, $w_{1.2}^1$, $w_{2.1}^1$, $w_{2.2}^1$, $w_{1.1}^2$, $w_{1.2}^2$ and $w_{1.0}^1$, $w_{2.0}^1$, $w_{1.0}^2$, which minimize the error.

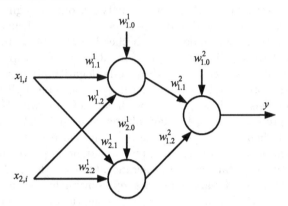

FIGURE 7.36. Structure of the neural network for solving the XOR problem

$$Q = \frac{1}{4} \sum_{i=1}^{4} (d_i - y_i)^2 .$$

The output signal is given by the formula

$$y_i = \left(1/\left(1+\exp\left((-\beta)\left(w_{1.1}^2\left(1/\left(1+\exp((-\beta)\left(w_{1.1}^1, x_{1,i}+w_{1.2}^1, x_{2,i}+w_{1.0}^1\right)\right)\right)\right)\right.\right.\right.$$
$$\left.\left.\left.+w_{1.2}^2\left(1/\left(1+\exp\left((-\beta)\left(w_{2.1}^1 x_{1,i}+w_{2.2}^1 x_{2,i}+w_{2.0}^1\right)\right)\right)\right)+w_{1.0}^2\right)\right)\right)$$

The task comes down to finding the value of the nine above-listed weights for which the network will learn how to correctly perform the XOR task. Information in the chromosome will be stored analogically to what is shown in Fig. 7.35. In order to determine the optimal weights, we are going to use the classical genetic algorithm implemented in *FlexTool* [54] program.

A population consists of 31 individuals. The roulette wheel method was chosen as the selection method. Each value of the weight, being an integer number in interval $[-10, 10]$, undergoes binary coding with the use of five bits. The probability of crossover and mutation is respectively $p_c = 0.77$ and $p_m = 0.0077$, while crossover is performed on two points. The genetic algorithm operates during 200 generations.

The result of the operation of the genetic algorithm is to find weights having values: $w_{1.1}^1 = -10$, $w_{1.2}^1 = 9$, $w_{2.1}^1 = 8$, $w_{2.2}^1 = -9$, $w_{1.1}^2 = 10$, $w_{1.2}^2 = 10$ and $w_{1.0}^1 = -4$, $w_{2.0}^1 = -4$, $w_{1.0}^2 = -4$. The mean squared error for these values of weights is $Q = 2.7260 \cdot 10^{-4}$. As it is shown, the network performs its task with quite low value of this error. Figure 7.37 presents the values of the mean squared error for subsequent generations.

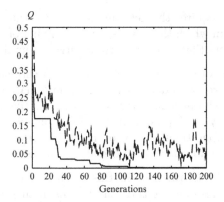

FIGURE 7.37. The best (continuous curve) and average (dotted curve) values of the fitness function of individuals in the 200 subsequent generations of the genetic algorithm in Example 7.24

7.5.2 Evolutionary algorithms for determining the topology of the neural network

In previous section concerning the evolutionary learning of the neural network it was assumed that the network architecture is previously determined and does not change during the process of weights evolution. However, the question how to select the network architecture is essential. It is known that architecture has a decisive impact on the processing of information by the neural network. Unfortunately, in the majority of cases it is created by experts using the trial and error method. It is worth considering an automatic method of designing the architecture of the neural network for solving a specific task. Such a method may be the evolutionary designing of architecture using the evolutionary algorithm.

As in the case of evolutionary learning, the first stage of the evolutionary designing of architecture is taking a decision concerning its appropriate representation in the chromosome. However, in this case the problem does not concern the choice between binary representation and real representation (real numbers), since we deal only with discrete values. At present, this issue is more connected with the concept of the representation structure, i.e. a matrix, graph, or certain general rules. Generally, the types of coding may be divided between *direct coding* and *indirect coding*. Direct coding consists in representing, in a chromosome, the smallest units that we select in order to specify the construction of the neural network. These may be connections, neurons (network nodes) or layers. Depending on the choice we may distinguish between the following network coding in chromosomes.

- **Connection-based encoding** – this is one of the first methods of representing the network structure. The chromosome is a chain of weights values or information on the occurrence of connections. Such an approach requires the determination of the maximum size of the architecture in which there is a specified number of connections, neurons and layers that can occur in the largest neural network.

- **Node-based encoding** – in this method a chromosome represents a chain or a tree of nodes. Each piece of information about the node coded in the chromosome may contain its relative position, connections with previous layers, activation function and others. When performing the operations of crossover and mutation we should ensure that the entire information on coded neurons is exchanged.

- **Layer-based encoding** – in this method, the basic information coded in the chromosome is the layer. This method can be applied to the designing of larger networks. The coding schema is a complicated description of connections between layers and thus, special genetic operators are required.

- **Pathway-based encoding** – the network is presented as a set of pathways from the input to the output of the neuron. This type of coding may be used for the designing of recurrence neural networks. Special operators of genetic algorithms are also used here.

The simplest method of direct coding is the coding of connections in the neural network with the use of connection matrix. Matrix C having a size of $n \times n$, $C = [c_{ij}]_{n \times n}$ may represent connections of the neural network having n nodes (neurons) where c_{ij} means the connection or the lack of connection between neurons i and j, i.e. $c_{ij} = 1$, if such a connection exists, and in case of the lack of such a connection $c_{ij} = 0$. The binary sequence (chromosome) representing connections in the neural network is simply a composition of rows (or columns) of matrix C. An example of such a method of coding for $n = 5$ is shown in Fig. 7.38. If n means the number of neurons in the network then connections between these neurons are presented in the form

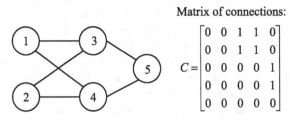

Matrix of connections:

$$C = \begin{bmatrix} 0 & 0 & 1 & 1 & 0 \\ 0 & 0 & 1 & 1 & 0 \\ 0 & 0 & 0 & 0 & 1 \\ 0 & 0 & 0 & 0 & 1 \\ 0 & 0 & 0 & 0 & 0 \end{bmatrix}$$

Chromosome:
[0011000110000010000100000]

FIGURE 7.38. Coding of connections in the neural network with the use of connection matrix C

of binary sequence having a length of n^2. An obvious disadvantage of such a coding schema is a rapid growth of the genotype's length along with an increase in the neural network. However, certain bounds may be easily connected to such a representation schema, which brings the shortening of the length of chromosomes [137]. For instance, we may consider only feed-forward connections, i.e. take into consideration only these elements of matrix C which concern a connection of a given node (neuron) with the next one. Then the chromosome from the example in Fig. 7.38 will be the following: 0110110011.

Direct coding has its advantages, namely, it is quite easy to assess the architecture of the neural network and the easiness of structure coding or decoding. This is quite convenient and effective in case of designing small neural networks. In case of coding large networks longer chromosomes are generated and the effectiveness of the operation of the genetic algorithm is reduced. A solution to the situation may be to apply indirect coding which consists in coding only the most important features instead of each connection of the neural network. It is generally assumed that this coding consists in searching useful common blocks which are repeated in the neural network construction. This representation uses, among other things, coding applied in genetic programming (see Section 7.3.5) or in other methods based on graphs. This schema is more justified biologically, since, following discoveries in the field of neurology, it is not possible for genetic information coded in chromosomes to determine the entire nervous system directly and independently. This results from the fact that, for instance, the genotype of a human being contains much less genes than the number of neurons the human brain contains.

The second phase of the evolutionary designing of the neural network architecture goes along the following steps, in accordance with a typical evolution cycle:

1) Decoding each individual of the current generation to the architecture resulting from the adopted coding schema.

2) Learning each neural network with an architecture obtained in step 1, with the use of a learning rule (certain parameters of the learning rule may be updated during the learning process). Learning should begin with randomly selected initial weights values and parameters of the learning rule, if they occur.

3) The assessment of the fitness of each individual (coded architecture) upon the basis of the above-mentioned learning results, i.e. the smallest total mean squared error or upon the basis of testing if more emphasis is put on generalization, the shortest learning time, the complexity of architecture (e.g. the smallest number of neurons and connections between them), etc.

4) The reproduction of individuals with the probability that is appropriate to their fitness or rank, depending on the selection method applied.

5) Using genetic operators, such as crossover, mutation and/or inversion and obtaining a new generation.

7.5.3 Evolutionary algorithms for learning weights and determining the topology of the neural network

The process of the evolution of architectures described in the previous section has many disadvantages. Learning requires a long time for computer calculations and its result depends on the initiation of connections' weights at the beginning. Another disadvantage is a so-called *permutation problem*: one phenotype may have many different equivalents in genotypes, i.e. a given neural network may be encoded in a chromosome in different ways. These disadvantages may be eliminated by combining the methods of coding weights and structures (based on connections coding), which are presented in Sections 7.5.1 and 7.5.2. The simplest method of representing the neural network is shown in Fig. 7.39. The connection of neurons in the network is determined, as in case of matrix representation, with the use of one bit. Additionally, the values of connections' weights after a binary coding were introduced. Alternatively, a two-level representation may be applied where the weights values are separated from connections. One part of a chromosome will contain the formula of connections, the other part – the weights values. The main characteristic of both methods is that if the gene responsible for connection is inactive (holds the value of zero) then the coded weights values are not taken into consideration when the structure is decoded.

Another method of the simultaneous coding of information about the weights and connections of the network uses the representation presented

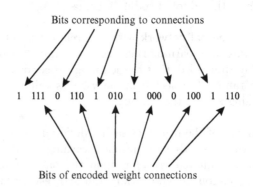

Bits corresponding to connections

1 111 0 110 1 010 1 000 0 100 1 110

Bits of encoded weight connections

FIGURE 7.39. Simultaneous coding of structure and weights of the neural network

in Fig. 7.35. The information on the existence of the connection is contained in the weight value itself. If this value is zero then the connection of this weight is not taken into consideration. This method needs to use an additional genetic operator which would remove or create, with certain probability, new connections (zeroing or drawing the values of connections weights). The cycle of the operation of the evolutionary algorithm applied to the simultaneous learning of weights and determining the neural network structure is similar to the one described in Section 7.5.2.

Example 7.25

We are going to demonstrate an example of using a modified genetic algorithm in order to find an appropriate architecture and the values of weights of the neural network connections. The network will be used to solve the XOR problem, we, thus, have two inputs and one output. We look for a network with a layer structure $(x; 1)$ where x means an unknown number of neurons in the first layer (which in this case is equivalent to the hidden layer). We are going to apply connection-based encoding (Section 7.5.2). Firstly, we are going to determine the maximum network size in order to be able to code all possible connections between neurons. We assume that the first layer may have maximum 5 neurons. The maximum structure of the network is shown in Fig. 7.40.

We begin the preparation of the genetic algorithm with determining the method of task coding. Weights will be coded using the binary method, each with the use of 8 bits $s(i) \in \{0, 1\}$, $i = 0, ..., 7$ in the following manner:

$$w = (-1)^{s(0)} \left(\sum_{i=1}^{7} 2^{i-2} s(i) \right).$$

We assume the first one of the bits $s(0)$ as a sign of the weight value. The application of the formula presented above allows to determine the range of search of the weights value from -31.5 to 31.5. As it is easy to verify, the maximum network structure has 17 connections between neurons and additionally, 6 connections through which the constant value is indicated.

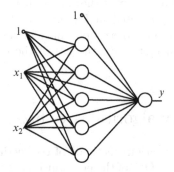

FIGURE 7.40. The maximal structure of the neural network in Example 7.25

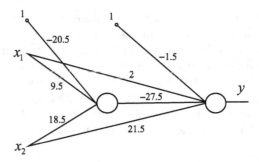

FIGURE 7.41. Final structure of the neural network obtained after 211 generations in Example 7.25

Therefore, a chromosome will represent 23 connection weights, each of them being coded with the use of 8 bits, which brings 184 genes. We use a two-point operator for crossover, but we are going to distinguish between two mutation operators. The operation of the first of them consists in drawing a new weight value. The second mutation operator will be responsible for the change of the network structure, since we assume that a connection exists where the value of this weight is different than zero. When all the coding bits have their weights equal to zero then a given connection is not taken into consideration. This way the lack of input or output connections in certain neuron results in omitting this neuron in the network structure. Owing to this, the architecture of the neural network is modified. The operation of the second mutation operator will consist in removing (we assume weight values equal to 0) or adding a connection (we draw the values of 8 genes). The task of the fitness function is the assessment of the network. We take into consideration the network size, while a reduction of the number of neurons is of great importance. Furthermore, an important coefficient is the mean squared error which we will obtain when testing the network. The result of the algorithm operation is illustrated in Fig. 7.41, where the network obtained after 211 generations is shown. The population contained 100 individuals. As it is shown, the network structure is quite different than the one presented in Example 7.24 as it consists of two neurons only. The mean squared error obtained for this network was 0.0372. The network may learn more with the use of the error backpropagation algorithm.

7.6   Evolutionary algorithms vs fuzzy systems

The literature describes various methods of connecting evolutionary algorithms and fuzzy systems. One of the possibilities consists in controlling the evolutionary algorithm operation with the use of fuzzy knowledge base.

Such a connection permits to steer the algorithm parameters and monitor its operation in order to avoid undesired behavior, such as premature convergence. Another option is to use fuzzy logic in the evolutionary algorithm itself; there is a possibility to define fuzzy genetic operations, and even fuzzy genes. One of the most frequently described hybrid methods consists in the use of evolutionary algorithms for optimization of fuzzy systems.

7.6.1   Fuzzy systems for evolution control

In point 7.4.1 we focused on the problem of exploration and exploitation of the evolutionary algorithm. It is particularly important that the algorithm should be appropriately balanced between those two extreme cases. If, for example, all individuals are similar to each other, yet we are still far from the optimal solution, we should focus on greater exploration of search space. And vice versa, excessive diversity of population results in inability to find the optimum. Thus, we could slightly increase the selective pressure and, consequently, cause the increase in exploitation. It is one of the possibilities to influence the evolutionary algorithm operation. Beside the selective pressure, we can change the crossover and mutation probabilities. While the algorithm is in operation, the expert checks its behavior. On the basis of appropriately selected statistics we can change algorithm parameters so that premature convergence could be avoided. Moreover, we can decide on the solution quality, depending on the time of the algorithm operation.

An alternative to the expert's intervention in algorithm operation are the so-called *adaptive evolutionary algorithms*, capable of setting their own parameters independently while being in operation so as to achieve the best results possible. To modify those parameters, fuzzy inference systems are used. The general schema of an adaptive evolutionary algorithm is presented in Fig. 7.42.

While the evolutionary algorithm is in operation, various statistics are generated. On the basis of this information, a fuzzy inference system checks correctness of the algorithm operation and dynamically influences its operation by changing some of the parameters (e.g. mutation probability). A fuzzy inference system has an implemented knowledge base, appropriately prepared by an expert in the form of linguistic rules. Information on the algorithm operation (statistics) is fed at the system input, next is subject to fuzzification, i.e. is converted into fuzzy sets. The result of a fuzzy system operation may be a decision to change some of the evolutionary algorithm parameters. The parameters controlled may be mutation and crossover probabilities, size of population, selective pressure or other parameters depending on the problem, which is being solved. Such control is performed systematically. The algorithm operation is then steered automatically in a manner permitting to retain an appropriate relationship between exploration and exploitation.

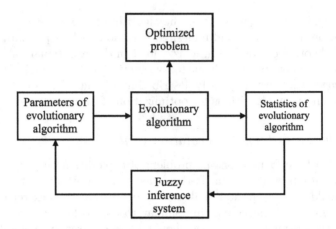

FIGURE 7.42. Block diagram of an adaptive evolutionary algorithm

The statistics of evolutionary algorithm on the basis of which a fuzzy driver makes decisions are checked every fixed number of generations. Those statistics may be divided into two groups. The first one refers to relations between genotypes of individuals in the population, e.g. examining diversity of individuals by means of a certain measure function. The other deals with examining the fitness of phenotypes. The maximum, average and minimum value of the fitness function in the population or the ratio of the best fitness value to the average one could be adopted as the measure. We could also use completely different statistics, for example the number of mutations which improved their fitness, compared to the number of all mutations (which leads to the same calculations as in the 1/5 success rule – point 7.3.3).

Example 7.26
We will show an example illustrating how the rule base looks by means of which we can steer the evolutionary algorithm parameters. The input data for a fuzzy driver will be four parameters; two of them are statistical data reflecting the algorithm behavior, i.e. the ratio of the average value of fitness function to the best value of that function and the ratio of the worst value of fitness function to the average value of that function. In addition, we must take into account the crossover probability p_c and the size of population. The following linguistic values shall be input into the rule base: *big, medium* and *small*. Let's denote the following:

$$\alpha = \frac{\text{average fitness}}{\text{best fitness}},$$

$$\beta = \frac{\text{worst fitness}}{\text{average fitness}}.$$

An example fragment of the rule base may look as follows [125]:

IF α is *big* **THEN** increase the population

IF β is *small* **THEN** decrease the population

IF p_c is *small* **AND** the population is *small* **THEN** increase the population

We can see that the steered parameter is the size of population through which we can influence the diversity of individuals.

7.6.2 Evolution of fuzzy systems

When designing a fuzzy system, we face the task of selecting an appropriate rule base. Like in the case of designing neural networks, it takes time, experience and expert knowledge. This process may be automated thanks to evolutionary algorithms, the flexibility of which and independence of the problem solved is used, among other things, at the following moments of designing a fuzzy system:

- When we have a fuzzy system at our disposal and strive for improving the efficiency of its operation. Evolutionary algorithms may be used for tuning of a membership function, i.e. changing their location or shape.

- When a set of membership functions of linguistic terms is defined, we may generate a rule base by means of evolutionary methods. Three approaches to solve this problem are used – Michigan approach, Pittsburgh approach and iterative rule learning, which stand out due to the method of coding and constructing a rule base.

The Michigan approach uses the concept of the so-called *classifier systems*. Each of the individuals represents one coded rule. All or only part of chromosomes of the population are treated as the rule base which is searched for.

The Pittsburgh approach is a coding method which corresponds rather to the operation of evolutionary algorithms. The whole rule base is coded in one chromosome. Thus, the solution searched for is found in the best fit individual.

The third approach to the search for a rule base is the so-called *iterative rule learning*. It combines the best features of the Michigan and Pittsburgh approaches. The concept of coding one rule per chromosome was used here, yet the creation of the entire base happens gradually. The final rule base is made up of the best individuals being the result of operation of subsequent activations of the evolutionary algorithm.

We will discuss the methods of evolutionary tuning of membership functions, in particular coding methods for those functions. Next, we will present the concept of the Michigan approach, the Pittsburgh approach as well as the iterative rule learning.

7.6.2.1. Tuning of membership functions

The effectiveness of the fuzzy system operation could be increased by appropriate tuning of the fuzzy sets, while the rule base remains unchanged. The evolutionary algorithm modifies membership functions by changing the location of characteristic points of their shapes. Most frequently, they are coordinates of vertices of the figures described by those functions. The information on vertices is coded in chromosomes. The shape of sets and coding are strictly interrelated. Several popular functions can be distinguished (Fig. 7.43):

- isosceles triangle – the functions are coded in a chromosome by means of two extreme points of the triangle: a and b (Fig. 7.43a).

- asymmetrical triangle – location and shape of the asymmetrical triangle is coded by means of three parameters: a, b and c (Fig. 7.43b). In comparison to the isosceles triangle, the location of the central vertex is additionally defined. Instead of coordinates of two extreme points, we can also give distances from the central point.

- trapezoid – a trapezoid is characterized by four points (Fig. 7.43c) but it should be remembered, like in the case of triangles, that the points: a, b, c and d, which represent the trapezoid, must fulfill the condition $a < b < c < d$.

- also other functions could be distinguished, e.g. the Gaussian function which could be described by means of two parameters: center $\bar{x}$ and width σ. The other functions used include the radial function and the sigmoidal function (for extreme fuzzy sets which represent the values of linguistic variables).

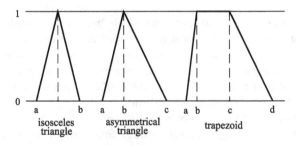

FIGURE 7.43. Different membership functions with characteristic points

Depending on the evolutionary algorithm applied, the characteristic points of membership functions are coded in the chromosome in the binary form or by means of real numbers. After the appropriate representation of fuzzy sets in the chromosome has been selected, the evolutionary algorithm operates on the population of individuals (of chromosomes containing coded shapes of fuzzy system membership functions) according to the evolutionary cycle which may comprise the following steps:

1) Decoding each of the individuals of the population consists in recreation of the set of membership functions and constructing an appropriate fuzzy system. The rule base is pre-defined.

2) The operation of a fuzzy system is evaluated on the basis of the difference (error) between the system's responses and the desired values. This error defines the individual's fitness.

3) Reproduction of individuals and application of genetic operators. Specific techniques depend on the evolutionary algorithm selected, as the genetic algorithm and evolution strategies are characterized by various selection and recombination mechanisms.

4) If the stopping criterion is not met, we proceed to point 1.

Example 7.27
Let us consider example fuzzy sets (linguistic values) of *cold, moderate* and *warm*, which may serve the purpose of describing the surrounding temperature (Fig. 7.44a). At first we assume that the term cold refers to temperatures below 5°C, *moderate* – when the temperature varies between 0 to 20°C, while *warm* – for temperatures above 15°C. The vertex corresponding to the temperature of –5°C (cold) and temperature of 25°C (warm) does not undergo any evolutionary changes. Membership functions have the shape of triangles which may be described by means of characteristic points in the following manner (Fig. 7.44b): two extreme sets by means of one point (*cold* – c, *warm* – d, the other vertices of the triangle are constant), while the central set *moderate* – by means of two points (a and b, vertex of the triangle is halfway between those two points). Let us try to code those fuzzy sets in the chromosome by placing characteristic points one by one next to each other (Fig. 7.44c). The first one refers to the linguistic value of *cold* and is defined by one value c, the fuzzy set of moderate by two points a and b, while the fuzzy set of *warm* by point d. Those values can be presented as real numbers or binary chains by applying one of the coding methods presented above (point 7.4.5).

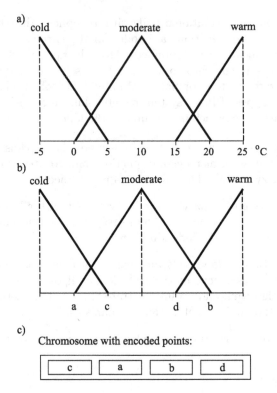

FIGURE 7.44. Coding of triangular membership functions in Example 7.27

Example 7.28

In this example we will show how to code a Mamdani fuzzy set with N rules and n inputs, with trapezoid membership functions. Let us consider the k-th rule of the form (Subchapter 4.9)

$$R^{(k)}: \textbf{IF } x_1 \text{ is } A_1^k \textbf{ AND } x_2 \text{ is } A_2^k...\textbf{AND } x_n \text{ is } A_n^k \textbf{ THEN } y \text{ is } B^k$$

All fuzzy sets A_i^k, $i = 1, ..., n$, $k = 1, ..., N$, are coded, as shown in Fig. 7.45. Those sets are represented by the following chromosomes:

$$C_{i,k} = (a_{i,k}, b_{i,k}, c_{i,k}, d_{i,k}).$$

Taking into consideration all fuzzy sets in the k-th rule, we obtain

$$C_k = (a_{1,k}, b_{1,k}, c_{1,k}, d_{1,k}, ..., a_{n,k}, b_{n,k}, c_{n,k}, d_{n,k}, a_k', b_k', c_k', d_k'),$$

while set B^k is also a trapezoid set described by means of points (a_k', b_k', c_k', d_k'). The final chromosome C, coding all rules, will be obtained by combining the subsequent fragments C_k, $k = 1, ..., N$, i.e.

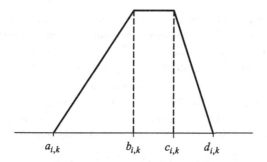

Chromosome with encoded points:

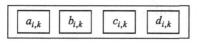

FIGURE 7.45. Coding of the trapezoidal membership function in Example 7.28

$$C = (C_1, C_2, ..., C_N).$$

The length of this chromosome equals $4N(n+1)$. Therefore, for a larger number of rules it is worth applying coding with the use of real numbers.

Example 7.29

We will demonstrate coding of the Takagi-Sugeno fuzzy system rules (see Chapter 9) in a chromosome. We assume that the system has N rules and n inputs, while membership functions of particular fuzzy sets are trapezoidal functions. Below is presented the k-th rule of the system

$$R^{(k)}: \textbf{IF } x_1 \text{ is } A_1^k ... \textbf{AND } x_n \text{ is } A_n^k \textbf{ THEN } y = c_0^{(k)} + c_1^{(k)} x_1 + ... + c_n^{(k)} x_n$$

As we can see, there occur functional dependencies (linear functions) with parameters $c_j^{(k)}$, $j = 0, ..., n$, $k = 1, ..., N$, in the rule consequents. Those parameters must also be coded in the chromosome. Two parts could be distinguished in a fragment of the individual describing this rule. The first of them describes coordinates of all membership functions of rule antecedents and looks as follows:

$$C_k^1 = (a_{1,k}, b_{1,k}, c_{1,k}, d_{1,k}, ..., a_{i,k}, b_{i,k}, c_{i,k}, d_{i,k}, ..., a_{n,k}, b_{n,k}, c_{n,k}, d_{n,k}).$$

In the second part we will place parameters of rule consequents

$$C_k^2 = \left(c_0^{(k)}, c_1^{(k)}, ..., c_n^{(k)} \right).$$

While encoding parameters $c_j^{(k)}$, we must know the range of values which decide what is essential, for instance, for binary coding. Unfortunately, most frequently we do not know this range. In this case we can apply a method known in the literature in English as *angular coding* [28]. Instead of direct coding of parameters $c_j^{(k)}$, we could code values $\alpha_{j,k} = \arctan c_j^{(k)}$, which are in the interval $\left(-\frac{\pi}{2}, \frac{\pi}{2}\right)$. This allows the choice of any coding method. Like in the previous example, the whole chromosome consists of pieces of the above coded rules. A chromosome with N rules could look as follows

$$C = (C_1^1, C_2^1, ..., C_N^1, C_1^2, C_2^2, ..., C_N^2).$$

7.6.2.2. Evolution of rules

As has been mentioned in the introduction to this subchapter, we can distinguish between three methods of using evolutionary algorithms in the generation process of fuzzy system rules: Michigan approach, Pittsburgh approach and *iterative learning process*. Let us have a look at individual methods.

7.6.2.2.1. Michigan approach

This approach has been developed at the University of Michigan. A characteristic feature of this approach is that particular rules are coded in separate chromosomes. The original Michigan method uses the concept of the so-called classifier system. We refer the interested reader to the literature on this subject [29, 35, 63]. Now we will only present the coding method of particular rules in separate chromosomes.

Example 7.30
Let us consider a fuzzy system with three inputs and one output. The linguistic variables take the following values:

1: *very little,*

2: *little,*

3: *medium,*

4: *much,*

5: *very much.*

Each of the terms will be coded by means of a corresponding digit. Let us also add a symbol denoting that a relevant linguistic value is missing: #. In Fig. 7.46 two rules have been coded (Parent 1 and Parent 2). For example, chromosome 4#12 denotes the following rule:

IF x_1 is *much* **AND** x_3 is *very little* **THEN** y is *little*

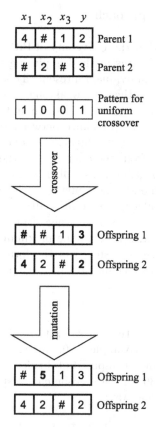

FIGURE 7.46. Genetic operations in the Michigan approach

Variable x_2, as we can see, does not occur, as in the relevant place of the chromosome the sign # is placed. Generating new individuals is performed by means of genetic operators already known to us – crossover and mutation. Example operations have been shown in Fig. 7.46. A homogenous crossover and a mutation consisting in drawing a new gene value have been used in this case. Selection consists in rejecting the worst fit individuals and replacing them with new ones, created by means of genetic operators. The literature specifies a number of methods of rule base automatic generation which are referred to as the Michigan approach. Most of them, however, have nothing in common with the operation algorithm of classifier systems from which the original Michigan method derives. What they have in common, however, is the concept of representation of one rule in the chromosome and regarding the whole population as a set of rules searched for.

7.6.2.2.2. Pittsburgh approach

This approach originated at the University of Pittsburgh, hence the name. A characteristic feature of this method is placing the whole solution in a single chromosome. Each individual represents a complete set of rules. Individuals compete with each other, the weak ones die, the strong survive and reproduce. It happens on the basis of selection and crossover and mutation operators. It is possible to keep balance between exploration of new solutions and exploitation of the best solution. The operation of the Pittsburgh approach does not differ from the operation of the classical genetic algorithm. Unfortunately, it has faults, as well. The chromosome coding all rules is much bigger than in the Michigan method, which extends the operation time of the evolutionary algorithm, additionally the chance to find a correct solution also decreases. In the simplest algorithm of searching for a fuzzy rule base we must determine a priori the maximum number of rules. The set of all fuzzy system rules is stored in the form of a chain of constant length with division into subchains in which particular rules are coded.

Example 7.31

We will show the coding method in the Pittsburgh method, considering a fuzzy system described in Example 7.30. In Fig. 7.47 two rule bases have been coded, each of them is composed of 3 rules (Parent 1 and Parent 2). Similarly to Example 7.30, the chromosome corresponding to Parent 1 can be decoded in the following way:

$$R^{(1)}: \textbf{IF } x_1 \text{ is } very\ little \textbf{ THEN } y \text{ is } very\ little$$

$$R^{(2)}: \textbf{IF } x_2 \text{ is } medium \textbf{ AND } x_3 \text{ is } very\ little \textbf{ THEN } y \text{ is } very\ little$$

$$R^{(3)}: \textbf{IF } x_1 \text{ is } much \textbf{ AND } x_2 \text{ is } medium \textbf{ AND }$$
$$x_3 \text{ is } very\ little \textbf{ THEN } y \text{ is } little$$

The operation of genetic operators is illustrated in Fig. 7.47. Those operators work in a similar way as in the Michigan method (Fig. 7.46), yet we can clearly see the difference in the chromosome length.

Obviously, it is a very simple algorithm of rules generation, however we can apply its numerous modifications, for instance a variable chromosome length, to generate a various number of rules. Additionally, by coding the shapes of membership functions in the manner specified in point 7.6.2.1, beside rules generation we can tune at the same time the parameters of membership functions of particular fuzzy sets.

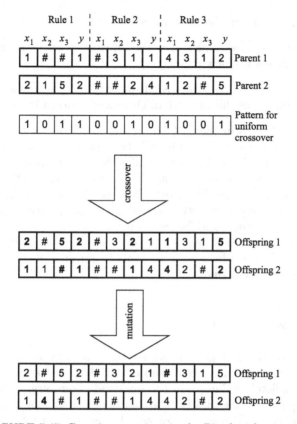

FIGURE 7.47. Genetic operations in the Pittsburgh approach

7.6.2.2.3. Iterative rule learning

Iterative rule learning was developed as an attempt to combine the best features of the previously described Michigan and Pittsburgh approaches. This method is characterized by the fact that each chromosome in the population represents a single rule, like in the Michigan approach. In contrast, what is characteristic for the Pittsburgh approach is the fact that only the best individual in the population is chosen. The complete rule base can be obtained by repeating the evolutionary algorithm many times, each time adding the best rule to the rule base until the entire solution is found. The rule base generation algorithm based on the iterative rule learning could be presented as follows:

1) Apply an evolutionary algorithm to find a rule. As the fitness of the chromosome coding a rule we can take e.g. the number of correctly classified learning data or the rule simplicity.

2) Attach the rule to the ultimate set of rules.

3) Evaluate the effectiveness of the set of rules, particularly taking into consideration the newly found rule.

4) If the set of generated rules is adequate to the problem solved, finish the operation of this algorithm. Otherwise proceed to point 1.

Generation of rules in the Pittsburgh and Michigan methods consisted in simultaneous obtaining the entire rule base. In the iterative rule learning particular rules develop independently of each other, without any information on the previously generated ones. It may result in repetition of rules or their mutual exclusion. Therefore, after the operation of the algorithm is finished, rule base simplification procedures are frequently applied. However, already when the algorithm is in operation, we can prevent identical rules from appearing. The method is based on removing such data from the learning sequence which were used for the purpose of correct learning of the previously found rules.

7.7 Notes

The pioneer of classical genetic algorithms was Alex Fraser, who published his first work [57] on this subject in 1957. His scientific activity is almost completely unknown in the worldwide literature. Goldberg [63], Holland [82] and Michalewicz [136] are authors of well-known monographs on evolutionary algorithms. In Poland, the monograph by Arabas [2] is well worth recommending. The authors of monographs [59, 60, 248, 260] discuss various issues concerning evolutionary algorithms and their applications. A popular program for simulation of genetic algorithms is Evolver [50]. In monographs [2] and [136] the authors describe the historic works by Ingo Rechenberg and Hans-Paul Schwefel on evolution strategies and by Lawrence Fogel on evolutionary programming.

The author of the concept of genetic programming is John Koza [120]. Monographs [24, 63, 82, 136] and work [175] discuss the schemata theorem, bricks (or building blocks) hypothesis as well as the theoretical basis for the operation of genetic algorithms. Various selection and scaling algorithms have been described in papers [38] and [75]. The advantages of Gray coding have been described in monographs [2] and [31]. A detailed description of crossover methods can be found in studies [35] and [63].

Evolutionary learning of small feed-forward neural networks has been described in work [256], while the use of evolutionary algorithms for learning of relatively large networks has been presented in work [230]. The description of the evolutionary design of the neural network architecture can be found in works [232, 263, 264]. Various methods of coding the weights of neural networks have been suggested in works [6, 7, 139]. Methods of direct

TABLE 7.13a. List of main features of evolutionary algorithms

	Genetic algorithm (GA)	Evolution strategies (ES)	Evolutionary programming (EP)	Genetic programming (GP)
Coding, represen- tation	In the classical GA the binary coding occurs. It could be replaced by Gray coding or another coding with fixed alphabet may be applied	Representation by means of vectors of floating point numbers, consisting of the values of variables of the task solved and information on the mutation range	Representation adequate to the problem solved, often similar to the representation in ES	Tree coding (originally use of the LISP language)
Mutation operator	Switch of the value of genes for the opposite one	At first random modification of the values of an individual's standard deviations occurs, then modification of values of independent variables	Similar to ES, self-adaptation of the mutation range is possible	Change of the contents of the terminal node or swap from the terminal node to the intermediate node. Swap from the intermediate node along with its subtree to the terminal node or to another intermediate node with a randomly generated subtree.
Mutation role	Secondary, p_m of order $[0, 0.1]$	Primary, self-adaptation of the mutation range is applied	The only genetic operator	Secondary

coding of neural networks structure have been discussed in paper [112]. Papers [110, 230, 263] describe the methods of indirect coding of neural networks with the use of graphs. Simultaneous coding of weights and connections between neural networks has been presented in papers [15, 250, 251]. Paper [22] describes the application of fuzzy systems for evolution control. The issue of using fuzzy logic in evolutionary algorithms has been

TABLE 7.13b. List of main features of evolutionary algorithms

	Genetic algorithm (GA)	Evolution strategies (ES)	Evolutionary programming (EP)	Genetic programming (GP)
Crossover (recombination) operator	One-point, multi-point (in particular two-point), homogenous	No occurrence in the classical version. Averaging crossover or crossover exchanging values of elements of parent vectors is possible.	None	Exchange of subtrees in chromosomes
Crossover role	Primary, p_c in the interval $[0.5, 1]$ and $p_c \gg p_m$	Secondary, lack of crossover possible	None	Primary
Selection of parents or creation of temporary population	Probability of choosing an individual for mating pool depends on the value of its fitness function. The roulette-wheel method is applied as well as the tournament, ranking and other methods	A temporary population is created (of size λ) by means of multiple sampling among μ individuals, the individuals of which are subjected to genetic operators.	Each of the parents creates one offspring, using a mutation operator	Probability of choosing an individual for mating pool depends on the value of its fitness function

described in papers [77] and [78]. Methods of coding various membership functions have been given in papers [29, 102, 103]. The Michigan approach along with the description of classifier systems are presented in monograph [29] and in paper [249]. Methods of coding the rule base with the use of the Michigan and Pittsburgh approaches have been described in paper [88]. A detailed description of the iterative rule learning can be found in monograph [29]. In papers [36, 37, 182] the authors discuss the applications of evolutionary algorithms for designing electric machines, in paper [75] for optimization of power grids, in paper [160] for production planning, while in monograph [153] for technical diagnostics.

The differences between the evolutionary algorithms discussed in this chapter increasingly diminish. At present those algorithms occur only rarely

TABLE 7.13c. List of main features of evolutionary algorithms

	Genetic algorithm (GA)	Evolution strategies (ES)	Evolutionary programming (EP)	Genetic programming (GP)
Creation of a new population	All individuals selected in the selection process are subjected to genetic operators and create a new population. In particular the elitist strategy is applied	Depending on the strategy: $(1+1)$ – choice of the best of two individuals, $(\mu + \lambda)$ – choice of μ individuals from the old and new populations, (μ, λ) – choice of μ from the new, bigger population	μ best individuals from the old and new populations are chosen (equivalent of ES $(\mu+\mu)$)	All individuals selected in the selection process are subjected to genetic operators and create a new population
Application	Combinatorial optimization	Optimization in case of continuous independent variables	Majority of optimization tasks	Generation of programs or optimization of problems described in the form of trees
Authors	Alex Fraser, the 1950s John Holland, the 1960s and 1970s	Ingo Rechenberg, Hans-Paul Schwefel, the 1960s	Lawrence Fogel, the 1960s	John Koza, the 1990s

in their original forms. They are most frequently used for comparative tests. It should be added that the term "genetic algorithms" is used both in a narrow sense to mean classical genetic algorithms or their minor modifications, and in a broad sense meaning evolutionary algorithms which significantly differ from the classical genetic algorithm.

To sum up this chapter, in Tables 7.13a, 7.13b and 7.13c we compare the most important features of evolutionary algorithms, such as: coding methods, crossover and mutation methods and their significance in the algorithm process, selection methods as well as the method of creating a new population. We have presented potential applications and the names of authors of various algorithms. The list enclosed may be helpful in choosing a proper type of algorithm to solve a specific problem.

8

Data clustering methods

8.1 Introduction

In daily life as well as in different fields of science we encounter big, sometimes enormous volume of information. One look is enough for humans to
distinguish the shapes of objects being of interest to us from a specific
image. Intelligent machines, however, are still incapable of prompt and
unerring distinguishing of objects in the image, due to the lack of universal
algorithms which would work in every situation.

The objective of data clustering is a partition of data set into clusters of
similar data. Objects in the data set may be e.g. bank customers, figures or
things in a photograph, sick and healthy persons. A human being may effectively group only one- and two-dimensional data, while three-dimensional
data may cause serious difficulties. The scale of the problem is intensified
by the fact that the number of samples in real tasks may amount to thousands and millions. In the light of those facts it would be very useful to
have algorithms for automatic data clustering. Operation of those algorithms would result in a fixed structure of data partition, i.e. location and
shape of the clusters and membership degrees of each sample to each cluster. Data clustering is a complicated issue as the structures hidden in the
data set may have any shapes and sizes. Moreover, the number of clusters
is usually unknown. Unfortunately, the literature so far does not provide
any algorithm which would work in the case of any shapes of clusters.

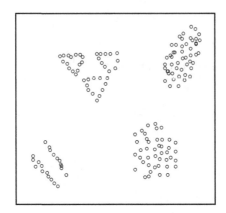

FIGURE 8.1. Various shapes of clusters in two-dimensional space

When choosing the proper clustering algorithm, we should use the knowledge of the problem described by the data set. Generally, data partition should have two features:

- *homogeneity in clusters*, i.e. data within a given cluster should be as similar to each other as possible,

- *heterogeneity between clusters*, i.e. data belonging to different clusters should be as different from each other as possible.

Similarity of data vectors may be defined in different ways, depending on the type of data being clustered. As data most often describe features of objects in a numerical form (as numbers), the most appropriate similarity measure is to measure the distance between objects. We may use e.g. the Euclidean norm which is the most frequently used method of measuring the similarity of objects. Clusters may be represented in different ways. Most frequently the cluster is represented by its central point in the data space. By using various similarity measures we can obtain different shapes of clusters, with the center represented by the central point. Figure 8.1 illustrates examples of various clusters in the two-dimensional space.

In the data clustering tasks we do not have at our disposal the so-called desired output signal, from the teacher. Thus, the process of data clustering may be equated with unsupervised learning. This chapter presents various methods of data partitioning and algorithms for automatic data clustering. Data clustering validity measures are also discussed in this chapter.

8.2   Hard and fuzzy partitions

Data subject to clustering will be represented by n-dimensional vectors $\mathbf{x}_k = [x_{k1}, \ldots, x_{kn}]^T, \mathbf{x}_k \in R^n, k = 1, \ldots, M$, which consist of numerical values describing the objects. The set of M vectors creates matrix $\mathbf{X}$ of dimension $n \times M$

$$
\mathbf{X} =
\begin{bmatrix}
x_{11} & x_{21} & \cdots & x_{M1} \\
x_{12} & x_{22} & \cdots & x_{M2} \\
\vdots & \vdots & \vdots & \vdots \\
x_{1n} & x_{2n} & \cdots & x_{Mn}
\end{bmatrix}.
\tag{8.1}
$$

In case of classification, matrix columns (8.1) are objects, and the rows are features (attributes). In the case of medical diagnostics, objects may be identified with patients, while features will be identified with symptoms of a disease or with results of laboratory analysis of those patients.

If clusters are represented by their centers, the objective of clustering algorithms is to obtain c vectors $\mathbf{v}_i = [v_{i1}, \ldots, v_{in}]$, $i = 1, \ldots, c$, which are representatives of particular clusters in the data space.

It must be emphasized that from the computational point of view it would be very hard to analyze all possible partitions of M objects into c clusters as their number equals [48]

$$
\frac{1}{c!} \sum_{i=1}^{c} \binom{c}{i} (-1)^{(c-i)} i^M.
\tag{8.2}
$$

Example 8.1

Let us consider the problem of partitioning 100 patients ($M = 100$) into 5 different clusters, characterizing particular pathological cases ($c = 5$). It is easy to check that by using formula (8.2), we obtain approximately $6.57 \cdot 10^{67}$ different partitions. Thus it is extremely important to find methods which would perform optimal partition without the necessity to analyze all possible results of clustering.

In the data clustering tasks it is essential to define the type of data partition. The literature distinguishes between hard, fuzzy and possibilistic partitions, where possibilistic partitions are treated as modification of fuzzy partitions.

In the hard data clustering the object entirely belongs or does not belong to a given cluster. The objective of data clustering is data partitioning into c clusters A_i so that

$$
\bigcup_{i=1}^{c} A_i = \mathbf{X},
\tag{8.3}
$$

$$
A_i \cap A_j = \emptyset, \quad 1 \le i \ne j \le c,
\tag{8.4}
$$

$$
\emptyset \subset A_i \subset \mathbf{X}, \quad 1 \le i \le c.
\tag{8.5}
$$

Assumption (8.3) means that the set of all clusters contains all data vectors, and each object belongs to exactly one cluster. The clusters are disjoint (condition (8.4)), and none of them is empty nor contains the whole data set $\mathbf{X}$ (condition (8.5)). In order to partition data into c clusters, it is comfortable to use the partition matrix $\mathbf{U}$ of the dimension $c \times M$, containing the membership degrees μ_{ik} of the k-th data $\mathbf{x}_k$ to the i-th cluster, $k = 1, ..., M$, $i = 1, ..., c$.

Definition 8.1
Let $\mathbf{X} = \{\mathbf{x}_1, ..., \mathbf{x}_M\}$ be a finite set. Let c, $2 \le c < M$, be an integer. *Hard partitioning space* of the set $\mathbf{X}$ is defined in the following way:

$$Z_1 = \left\{ \mathbf{U} \in R^{c \times M} \mid \mu_{ik} \in \{0, 1\}, \quad \forall i, k; \sum_{i=1}^{c} \mu_{ik} \right.$$

$$\left. = 1, \quad \forall k; \quad 0 < \sum_{k=1}^{M} \mu_{ik} < M, \quad \forall i \right\}.$$

(8.6)

The partition above assumes that the object belongs to one cluster only and there are no empty clusters or clusters containing all objects.

Example 8.2
Let us consider the data presented in Fig. 8.1. For such data hard partitioning into three clusters ($c = 3$) may be represented by the following matrix $\mathbf{U}$:

$$\mathbf{U} = \begin{bmatrix} 1 & 1 & 1 & 0 & 0 & 0 & 0 & 0 & 0 & 0 \\ 0 & 0 & 0 & 1 & 1 & 1 & 0 & 0 & 0 & 1 \\ 0 & 0 & 0 & 0 & 0 & 0 & 1 & 1 & 1 & 0 \end{bmatrix}.$$

(8.7)

Let us notice that the object $\mathbf{x}_{10}$ is assigned to cluster 2, although intuitively we would not include it in any of the clusters. However, hard partition makes it necessary for each of the objects to belong to one of the clusters.

The most frequently considered problems do not permit such an unambiguous data partition as in Definition 8.1, as the areas of clusters occurrence may overlap. What is helpful in such a case are algorithms which cause that objects may belong to many clusters with different membership degrees at the same time. It is a natural extension of the hard partition where, like in real problems, a given object may not always be classified unambiguously to one category. For example the boundaries between small, compact and big cars are not strictly defined. There are two types of soft partition: fuzzy and possibilistic. In both partitions the objects may belong to any number of clusters with a membership degree which is a number from the range $[0, 1]$. In the fuzzy partition there is additionally a constraint imposed on membership degrees of a particular object so that the sum of membership degrees of this object to each of c clusters equals 1. This constraint is analogous to the constraint occurring in the probabilistics, therefore this partition is also called *probabilistic partition*.

Definition 8.2

Let $\mathbf{X} = \{\mathbf{x}_1, \ldots, \mathbf{x}_M\}$ be a finite set. Let c, $2 \leq c < M$, be an integer. *Fuzzy partition* of the set $\mathbf{X}$ is defined in the following way:

$$Z_2 = \left\{ \mathbf{U} \in R^{c \times M} \mid \mu_{ik} \in [0,1], \quad \forall i, k; \sum_{i=1}^{c} \mu_{ik} \right. \tag{8.8}$$

$$\left. = 1, \quad \forall k; \quad 0 < \sum_{k=1}^{M} \mu_{ik} < M, \quad \forall i \right\}.$$

The partition above assumes that the object may at the same time belong to all clusters with a certain membership degree but the sum of all membership degrees must equal 1. Moreover, there may be no empty clusters or clusters containing all data.

Example 8.3

Let us consider the data presented in Fig. 8.2. For such data the fuzzy partition into three clusters may be represented by the following matrix $\mathbf{U}$:

$$\mathbf{U} = \begin{bmatrix} 0 & 0.06 & 0.02 & 0.98 & 0.98 & 0.99 & 0.01 & 0.01 & 0 & 0.29 \\ 1 & 0.89 & 0.93 & 0.01 & 0.01 & 0.00 & 0.01 & 0.01 & 0 & 0.33 \\ 0 & 0.05 & 0.05 & 0.01 & 0.01 & 0.01 & 0.98 & 0.98 & 1 & 0.38 \end{bmatrix}. \tag{8.9}$$

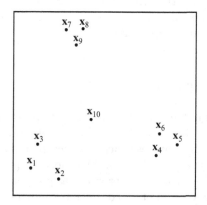

FIGURE 8.2. A dataset in Example 8.3

Let us notice that object $\mathbf{x}_{10}$ is characterized by similar degrees of membership to all three clusters, which corresponds to its almost equal distance from the centers of those clusters. This object may be identified as an outlier (noise). Intuitively we would assign very low membership degrees to noise $\mathbf{x}_{10}$, equal to e.g. 0.1, to all three clusters. However, then the condition that the sum of all membership degrees of a given object must equal 1 would not be met.

Historically in the literature the next partition was a possibilistic partition getting rid of the restriction that the sum of membership degrees is equal one. The only restriction for the object is to belong at least to one cluster. In practice it is not a big inconvenience as the low value of membership degree may be regarded as the lack of membership.

Definition 8.3
Let $\mathbf{X} = \{\mathbf{x}_1, \ldots, \mathbf{x}_M\}$ be a finite set. Let c, $2 \leq c < M$, be an integer. *Possibilistic partition* of the set $\mathbf{X}$ is defined in the following way:

$$Z_3 = \left\{ \mathbf{U} \in R^{c \times M} \mid \mu_{ik} \in [0,1], \ \forall i, k; \ \forall k, \ \exists i, \ \mu_{ik} > 0; \ 0 < \sum_{k=1}^{M} \mu_{ik} < M, \ \forall i \right\}. \quad (8.10)$$

Example 8.4
Let us consider the data presented in Fig. 8.2. For such data the possibilistic partition into three clusters may be represented by the following matrix $\mathbf{U}$:

$$\mathbf{U} = \begin{bmatrix} 0.01 & 0.02 & 0.01 & 0.52 & 0.39 & 0.87 & 0.01 & 0.01 & 0.01 & 0.03 \\ 0.87 & 0.44 & 0.79 & 0.04 & 0.03 & 0.03 & 0.05 & 0.04 & 0.05 & 0.12 \\ 0.01 & 0.01 & 0.02 & 0.01 & 0.01 & 0.01 & 0.53 & 0.63 & 0.79 & 0.03 \end{bmatrix}$$

$$(8.11)$$

Currently the condition that the sum of membership degrees is equal to one does not have to be met. Therefore the noise $\mathbf{x}_{10}$ belongs to all clusters but with a small membership degree.

8.3   Distance measures

An important factor influencing the result of data partition is the method of determining distances between objects. In case of data clustering we measure the distance in the features space in which there are clustered objects and centers (prototypes) of clusters. The most frequently used distance measure is *the Euclidean norm*, interpreted as geometric distance between two points in the space $\mathbf{X}$. Let us consider two points $\mathbf{x}_d = [x_{d1}, \ldots, x_{dn}]^T$ and $\mathbf{v}_i = [v_{i1}, \ldots, v_{in}]^T$. *The Euclidean distance* between those points is defined in the following way:

$$D_{id} = \sqrt{\sum_{j=1}^{n} (x_{dj} - v_{ij})^2} = \|\mathbf{x}_d - \mathbf{v}_i\|_2, \quad (8.12)$$

and in the vector notation

$$D_{id} = \left[(\mathbf{v}_i - \mathbf{x}_d)^T (\mathbf{v}_i - \mathbf{x}_d) \right]^{\frac{1}{2}}. \quad (8.13)$$

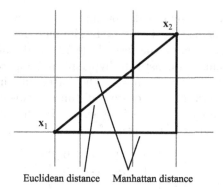

Euclidean distance   Manhattan distance

FIGURE 8.3. Illustration of the Euclidean and Manhattan distance

This measure is a generalization of the *Minkowski metric*

$$D_{id} = \left(\sum_{j=1}^{n} |x_{dj} - v_{ij}|^r \right)^{\frac{1}{r}}. \qquad (8.14)$$

For different values of the parameter r we may obtain other than the Euclidean norm distance measures. For example for $r = 1$ we obtain *the Manhattan distance* (also called *the city block* measure). The interpretation of this measure may be identified with moving along city streets where we are forced to keep to the network of streets and only 90-degree turns are allowed. Figure 8.3 illustrates the interpretation of the Euclidean norm and Manhattan distance. In case of binary variables the Manhattan distance is called *the Hamming distance*. This measure gives the number of bits by which two bit strings differ. Those strings may represent for example black and white images.

Minkowski measures are susceptible to differences in size (scale) of particular variables. High value variables will dominate low value variables which are for example in a different scale. The method to avoid this problem is variables scaling, which leads to the weighted Euclidean norm

$$D_{id} = \sqrt{\sum_{j=1}^{n} w_j \left(x_{dj} - v_{ij} \right)^2}, \qquad (8.15)$$

where w_j is the weight of a given dimension. Assigning the weights to particular variables is useful if we want to obtain the same importance of variables without scaling the data set, or to impose another dimension hierarchy.

If we introduce an additional matrix **A** to the Euclidean norm, the clusters may take the shape of ellipses of any orientation. Then we obtain the family of norms induced by a scalar product. In the simplest case the matrix $\mathbf{A}$ is an identity matrix, i.e. $\mathbf{A} = \mathbf{I}$. The measure then becomes the Euclidean distance given by the formula (8.12), and from the geometric point of view the clusters constitute hyperspheres. Figure 8.4 shows how the Euclidean norm operates for $n = 2$. The dotted lines have been used to mark circles characterized by a constant distance between the points lying on those circles from the central point (center). In general the matrix $\mathbf{A}$ is an $n \times n$ diagonal matrix of the form of

$$
\mathbf{A} = \begin{bmatrix} c_1 & 0 & \cdots & 0 \\ 0 & c_2 & \cdots & 0 \\ \vdots & \vdots & \ddots & \vdots \\ 0 & 0 & \cdots & c_n \end{bmatrix}, \tag{8.16}
$$

where $c_i > 0$, $i = 1, ..., n$. The clusters generated by the norm with such a matrix are hyperellipses with main diagonals perpendicular to the axis of data space, which is illustrated in Fig. 8.5. The dotted lines have been used to mark ellipses characterized by a constant distance between the points lying on those ellipses from the central point.

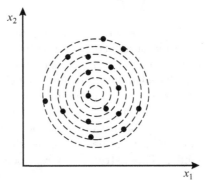

FIGURE 8.4. Euclidean norm

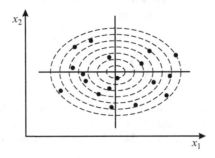

FIGURE 8.5. Diagonal norm

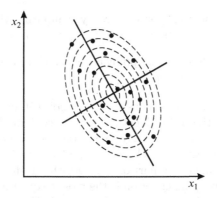

FIGURE 8.6. Mahalanobis norm

Now we will show another method of creation of the matrix $\mathbf{A}$. Let us define the covariance matrix of data from the set $\mathbf{X}$

$$\mathbf{R} = \frac{1}{M} \sum_{k=1}^{M} (\mathbf{x}_k - \overline{\mathbf{x}}) (\mathbf{x}_k - \overline{\mathbf{x}})^T , \qquad (8.17)$$

where $\overline{\mathbf{x}}$ means the average of data $\mathbf{x}_k$, $k = 1, ..., M$. Matrix $\mathbf{A}$ is defined in the following way:

$$\mathbf{A} = \mathbf{R}^{-1}. \qquad (8.18)$$

Matrix $\mathbf{A}$ created in this way induces *the Mahalanobis norm* in the space R^n, and the clusters are now hyperellipses with any shape and orientation, which is illustrated in Fig. 8.6.

8.4   HCM algorithm

The HCM algorithm (*Hard C-Means*) unambiguously partitions the data contained in the matrix $\mathbf{X}$ into c clusters. When executing this algorithm, we compute the distance between each vector $\mathbf{x}_k \in R^n$, $k = 1, \ldots, M$ and the cluster center $\mathbf{v}_i$, $i = 1, ..., c$. *The cluster center* is the average of the location of all objects belonging to this cluster. It is convenient to describe the membership in a cluster by means of matrix $\mathbf{U} = [\mu_{ik}] \in Z_1$ (see Definition 8.1). Elements of this matrix are zeros and ones saying that the object $\mathbf{x}_k$ belongs to the i-th cluster. The algorithm is performed in the following stages:

1. Algorithm initialization.

2. Determining the membership of objects on the basis of their distance from the cluster centers.

3. Determining new cluster centers by computing the average of the location of the objects belonging to a given cluster.

4. Checking the algorithm stopping criterion. If the condition is not met, then we proceed to step 2.

The algorithm initialization consists in the choice of the number of clusters c and determining the initial location of their centers. This location may be chosen at random. Alternatively, the initial location of the centers may be identical with c vectors $\mathbf{x}_k$ chosen at random or with first c objects in the data set. A detailed flowchart of the algorithm is illustrated in Fig. 8.7. The algorithm stopping criterion is the most frequently an appropriately small change of the value of elements of the matrix $\mathbf{U}$, that is $\left\|\mathbf{U}^{(t+1)} - \mathbf{U}^{(t)}\right\| < \varepsilon$, where ε is a fixed constant. Alternatively, we may check the change

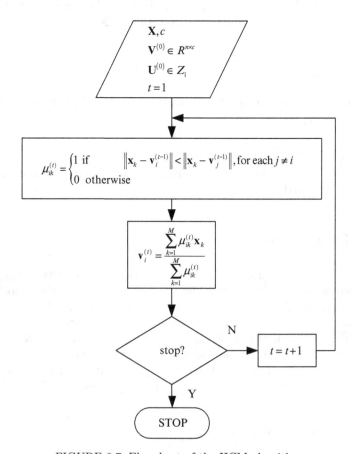

FIGURE 8.7. Flowchart of the HCM algorithm

of cluster centers location, i.e. $\left\| \mathbf{V}^{(t+1)} - \mathbf{V}^{(t)} \right\| < \varepsilon$. The HCM algorithm may give various results, depending on the initial location of the cluster centers.

8.5 FCM algorithm

Now let us present the FCM algorithm (*Fuzzy C-Means*) which allows assigning the same objects to various clusters with appropriate membership degrees. The FCM algorithm is the most frequently used algorithm of fuzzy clustering. It detects clusters with prototypes which are points in the data space. All clusters have the same shape dependent on the norm chosen in advance since the algorithm has no possibility to adjust the matrix $\mathbf{A}$ to existing data. This algorithm is derived by minimization of the criterion

$$J\left(\mathbf{X}; \mathbf{U}, \mathbf{V}\right) = \sum_{i=1}^{c} \sum_{k=1}^{M} \left(\mu_{ik}\right)^{m} \left\| \mathbf{x}_k - \mathbf{v}_i \right\|_{\mathbf{A}}^{2}, \tag{8.19}$$

where

$$\mathbf{U} = [\mu_{ik}] \in Z_2 \tag{8.20}$$

is the matrix of the set $\mathbf{X}$ partition, whereas

$$\mathbf{V} = [\mathbf{v}_1, \mathbf{v}_2, \ldots, \mathbf{v}_c] \tag{8.21}$$

is the vector of centers which are to be defined as a result of the algorithm operation, $\mathbf{v}_i \in R^n$, $i = 1, ..., c$. The following term appearing in formula (8.19)

$$D_{ik\mathbf{A}}^2 = \left\| \mathbf{x}_k - \mathbf{v}_i \right\|_{\mathbf{A}}^2 = \left(\mathbf{x}_k - \mathbf{v}_i\right)^T \mathbf{A} \left(\mathbf{x}_k - \mathbf{v}_i\right) \tag{8.22}$$

permits to compute the distance between vector $\mathbf{x}_k$ and cluster center $\mathbf{v}_i$, and $m \in (1, \infty)$ is a coefficient indicating the fuzziness degree of formed clusters. When $m \to 1$, the partition becomes less and less fuzzy. When $m \to \infty$, the partition becomes more and more fuzzy (then $\mu_{ik} = 1/c$). In practice the value $m = 2$ is chosen. In order to execute the algorithm, having a given data set $\mathbf{X}$, we must choose the number of clusters c, fuzziness degree m, parameter ε in the algorithm stopping criterion and initiate at random matrix $\mathbf{U}^{(0)} \in Z_2$ and vector of clusters prototypes $\mathbf{V}^{(0)}$. The algorithm stopping criterion is the same as in case of the HCM algorithm. The FCM algorithm, like HCM, may give various results depending on the initialization. The shape of clusters depends on the adopted distance measure. The flowchart of the FCM algorithm operation is illustrated in Fig. 8.8.

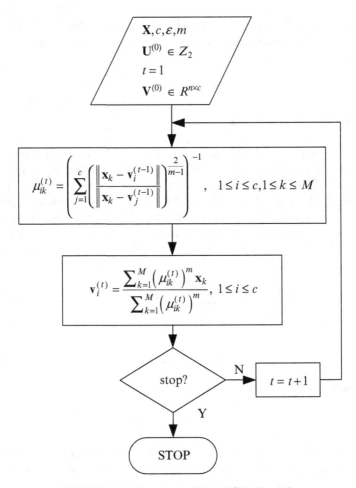

FIGURE 8.8. Flowchart of the FCM algorithm

8.6   PCM algorithm

When deriving the FCM algorithm it is assumed that the sum of the membership degrees of a given object to each of the clusters always equals 1. This restriction may cause undesirable shift of cluster centers in a situation when single incidental objects (noise) occur which sometimes lie far away from the proper clusters. Giving up this constraint, we will get the PCM algorithm (*Possibilistic C-Means*) which may be obtained as a result of minimization of the following objective function:

$$J\left(\mathbf{X}, \eta; \mathbf{U}, \mathbf{V}\right) = \sum_{i=1}^{c}\sum_{k=1}^{M}(\mu_{ik})^{m}\left\|\mathbf{x}_k - \mathbf{v}_i\right\|_{\mathbf{A}}^2 + \sum_{i=1}^{c}\eta_i\sum_{k=1}^{M}(1 - \mu_{ik})^{m}, \quad (8.23)$$

where η_i is a certain positive constant. The first term of criterion (8.23) is the same as in criterion (8.19) concerning the FCM algorithm. The second term, however, makes it necessary for membership degrees to be as big as possible, without which the solution would be achieved for matrix $\mathbf{U}$ with elements equal 0. Such a solution would result from giving up the assumption saying that the sum of membership degrees of a given object to each of the clusters always equals 1. It is easy to notice that the global objective function (8.23) can be decomposed into c objective functions for particular clusters. As a result of minimization we get

$$\mu_{ik} = \left(1 + \left(\frac{D_{ik\mathbf{A}}}{\eta_i} \right)^{\frac{2}{m-1}} \right)^{-1}, \tag{8.24}$$

where distance $D_{ik\mathbf{A}}$ is given by (8.22). The coefficient η_i defines the so-called *width of resulting possibilistic distribution*. We can choose the same value of the coefficient η_i for all clusters or compute it separately for each of the clusters, proportionally to the average distance of the objects from the center of a given cluster, i.e.

$$\eta_i = \frac{\sum_{k=1}^{M} (\mu_{ik})^m D_{ik\mathbf{A}}^2}{\sum_{k=1}^{M} (\mu_{ik})^m}. \tag{8.25}$$

The algorithm stopping criterion is chosen in the same way as in case of the HCM algorithm. We must note the fact that improper initialization of the PCM algorithm may lead to partitioning in which all membership degrees are equal. Therefore the initial partitioning in the PCM algorithm usually takes place with use of the FCM algorithm. The flowchart of the PCM algorithm is illustrated in Fig. 8.9.

8.7 Gustafson-Kessel algorithm

In the algorithms presented so far the type of norm must be defined in advance. Therefore we must know what cluster shapes occur in the data. The main disadvantage of the algorithms with a constant norm is searching for clusters with the shape which may not occur in the data set. The Gustafson-Kessel algorithm (GK) is a modification of the FCM algorithm. In this algorithm each cluster is associated with a separate matrix $\mathbf{A}_i$, and the distance between object $\mathbf{x}_k$ and the cluster center $\mathbf{v}_i$ equals

$$D_{ik}^2 = (\mathbf{x}_k - \mathbf{v}_i)^T \mathbf{A}_i (\mathbf{x}_k - \mathbf{v}_i). \tag{8.26}$$

During the algorithm operation also matrices $\mathbf{A}_i$, inducing the distance measure, $i = 1, ..., c$, are modified. The objective function in the GK algorithm is defined in the same way as in the FCM algorithm (8.19),

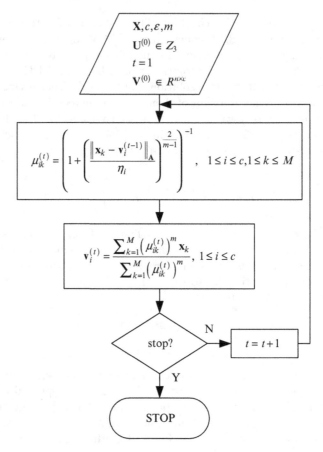

FIGURE 8.9. Flowchart of the PCM algorithm

but now distance measure (8.26) is used. Thus the objective function takes the form of

$$J\left(\mathbf{X}; \mathbf{U}, \mathbf{V}, \mathbf{A}\right) = \sum_{i=1}^{c} \sum_{k=1}^{M} \left(\mu_{ik}\right)^{m} D_{ik}^{2}, \qquad (8.27)$$

where $\mathbf{A} = (\mathbf{A}_1, \mathbf{A}_2, \ldots, \mathbf{A}_c)$. Let us notice that the direct minimization of criterion (8.27) does not lead to an effective solution as the value of this criterion may have any small value, e.g. for matrix $\mathbf{A}_i$ with almost exclusively zero-value elements. In order to obtain a correct result, matrices $\mathbf{A}_i$ must be constrained, e.g. by setting the values of their determinants, i.e.

$$\det\left(\mathbf{A}_i\right) = \rho_i, \quad \rho_i > 0, \quad \forall i, \quad i = 1, \ldots, c, \qquad (8.28)$$

where ρ_i is a chosen constant reflecting the information of data subject to clustering. In case of lack of such information it is assumed that $\rho_i = 1$ for

$i = 1, ..., c$. The constraint (8.28) causes that the volumes of clusters are constant, and we only permit a change of the shape of clusters. As a result of the minimization of criterion (8.27) with respect to matrix $\mathbf{A}_i$ we get

$$\mathbf{A}_i = [\rho_i \det (\mathbf{F}_i)]^{\frac{1}{n}} \mathbf{F}_i^{-1}, \tag{8.29}$$

where $\mathbf{F}_i$ is the so-called *fuzzy covariance matrix* of the i-th cluster

$$\mathbf{F}_i = \frac{\sum_{k=1}^{M} (\mu_{ik})^m (\mathbf{x}_k - \mathbf{v}_i)(\mathbf{x}_k - \mathbf{v}_i)^T}{\sum_{k=1}^{M} (\mu_{ik})^m}. \tag{8.30}$$

The algorithm initialization requires determining the same parameters as in the FCM algorithm, and additionally coefficients ρ_i defining the volumes of particular clusters (if we do not have the knowledge on the problem, we may assume that $\rho_i = 1$). The GK algorithm finds clusters of any shapes but requires more computations than the FCM algorithm due to the necessity to compute the determinant and inverse of the matrix $\mathbf{F}_i$. The flowchart of the GK algorithm is illustrated in Fig. 8.10.

8.8 FMLE algorithm

In the FMLE clustering algorithm (*Fuzzy Maximum Likelihood Estimates*) the distance measure refers to the form of maximum likelihood estimates. This measure is given by the following formula:

$$D_{ik\mathbf{G}_i} = \frac{[\det (\mathbf{G}_i)]^{\frac{1}{2}}}{P_i} \exp \left[\frac{1}{2} (\mathbf{x}_k - \mathbf{v}_i)^T \mathbf{G}_i^{-1} (\mathbf{x}_k - \mathbf{v}_i) \right], \tag{8.31}$$

where $\mathbf{G}_i$ is the covariance matrix of the i-th cluster

$$\mathbf{G}_i = \frac{\sum_{k=1}^{M} \mu_{ik} (\mathbf{x}_k - \mathbf{v}_i)(\mathbf{x}_k - \mathbf{v}_i)^T}{\sum_{k=1}^{M} \mu_{ik}}, \tag{8.32}$$

and P_i is the *a priori* probability of choosing the i-th cluster

$$P_i = \frac{1}{M} \sum_{k=1}^{M} \mu_{ik}. \tag{8.33}$$

Membership degree μ_{ik} *may be interpreted as the probability* of assigning object $\mathbf{x}_k$ to the i-th cluster. Convergence of the FMLE algorithm strongly depends on the initialization as it often gets stuck in the local minimum. Contrary to the GK algorithm, realization of the FMLE algorithm does not require the knowledge or arbitrary assumption of the value of parameter ρ_i for $i = 1, ..., c$. The algorithm flowchart is illustrated in Fig. 8.11.

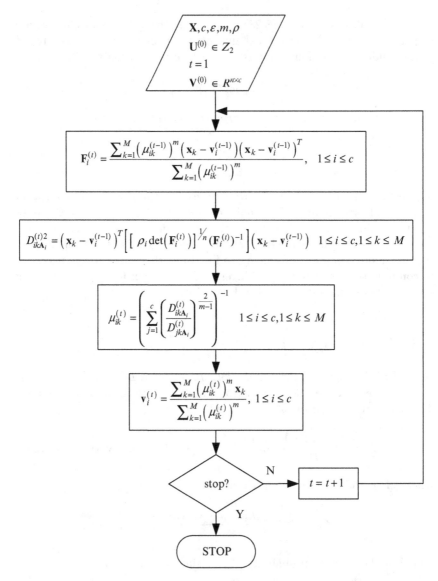

FIGURE 8.10. Flowchart of the Gustafson-Kessel algorithm

8.9   Clustering validity measures

The number of clusters is an important factor influencing clustering validity. It should reflect the actual number of clusters of objects similar to each other in the set $\mathbf{X}$. The proper number of clusters may be found by clustering the data set for a different number of clusters and different values

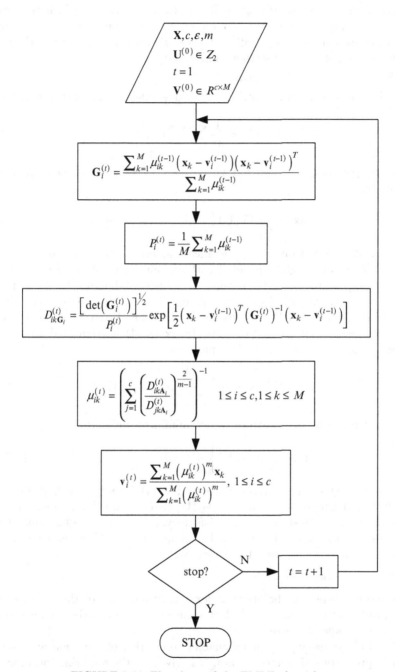

FIGURE 8.11. Flowchart of the FMLE algorithm

of parameters (e.g. parameter m in the FCM algorithm). Each time we must also evaluate the partition obtained. Such evaluation is performed by means of special indexes called *clustering validity indexes*. Below is the list of several best-known clustering validity indexes.

a) Fuzziness in partition matrix **U**

It is the simplest index measuring the fuzziness degree of a partition matrix

$$V_1\left(\mathbf{U}\right) = \frac{1}{M} \sum_{i=1}^{c} \sum_{k=1}^{M} \left(\mu_{ik}\right)^2. \tag{8.34}$$

The best partition is a partition where index $V_1\left(\mathbf{U}\right)$ reaches the maximum value, that is

$$\max_{c}\left\{\max_{Z_2} V_1\left(\mathbf{U}\right)\right\}, \quad c = 2, \ldots, M-1. \tag{8.35}$$

Coefficient $V_1\left(\mathbf{U}\right)$ evaluates the distance of all objects to the cluster centers. If each data is strongly connected with one cluster only, i.e. if for each k membership degree μ_{ik} is big for only one cluster i, the uncertainty of data is low, and consequently $V_1\left(\mathbf{U}\right)$ takes a high value. It is easy to notice that the value of index (8.34) depends on the distance of particular objects $\mathbf{x}_k$ from the centers of created clusters. Index (8.34) is connected with the index defining the entropy of data partition

$$V_2\left(\mathbf{U}\right) = -\frac{1}{M} \sum_{i=1}^{c} \sum_{k=1}^{M} \mu_{ik} \ln\left(\mu_{ik}\right). \tag{8.36}$$

The best partition is a partition which minimizes index (8.36) that is

$$\min_{c}\left\{\min_{Z_2} V_2\left(\mathbf{U}\right)\right\} \quad c = 2, \ldots, M-1. \tag{8.37}$$

When all degrees have values close to $1/c$, which means a high degree of clusters fuzziness, then measure $V_2\left(\mathbf{U}\right)$ takes high values, which means that the result of clustering is unsatisfactory. By analogy, if all membership degrees μ_{ik} take values close to 0 or 1, then measure $V_2\left(\mathbf{U}\right)$ takes low values, which indicates a good result of clustering.

b) Fukuyama-Sugeno index

The inconvenience of the above indexes is dependence of their values on the number of clusters c and lack of connection between those values and geometric shape of clusters.

The Fukuyama-Sugeno index enables connection of partition with geometric properties of clustered data. It is given by the following formula:

$$V_3\left(\mathbf{U}, \mathbf{V}; \mathbf{X}\right) = \sum_{i=1}^{c} \sum_{k=1}^{M} \left(\mu_{ik}\right)^m \left(\|\mathbf{x}_k - \mathbf{v}_i\|_{\mathbf{A}}^2 - \|\mathbf{x}_k - \overline{\mathbf{v}}\|_{\mathbf{A}}^2\right), \tag{8.38}$$

where $\bar{\mathbf{v}}$ is an average of all points in the data set, i.e.

$$\bar{\mathbf{v}} = \frac{1}{M} \sum_{k=1}^{M} x_k. \tag{8.39}$$

The optimal data partition minimizes index V_3.

c) Xie-Beni index

Xie-Beni index is given by the formula

$$V_4\left(\mathbf{U}, \mathbf{V}; \mathbf{X}\right) = \frac{\sum_{i=1}^{c} \sum_{k=1}^{M} \left(\mu_{ik}\right)^m \|\mathbf{x}_k - \mathbf{v}_i\|^2}{M\left(\min_{i,j}\left\{\|v_i - v_j\|\right\}^2\right)}, \tag{8.40}$$

and the optimal selection of the number of classes is given by the formula

$$\min_{c}\left\{\min_{M_2} V_4\left(\mathbf{U}\right)\right\} \quad c = 2, \ldots, M - 1. \tag{8.41}$$

The best partition minimizes index (8.40) which is a quotient of the average of all distances between clusters and objects and the smallest distance between clusters. The proper clustering procedure should result in a situation in which all data will be as close to the centers of the respective clusters as possible and all centers will be as far from each other as possible.

8.10 Illustration of operation of data clustering algorithms

The most frequently used algorithm of data clustering is the FCM algorithm. Therefore in this chapter we will perform a simulation of this algorithm and compare it with the HCM and PCM algorithms.

Example 8.5
Figure 8.12 presents an exemplary data set composed of 9 two-dimensional objects, i.e. $M = 9$ and $n = 2$. Matrix $\mathbf{X}$ corresponding to this set is in the form of

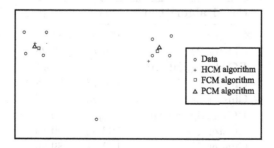

FIGURE 8.12. Comparison of three clustering algorithms

$$\mathbf{X} = \begin{bmatrix} 98 & 97 & 111 & 109 & 178 & 178 & 190 & 189 & 143 \\ 86 & 99 & 99 & 85 & 85 & 95 & 97 & 85 & 46 \end{bmatrix}. \qquad (8.42)$$

It is easy to notice two separate groups of objects and one object number 9 which "does not fit" in those groups. Symbol "+" has been used to mark the cluster centers obtained by means of the HCM algorithm. As one can see, object 9 has been qualified to cluster 2 and influenced the position of this cluster's center, by "drawing" this center towards itself. As a result of the HCM algorithm operation, the following partition matrix has been created:

$$\mathbf{U}_{\mathrm{HCM}} = \begin{bmatrix} 0 & 0 & 0 & 0 & 1 & 1 & 1 & 1 & 1 \\ 1 & 1 & 1 & 1 & 0 & 0 & 0 & 0 & 0 \end{bmatrix}. \qquad (8.43)$$

The FCM algorithm assigned object number 9 to both groups with the same membership degree equal to 0.5. In such a case the centers of both clusters are moved towards object number 9. As a result of clustering using the FCM algorithm, the following partition matrix has been created:

$$\mathbf{U}_{\mathrm{FCM}} = \begin{bmatrix} 0.99 & 0.98 & 0.98 & 0.99 & 0.00 & 0.01 & 0.02 & 0.01 & 0.50 \\ 0.01 & 0.02 & 0.02 & 0.01 & 1.00 & 0.99 & 0.98 & 0.99 & 0.50 \end{bmatrix}.$$

$$(8.44)$$

The problem of noise has been best dealt with by the PCM algorithm which assigned membership degrees equal to 0.04 and 0.03 to the object number 9, which can be seen when analyzing the partition matrix created

$$\mathbf{U}_{\mathrm{PCM}} = \begin{bmatrix} 0.76 & 0.56 & 0.53 & 0.73 & 0.03 & 0.03 & 0.02 & 0.02 & 0.04 \\ 0.02 & 0.02 & 0.02 & 0.02 & 0.74 & 0.73 & 0.54 & 0.65 & 0.03 \end{bmatrix}.$$

$$(8.45)$$

The centers of both clusters have only slightly been moved comparing to the results of the HCM and PCM algorithms. Table 8.1 shows the coordinates of cluster centers which were created as a result of clustering set (8.42).

TABLE 8.1. Coordinates of cluster centers which were created as a result of clustering the set (8.42)

HCM algorithm	x_1	x_2
Cluster 1	175.60	81.60
Cluster 2	103.75	92.25
FCM algorithm	x_1	x_2
Cluster 1	106.13	89.38
Cluster 2	181.20	87.73
PCM algorithm	x_1	x_2
Cluster 1	103.58	90.19
Cluster 2	182.41	89.86

8.11   Notes

This chapter presents only basic methods of data clustering. In order to illustrate their operation we compared the HCM, FCM and PCM algorithms. The most frequently used FCM method is sensitive to occurrence of noisy data. This method may serve as an initialization of the PCM algorithm, which is resistant to noise and outliers. It is also applied for preliminary setup of membership functions during design of neural and fuzzy systems (Chapters 9 and 10). Data clustering methods constitute an extremely important research tool in computational intelligence and have numerous applications. Both, basic algorithms presented in this chapter as well as more advanced methods, e.g. oriented at detection of clusters of specific shapes, have been discussed in detail in the literature [3, 9, 23, 34, 83]. It is worth noting that pioneers in the field of data clustering methods are James C. Bezdek and Enrique H. Ruspini whose original works have been reprinted in the book [10].

9
Neuro-fuzzy systems of Mamdani, logical and Takagi-Sugeno type

9.1 Introduction

Within the last dozen of years, different structures of neuro-fuzzy networks have been presented, often referred to in the world literature as neuro-fuzzy systems. They combine the advantages of neural networks and classic fuzzy systems. In particular, the neuro-fuzzy networks are characterized – in contrast with neural networks – by a interpretable representation of knowledge represented by fuzzy rules. As generally known, the knowledge in neural networks is represented by the values of synaptic weights, and therefore is completely not interpretable, for instance, for a user of a medical expert system that uses neural networks. Moreover, neuro-fuzzy networks can be trained, using the idea of error backpropagation method, which is the basis of learning of multilayer neural networks. The learning usually applies to membership function parameters of the **IF** and **THEN** part of the fuzzy rules. As shown in Chapter 7, there is also the possibility to apply the evolutionary algorithms to learn not only the parameters of the membership functions but also the fuzzy rules themselves. The above discussed advantages of the neuro-fuzzy networks are the reason for their common application in classification, approximation and prediction problems. Most of neuro-fuzzy structures described in the world literature utilizes the Mamdani type inference or the Takagi-Sugeno schema. As mentioned in Chapter 4, the Mamdani type inference consists in connecting the antecedents and the consequents of rules using a t-norm (most often the t-norm of the min type or of the product type). Then the aggregation of particular rules is made using a t-conorm. In case of the Takagi-Sugeno schema, the consequents of the rules are not fuzzy in nature, but are functions of the input variables. Less often the logical inference is applied, which consists in connecting the antecedents and the consequents of rules using a fuzzy implication that satisfies the conditions of Definition 4.47. In case of an inference of logical type the aggregation of particular rules is made using a t-conorm. It is obvious that the designers and users of neuro-fuzzy systems would like to obtain a possibly high accuracy of these systems operation in the sense of the chosen quality criterion. In approximation and prediction problems, such quality criterion is the mean squared error, and in classification problems – the number of erroneously classified samples. In both problems, the experiments are made on learning sequences and testing sequences. It should be stressed that the satisfactory results obtained on a learning sequence do not guarantee a correct system operation on a testing sequence. In other words, the neuro-fuzzy system should have good properties of the so-called generalization. In particular, neuro-fuzzy systems designed using both the membership function and the weights describing the importance of rules and importance of linguistic variables in individual rules should be characterized by an appropriate number of all parameters which are to be subject of learning. A big number of parameters ensures a small learning error, but usually leads to wrong generalization. On the other hand, a small number of parameters in the system leads to a larger learning error. In this chapter, we will present the Mamdani, logical and Takagi-Sugeno systems, their learning algorithms and we will make a comparative analysis of their effectiveness. We will solve the issue of designing neuro-fuzzy systems, which are a compromise between accuracy and the number of parameters describing this system.

9.2   Description of simulation problems used

Neuro-fuzzy system discussed in this and the next chapter will be tested using standard testing problems (*benchmarks*).

Table 9.1 presents the name of the problem, number of input data, length of the learning sequence and length of the testing sequence.

Below, we present a detailed description of the problems listed in Table 9.1. Information on the number of rules and the number of epochs relates to the simulations performed in this chapter (problems 9.2.1 - 9.2.4).

9.2.1   Polymerization

We consider the problem of modeling the polymer manufacturing process. The device produces polymers (macromolecular compounds obtained from monomers, i.e. small-molecule compounds) as a result of chemical reaction

TABLE 9.1. Simulation problems

No.	Name of problem	Number of inputs	Length of the learning sequence	Length of the testing sequence
1	Polymerization	3	70	20
2	HANG (modeling a static nonlinear function)	2	50	20
3	NPD (modeling a dynamic nonlinear function)	2	1000	200
4	Modeling the taste of rice	5	75	30
5	Distinguishing of the brand of wine	13	125	53
6	Classification of iris flower	4	90	60

called polymerization, during which many small molecules of the same compound connect spontaneously (or under the influence of catalytic agents). In order to model the system, three continuous input variables are selected. They include: monomer concentration, change of monomer concentration and its current flow rate. Based on the values of input variables, the next value of the monomer flow rate should be determined. Simulation tests of systems made of 3 inputs, one output and 6 rules have been performed. The experiment was repeated many times for 6000 epochs (420 000 iterations) and its results were averaged.

9.2.2   Modeling a static non-linear function

It is an issue of approximation of a non-linear function – HANG, described by the formula

$$y(x_1, x_2) = \left(1 + x_1^{-2} + x_2^{-1.5}\right)^2,$$ (9.1)

where $x_1, x_2 \in [1, 5]$. The learning sequence consists of 50 input data vectors and the corresponding function values. Simulation tests of systems made of 2 inputs, one output and 8 rules were performed. The experiment was repeated many times for 8000 epochs (400 000 iterations) and its results were averaged.

9.2.3   Modeling a non-linear dynamic object (Nonlinear Dynamic Problem - NDP)

It is the problem of modeling a *nonlinear dynamic object* the behavior of which is described by the formula

$$y(t) = g\left(y(t-1), y(t-2)\right) + u(t),$$ (9.2)

where

$$g\left(u\left(t-1\right),y\left(t-2\right)\right) = \frac{y\left(t-1\right)y\left(t-2\right)\left(y\left(t-1\right)-0.5\right)}{1+y^2\left(t-1\right)+y^2\left(t-2\right)}, \qquad (9.3)$$

and $u\left(t\right)$ is the output signal.

For the purpose of learning neuro-fuzzy systems, a sequence of model states of the objects for a random input signal with uniform distribution is used (first 500 samples) and for a sinusoidal input signal $u\left(t\right) = \sin\left(2\pi t/25\right)$ (next 500 samples). The sequence has been generated for a zero initial state. Simulation tests of systems made of 3 inputs, one output and 6 rules were performed. The experiment was repeated many times for 500 epochs (500 000 iterations) and its results were averaged.

9.2.4 Modeling the taste of rice

The problem to be solved in this example is to find a nonlinear dependency between input data, characterizing the rice samples, and the output signal containing the interpretation of the taste of rice. Data consist of 105 cases. Each sample has been described by 5 features: flavor, appearance, taste, viscosity and hardness, constituting the system input data. The system output is a general assessment of the taste of rice. Input and output data have been normalized to the interval $[0, 1]$. Simulation tests of systems made of 2 inputs, one output and 6 rules were performed. The experiment was repeated many times for 5000 epochs (375 000 iterations) and its results were averaged.

9.2.5 Distinguishing of the brand of wine

The problem to be solved is the correct classification of wine samples. Data in the problem of wine distinguishing consist of chemical analysis of 178 wines from same region of Italy, but from three different vineyards. The input data consist of 13 continuous attributes which include among other thing: alcohol contents, malic acid contents, sediment, sediment alkalinity, magnesium contents, total phenol contents, color intensity and shade. In the experiment discussed, all the data have been divided into a learning sequence (125 samples) and a testing sequence (53 samples).

9.2.6 Classification of iris flower

The problem consist in the classification of the Iris flower based on the length of the leaf in cm, width of the leaf in cm, length of the petal in cm, width of the petal in cm. We distinguish three classes: *Iris setosa*, *Iris Versicolor* and *Iris Virginica*. Data include 150 sets, which were divided at random into the learning sequence (90 sets) and the testing sequence (60 sets).

Remark 9.1
Gradient algorithms of the momentum type with learning coefficient $\eta =$ 0.25 and with momentum coefficient 0.1 have been used for learning of all the neuro-fuzzy systems presented in this chapter. These algorithms have been derived in Subchapter 9.6 without taking account of the momentum term in particular iteration procedures. In all neuro-fuzzy systems considered in this chapter, the following principle has been adopted:

- particular rules are aggregated using a t-conorm of the max type in case of the Mamdani system and a t-norm of the min type in case of a logical system,

- the antecedents of rules are aggregated by means of t-norm of the product type.

The basis for assessment of neuro-fuzzy systems will be the value of error (mean squared error in case of approximation issues or number of erroneously classified samples in case of classification issues). At first, the mean error in particular epochs is determined, and then the minimum error is found among these errors.

9.3 Neuro-fuzzy systems of Mamdani type

Let us consider two types of neuro-fuzzy systems of Mamdani type, the so-called A type and B type systems. In both cases, the antecedents and the consequents of rules are connected by means of a t-norm. In A type systems at the inference block output we have N fuzzy sets, while in B-type systems at the block output we have one fuzzy set which is the result of aggregation of inference results in particular rules.

9.3.1   A-type systems

In A-type systems, the defuzzification is realized using the dependency:

$$\bar{y} = \frac{\sum_{r=1}^{N} \bar{y}^r \cdot \mu_{\overline{B}^r}(\bar{y}^r)}{\sum_{r=1}^{N} \mu_{\overline{B}^r}(\bar{y}^r)}. \tag{9.4}$$

The membership functions of fuzzy sets $\overline{B}^r$, $r = 1, 2, \ldots, N$, are defined using the following formula:

$$\mu_{\overline{B}^r}(y) = \sup_{\mathbf{x} \in \mathbf{X}} \left\{ \mu_{A^r}(\mathbf{x}) \overset{T}{*} \mu_{A^r \to B^r}(\mathbf{x}, y) \right\}. \tag{9.5}$$

With singleton type fuzzification, formula (9.5) takes the form

$$\mu_{\overline{B}^r}(y) = \mu_{A^r \to B^r}(\overline{\mathbf{x}}, y) = T\left(\mu_{A^r}(\overline{\mathbf{x}}), \mu_{B^r}(y)\right). \tag{9.6}$$

Since

$$\mu_{A^r}(\overline{\mathbf{x}}) = \mathop{T}_{i=1}^{n}\left(\mu_{A_i^r}(\overline{x}_i)\right),$$ (9.7)

we have

$$\mu_{\overline{B}^r}(y) = \mu_{A^r \to B^r}(\overline{\mathbf{x}}, y) = T\left[\mathop{T}_{i=1}^{n}\left(\mu_{A_i^r}(\overline{x}_i)\right), \mu_{B^r}(y)\right],$$ (9.8)

where T is any t-norm. Owing to the fact that

$$\mu_{B^r}(\overline{y}^r) = 1$$ (9.9)

and

$$T(a, 1) = a,$$ (9.10)

we obtain the following dependency:

$$\mu_{\overline{B}^r}(\overline{y}^r) = \mathop{T}_{i=1}^{n}\left(\mu_{A_i^r}(\overline{x}_i)\right).$$ (9.11)

By substituting dependency (9.11) to formula (9.4), we get

$$\overline{y} = \frac{\sum_{r=1}^{N} \overline{y}^r \cdot T_{i=1}^{n}\left(\mu_{A_i^r}(\overline{x}_i)\right)}{\sum_{r=1}^{N} T_{i=1}^{n}\left(\mu_{A_i^r}(\overline{x}_i)\right)}.$$ (9.12)

In A-type systems, separate inference is made within each rule and $\mu_{\overline{B}^r}(\overline{y}^r)$, $r = 1, 2, \ldots, N$, is computed. Let us assume that input and output linguistic variables are described by means of Gaussian membership functions, that is

$$\mu_{A_i^r}(x_i) = \exp\left[-\left(\frac{x_i - \overline{x}_i^r}{\sigma_i^r}\right)^2\right],$$ (9.13)

$$\mu_{B^r}(y) = \exp\left[-\left(\frac{y - \overline{y}^r}{\sigma^r}\right)^2\right].$$ (9.14)

By substituting the above dependencies to formula (9.4) and applying the Larsen rule, we will get the following formula:

$$\overline{y} = \frac{\sum_{r=1}^{N} \overline{y}^r \left(\prod_{i=1}^{n} \exp\left[-\left(\frac{\overline{x}_i - \overline{x}_i^r}{\sigma_i^r}\right)^2\right]\right)}{\sum_{r=1}^{N} \left(\prod_{i=1}^{n} \exp\left[-\left(\frac{\overline{x}_i - \overline{x}_i^r}{\sigma_i^r}\right)^2\right]\right)}.$$ (9.15)

Let us notice that in dependency (9.15), there is no parameter σ^r of the output fuzzy set B^r, $r = 1, 2, \ldots, N$. Figure 9.1 presents a block schema of the structure reflecting dependency (9.15). As we can see, it is a multilayer network structure. Such a structure is called a neuro-fuzzy network. To train it, the idea of error backpropagation method may be applied.

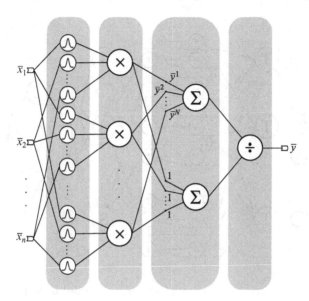

FIGURE 9.1. Network structure of a system described by formula (9.15)

9.3.2   B-type systems

In B-type systems, the defuzzification is made using the dependency

$$\overline{y} = \frac{\sum_{r=1}^{N} \overline{y}^r \cdot \mu_{B'}\left(\overline{y}^r\right)}{\sum_{r=1}^{N} \cdot \mu_{B'}\left(\overline{y}^r\right)}. \tag{9.16}$$

In these systems, aggregation of particular fuzzy sets $\overline{B}^k$ given by formula (9.6) is made, which means that the fuzzy set B' is determined through operation of union of fuzzy sets $\overline{B}^k$

$$B' = \bigcup_{k=1}^{N} \overline{B}^k. \tag{9.17}$$

The membership function of fuzzy set B' is determined using a t-conorm, i.e.

$$\mu_{B'}\left(y\right) = \mathop{S}_{k=1}^{N} \left\{\mu_{\overline{B}^k}\left(y\right)\right\}. \tag{9.18}$$

Therefore

$$\mu_{B'}\left(\overline{y}^r\right) = \mathop{S}_{k=1}^{N} \left\{\mu_{\overline{B}^k}\left(\overline{y}^r\right)\right\} = \mathop{S}_{k=1}^{N} \left\{T\left(\mu_{A^k}\left(\mathbf{x}\right), \mu_{B^k}\left(\overline{y}^r\right)\right)\right\} \tag{9.19}$$

$$= \mathop{S}_{k=1}^{N} \left\{T\left(\mathop{T}_{i=1}^{n} \mu_{A_i^k}\left(\overline{x}_i\right), \mu_{B^k}\left(\overline{y}^r\right)\right)\right\}.$$

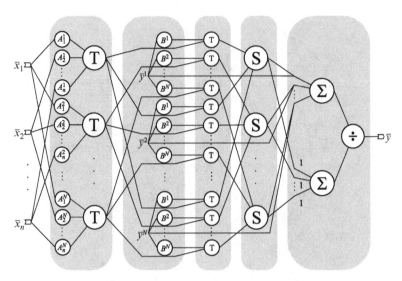

FIGURE 9.2. Network structure of a system described by formula (9.20)

By substituting formula (9.19) to dependency (9.16), we get

$$\overline{y} = \frac{\sum_{r=1}^{N} \overline{y}^r \cdot S_{k=1}^{N} \left\{ T \left(T_{i=1}^{n} \left\{ \mu_{A_i^k} (\overline{x}_i) \right\}, \mu_{B^k} (\overline{y}^r) \right) \right\}}{\sum_{r=1}^{N} S_{k=1}^{N} \left\{ T \left(T_{i=1}^{n} \left\{ \mu_{A_i^k} (\overline{x}_i) \right\}, \mu_{B^k} (\overline{y}^r) \right) \right\}}. \tag{9.20}$$

In Fig. 9.2 the network structure of the system described by formula (9.20) is presented.

In B-type systems, separate inference is also made within each rule, but next, the aggregation of inference results is made in individual rules and only then $\mu_{B'} (\overline{y}^r)$, $r = 1, 2, \ldots, N$, is computed.

9.3.3   Mamdani type systems in modeling problems

Mamdani type systems will be applied to modeling problems. These problems were described in detail in Subchapter 9.2. We will assume that fuzzy sets A_i^r and B^r are characterized by Gaussian membership functions given by formula (9.13) and (9.14).

9.3.3.1. M1-type systems

Let us consider Mamdani type systems which are constructed using definitions of triangular norms without taking the weights into account. Using dependency (9.20) and min type Mamdani rule, we obtain the following description of the neuro-fuzzy system:

$$\bar{y} = \frac{\sum_{r=1}^{N} \bar{y}^r \cdot S_{k=1}^{N} \left\{ \min \left(T_{i=1}^n \left\{ \mu_{A_i^k}(\bar{x}_i) \right\}, \mu_{B^k}(\bar{y}^r) \right) \right\}}{\sum_{r=1}^{N} S_{k=1}^{N} \left\{ \min \left(T_{i=1}^n \left\{ \mu_{A_i^k}(\bar{x}_i) \right\}, \mu_{B^k}(\bar{y}^r) \right) \right\}}. \tag{9.21}$$

Substituting dependencies (9.13) and (9.14) to formula (9.21) and using the contents of Remark 9.1, we obtain

$$\bar{y} = \frac{\sum_{r=1}^{N} \bar{y}^r \cdot S_{k=1}^{N} \left\{ \min \left(T_{i=1}^n \left\{ \exp \left[-\left(\frac{\bar{x}_i - \bar{x}_i^k}{\sigma_i^k} \right)^2 \right] \right\}, \exp \left[-\left(\frac{\bar{y}^r - \bar{y}^k}{\sigma^k} \right)^2 \right] \right) \right\}}{\sum_{r=1}^{N} S_{k=1}^{N} \left\{ \min \left(T_{i=1}^n \left\{ \exp \left[-\left(\frac{\bar{x}_i - \bar{x}_i^k}{\sigma_i^k} \right)^2 \right] \right\}, \exp \left[-\left(\frac{\bar{y}^r - \bar{y}^k}{\sigma^k} \right)^2 \right] \right) \right\}}$$

$$= \frac{\sum_{r=1}^{N} \bar{y}^r \cdot \max_{1 \leq k \leq N} \left\{ \min \left(\prod_{i=1}^n \exp \left[-\left(\frac{\bar{x}_i - \bar{x}_i^k}{\sigma_i^k} \right)^2 \right] \right) \cdot \exp \left[-\left(\frac{\bar{y}^r - \bar{y}^k}{\sigma^k} \right)^2 \right] \right\}}{\sum_{r=1}^{N} \max_{1 \leq k \leq N} \left\{ \min \left(\prod_{i=1}^n \exp \left[-\left(\frac{\bar{x}_i - \bar{x}_i^k}{\sigma_i^k} \right)^2 \right] \right) \cdot \exp \left[-\left(\frac{\bar{y}^r - \bar{y}^k}{\sigma^k} \right)^2 \right] \right\}}. \tag{9.22}$$

Using dependency (9.20) and product type Mamdani rule (known as Larsen rule), we obtain the following description of the neuro-fuzzy system:

$$\bar{y} = \frac{\sum_{r=1}^{N} \bar{y}^r \cdot S_{k=1}^{N} \left\{ T_{i=1}^n \left\{ \mu_{A_i^k}(\bar{x}_i) \right\} \cdot \mu_{B^k}(\bar{y}^r) \right\}}{\sum_{r=1}^{N} S_{k=1}^{N} \left\{ T_{i=1}^n \left\{ \mu_{A_i^k}(\bar{x}_i) \right\} \cdot \mu_{B^k}(\bar{y}^r) \right\}}. \tag{9.23}$$

Substituting dependencies (9.13) and (9.14) to formula (9.23) and using the contents of Remark 9.1, we obtain

$$\bar{y} = \frac{\sum_{r=1}^{N} \bar{y}^r \cdot S_{k=1}^{N} \left\{ T_{i=1}^n \left\{ \exp \left[-\left(\frac{\bar{x}_i - \bar{x}_i^k}{\sigma_i^k} \right)^2 \right] \right\} \cdot \exp \left[-\left(\frac{\bar{y}^r - \bar{y}^k}{\sigma^k} \right)^2 \right] \right\}}{\sum_{r=1}^{N} S_{k=1}^{N} \left\{ T_{i=1}^n \left\{ \exp \left[-\left(\frac{\bar{x}_i - \bar{x}_i^k}{\sigma_i^k} \right)^2 \right] \right\} \cdot \exp \left[-\left(\frac{\bar{y}^r - \bar{y}^k}{\sigma^k} \right)^2 \right] \right\}}$$

$$= \frac{\sum_{r=1}^{N} \bar{y}^r \cdot \max_{1 \leq k \leq N} \left\{ \prod_{i=1}^n \exp \left[-\left(\frac{\bar{x}_i - \bar{x}_i^k}{\sigma_i^k} \right)^2 \right] \cdot \exp \left[-\left(\frac{\bar{y}^r - \bar{y}^k}{\sigma^k} \right)^2 \right] \right\}}{\sum_{r=1}^{N} \max_{1 \leq k \leq N} \left\{ \prod_{i=1}^n \exp \left[-\left(\frac{\bar{x}_i - \bar{x}_i^k}{\sigma_i^k} \right)^2 \right] \cdot \exp \left[-\left(\frac{\bar{y}^r - \bar{y}^k}{\sigma^k} \right)^2 \right] \right\}}. \tag{9.24}$$

Further in this chapter, we will not remind the wording of Remark 9.1. We should however remember that particular rules are aggregated using a t-conorm of the max type in case of the Mamdani system and a t-norm of the min type in case of a logic system and that the antecedents of the rules are aggregated using a t-norm of the product type. Neuro-fuzzy systems (9.22) and (9.24) are special cases of B-type system described in point 9.3.2. In systems (9.22) and (9.24) the following parameters of the membership functions are subject to learning; $\overline{x}_i^k, \sigma_i^k, \overline{y}^k, \sigma^k$, $k = 1, 2, \ldots, N$. One of subjects of studies is also A-type Mamdani system described in point 9.3.1, the description of which, for the Reader's convenience, is recalled below:

$$
\overline{y} = \frac{\sum_{r=1}^{N} \overline{y}^r \cdot \left[\prod_{i=1}^{n} \left(\exp \left[- \left(\frac{\overline{x}_i - \overline{x}_i^r}{\sigma_i^r} \right)^2 \right] \right) \right]}{\sum_{r=1}^{N} \left[\prod_{i=1}^{n} \left(\exp \left[- \left(\frac{\overline{x}_i - \overline{x}_i^r}{\sigma_i^r} \right)^2 \right] \right) \right]}.
\tag{9.25}
$$

System (9.25) has been called a simplified Larsen structure. In this system, the parameters $\overline{x}_i^r$, σ_i^r, $\overline{y}^r$, $r = 1, 2, \ldots, N$, are subject to learning. It may be shown that system (9.25) is a special case of system (9.24) Neuro-fuzzy systems (9.22), (9.24) and (9.25) have been used to solve four problems specified in Table 9.1: polymerization, HANG, NDP and modeling the taste of rice. All the parameters of the neuro-fuzzy systems have been trained using error backpropagation method: centers and widths of Gaussian functions were trained. In case of structure (9.25), there are no widths of consequents of the Gaussian function.

9.3.3.1.1. Polymerization

Table 9.2 presents the smallest error for individual structures and the number of epochs corresponding to this error. Table 9.3 presents three desired error values and the number of epochs, after which this error was obtained.

As it may be inferred from Table 9.3, for the Larsen structure, it was impossible to train the system with error 0.0045.

9.3.3.1.2. HANG

Table 9.4 presents the smallest error for individual structures and the number of epochs corresponding to this error. Table 9.5 presents three

TABLE 9.2. The smallest error obtained as a result of learning

POLYMERIZATION		
Structure	The smallest error	Number of epochs
Mamdani	0.0041	3734
Larsen	0.0049	5984
Larsen (simplified)	0.0042	4689

TABLE 9.3. Number of epochs required to train the system which is characterized by a definite error

POLYMERIZATION			
Structure	Value of error		
	0.0055	0.0050	0.0045
Mamdani	1086	1479	1943
Larsen	3621	5984	–
Larsen (simplified)	807	2718	4454

TABLE 9.4. The smallest error obtained as a result of learning

HANG		
Structure	The smallest error	Number of epochs
Mamdani	0.0340	7848
Larsen	0.0387	8000
Larsen (simplified)	0.0240	7102

TABLE 9.5. Number of epochs required to train the system which is characterized by a given error

HANG			
POLYMERIZATION			
Structure	Value of error		
	0.028	0.026	0.024
Mamdani	–	–	–
Larsen	–	–	–
Larsen (simplified)	4071	6024	7102

desired values of error and the number of epochs, after which this error was obtained.

As it may be inferred from Table 9.5, neither the Mamdani nor the Larsen structure did achieve any of the desired values of error.

9.3.3.1.3. NDP

Table 9.6 presents the smallest error for individual structures and the number of epochs corresponding to this error.

Table 9.7 presents three desired values of error and the number of epochs, after which this error was obtained.

As it may be inferred from Table 9.7, the Mamdani structure did not attain any of the desired values of error.

TABLE 9.6. The smallest error obtained as a result of learning

NDP		
Structure	The smallest error	Number of epochs
Mamdani	0.0263	436
Larsen	0.0176	433
Larsen (simplified)	0.0140	393

TABLE 9.7. Number of epochs required to train the system which is characterized by a given error

NDP			
Structure	Value of error		
	0.026	0.023	0.020
Mamdani	–	–	–
Larsen	172	233	302
Larsen (simplified)	74	82	93

9.3.3.1.4. Modeling the taste of rice

Table 9.8 presents the smallest error for individual structures and the number of epochs corresponding to this error.

As it may be inferred from Table 9.9, only the simplified Larsen structure obtained all the desired values of error.

9.3.3.2. M2-type systems

Let us consider Mamdani type systems which are constructed using definitions of triangular norms taking into account the weights w_k, characterizing the importance of particular rules. Using the definition of weighted t-conorm and dependency (9.22), (9.24) and (9.25) we obtain the following description of the neuro-fuzzy systems:

TABLE 9.8. The smallest error obtained as a result of learning

MODELING THE TASTE OF RICE		
Structure	The smallest error	Number of epochs
Mamdani	0.0244	4459
Larsen	0.0252	2501
Larsen (simplified)	0.0205	3888

TABLE 9.9. Number of epochs required to train the system which is characterized by a definite error

MODELING THE TASTE OF RICE			
Structure	Value of error		
	0.028	0.025	0.022
Mamdani	233	1978	–
Larsen	506	–	–
Larsen (simplified)	67	451	2936

a) Mamdani system with weights of rules

$$\bar{y} = \frac{\sum_{r=1}^{N} \bar{y}^r \cdot S_{k=1}^{*N} \left\{ \min \left(T_{i=1}^{n} \left\{ \mu_{A_i^k}(\bar{x}_i) \right\}, \mu_{B^k}(\bar{y}^r) \right), w_k \right\}}{\sum_{r=1}^{N} S_{k=1}^{*N} \left\{ \min \left(T_{i=1}^{n} \left\{ \mu_{A_i^k}(\bar{x}_i) \right\}, \mu_{B^k}(\bar{y}^r) \right), w_k \right\}}$$ (9.26)

$$= \frac{\sum_{r=1}^{N} \bar{y}^r \cdot S_{k=1}^{*N} \left\{ \min \left(T_{i=1}^{n} \left\{ \exp\left[-\left(\frac{\bar{x}_i - \bar{x}_i^k}{\sigma_i^k}\right)^2 \right] \right\}, \exp\left[-\left(\frac{\bar{y}^r - \bar{y}^k}{\sigma^k}\right)^2 \right] \right), w_k \right\}}{\sum_{r=1}^{N} S_{k=1}^{*N} \left\{ \min \left(T_{i=1}^{n} \left\{ \exp\left[-\left(\frac{\bar{x}_i - \bar{x}_i^k}{\sigma_i^k}\right)^2 \right] \right\}, \exp\left[-\left(\frac{\bar{y}^r - \bar{y}^k}{\sigma^k}\right)^2 \right] \right), w_k \right\}}.$$

b) Larsen system with weights of rules

$$\bar{y} = \frac{\sum_{r=1}^{N} \bar{y}^r \cdot S_{k=1}^{*N} \left\{ T_{i=1}^{n} \left(\left\{ \mu_{A_i^k}(\bar{x}_i) \right\} \cdot \mu_{B^k}(\bar{y}^r) \right), w_k \right\}}{\sum_{r=1}^{N} S_{k=1}^{*N} \left\{ T_{i=1}^{n} \left(\left\{ \mu_{A_i^k}(\bar{x}_i) \right\} \cdot \mu_{B^k}(\bar{y}^r) \right), w_k \right\}}$$ (9.27)

$$= \frac{\sum_{r=1}^{N} \bar{y}^r \cdot S_{k=1}^{*N} \left\{ T_{i=1}^{n} \left\{ \exp\left[-\left(\frac{\bar{x}_i - \bar{x}_i^k}{\sigma_i^k}\right)^2 \right] \right\} \cdot \exp\left[-\left(\frac{\bar{y}^r - \bar{y}^k}{\sigma^k}\right)^2 \right], w_k \right\}}{\sum_{r=1}^{N} S_{k=1}^{*N} \left\{ T_{i=1}^{n} \left\{ \exp\left[-\left(\frac{\bar{x}_i - \bar{x}_i^k}{\sigma_i^k}\right)^2 \right] \right\} \cdot \exp\left[-\left(\frac{\bar{y}^r - \bar{y}^k}{\sigma^k}\right)^2 \right], w_k \right\}}.$$

In both a) and b) systems, the parameters of membership function, i.e. $\bar{x}_i^k, \sigma_i^k, \bar{y}^k, \sigma^k$ and weights w_k are subject to learning.

c) Simplified Larsen system with weights of rules

$$\bar{y} = \frac{\sum_{r=1}^{N} \bar{y}^r \cdot w_r \cdot \prod_{i=1}^{n}\left(\mu_{A_i^r}\left(\bar{x}_i\right)\right)}{\sum_{r=1}^{N} w_r \cdot \prod_{i=1}^{n}\left(\mu_{A_i^r}\left(\bar{x}_i\right)\right)} \tag{9.28}$$

$$= \frac{\sum_{r=1}^{N} \bar{y}^r \cdot w_r \cdot \prod_{i=1}^{n}\left(\exp\left[-\left(\frac{\bar{x}_i - \bar{x}_i^r}{\sigma_i^r}\right)^2\right]\right)}{\sum_{r=1}^{N} w_r \cdot \prod_{i=1}^{n}\left(\exp\left[-\left(\frac{\bar{x}_i - \bar{x}_i^r}{\sigma_i^r}\right)^2\right]\right)}.$$

In c) system, the parameters of membership function $\bar{x}_i^r, \sigma_i^r, \bar{y}^r$ and weights w_r. are subject to learning. Neuro-fuzzy systems (9.26), (9.27) and (9.28), have been used to solve four problems specified in Table 9.1.

9.3.3.2.1. Polymerization

Table 9.10 presents the smallest error for individual structures and the number of epochs corresponding to this error.

Table 9.11 presents three desired values of error and the number of epochs, after which this error was obtained.

9.3.3.2.2. HANG

Table 9.12 presents the smallest error for individual structures and the number of epochs corresponding to this error. Table 9.13 presents three

TABLE 9.10. The smallest error obtained as a result of learning

POLYMERIZATION		
Structure	The smallest error	Number of epochs
Mamdani with weights	0.0039	4088
Larsen with weights	0.0043	4501
Larsen (simplified) weights	0.0039	3691

TABLE 9.11. Number of epochs required to train the system which is characterized by a definite error

POLYMERIZATION			
Structure	Value of error		
	0.0055	0.0050	0.0045
Mamdani with weights	26	44	2440
Larsen with weights	2646	3154	4099
Larsen (simplified) with weights	1633	1633	3443

TABLE 9.12. The smallest error obtained as a result of learning

HANG		
Structure	The smallest error	Number of epochs
Mamdani with weights	0.0318	7848
Larsen with weights	0.0353	6773
Larsen (simplified) with weights	0.0183	1955

TABLE 9.13. Number of epochs required to train the system which is characterized by a definite error

HANG			
Structure	Value of error		
	0.028	0.026	0.024
Mamdani with weights	–	–	–
Larsen with weights	–	–	–
Larsen (simplified) with weights	191	366	632

TABLE 9.14. The smallest error obtained as a result of learning

NDP		
Structure	The smallest error	Number of epochs
Mamdani with weights	0.0238	389
Larsen with weights	0.0164	495
Larsen (simplified) with weights	0.0136	487

desired values of error and the number of epochs, after which this error was obtained.

As it may be inferred from Table 9.13, neither for the Mamdani nor for the Larsen structure the system was able to learn as to obtain the desired values of error.

9.3.3.2.3. NDP

Table 9.14 presents the smallest error for individual structures and the number of epochs corresponding to this error. Table 9.15 presents three desired values of error and the number of epochs, after which this error was obtained.

As it may be inferred from Table 9.15, for the Mamdani structure with weights of rules, the system was unable to learn as to obtain the error 0.020 and 0.023.

9.3.3.2.4. Modeling the taste of rice

Table 9.16 presents the smallest error for individual structures and the number of epochs corresponding to this error. Table 9.17 presents three desired values of error and the number of epochs, after which this error was obtained.

As it may be inferred from Table 9.17, for the Larsen structure with weights of rules, the system was unable to learn as to obtain the error 0.022.

9.3.3.3. M3-type systems

Let us consider Mamdani type systems which are constructed using definitions of triangular norms taking into account the weights w_k, characterizing

TABLE 9.15. Number of epochs required to train the system which is characterized by a definite error

NDP			
Structure	Value of error		
	0.028	0.020	0.023
Mamdani with weights	157	–	–
Larsen with weights	121	180	272
Larsen (simplified) with weights	55	59	90

TABLE 9.16. The smallest error obtained as a result of learning

MODELING THE TASTE OF RICE		
Structure	The smallest error	Number of epochs
Mamdani with weights	0.0178	4978
Larsen with weights	0.0229	4154
Larsen (simplified) with weights	0.0199	4935

TABLE 9.17. Number of epochs required to train the system which is characterized by a definite error

MODELING THE TASTE OF RICE			
Structure	Value of error		
	0.028	0.025	0.022
Mamdani with weights	335	716	1751
Larsen with weights	562	1479	–
Larsen (simplified) with weights	421	852	3264

the importance of particular rules, and the weights $w_{i,k}$, characterizing the importance of particular input linguistic variables. Using the definition of weighted t-conorm and dependencies (9.22), (9.24) and (9.25), we obtain the following description of the neuro-fuzzy systems:

a) *Mamdani system with weights of inputs and rules*

$$
\bar{y} = \frac{\sum_{r=1}^{N} \bar{y}^r \cdot S_{k=1}^{*N} \left\{ \min\left(T_{i=1}^{*n} \left\{ \mu_{A_i^k}(\bar{x}_i), w_{i,k} \right\}, \mu_{B^k}(\bar{y}^r) \right), w_k \right\}}{\sum_{r=1}^{N} S_{k=1}^{*N} \left\{ \min\left(T_{i=1}^{*n} \left\{ \mu_{A_i^k}(\bar{x}_i), w_{i,k} \right\}, \mu_{B^k}(\bar{y}^r) \right), w_k \right\}}
\tag{9.29}
$$

$$
= \frac{\sum_{r=1}^{N} \bar{y}^r \cdot S_{k=1}^{*N} \left\{ \min\left(\begin{array}{l} T_{i=1}^{*n} \left\{ \exp\left[-\left(\frac{\bar{x}_i - \bar{x}_i^k}{\sigma_i^k} \right)^2 \right], w_{i,k} \right\}, \\ \exp\left[-\left(\frac{\bar{y}^r - \bar{y}^k}{\sigma^k} \right)^2 \right] \end{array} \right), w_k \right\}}{\sum_{r=1}^{N} S_{k=1}^{*N} \left\{ \min\left(\begin{array}{l} T_{i=1}^{*n} \left\{ \exp\left[-\left(\frac{\bar{x}_i - \bar{x}_i^k}{\sigma_i^k} \right)^2 \right], w_{i,k} \right\}, \\ \exp\left[-\left(\frac{\bar{y}^r - \bar{y}^k}{\sigma^k} \right)^2 \right] \end{array} \right), w_k \right\}}.
$$

b) *Larsen system with weights of inputs and rules*

$$
\bar{y} = \frac{\sum_{r=1}^{N} \bar{y}^r \cdot S_{k=1}^{*N} \left\{ T_{i=1}^{*n} \left\{ \mu_{A_i^k}(\bar{x}_i), w_{i,k} \right\} \cdot \mu_{B^k}(\bar{y}^r), w_k \right\}}{\sum_{r=1}^{N} S_{k=1}^{*N} \left\{ T_{i=1}^{*n} \left\{ \mu_{A_i^k}(\bar{x}_i), w_{i,k} \right\} \cdot \mu_{B^k}(\bar{y}^r), w_k \right\}}
\tag{9.30}
$$

$$
= \frac{\sum_{r=1}^{N} \bar{y}^r \cdot S_{k=1}^{*N} \left\{ T_{i=1}^{*n} \left\{ \exp\left[-\left(\frac{\bar{x}_i - \bar{x}_i^k}{\sigma_i^k} \right)^2 \right], w_{i,k} \right\} \cdot \exp\left[-\left(\frac{\bar{y}^r - \bar{y}^k}{\sigma^k} \right)^2 \right], w_k \right\}}{\sum_{r=1}^{N} S_{k=1}^{*N} \left\{ T_{i=1}^{*n} \left\{ \exp\left[-\left(\frac{\bar{x}_i - \bar{x}_i^k}{\sigma_i^k} \right)^2 \right], w_{i,k} \right\} \cdot \exp\left[-\left(\frac{\bar{y}^r - \bar{y}^k}{\sigma^k} \right)^2 \right], w_k \right\}}.
$$

In both a) and b) systems, the parameters of membership function, i.e. $\bar{x}_i^k, \sigma_i^k, \bar{y}^k, \sigma^k$ and weights $w_{i,k}$ and w_k are subject to learning.

c) *Simplified Larsen system with weights of inputs and rules*

$$
\bar{y} = \frac{\sum_{r=1}^{N} \bar{y}^r \cdot w_r \left[T_{i=1}^{n} \left\{ 1 - w_{i,r} \left(1 - \mu_{A_i^r}(\bar{x}_i) \right) \right\} \right]}{\sum_{r=1}^{N} w_r \left[T_{i=1}^{n} \left\{ 1 - w_{i,r} \left(1 - \mu_{A_i^r}(\bar{x}_i) \right) \right\} \right]}
\tag{9.31}
$$

$$
= \frac{\sum_{r=1}^{N} \bar{y}^r \cdot w_r \left[T_{i=1}^{n} \left\{ 1 - w_{i,r} \left(1 - \left(\exp\left[-\left(\frac{\bar{x}_i - \bar{x}_i^r}{\sigma_i^r} \right)^2 \right] \right) \right) \right\} \right]}{\sum_{r=1}^{N} w_r \left[T_{i=1}^{n} \left\{ 1 - w_{i,r} \left(1 - \left(\exp\left[-\left(\frac{\bar{x}_i - \bar{x}_i^r}{\sigma_i^r} \right)^2 \right] \right) \right) \right\} \right]}.
$$

In system c) the parameters of membership function, i.e. $\overline{x}_i^r, \sigma_i^r, \overline{y}^r$ and weights $w_{i,r}$ and w_r are subject to learning. Neuro-fuzzy systems (9.29), (9.30) and (9.31) have been used to solve four problems specified in Table 9.1.

9.3.3.3.1. Polymerization

Table 9.18 presents the smallest error for individual structures and the number of epochs corresponding to this error. Table 9.19 presents three desired values of error and the number of epochs, after which this error was obtained.

9.3.3.3.2. HANG

Table 9.20 presents the smallest error for individual structures and the number of epochs corresponding to this error. Table 9.21 presents three desired values of error and the number of epochs, after which this error was obtained.

As it may be inferred from Table 9.21, the desired error values could not be obtained for the Larsen structure with weights of inputs and rules.

TABLE 9.18. The smallest error obtained as a result of learning

POLYMERIZATION		
Structure	The smallest error	Number of epochs
Mamdani with weights of inputs and rules	0.0034	4704
Larsen with weights of inputs and rules	0.0035	3822
Larsen (simplified) with weights of inputs and rules	0.0031	2953

TABLE 9.19. Number of epochs required to train the system which is characterized by a definite error

POLYMERIZATION			
Structure	Value of error		
	0.0055	0.0050	0.0045
Mamdani with weights of inputs and rules	1915	2303	2549
Larsen with weights of inputs and rules	1	1	1
Larsen (simplified) with weights of inputs and rules	1	6	13

TABLE 9.20. The smallest error obtained as a result of learning

HANG		
Structure	The smallest error	Number of epochs
Mamdani with weights of inputs and rules	0.0209	5474
Larsen with weights of inputs and rules	0.0346	1541
Larsen (simplified) with weights of inputs and rules	0.0124	4252

TABLE 9.21. Number of epochs required to train the system which is characterized by a definite error

HANG			
Structure	Value of error		
	0.028	0.026	0.024
Mamdani with weights of inputs and rules	4213	5474	5474
Larsen with weights of inputs and rules	–	–	–
Larsen (simplified) with weights of inputs and rules	628	750	750

TABLE 9.22. The smallest error obtained as a result of learning

NDP		
Structure	The smallest error	Number of epochs
Mamdani with weights of inputs and rules	0.0181	498
Larsen with weights of inputs and rules	0.0146	500
Larsen (simplified) with weights of inputs and rules	0.0188	484

9.3.3.3.3. NDP

Table 9.22 presents the smallest error for individual structures and the number of epochs corresponding to this error. Table 9.23 presents three

TABLE 9.23. Number of epochs required to train the system which is characterized by a definite error

Structure	NDP		
	Value of error		
	0.026	0.023	0.020
Mamdani with weights of inputs and rules	60	126	294
Larsen with weights of inputs and rules	24	31	74
Larsen (simplified) with weights of inputs and rules	–	–	–

TABLE 9.24. The smallest error obtained as a result of learning

MODELING THE TASTE OF RICE		
Structure	The smallest error	Number of epochs
Mamdani with weights of inputs and rules	0.0168	2218
Larsen with weights of inputs and rules	0.0218	2325
Larsen (simplified) with weights of inputs and rules	0.0190	4975

desired values of error and the number of epochs, after which this error was obtained.

As it may be inferred from Table 9.23, the desired error values could not be obtained for the simplified Larsen structure with weights of inputs and rules.

9.3.3.3.4. Modeling the taste of rice

Table 9.24 presents the smallest error for individual structures and the number of epochs corresponding to this error.

Table 9.25 presents three desired values of error and the number of epochs, after which this error was obtained.

9.4 Neuro-fuzzy systems of logical type

In the previous subchapter, we have discussed neuro-fuzzy systems with Mamdani type inference. Currently, we will consider systems in which the

TABLE 9.25. Number of epochs required to train the system which is characterized by a definite error

MODELING THE TASTE OF RICE			
Structure	Value of error		
	0.028	0.025	0.022
Mamdani with weights of inputs and rules	1	3	3
Larsen with weights of inputs and rules	295	528	2325
Larsen (simplified) with weights of inputs and rules	1	1	5

antecedents and the consequents of rules are connected with each other using a fuzzy implication.

In logical type systems, the defuzzification is made by means of dependency

$$\bar{y} = \frac{\sum_{r=1}^{N} \bar{y}^r \cdot \mu_{B'}(\bar{y}^r)}{\sum_{r=1}^{N} \mu_{B'}(\bar{y}^r)}. \tag{9.32}$$

In these systems, the fuzzy set B' is created as a result of intersection of fuzzy sets $\overline{B}^k$, i.e.

$$B' = \bigcap_{k=1}^{N} \overline{B}^k. \tag{9.33}$$

The membership function of fuzzy set B' is determined using a t-norm, which shall be notated as follows:

$$\mu_{B'}(y) = \underset{k=1}{\overset{N}{T}} \left\{ \mu_{\overline{B}^k}(y) \right\}. \tag{9.34}$$

Using formulas (9.34), (9.6) and (9.7), we have

$$\mu_{B'}(\bar{y}^r) = \underset{k=1}{\overset{N}{T}} \left\{ \mu_{\overline{B}^k}(\bar{y}^r) \right\} = \underset{k=1}{\overset{N}{T}} \left\{ I\left(\mu_{A^k}(\bar{\mathbf{x}}), \mu_{B^k}(\bar{y}^r) \right) \right\} \tag{9.35}$$

$$= \underset{k=1}{\overset{N}{T}} \left\{ I\left(\underset{i=1}{\overset{N}{T}} \mu_{A_i^k}(\bar{x}_i), \mu_{B^k}(\bar{y}^r) \right) \right\},$$

where I is a fuzzy implication defined in point 4.8.4. By substituting formula (9.35) to dependency (9.32), we obtain

$$\bar{y} = \frac{\sum_{r=1}^{N} \bar{y}^r \cdot T_{k=1}^{N} \left\{ I\left(T_{i=1}^{n} \left\{ \mu_{A_i^k}(\bar{x}_i) \right\}, \mu_{B^k}(\bar{y}^r) \right) \right\}}{\sum_{r=1}^{N} T_{k=1}^{N} \left\{ I\left(T_{i=1}^{n} \left\{ \mu_{A_i^k}(\bar{x}_i) \right\}, \mu_{B^k}(\bar{y}^r) \right) \right\}}. \tag{9.36}$$

The specific form of formula (9.36) depends on the chosen definition of I function. Logical type systems will be applied to solve modeling problems. We will consider M1 systems (without weights), M2 systems (with weights of rules) and M3 systems (with weights of rules and weights of input linguistic variables). We will apply the Łukasiewicz, binary, Reichenbach, Zadeh and Willmott fuzzy implications. Moreover we will present and test simplified neuro-fuzzy structures using Łukasiewicz and Zadeh implications.

9.4.1 M1-type systems

Let us consider logical type systems which are constructed using definitions of triangular norms without taking the weights into account. First, we will use Łukasiewicz implication. As a result of applying this implication, we will obtain the following dependency:

$$\mu_{A^k \to B^k}(\overline{\mathbf{x}}, y) = I\left(\mu_{A^k}(\overline{\mathbf{x}}), \mu_{B^k}(y)\right) = I\left(\underset{k=1}{\overset{n}{T}}\left(\mu_{A_i^k}(\overline{x}_i)\right), \mu_{B^k}(y)\right) \quad (9.37)$$

$$= \min\left[1, 1 - \underset{i=1}{\overset{n}{T}}\left(\mu_{A_i^k}(\overline{x}_i)\right) + \mu_{B^k}(y)\right].$$

By substituting dependency (9.37) to formula (9.36), we obtain

$$\overline{y} = \frac{\sum_{r=1}^{N} \overline{y}^r T_{k=1}^{N}\left\{\min\left[1, 1 - T_{i=1}^{n}\left(\mu_{A_i^k}(\overline{x}_i)\right) + \mu_{B^k}(\overline{y}^r)\right]\right\}}{\sum_{r=1}^{N} T_{k=1}^{N}\left\{\min\left[1, 1 - T_{i=1}^{n}\left(\mu_{A_i^k}(\overline{x}_i)\right) + \mu_{B^k}(\overline{y}^r)\right]\right\}}. \quad (9.38)$$

By substituting dependencies (9.13) and (9.14) to formula (9.38), we obtain

$$\overline{y} = \frac{\sum_{r=1}^{N} \overline{y}^r T_{k=1}^{N}\left\{\min\left[\begin{array}{c} 1, 1 - T_{i=1}^{n}\left(\exp\left[-\left(\frac{\overline{x}_i - \overline{x}_i^k}{\sigma_i^k}\right)^2\right]\right) \\ + \exp\left[-\left(\frac{\overline{y}^r - \overline{y}^k}{\sigma^k}\right)^2\right] \end{array}\right]\right\}}{\sum_{r=1}^{N} T_{k=1}^{N}\left\{\min\left[\begin{array}{c} 1, 1 - T_{i=1}^{n}\left(\exp\left[-\left(\frac{\overline{x}_i - \overline{x}_i^k}{\sigma_i^k}\right)^2\right]\right) \\ + \exp\left[-\left(\frac{\overline{y}^r - \overline{y}^k}{\sigma^k}\right)^2\right] \end{array}\right]\right\}}. \quad (9.39)$$

By applying the binary fuzzy implication, we obtain

$$\mu_{A^k \to B^k}(\overline{\mathbf{x}}, y) = I\left(\mu_{A^k}(\overline{\mathbf{x}}), \mu_{B^k}(y)\right) = I\left(\underset{i=1}{\overset{n}{T}}\left(\mu_{A_i^k}(\overline{x}_i)\right), \mu_{B^k}(y)\right) \quad (9.40)$$

$$= \max\left[1 - \underset{i=1}{\overset{n}{T}}\left(\mu_{A_i^k}(\overline{x}_i)\right), \mu_{B^k}(y)\right].$$

By substituting dependency (9.40) to formula (9.36), we obtain

$$\bar{y} = \frac{\sum_{r=1}^{N} \bar{y}^r T_{k=1}^{N} \left\{ \max \left[1 - T_{i=1}^n \left(\mu_{A_i^k} \left(\bar{x}_i \right) \right), \mu_{B^k} \left(\bar{y}^r \right) \right] \right\}}{\sum_{r=1}^{N} T_{k=1}^{N} \left\{ \max \left[1 - T_{i=1}^n \left(\mu_{A_i^k} \left(\bar{x}_i \right) \right), \mu_{B^k} \left(\bar{y}^r \right) \right] \right\}}. \tag{9.41}$$

By substituting dependencies (9.13) and (9.14) to formula (9.41), we obtain

$$\bar{y} = \frac{\sum_{r=1}^{N} \bar{y}^r T_{k=1}^{N} \left\{ \max \left[\begin{array}{c} 1 - T_{i=1}^n \left(\exp \left[- \left(\frac{\bar{x}_i - \bar{x}_i^k}{\sigma_i^k} \right)^2 \right] \right), \\ \exp \left[- \left(\frac{\bar{y}^r - \bar{y}^k}{\sigma^k} \right)^2 \right] \end{array} \right] \right\}}{\sum_{r=1}^{N} T_{k=1}^{N} \left\{ \max \left[\begin{array}{c} 1 - T_{i=1}^n \left(\exp \left[- \left(\frac{\bar{x}_i - \bar{x}_i^k}{\sigma_i^k} \right)^2 \right] \right), \\ \exp \left[- \left(\frac{\bar{y}^r - \bar{y}^k}{\sigma^k} \right)^2 \right] \end{array} \right] \right\}}. \tag{9.42}$$

Applying Reichenbach fuzzy implication we get:

$$\mu_{A^k \to B^k} \left(\bar{\mathbf{x}}, y \right) = I \left(\mu_{A^k} \left(\bar{\mathbf{x}} \right), \mu_{B^k} \left(y \right) \right) = I \left(\mathop{T}_{i=1}^{n} \left(\mu_{A_i^k} \left(\bar{x}_i \right) \right), \mu_{B^k} \left(y \right) \right) \tag{9.43}$$

$$= 1 - \mathop{T}_{i=1}^{n} \left(\mu_{A_i^k} \left(\bar{x}_i \right) \right) \left(1 - \mu_{B^k} \left(y \right) \right).$$

By substituting dependency (9.43) to formula (9.36), we obtain

$$\bar{y} = \frac{\sum_{r=1}^{N} \bar{y}^r T_{k=1}^{N} \left\{ 1 - T_{i=1}^n \left(\mu_{A_i^k} \left(\bar{x}_i \right) \right) \left(1 - \mu_{B^k} \left(\bar{y}^r \right) \right) \right\}}{\sum_{r=1}^{N} T_{k=1}^{N} \left\{ 1 - T_{i=1}^n \left(\mu_{A_i^k} \left(\bar{x}_i \right) \right) \left(1 - \mu_{B^k} \left(\bar{y}^r \right) \right) \right\}}. \tag{9.44}$$

By substituting dependencies (9.13) and (9.14) to formula (9.44), we have

$$\bar{y} = \frac{\sum_{r=1}^{N} \bar{y}^r T_{k=1}^{N} \left\{ \begin{array}{c} 1 - T_{i=1}^n \left(\exp \left[- \left(\frac{\bar{x}_i - \bar{x}_i^k}{\sigma_i^k} \right)^2 \right] \right) \\ \left(1 - \exp \left[- \left(\frac{\bar{y}^r - \bar{y}^k}{\sigma^k} \right)^2 \right] \right) \end{array} \right\}}{\sum_{r=1}^{N} T_{k=1}^{N} \left\{ \begin{array}{c} 1 - T_{i=1}^n \left(\exp \left[- \left(\frac{\bar{x}_i - \bar{x}_i^k}{\sigma_i^k} \right)^2 \right] \right) \\ \left(1 - \exp \left[- \left(\frac{\bar{y}^r - \bar{y}^k}{\sigma^k} \right)^2 \right] \right) \end{array} \right\}}. \tag{9.45}$$

Applying Zadeh fuzzy implication we get:

$$\mu_{A^k \to B^k}(\overline{\mathbf{x}}, y) = I\left(\mu_{A^k}(\overline{\mathbf{x}}), \mu_{B^k}(y)\right) = I\left(\underset{i=1}{\overset{n}{T}}\left(\mu_{A_i^k}(\overline{x}_i)\right), \mu_{B^k}(y)\right) \quad (9.46)$$

$$= \max\left\{\min\left[\underset{i=1}{\overset{n}{T}}\left(\mu_{A_i^k}(\overline{x}_i)\right), \mu_{B^k}(y)\right], 1 - \underset{i=1}{\overset{n}{T}}\left(\mu_{A_i^k}(\overline{x}_i)\right)\right\}.$$

By substituting dependency (9.46) to formula (9.36), we obtain

$$\overline{y} = \frac{\sum_{r=1}^{N} \overline{y}^r T \left\{ \underset{\substack{k=1 \\ k \neq r}}{\overset{N}{T}} \max\left\{ \begin{array}{l} \max\left\{\underset{i=1}{\overset{n}{T}}\left\{\mu_{A_i^r}(\overline{x}_i)\right\}, 1 - \underset{i=1}{\overset{n}{T}}\left\{\mu_{A_i^r}(\overline{x}_i)\right\}\right\}, \\ \left\{ \begin{array}{l} \min\left[\underset{i=1}{\overset{n}{T}}\left\{\mu_{A_i^k}(\overline{x}_i)\right\}, \mu_{B^k}(\overline{y}^r)\right], \\ 1 - \underset{i=1}{\overset{n}{T}}\left\{\mu_{A_i^k}(\overline{x}_i)\right\} \end{array}\right\} \end{array}\right\}\right\}}{\sum_{r=1}^{N} T \left\{ \underset{\substack{k=1 \\ k \neq r}}{\overset{N}{T}} \max\left\{ \begin{array}{l} \max\left\{\underset{i=1}{\overset{n}{T}}\left\{\mu_{A_i^r}(\overline{x}_i)\right\}, 1 - \underset{i=1}{\overset{n}{T}}\left\{\mu_{A_i^r}(\overline{x}_i)\right\}\right\}, \\ \left\{ \begin{array}{l} \min\left[\underset{i=1}{\overset{n}{T}}\left\{\mu_{A_i^k}(\overline{x}_i)\right\}, \mu_{B^k}(\overline{y}^r)\right], \\ 1 - \underset{i=1}{\overset{n}{T}}\left\{\mu_{A_i^k}(\overline{x}_i)\right\} \end{array}\right\} \end{array}\right\}\right\}}.$$

$$(9.47)$$

By substituting dependencies (9.13) and (9.14) to formula (9.47), we get:

$$\overline{y} = \frac{\sum_{r=1}^{N} \overline{y}^r T \left\{ \underset{\substack{k=1 \\ k \neq r}}{\overset{N}{T}} \max\left\{ \begin{array}{l} \max\left\{ \begin{array}{l} \underset{i=1}{\overset{n}{T}}\left\{\exp\left[-\left(\frac{\overline{x}_i - \overline{x}_i^r}{\sigma_i^r}\right)^2\right]\right\}, \\ 1 - \underset{i=1}{\overset{n}{T}}\left\{\exp\left[-\left(\frac{\overline{x}_i - \overline{x}_i^r}{\sigma_i^r}\right)^2\right]\right\} \end{array}\right\}, \\ \left\{ \begin{array}{l} \min\left\{ \begin{array}{l} \underset{i=1}{\overset{n}{T}}\left\{\exp\left[-\left(\frac{\overline{x}_i - \overline{x}_i^k}{\sigma_i^k}\right)^2\right]\right\}, \\ \exp\left[-\left(\frac{\overline{y}^r - \overline{y}^k}{\sigma^k}\right)^2\right] \end{array}\right\}, \\ 1 - \underset{i=1}{\overset{n}{T}}\left\{\exp\left[-\left(\frac{\overline{x}_i - \overline{x}_i^k}{\sigma_i^k}\right)^2\right]\right\} \end{array}\right\} \end{array}\right\}\right\}}{\sum_{r=1}^{N} T \left\{ \underset{\substack{k=1 \\ k \neq r}}{\overset{N}{T}} \max\left\{ \begin{array}{l} \max\left\{ \begin{array}{l} \underset{i=1}{\overset{n}{T}}\left\{\exp\left[-\left(\frac{\overline{x}_i - \overline{x}_i^r}{\sigma_i^r}\right)^2\right]\right\}, \\ 1 - \underset{i=1}{\overset{n}{T}}\left\{\exp\left[-\left(\frac{\overline{x}_i - \overline{x}_i^r}{\sigma_i^r}\right)^2\right]\right\} \end{array}\right\}, \\ \left\{ \begin{array}{l} \min\left\{ \begin{array}{l} \underset{i=1}{\overset{n}{T}}\left\{\exp\left[-\left(\frac{\overline{x}_i - \overline{x}_i^k}{\sigma_i^k}\right)^2\right]\right\}, \\ \exp\left[-\left(\frac{\overline{y}^r - \overline{y}^k}{\sigma^k}\right)^2\right] \end{array}\right\}, \\ 1 - \underset{i=1}{\overset{n}{T}}\left\{\exp\left[-\left(\frac{\overline{x}_i - \overline{x}_i^k}{\sigma_i^k}\right)^2\right]\right\} \end{array}\right\} \end{array}\right\}\right\}}.$$

$$(9.48)$$

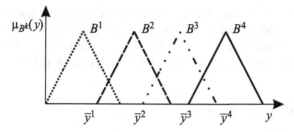

FIGURE 9.3. Example fuzzy sets satisfying the assumption $\mu_{B^k}(\overline{y}^r) \approx 0$

The simplified systems were also studied which are characterized by a small coincidence or total separation one from another of output fuzzy sets B^k. In this situation, the condition $\mu_{B^k}(\overline{y}^r) \approx 0$ is satisfied, which is illustrated by Fig. 9.3.

If $\mu_{B^k}(\overline{y}^r) \approx 0$, then we will obtain from dependency (9.39) a simplified Łukasiewicz structure of the following form

$$
\overline{y} = \frac{\sum\limits_{r=1}^{N} \overline{y}^r \mathop{T}\limits_{\substack{k=1 \\ k \neq r}}^{N} \left\{ 1 - \mathop{T}\limits_{i=1}^{n} \left\{ \exp\left[-\left(\frac{\overline{x}_i - \overline{x}_i^k}{\sigma_i^k} \right)^2 \right] \right\} \right\}}{\sum\limits_{r=1}^{N} \mathop{T}\limits_{\substack{k=1 \\ k \neq r}}^{N} \left\{ 1 - \mathop{T}\limits_{i=1}^{n} \left\{ \exp\left[-\left(\frac{\overline{x}_i - \overline{x}_i^k}{\sigma_i^k} \right)^2 \right] \right\} \right\}}.
\tag{9.49}
$$

Similarly, if $\mu_{B^k}(\overline{y}^r) \approx 0$, then we will obtain from dependency (9.48) a simplified Zadeh structure given by the formula

$$
\overline{y} = \frac{\sum\limits_{r=1}^{N} \overline{y}^r T \left\{ \max \left\{ \begin{array}{c} \mathop{T}\limits_{i=1}^{n} \left\{ \exp\left[-\left(\frac{\overline{x}_i - \overline{x}_i^r}{\sigma_i^r} \right)^2 \right] \right\}, \\ 1 - \mathop{T}\limits_{i=1}^{n} \left\{ \exp\left[-\left(\frac{\overline{x}_i - \overline{x}_i^r}{\sigma_i^r} \right)^2 \right] \right\} \end{array} \right\}, \mathop{T}\limits_{\substack{k=1 \\ k \neq r}}^{N} \left\{ 1 - \mathop{T}\limits_{i=1}^{n} \left\{ \exp\left[-\left(\frac{\overline{x}_i - \overline{x}_i^k}{\sigma_i^k} \right)^2 \right] \right\} \right\} \right\}}{\sum\limits_{r=1}^{N} T \left\{ \max \left\{ \begin{array}{c} \mathop{T}\limits_{i=1}^{n} \left\{ \exp\left[-\left(\frac{\overline{x}_i - \overline{x}_i^r}{\sigma_i^r} \right)^2 \right] \right\}, \\ 1 - \mathop{T}\limits_{i=1}^{n} \left\{ \exp\left[-\left(\frac{\overline{x}_i - \overline{x}_i^r}{\sigma_i^r} \right)^2 \right] \right\} \end{array} \right\}, \mathop{T}\limits_{\substack{k=1 \\ k \neq r}}^{N} \left\{ 1 - \mathop{T}\limits_{i=1}^{n} \left\{ \exp\left[-\left(\frac{\overline{x}_i - \overline{x}_i^k}{\sigma_i^k} \right)^2 \right] \right\} \right\} \right\}}.
\tag{9.50}
$$

In systems (9.39), (9.42), (9.45) and (9.48) the following parameters of the membership functions are subject to learning: $\overline{x}_i^k, \sigma_i^k, \overline{y}^k, \sigma^k$. In simplified systems (9.49) and (9.50), the parameters $\bar{x}_i^k, \sigma_i^k, \bar{y}^k$ are subject to learning. We will solve the modeling problems using Łukasiewicz structure, binary structure, Reichenbach structure, Łukasiewicz simplified structure, Zadeh structure, Willmott structure and Zadeh simplified structure.

9.4.1.1. Polymerization

Table 9.26 presents the smallest error for individual structures and the number of epochs corresponding to this error.

As it may be inferred from Table 9.26, the smallest error was 0.0038 and was obtained for Zadeh structure.

Table 9.27 presents three desired values of error and the number of epochs, after which this error was obtained. As it may be inferred from this table, not all the structures were able to obtain the desired error value.

TABLE 9.26. The smallest error obtained as a result of learning

POLYMERIZATION		
Structure	The smallest error	Number of epochs
Łukasiewicz	0.0065	5863
Binary	0.0063	5980
Reichenbach	0.0040	5494
Łukasiewicz simplified	0.0059	3385
Zadeh	0.0038	3648
Willmott	0.0056	5918
Zadeh simplified	0.0049	3432

TABLE 9.27. Number of epochs required to train the system which is characterized by a definite error

POLYMERIZATION			
Structure	Value of error		
	0.0055	0.0050	0.0040
Łukasiewicz	–	–	–
Binary	–	–	–
Reichenbach	1627	1903	2783
Łukasiewicz simplified	–	–	–
Zadeh	949	1022	1809
Willmott	–	–	–
Zadeh simplified	2844	3432	–

9.4.1.2. HANG

Table 9.28 presents the smallest error for individual structures and the number of epochs corresponding to this error.

As it may be inferred from Table 9.28, the smallest error was 0.0177 and was obtained for binary structure. Table 9.29 presents three desired values of error and the number of epochs, after which this error was obtained.

As it may be inferred from Table 9.29, not all the structures were able to obtain the desired value of error.

9.4.1.3. NDP

Table 9.30 presents the smallest error for individual structures and the number of epochs corresponding to this error.

TABLE 9.28. The smallest error obtained as a result of learning

HANG		
Structure	The smallest error	Number of epochs
Łukasiewicz	0.0289	3908
Binary	0.0177	7773
Reichenbach	0.0320	7989
Łukasiewicz simplified	0.0361	6536
Zadeh	0.0216	5288
Willmott	0.0366	7327
Zadeh simplified	0.0265	2317

TABLE 9.29. Number of epochs required to train the system which is characterized by a definite error

HANG			
Structure	Value of error		
	0.028	0.026	0.024
Łukasiewicz	–	–	–
Binary	1795	2996	3762
Reichenbach	–	–	–
Łukasiewicz simplified	–	–	–
Zadeh	3382	3875	4218
Willmott	–	–	–
Zadeh simplified	1787	–	–

TABLE 9.30. The smallest error obtained as a result of learning

	NDP	
Structure	The smallest error	Number of epochs
Łukasiewicz	0.0166	457
Binary	0.0149	437
Reichenbach	0.0157	454
Łukasiewicz simplified	0.0229	497
Zadeh	0.0156	498
Willmott	0.0180	488
Zadeh simplified	0.0156	496

TABLE 9.31. Number of epochs required to train the system which is characterized by a definite error

	NDP		
Structure	Value of error		
	0.026	0.023	0.020
Łukasiewicz	255	276	313
Binary	74	111	166
Reichenbach	166	173	251
Łukasiewicz simplified	190	497	–
Zadeh	38	59	109
Willmott	121	171	303
Zadeh simplified	39	83	147

As it may be inferred from Table 9.30, the smallest error was 0.0149 and was obtained for the binary structure. Table 9.31 presents three desired values of error and the number of epochs, after which this error was obtained. As it may be inferred from Table 9.31, for the simplified Łukasiewicz structure, it was impossible to obtain the error equal to 0.020.

9.4.1.4. Modeling the taste of rice

Table 9.32 presents the smallest error for individual structures and the number of epochs corresponding to this error.

As it may be inferred from Table 9.32, the smallest error was 0.0211 and was obtained for Willmott structure. Table 9.33 presents three desired values of error and the number of epochs, after which this error was obtained.

As it may be inferred from Table 9.33, not all the structures were able to obtain the desired value of error equal to 0.022.

TABLE 9.32. The smallest error obtained as a result of learning

MODELING THE TASTE OF RICE		
Structure	The smallest error	Number of epochs
Łukasiewicz	0.0221	4048
Binary	0.0230	3201
Reichenbach	0.0212	4575
Łukasiewicz simplified	0.0243	2328
Zadeh	0.0219	4534
Willmott	0.0211	2605
Zadeh simplified	0.0246	4588

TABLE 9.33. Number of epochs required to train the system which is characterized by a definite error

MODELING THE TASTE OF RICE			
Structure	Value of error		
	0.028	0.025	0.022
Łukasiewicz	408	2327	–
Binary	286	817	–
Reichenbach	313	938	2354
Łukasiewicz simplified	1030	1850	–
Zadeh	344	1484	4534
Willmott	134	517	2605
Zadeh simplified	998	4588	–

9.4.2   M2-type systems

Let us consider logical type systems which are constructed using definitions of triangular norms taking into account the weights w_k, characterizing the importance of particular rules. Using the definition of weighted t-norm and dependencies (9.39), (9.42), (9.45), and (9.48) - (9.50), we obtain the following neuro-fuzzy systems:

a) Neuro-fuzzy system with weights of rules and the Łukasiewicz implication

$$\bar{y} = \frac{\sum_{r=1}^{N} \bar{y}^r T_{k=1}^{*N} \left\{ \min \left[1, 1 - \underset{i=1}{\overset{n}{T}} \left(\exp\left[-\left(\frac{\bar{x}_i - \bar{x}_i^k}{\sigma_i^k} \right)^2 \right] \right) + \exp\left[-\left(\frac{\bar{y}^r - \bar{y}^k}{\sigma^k} \right)^2 \right] \right], w_k \right\}}{\sum_{r=1}^{N} T_{k=1}^{*N} \left\{ \min \left[1, 1 - \underset{i=1}{\overset{n}{T}} \left(\exp\left[-\left(\frac{\bar{x}_i - \bar{x}_i^k}{\sigma_i^k} \right)^2 \right] \right) + \exp\left[-\left(\frac{\bar{y}^r - \bar{y}^k}{\sigma^k} \right)^2 \right] \right], w_k \right\}}. \quad (9.51)$$

b) Neuro-fuzzy system with weights of rules and the binary implication

$$\bar{y} = \frac{\sum_{r=1}^{N} \bar{y}^r\, T_{k=1}^{*N} \left\{ \max\left[\begin{array}{c} 1 - \mathop{T}\limits_{i=1}^{n}\left(\exp\left[-\left(\frac{\bar{x}_i - \bar{x}_i^k}{\sigma_i^k}\right)^2\right]\right), \\ \exp\left[-\left(\frac{\bar{y}^r - \bar{y}^k}{\sigma^k}\right)^2\right] \end{array} \right], w_k \right\}}{\sum_{r=1}^{N} T_{k=1}^{*N} \left\{ \max\left[\begin{array}{c} 1 - \mathop{T}\limits_{i=1}^{n}\left(\exp\left[-\left(\frac{\bar{x}_i - \bar{x}_i^k}{\sigma_i^k}\right)^2\right]\right), \\ \exp\left[-\left(\frac{\bar{y}^r - \bar{y}^k}{\sigma^k}\right)^2\right] \end{array} \right], w_k \right\}}. \qquad (9.52)$$

c) Neuro-fuzzy system with weights of rules and the Reichenbach implication

$$\bar{y} = \frac{\sum_{r=1}^{N} \bar{y}^r\, T_{k=1}^{*N} \left\{ \begin{array}{c} 1 - \mathop{T}\limits_{i=1}^{n}\left(\exp\left[-\left(\frac{\bar{x}_i - \bar{x}_i^k}{\sigma_i^k}\right)^2\right]\right) \\ \left(1 - \exp\left[-\left(\frac{\bar{y}^r - \bar{y}^k}{\sigma^k}\right)^2\right]\right), w_k \end{array} \right\}}{\sum_{r=1}^{N} T_{k=1}^{*N} \left\{ \begin{array}{c} 1 - \mathop{T}\limits_{i=1}^{n}\left(\exp\left[-\left(\frac{\bar{x}_i - \bar{x}_i^k}{\sigma_i^k}\right)^2\right]\right) \\ \left(1 - \exp\left[-\left(\frac{\bar{y}^r - \bar{y}^k}{\sigma^k}\right)^2\right]\right), w_k \end{array} \right\}}. \qquad (9.53)$$

d) Neuro-fuzzy system with weights of rules and a fuzzy Zadeh implication

$$\bar{y} = \frac{\displaystyle\sum_{r=1}^{N} \bar{y}^r\, T^* \left\{ \begin{array}{c} \max\left\{ \begin{array}{c} \mathop{T}\limits_{i=1}^{n}\left\{\exp\left[-\left(\frac{\bar{x}_i - \bar{x}_i^r}{\sigma_i^r}\right)^2\right]\right\}, \\ 1 - \mathop{T}\limits_{i=1}^{n}\left\{\exp\left[-\left(\frac{\bar{x}_i - \bar{x}_i^r}{\sigma_i^r}\right)^2\right]\right\} \end{array}, w_r \right\} \\ \mathop{T}\limits_{\substack{k=1 \\ k\neq r}}^{N} \left\{ \max\left\{ \begin{array}{c} \min\left\{ \begin{array}{c} \mathop{T}\limits_{i=1}^{n}\left\{\exp\left[-\left(\frac{\bar{x}_i - \bar{x}_i^k}{\sigma_i^k}\right)^2\right]\right\}, \\ \exp\left[-\left(\frac{\bar{y}^r - \bar{y}^k}{\sigma^k}\right)^2\right] \end{array} \right\}, \\ 1 - \mathop{T}\limits_{i=1}^{n}\left\{\exp\left[-\left(\frac{\bar{x}_i - \bar{x}_i^k}{\sigma_i^k}\right)^2\right]\right\} \end{array} \right\}, w_k \right\} \end{array} \right\}}{\displaystyle\sum_{r=1}^{N} T^* \left\{ \begin{array}{c} \max\left\{ \begin{array}{c} \mathop{T}\limits_{i=1}^{n}\left\{\exp\left[-\left(\frac{\bar{x}_i - \bar{x}_i^r}{\sigma_i^r}\right)^2\right]\right\}, \\ 1 - \mathop{T}\limits_{i=1}^{n}\left\{\exp\left[-\left(\frac{\bar{x}_i - \bar{x}_i^r}{\sigma_i^r}\right)^2\right]\right\} \end{array}, w_r \right\} \\ \mathop{T}\limits_{\substack{k=1 \\ k\neq r}}^{N} \left\{ \max\left\{ \begin{array}{c} \min\left\{ \begin{array}{c} \mathop{T}\limits_{i=1}^{n}\left\{\exp\left[-\left(\frac{\bar{x}_i - \bar{x}_i^k}{\sigma_i^k}\right)^2\right]\right\}, \\ \exp\left[-\left(\frac{\bar{y}^r - \bar{y}^k}{\sigma^k}\right)^2\right] \end{array} \right\}, \\ 1 - \mathop{T}\limits_{i=1}^{n}\left\{\exp\left[-\left(\frac{\bar{x}_i - \bar{x}_i^k}{\sigma_i^k}\right)^2\right]\right\} \end{array} \right\}, w_k \right\} \end{array} \right\}}.$$

$$(9.54)$$

e) Simplified neuro-fuzzy system with weights of rules and a fuzzy Łukasiewicz implication

$$\bar{y} = \frac{\displaystyle\sum_{r=1}^{N} \bar{y}^r \, \underset{\substack{k=1 \\ k \neq r}}{\overset{N}{T}} \left\{ 1 - \underset{i=1}{\overset{n}{T}} \left\{ \exp\left[-\left(\frac{\bar{x}_i - \bar{x}_i^k}{\sigma_i^k} \right)^2 \right] \right\}, w_k \right\}}{\displaystyle\sum_{r=1}^{N} \underset{\substack{k=1 \\ k \neq r}}{\overset{N}{T}} \left\{ 1 - \underset{i=1}{\overset{n}{T}} \left\{ \exp\left[-\left(\frac{\bar{x}_i - \bar{x}_i^k}{\sigma_i^k} \right)^2 \right] \right\}, w_k \right\}}. \tag{9.55}$$

f) Simplified neuro-fuzzy system with weights of rules and the Zadeh implication

$$\bar{y} = \frac{\displaystyle\sum_{r=1}^{N} \bar{y}^r T^* \left\{ \left(\max \left\{ \begin{array}{l} \underset{i=1}{\overset{n}{T}}\left\{ \exp\left[-\left(\frac{\bar{x}_i - \bar{x}_i^r}{\sigma_i^r}\right)^2\right]\right\}, \\ 1 - \underset{i=1}{\overset{n}{T}}\left\{ \exp\left[-\left(\frac{\bar{x}_i - \bar{x}_i^r}{\sigma_i^r}\right)^2\right]\right\} \end{array} \right\}, w_r, \right\} \underset{\substack{k=1 \\ k \neq r}}{\overset{N}{T^*}}\left\{ 1 - \underset{i=1}{\overset{n}{T}}\left\{ \exp\left[-\left(\frac{\bar{x}_i - \bar{x}_i^k}{\sigma_i^k}\right)^2\right]\right\}, w_k \right\} \right\}}{\displaystyle\sum_{r=1}^{N} T^* \left\{ \left(\max \left\{ \begin{array}{l} \underset{i=1}{\overset{n}{T}}\left\{ \exp\left[-\left(\frac{\bar{x}_i - \bar{x}_i^r}{\sigma_i^r}\right)^2\right]\right\}, \\ 1 - \underset{i=1}{\overset{n}{T}}\left\{ \exp\left[-\left(\frac{\bar{x}_i - \bar{x}_i^r}{\sigma_i^r}\right)^2\right]\right\} \end{array} \right\}, w_r, \right\} \underset{\substack{k=1 \\ k \neq r}}{\overset{N}{T^*}}\left\{ 1 - \underset{i=1}{\overset{n}{T}}\left\{ \exp\left[-\left(\frac{\bar{x}_i - \bar{x}_i^k}{\sigma_i^k}\right)^2\right]\right\}, w_k \right\} \right\}}. \tag{9.56}$$

In systems (9.51) - (9.54) the parameters of the membership functions, i.e. $\bar{x}_i^k, \sigma_i^k, \bar{y}^k, \sigma^k$ and weights w_k are subject to learning. In systems (9.55) and (9.56), the parameters of membership function $\bar{x}_i^k, \sigma_i^k, \bar{y}^k$ and weights w_k are subject to learning.

9.4.2.1. Polymerization

Table 9.34 presents the smallest error for individual structures and the number of epochs corresponding to this error. As it may be inferred from the table, the smallest error was 0.0030 and was obtained for Zadeh structure with weights of rules. Table 9.35 presents three desired values of error and the number of epochs, after which this error was obtained.

As it may be inferred from the table, not all the structures were able to obtain the desired value of error.

9.4.2.2. HANG

Table 9.36 presents the smallest error for individual structures and the number of epochs corresponding to this error.

TABLE 9.34. The smallest error obtained as a result of learning

POLYMERIZATION		
Structure	The smallest error	Number of epochs
Łukasiewicz with weights of rules	0.0041	4765
Binary with weights of rules	0.0054	5980
Reichenbach with weights of rules	0.0037	4653
Łukasiewicz simplified with weights of rules	0.0039	4694
Zadeh with weights of rules	0.0030	5650
Willmott with weights of rules	0.0047	5539
Zadeh simplified with weights of rules	0.0041	5151

TABLE 9.35. Number of epochs required to train the system which is characterized by a definite error

POLYMERIZATION			
Structure	Value of error		
	0.0055	0.0050	0.0040
Łukasiewicz with weights of rules	1258	1662	3266
Binary with weights of rules	5980	–	–
Reichenbach with weights of rules	1385	1385	2521
Łukasiewicz simplified with weights of rules	4	4	209
Zadeh with weights of rules	1497	1497	2726
Willmott with weights of rules	1405	3084	–
Zadeh simplified with weights of rules	367	701	3103

As it may be inferred from Table 9.36, the smallest error was 0.0115 and was obtained for Reichenbach structure with weights of rules. Table 9.37 presents three desired values of error and the number of epochs, after which this error was obtained.

As it may be inferred from Table 9.37, not all the structures were able to obtain the desired value of error.

9.4.2.3. NDP

Table 9.38 presents the smallest error for individual structures and the number of epochs corresponding to this error.

TABLE 9.36. The smallest error obtained as a result of learning

HANG		
Structure	The smallest error	Number of epochs
Łukasiewicz with weights of rules	0.0247	6500
Binary with weights of rules	0.0161	6525
Reichenbach with weights of rules	0.0115	7580
Łukasiewicz simplified with weights of rules	0.0350	1840
Zadeh with weights of rules	0.0202	5290
Willmott with weights of rules	0.0335	7977
Zadeh simplified with weights of rules	0.0231	7935

TABLE 9.37. Number of epochs required to train the system which is characterized by a definite error

HANG			
Structure	Value of error		
	0.028	0.026	0.024
Łukasiewicz with weights of rules	3771	3908	–
Binary with weights of rules	1320	1320	1929
Reichenbach with weights of rules	506	660	660
Łukasiewicz simplified with weights of rules	–	–	–
Zadeh with weights of rules	3380	3393	4089
Willmott with weights of rules	–	–	–
Zadeh simplified with weights of rules	2483	2483	4139

As it may be inferred from Table 9.38, the smallest error was 0.0131 and was obtained for binary structure with weights of rules. Table 9.39 presents three desired values of error and the number of epochs, after which this error was obtained.

9.4.2.4. Modeling the taste of rice

Table 9.40 presents the smallest error for individual structures and the number of epochs corresponding to this error.

As it may be inferred from Table 9.40, the smallest error was 0.0199 and was obtained for Willmott structure with weights of rules. Table 9.41

TABLE 9.38. The smallest error obtained as a result of learning

NDP		
Structure	The smallest error	Number of epochs
Łukasiewicz with weights of rules	0.0161	492
Binary with weights of rules	0.0131	498
Reichenbach with weights of rules	0.0140	489
Łukasiewicz simplified with weights of rules	0.0177	459
Zadeh with weights of rules	0.0148	499
Willmott with weights of rules	0.0165	486
Zadeh simplified with weights of rules	0.0142	448

TABLE 9.39. Number of epochs required to train the system which is characterized by a definite error

NDP			
Structure	Value of error		
	0.026	0.023	0.020
Łukasiewicz with weights of rules	170	218	274
Binary with weights of rules	101	119	151
Reichenbach with weights of rules	139	153	186
Łukasiewicz simplified with weights of rules	267	277	364
Zadeh with weights of rules	222	281	352
Willmott with weights of rules	86	121	331
Zadeh simplified with weights of rules	58	78	212

TABLE 9.40. The smallest error obtained as a result of learning

MODELING THE TASTE OF RICE		
Structure	The smallest error	Number of epochs
Łukasiewicz with weights of rules	0.0207	3257
Binary with weights of rules	0.0219	3897
Reichenbach with weights of rules	0.0205	3800
Łukasiewicz simplified with weights of rules	0.0222	2841
Zadeh with weights of rules	0.0205	3531
Willmott with weights of rules	0.0199	3805
Zadeh simplified with weights of rules	0.0227	4432

TABLE 9.41. Number of epochs required to train the system which is characterized by a definite error

MODELING THE TASTE OF RICE			
Structure	Value of error		
	0.028	0.025	0.022
Łukasiewicz with weights of rules	1	16	462
Binary with weights of rules	143	1117	3387
Reichenbach with weights of rules	38	185	394
Łukasiewicz simplified with weights of rules	397	1152	–
Zadeh with weights of rules	108	314	879
Willmott with weights of rules	22	78	374
Zadeh simplified with weights of rules	461	696	–

presents three desired values of error and the number of epochs, after which
this error was obtained.

As it may be inferred from Table 9.41, the error of 0.022 could not be
obtained for the simplified Łukasiewicz structure with weights of rules and
Zadeh structure with weights of rules.

9.4.3 M3-type systems

Let us consider logical type systems which are constructed using definitions
of triangular norms taking into account the weights w_k, characterizing the
importance of particular rules, and the weights $w_{i,k}$, characterizing the
importance of particular input linguistic variables. Using the definition of
weighted t-norm and dependencies (9.37), (9.40), (9.43), (9.46), (9.49) and
(9.50), we obtain the following neuro-fuzzy systems:

*a) Neuro-fuzzy system with weights of inputs and rules and the
Łukasiewicz implication*

$$\overline{y} = \frac{\sum_{r=1}^{N} \overline{y}^r T_{k=1}^{*N} \left\{ \min \left[1, 1 - T_{i=1}^{*n} \left(\exp\left[-\left(\frac{\overline{x}_i - \overline{x}_i^k}{\sigma_i^k} \right)^2 \right], w_{i,k} \right) + \exp\left[-\left(\frac{\overline{y}^r - \overline{y}^k}{\sigma^k} \right)^2 \right] \right], w_k \right\}}{\sum_{r=1}^{N} T_{k=1}^{*N} \left\{ \min \left[1, 1 - T_{i=1}^{*n} \left(\exp\left[-\left(\frac{\overline{x}_i - \overline{x}_i^k}{\sigma_i^k} \right)^2 \right], w_{i,k} \right) + \exp\left[-\left(\frac{\overline{y}^r - \overline{y}^k}{\sigma^k} \right)^2 \right] \right], w_k \right\}}. \quad (9.57)$$

b) Neuro-fuzzy system with weights of inputs and rules and the binary implication

$$\bar{y} = \frac{\sum_{r=1}^{N} \bar{y}^r T_{k=1}^{*N} \left\{ \max\left[\begin{array}{c} 1 - T_{i=1}^{*n}\left(\exp\left[-\left(\frac{\bar{x}_i - \bar{x}_i^k}{\sigma_i^k}\right)^2 \right], w_{i,k} \right), \\ \exp\left[-\left(\frac{\bar{y}^r - \bar{y}^k}{\sigma^k}\right)^2 \right] \end{array} \right], w_k \right\} }{\sum_{r=1}^{N} T_{k=1}^{*N} \left\{ \max\left[\begin{array}{c} 1 - T_{i=1}^{*n}\left(\exp\left[-\left(\frac{\bar{x}_i - \bar{x}_i^k}{\sigma_i^k}\right)^2 \right], w_{i,k} \right), \\ \exp\left[-\left(\frac{\bar{y}^r - \bar{y}^k}{\sigma^k}\right)^2 \right] \end{array} \right], w_k \right\} }. \tag{9.58}$$

c) Neuro-fuzzy system with weights of inputs and rules and the Reichenbach implication

$$\bar{y} = \frac{\sum_{r=1}^{N} \bar{y}^r T_{k=1}^{*N} \left\{ \begin{array}{c} 1 - T_{i=1}^{*n}\left(\exp\left[-\left(\frac{\bar{x}_i - \bar{x}_i^k}{\sigma_i^k}\right)^2 \right], w_{i,k} \right) \\ \left(1 - \exp\left[-\left(\frac{\bar{y}^r - \bar{y}^k}{\sigma^k}\right)^2 \right]\right), w_k \end{array} \right\} }{\sum_{r=1}^{N} T_{k=1}^{*N} \left\{ \begin{array}{c} 1 - T_{i=1}^{*n}\left(\exp\left[-\left(\frac{\bar{x}_i - \bar{x}_i^k}{\sigma_i^k}\right)^2 \right], w_{i,k} \right) \\ \left(1 - \exp\left[-\left(\frac{\bar{y}^r - \bar{y}^k}{\sigma^k}\right)^2 \right]\right), w_k \end{array} \right\} }. \tag{9.59}$$

d) Neuro-fuzzy system with weights of inputs and rules and the Zadeh implication

$$\bar{y} = \frac{\sum_{r=1}^{N} \bar{y}^r T^* \left\{ \max\left\{ \begin{array}{c} T_{i=1}^{*n}\left\{ \exp\left[-\left(\frac{\bar{x}_i - \bar{x}_i^r}{\sigma_i^r}\right)^2 \right], w_{i,r} \right\}, \\ 1 - T_{i=1}^{*n}\left\{ \exp\left[-\left(\frac{\bar{x}_i - \bar{x}_i^r}{\sigma_i^r}\right)^2 \right], w_{i,r} \right\} \end{array} \right\}, w_r, \\ T^*_{\substack{k=1 \\ k \neq r}}^{N} \left\{ \max\left\{ \begin{array}{c} \min T_{i=1}^{*n}\left\{ \exp\left[-\left(\frac{\bar{x}_i - \bar{x}_i^k}{\sigma_i^k}\right)^2 \right], w_{i,k} \right\}, \\ \exp\left[-\left(\frac{\bar{y}^r - \bar{y}^k}{\sigma^k}\right)^2 \right], \\ 1 - T_{i=1}^{*n}\left\{ \exp\left[-\left(\frac{\bar{x}_i - \bar{x}_i^k}{\sigma_i^k}\right)^2 \right], w_{i,k} \right\} \end{array} \right\}, w_k \right\} \right\}}{\sum_{r=1}^{N} T^* \left\{ \max\left\{ \begin{array}{c} T_{i=1}^{*n}\left\{ \exp\left[-\left(\frac{\bar{x}_i - \bar{x}_i^r}{\sigma_i^r}\right)^2 \right], w_{i,r} \right\}, \\ 1 - T_{i=1}^{*n}\left\{ \exp\left[-\left(\frac{\bar{x}_i - \bar{x}_i^r}{\sigma_i^r}\right)^2 \right], w_{i,r} \right\} \end{array} \right\}, w_r, \\ T^*_{\substack{k=1 \\ k \neq r}}^{N} \left\{ \max\left\{ \begin{array}{c} \min\left\{ \begin{array}{c} T_{i=1}^{*n}\left\{ \exp\left[-\left(\frac{\bar{x}_i - \bar{x}_i^k}{\sigma_i^k}\right)^2 \right], w_{i,k} \right\}, \\ \exp\left[-\left(\frac{\bar{y}^r - \bar{y}^k}{\sigma^k}\right)^2 \right] \end{array} \right\}, \\ 1 - T_{i=1}^{*n}\left\{ \exp\left[-\left(\frac{\bar{x}_i - \bar{x}_i^k}{\sigma_i^k}\right)^2 \right], w_{i,k} \right\} \end{array} \right\}, w_k \right\} \right\}}. \tag{9.60}$$

e) Simplified neuro-fuzzy system with weights of inputs and rules and the Łukasiewicz implication

$$\overline{y} = \frac{\sum_{r=1}^{N} \overline{y}^r \underset{\substack{k=1 \\ k\neq r}}{\overset{N}{T^*}} \left\{ 1 - \underset{i=1}{\overset{*n}{T}} \left\{ \exp\left[-\left(\frac{\overline{x}_i - \overline{x}_i^k}{\sigma_i^k} \right)^2 \right], w_{i,k} \right\}, w_k \right\}}{\sum_{r=1}^{N} \underset{\substack{k=1 \\ k\neq r}}{\overset{N}{T^*}} \left\{ 1 - \underset{i=1}{\overset{*n}{T}} \left\{ \exp\left[-\left(\frac{\overline{x}_i - \overline{x}_i^k}{\sigma_i^k} \right)^2 \right], w_{i,k} \right\}, w_k \right\}}. \tag{9.61}$$

f) Simplified neuro-fuzzy system with weights of inputs and rules and the Zadeh implication

$$\overline{y} = \frac{\sum_{r=1}^{N} \overline{y}^r T^* \left\{ \max \left\{ \begin{matrix} \underset{i=1}{\overset{*n}{T}} \left\{ \exp\left[-\left(\frac{\overline{x}_i - \overline{x}_i^r}{\sigma_i^r} \right)^2 \right], w_{i,r} \right\}, \\ 1 - \underset{i=1}{\overset{*n}{T}} \left\{ \exp\left[-\left(\frac{\overline{x}_i - \overline{x}_i^r}{\sigma_i^r} \right)^2 \right], w_{i,r} \right\} \end{matrix} \right\}, w_r, \\ \underset{\substack{k=1 \\ k\neq r}}{\overset{N}{T^*}} \left\{ 1 - \underset{i=1}{\overset{*n}{T}} \left\{ \exp\left[-\left(\frac{\overline{x}_i - \overline{x}_i^k}{\sigma_i^k} \right)^2 \right], w_{i,k} \right\}, w_k \right\} \right\}}{\sum_{r=1}^{N} T^* \left\{ \max \left\{ \begin{matrix} \underset{i=1}{\overset{*n}{T}} \left\{ \exp\left[-\left(\frac{\overline{x}_i - \overline{x}_i^r}{\sigma_i^r} \right)^2 \right], w_{i,r} \right\}, \\ 1 - \underset{i=1}{\overset{*n}{T}} \left\{ \exp\left[-\left(\frac{\overline{x}_i - \overline{x}_i^r}{\sigma_i^r} \right)^2 \right], w_{i,r} \right\} \end{matrix} \right\}, w_r, \\ \underset{\substack{k=1 \\ k\neq r}}{\overset{N}{T^*}} \left\{ 1 - \underset{i=1}{\overset{*n}{T}} \left\{ \exp\left[-\left(\frac{\overline{x}_i - \overline{x}_i^k}{\sigma_i^k} \right)^2 \right], w_{i,k} \right\}, w_k \right\} \right\}}. \tag{9.62}$$

In systems (9.57) - (9.60), the parameters of membership function, i.e. $\overline{x}_i^k$, $\sigma_i^k, \overline{y}^k, \sigma^k$ and weights $w_{i,k}$ and w_k are subject to learning. In systems (9.61) and (9.62), the parameters of membership function $\overline{x}_i^k, \sigma_i^k, \overline{y}^k$ and weights $w_{i,k}$ and w_k are subject to learning. Neuro-fuzzy systems (9.57) - (9.62) have been used to solve four problems specified in Table 9.1.

9.4.3.1. Polymerization

Table 9.42 presents the smallest error for individual structures and the number of epochs corresponding to this error.

As it may be inferred from Table 9.42, the smallest error was 0.0028 and was obtained for Zadeh structure with weights of inputs and rules. Table 9.43 presents three desired values of error and the number of epochs, after which this error was obtained.

TABLE 9.42. The smallest error obtained as a result of learning

POLYMERIZATION		
Structure	The smallest error	Number of epochs
Łukasiewicz with weights of inputs and rules	0.0038	4773
Binary with weights of inputs and rules	0.0036	4896
Reichenbach with weights of inputs and rules	0.0034	4704
Łukasiewicz simplified with weights of inputs and rules	0.0037	4815
Zadeh with weights of inputs and rules	0.0028	5064
Willmott with weights of inputs and rules	0.0039	4810
Zadeh simplified with weights of inputs and rules	0.0038	5515

TABLE 9.43. Number of epochs required to train the system which is characterized by a definite error

POLYMERIZATION			
Structure	Value of error		
	0.0055	0.0050	0.0040
Łukasiewicz with weights of inputs and rules	1	9	867
Binary with weights of inputs and rules	2305	2386	2798
Reichenbach with weights of inputs and rules	1915	2303	2549
Łukasiewicz simplified with weights of inputs and rules	2502	2821	3225
Zadeh with weights of inputs and rules	1	1	6
Willmott with weights of inputs and rules	11	90	1341
Zadeh simplified with weights of inputs and rules	2	2	206

9.4.3.2. HANG

Table 9.44 presents the smallest error for individual structures and the number of epochs corresponding to this error. Table 9.45 presents the results analogous to those given in Table 9.43.

TABLE 9.44. The smallest error obtained as a result of learning

HANG		
Structure	The smallest error	Number of epochs
Łukasiewicz with weights of inputs and rules	0.0207	6502
Binary with weights of inputs and rules	0.0110	7882
Reichenbach with weights of inputs and rules	0.0092	7390
Łukasiewicz simplified with weights of inputs and rules	0.0203	7996
Zadeh with weights of inputs and rules	0.0105	5533
Willmott with weights of inputs and rules	0.0300	6545
Zadeh simplified with weights of inputs and rules	0.0178	8000

TABLE 9.45. Number of epochs required to train the system which is characterized by a definite error

HANG			
Structure	Value of error		
	0.028	0.026	0.024
Łukasiewicz with weights of inputs and rules	3724	3771	3771
Binary with weights of inputs and rules	556	608	608
Reichenbach with weights of inputs and rules	603	678	978
Łukasiewicz simplified with weights of inputs and rules	7992	7992	7992
Zadeh with weights of inputs and rules	666	666	1115
Willmott with weights of inputs and rules	–	–	–
Zadeh simplified with weights of inputs and rules	3943	4408	5407

TABLE 9.46. The smallest error obtained as a result of learning

NDP		
Structure	The smallest error	Number of epochs
Łukasiewicz with weights of inputs and rules	0.0140	498
Binary with weights of inputs and rules	0.0121	479
Reichenbach with weights of inputs and rules	0.0133	497
Łukasiewicz simplified with weights of inputs and rules	0.0162	457
Zadeh with weights of inputs and rules	0.0140	4
Willmott with weights of inputs and rules	0.0141	496
Zadeh simplified with weights of inputs and rules	0.0135	496

9.4.3.3. NDP

Table 9.46 presents the smallest error for individual structures and the number of epochs corresponding to this error.

As it may be inferred from Table 9.46, the smallest error was 0.0121 and was obtained for binary structure with weights of inputs and rules. Table 9.47 presents three desired values of error and the number of epochs, after which this error was obtained

9.4.3.4. Modeling the taste of rice

Table 9.48 presents the smallest error for individual structures and the number of epochs corresponding to this error.

As it may be inferred from Table 9.48 the smallest error was 0.0164 and was obtained for Zadeh structure with weights of inputs and rules. Table 9.49 presents three desired values of error and the number of epochs, after which this error was obtained.

9.5   Neuro-fuzzy systems of Takagi-Sugeno type

In the fuzzy Takagi-Sugeno type model [246], the base of rules is of a fuzzy character only in the **IF** part, whereas in the **THEN** part, there are functional dependencies

$$R^{(r)} : \textbf{IF} \ (x_1 \text{ is } A_i^r \ \textbf{AND} \ x_2 \text{ is } A_2^r...\textbf{AND} \ x_n \text{ is } A_n^r) \\ \textbf{THEN} \ y_r = f^{(r)}(x_1, x_2, ..., x_n) \tag{9.63}$$

TABLE 9.47. Number of epochs required to train the system which is character-ized by a definite error

NDP			
Structure	Value of error		
	0.026	0.023	0.020
Łukasiewicz with weights of inputs and rules	279	295	368
Binary with weights of inputs and rules	61	80	94
Reichenbach with weights of inputs and rules	107	176	237
Łukasiewicz simplified with weights of inputs and rules	285	299	315
Zadeh with weights of inputs and rules	95	109	142
Willmott with weights of inputs and rules	80	109	150
Zadeh simplified with weights of inputs and rules	60	82	170

TABLE 9.48. The smallest error obtained as a result of learning

MODELING THE TASTE OF RICE		
Structure	The smallest error	Number of epochs
Łukasiewicz with weights of inputs and rules	0.0192	3031
Binary with weights of inputs and rules	0.0194	4164
Reichenbach with weights of inputs and rules	0.0191	4460
Łukasiewicz simplified with weights of inputs and rules	0.0201	4804
Zadeh with weights of inputs and rules	0.0164	3994
Willmott with weights of inputs and rules	0.0187	3916
Zadeh simplified with weights of inputs and rules	0.0186	3646

If we assume that the input of the fuzzy system is signal $\overline{x} = (\overline{x}_1, \overline{x}_2, ..., \overline{x}_n)$, then in order to obtain the output signal $\overline{y}$ of the system, first we will determine

$$T\left(\mu_{A_1^r}\left(\overline{x}_1\right), \mu_{A_2^r}\left(\overline{x}_2\right), ..., \mu_{A_n^r}\left(\overline{x}_n\right)\right), \quad r = 1, ..., N. \tag{9.64}$$

TABLE 9.49. Number of epochs required to train the system which is characterized by a definite error

MODELING THE TASTE OF RICE			
Structure	Value of error		
	0.028	0.025	0.022
Łukasiewicz with weights of inputs and rules	74	331	2045
Binary with weights of inputs and rules	143	317	1679
Reichenbach with weights of inputs and rules	2	3	8
Łukasiewicz simplified with weights of inputs and rules	165	450	1702
Zadeh with weights of inputs and rules	40	143	197
Willmott with weights of inputs and rules	1	1	37
Zadeh simplified with weights of inputs and rules	76	202	404

The next step is to compute

$$\overline{y}_r = f^{(r)}\left(\overline{x}_1, \overline{x}_2, ..., \overline{x}_n\right), \quad r = 1, ..., N. \tag{9.65}$$

The output signal of the fuzzy Takagi-Sugeno system is a normalized weighted sum of particular inputs $\overline{y}_1, ..., \overline{y}_N$, i.e.

$$\overline{y} = \frac{\sum_{r=1}^{N} \overline{y}_r T_{i=1}^n \left\{\mu_{A_i^r}\left(\overline{x}_i\right)\right\}}{\sum_{r=1}^{N} T_{i=1}^n \left\{\mu_{A_i^r}\left(\overline{x}_i\right)\right\}}. \tag{9.66}$$

In the following part of this subchapter, we will consider the Takagi-Sugeno systems with linear dependencies in consequents of the base of rules, i.e.

$$\begin{aligned} R^{(r)} : \textbf{IF } & (x_1 \text{ is } A_i^r \textbf{ AND } x_2 \text{ is } A_2^r ... \textbf{AND } x_n \text{ is } A_n^r) \\ \textbf{THEN } & y_r = c_0^{(r)} + c_1^{(r)} x_1 + ... + c_n^{(r)} x_n \end{aligned} \tag{9.67}$$

for $r = 1, ..., N$. It should be noted that if $c_i^{(r)} = 0$, $i = 1, ..., n$, then system (9.66) is reduced to a simplified Mamdani system given by formula (9.12), and then $c_0^{(r)} = \overline{y}^r$, $r = 1, ..., N$.

The systems of Takagi-Sugeno type have been used to solve approximation and identification problems (polymerization, HANG, NDP, modeling the taste of rice). Like in case of Mamdani type structures and logical type structures, we will consider three types of systems, i.e. without weights, with weights of rules and with weights of rules and weights of inputs reflecting the importance of individual linguistic variables.

9.5.1   M1-type systems

To construct a neuro-fuzzy system, Gaussian membership functions and the assumption that the antecedents in each rule are connected by a t-norm of the product type have been used. In this situation, dependency (9.66) takes the following form

$$
\bar{y} = \frac{\sum_{r=1}^{N} T_{i=1}^{n} \left\{ \mu_{A_i^r} (\bar{x}_i) \right\} \left(c_0^{(r)} + c_1^{(r)} x_1 + \ldots + c_n^{(r)} x_n \right)}{\sum_{r=1}^{N} T_{i=1}^{n} \left\{ \mu_{A_i^r} (\bar{x}_i) \right\}} \tag{9.68}
$$

$$
= \frac{\sum_{r=1}^{N} T_{i=1}^{n} \left\{ \exp\left[-\left(\frac{\bar{x}_i - \bar{x}_i^r}{\sigma_i^r} \right)^2 \right] \right\} \left(c_0^{(r)} + c_1^{(r)} x_1 + \ldots + c_n^{(r)} x_n \right)}{\sum_{r=1}^{N} T_{i=1}^{n} \left\{ \exp\left[-\left(\frac{\bar{x}_i - \bar{x}_i^r}{\sigma_i^r} \right)^2 \right] \right\}}
$$

$$
= \frac{\sum_{r=1}^{N} \left[\prod_{i=1}^{n} \left(\exp\left[-\left(\frac{\bar{x}_i - \bar{x}_i^r}{\sigma_i^r} \right)^2 \right] \right) \right] \left(c_0^{(r)} + c_1^{(r)} x_1 + \ldots + c_n^{(r)} x_n \right)}{\sum_{r=1}^{N} \left[\prod_{i=1}^{n} \left(\exp\left[-\left(\frac{\bar{x}_i - \bar{x}_i^r}{\sigma_i^r} \right)^2 \right] \right) \right]}.
$$

All the parameters of the neuro-fuzzy systems have been subject to learning using error backpropagation method: centers and widths of Gaussian functions and function parameters $c_0^{(r)}, ..., c_n^{(r)}, r = 1, ..., N$.

9.5.1.1. Polymerization

The smallest error for the Takagi-Sugeno structure was 0.0034 and was obtained in the 3430th epoch. Table 9.50 presents three desired error values and the number of epochs, after which this error was obtained.

9.5.1.2. HANG

The smallest error for the Takagi-Sugeno structure was 0.0197 and was obtained in the 7551st epoch. Table 9.51 presents three desired values of error and the number of epochs, after which this error was obtained.

TABLE 9.50. Number of epochs required to train the system which is characterized by a definite error

POLYMERIZATION			
Structure	Value of error		
	0.0055	0.0050	0.0045
Takagi-Sugeno	72	83	83

9.5.1.3. NDP

The smallest error for the Takagi-Sugeno structure was 0.0156 and was obtained in the 481st epoch. Table 9.52 presents three desired values of error and the number of epochs, after which this error was obtained.

9.5.1.4. Modeling the taste of rice

The smallest error for the Takagi-Sugeno structure was 0.0176 and was obtained in the 1264th epoch. Table 9.53 presents three desired values of error and the number of epochs, after which this error was obtained.

9.5.2   M2-type systems

By introducing to system (9.68) the weights specifying the importance of particular rules, we will obtain the following dependency:

TABLE 9.51. Number of epochs required to train the system which is characterized by a definite error

HANG		
Structure	Value of error	
	0.028 0.026 0.024	
Takagi-Sugeno	4280 5159 5593	

TABLE 9.52. Number of epochs required to train the system which is characterized by a definite error

NDP		
Structure	Value of error	
	0.026 0.023 0.020	
Takagi-Sugeno	36 66 122	

TABLE 9.53. Number of epochs required to train the system which is characterized by a definite error

MODELING THE TASTE OF RICE		
Structure	Value of error	
	0.028 0.025 0.022	
Takagi-Sugeno	546 1060 1951	

$$\bar{y} = \frac{\sum_{r=1}^{N} w_r T_{i=1}^{n}\left\{\exp\left[-\left(\frac{\overline{x}_i - \overline{x}_i^r}{\sigma_i^r}\right)^2\right]\right\}\left(c_0^{(r)} + c_1^{(r)}x_1 + \ldots + c_n^{(r)}x_n\right)}{\sum_{r=1}^{N} w_r T_{i=1}^{n}\left\{\exp\left[-\left(\frac{\overline{x}_i - \overline{x}_i^r}{\sigma_i^r}\right)^2\right]\right\}} \quad (9.69)$$

$$= \frac{\sum_{r=1}^{N} w_r \left[\prod_{i=1}^{n}\left(\exp\left[-\left(\frac{\overline{x}_i - \overline{x}_i^r}{\sigma_i^r}\right)^2\right]\right)\right]\left(c_0^{(r)} + c_1^{(r)}x_1 + \ldots + c_n^{(r)}x_n\right)}{\sum_{r=1}^{N} w_r \left[\prod_{i=1}^{n}\left(\exp\left[-\left(\frac{\overline{x}_i - \overline{x}_i^r}{\sigma_i^r}\right)^2\right]\right)\right]}.$$

All the parameters of the neuro-fuzzy systems have been subject to learning using the error backpropagation method: centers and widths of Gaussian functions, weights of rules and function parameters $c_0^{(r)}, \ldots, c_n^{(r)}, r = 1, \ldots, N$.

9.5.2.1. Polymerization

The smallest error for the Takagi-Sugeno structure was 0.0031 and was obtained in the 3098th epoch. Table 9.54 presents three desired values of error and the number of epochs, after which this error was obtained.

9.5.2.2. HANG

The smallest error for the Takagi-Sugeno structure with weights of rules was 0.0145 and was obtained in the 3008th epoch. Table 9.55 presents three desired values of error and the number of epochs, after which this error was obtained.

9.5.2.3. NDP

The smallest error for the Takagi-Sugeno structure with weights of rules was 0.0140 and was obtained in the 497th epoch. Table 9.56 presents three desired values of error and the number of epochs, after which this error was obtained.

TABLE 9.54. Number of epochs required to train the system which is characterized by a definite error

POLYMERIZATION			
Structure	Value of error		
	0.0055	0.0050	0.0045
Takagi-Sugeno with weights of rules	56	57	95

TABLE 9.55. Number of epochs required to train the system which is characterized by a definite error

HANG			
Structure	Value of error		
	0.028	0.026	0.024
Takagi-Sugeno with weights of rules	478	779	1132

TABLE 9.56. Number of epochs required to train the system which is characterized by a definite error

HANG			
Structure	Value of error		
	0.026	0.023	0.020
Takagi-Sugeno with weights of rules	20	40	79

TABLE 9.57. Number of epochs required to train the system which is characterized by a definite error

MODELING THE TASTE OF RICE			
Structure	Value of error		
	0.028	0.025	0.022
Takagi-Sugeno with weights of rules	8	35	67

9.5.2.4. Modeling the taste of rice

The smallest error for the Takagi-Sugeno structure with weights of rules was 0.0149 and was obtained in the 1620th epoch. Table 9.57 presents three desired values of error and the number of epochs, after which this error was obtained.

9.5.3 M3-type systems

By introducing to system (9.69) the weights specifying the importance of particular linguistic variables in each rule, we will obtain the following dependency:

$$\bar{y} = \frac{\sum_{r=1}^{N} \left(\begin{array}{c} w_r \left[T_{i=1}^n \left\{ 1 - w_{i,r} \left(1 - \mu_{A_i^r} \left(\bar{x}_i \right) \right) \right\} \right] \cdot \\ \cdot \left(c_0^{(r)} + c_1^{(r)} x_1 + \ldots + c_n^{(r)} x_n \right) \end{array} \right)}{\sum_{r=1}^{N} w_r \left[T_{i=1}^n \left\{ 1 - w_{i,r} \left(1 - \mu_{A_i^r} \left(\bar{x}_i \right) \right) \right\} \right]} \tag{9.70}$$

$$
= \frac{\sum_{r=1}^{N} w_r \left[T_{i=1}^{n} \left\{ \cdot \left(1 - \exp\left[-\left(\frac{\overline{x}_i - \overline{x}_i^r}{\sigma_i^r} \right)^2 \right] \right) \right\} \cdot \right] \cdot \left(c_0^{(r)} + c_1^{(r)} x_1 + \ldots + c_n^{(r)} x_n \right)}{\sum_{r=1}^{N} w_r \left[T_{i=1}^{n} \left\{ 1 - w_{i,r} \left(1 - \exp \cdot \left[-\left(\frac{\overline{x}_i - \overline{x}_i^r}{\sigma_i^r} \right)^2 \right] \right) \right\} \right]}.
$$

All the parameters of the neuro-fuzzy systems have been subject to learning using the error backpropagation method: centers and widths of Gaussian functions, weights of inputs and rules and function parameters $c_0^{(r)}, \ldots, c_n^{(r)}$, $r = 1, \ldots, N$.

9.5.3.1. Polymerization

The smallest error for the Takagi-Sugeno structure with weights of rules was 0.0030 and was obtained in the 4859th epoch. Table 9.58 presents three desired values of error and the number of epochs, after which this error was obtained.

9.5.3.2. HANG

The smallest error for the Takagi-Sugeno structure with weights of rules was 0.0116 and was obtained in the 2381st epoch. Table 9.59 presents three

TABLE 9.58. Number of epochs required to train the system which is characterized by a definite error

POLYMERIZATION		
Structure	Value of error	
	0.0055 0.0050 0.0045	
Takagi-Sugeno with weights of inputs and rules	36 110 324	

TABLE 9.59. Number of epochs required to train the system which is characterized by a definite error

HANG		
Structure	Value of error	
	0.028 0.026 0.024	
Takagi-Sugeno with weights of inputs and rules	265 465 478	

TABLE 9.60. Number of epochs required to train the system which is characterized by a definite error

NDP		
Structure	Value of error		
	0.026	0.023	0.020
Takagi-Sugeno with weights of inputs and rules	60	77	111

TABLE 9.61. Number of epochs required to train the system which is characterized by a definite error

MODELING THE TASTE OF RICE		
Structure	Value of error		
	0.028	0.025	0.022
Takagi-Sugeno with weights of inputs and rules	1	1	1

desired values of error and the number of epochs, after which this error was obtained.

9.5.3.4. Modeling the taste of rice

The smallest error for the Takagi-Sugeno structure with weights of rules was 0.0129 and was obtained in the 4008th epoch. Table 9.61 presents three desired values of error and the number of epochs, after which this error was obtained.

9.5.3.3. NDP

The smallest error for the Takagi-Sugeno structure with weights of rules was 0.0085 and was obtained in the 495th epoch. Table 9.60 presents three desired values of error and the number of epochs, after which this error was obtained.

9.6   Learning algorithms of neuro-fuzzy systems

In Subchapters 9.3, 9.4 and 9.5 we have discussed the neuro-fuzzy systems of the Takagi-Sugeno, Mamdani and logical type. In this subchapter, we will derive the learning algorithms of the above specified systems. They have been used in simulation examples (Subchapters 9.3-9.5).

We will use the idea of error backpropagation method, which is the basic learning method of neural networks. Learning of the neuro-fuzzy systems will come down to application of gradient algorithms, minimizing the appropriately formulated quality criterion. By $\overline{\mathbf{x}}(t) \in \mathbf{R}^n$ and $d(t) \in \mathbf{R}$ we will notate, a sequence of input and desired (at the output of the neuro-fuzzy system) signals, respectively. The problem of learning of those systems comes down to determining, based on the learning sequence

$$(\overline{\mathbf{x}}(1), d(1)), (\overline{\mathbf{x}}(2), d(2)), \ldots \tag{9.71}$$

all the parameters of the membership function and weights (weights describing the importance of rules and importance of particular linguistic variables in each rule) so as to minimize the criterion

$$Q(t) = \frac{1}{2} [f(\overline{\mathbf{x}}(t)) - d(t)]^2, \tag{9.72}$$

where

$$\overline{y} = f(\overline{\mathbf{x}}(t)) \tag{9.73}$$

is the output of the neuro-fuzzy systems of the Mamdani, logical and Takagi-Sugeno type presented in previous subchapters. For example, in the Mamdani and logical type systems, the parameter $\overline{y}^r$, $r = 1, \ldots, N$, may be determined using the gradient algorithm

$$\overline{y}^r(t+1) = \overline{y}^r(t) - \eta \frac{\partial Q(t)}{\partial \overline{y}^r(t)}. \tag{9.74}$$

The direct determination of gradient $\frac{\partial Q(t)}{\partial \overline{y}^r(t)}$ in the above procedure is complicated from a computational point of view. That is why an analogy between the neural networks and the neuro-fuzzy networks has been used, considering the fact that the latter also have a multilayer structure. Therefore, the error backpropagation method may be applied to learning of neuro-fuzzy networks. The notation used in this subchapter shall be explained on the example of a single neuron described by formula

$$y = f(s), \quad s = \sum_{i=0}^{n} x_i w_i, \tag{9.75}$$

where f is a sigmoidal function, x_i and w_i, $i = 0, \ldots, n$ are inputs and weights of the neuron. Let d be the desired signal at the neuron output. Then

$$\varepsilon^f = \varepsilon = y - d \tag{9.76}$$

is the error at neuron output and the expression

$$\varepsilon^s = \varepsilon^f \{s\} = \varepsilon^f \frac{\partial f(s)}{\partial s} = (y - d) f'(s) \tag{9.77}$$

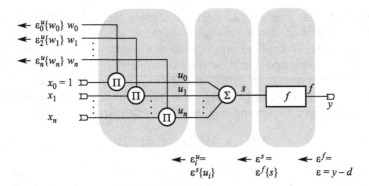

FIGURE 9.4. Flow of signals and errors in a single neuron

describes the error propagated from the functional block f to the summation block s. Figure 9.4 presents the flow of signals and errors in a single neuron.

At first, we will derive the learning algorithm for the Takagi-Sugeno system and next, for the Mamdani and logical type systems. We will modify the symbols used before to describe these systems, so that it would be possible to clearly present the flow of errors through particular blocks of the specified systems. The output signal of the Takagi-Sugeno system may be described as follows:

$$
\bar{y} = \frac{\sum_{r=1}^{N} \left(w_r^{\text{def}} \cdot T^* \left\{ \begin{array}{c} \mu_{A_1^r}(\bar{x}_1), \mu_{A_2^r}(\bar{x}_2), \dots, \mu_{A_n^r}(\bar{x}_n); \\ w_{1,r}^{\tau}, w_{2,r}^{\tau}, \dots, w_{n,r}^{\tau} \end{array} \right\} \cdot \left(c_{0,r}^f + \sum_{i=1}^{n} c_{i,r}^f \cdot \bar{x}_i \right) \right)}{\sum_{r=1}^{N} \left(w_r^{\text{def}} \cdot T^* \left\{ \begin{array}{c} \mu_{A_1^r}(\bar{x}_1), \mu_{A_2^r}(\bar{x}_2), \dots, \mu_{A_n^r}(\bar{x}_n); \\ w_{1,r}^{\tau}, w_{2,r}^{\tau}, \dots, w_{n,r}^{\tau} \end{array} \right\} \right)}, \qquad (9.78)
$$

where $w_{i,r}^{\tau} \in [0, 1]$, $i = 1, 2, \dots, n$, $r = 1, 2, \dots, N$, mean the weights of antecedents of rules and $w_r^{\text{def}} \in [0, 1]$, $r = 1, 2, \dots, N$, mean the weights of rules. By substituting

$$
T^* \left\{ \begin{array}{c} \mu_{A_1^r}(\bar{x}_1), \mu_{A_2^r}(\bar{x}_2), \dots, \mu_{A_n^r}(\bar{x}_n); \\ w_{1,r}^{\tau}, w_{2,r}^{\tau}, \dots, w_{n,r}^{\tau} \end{array} \right\} = \tau_r(\bar{\mathbf{x}}) \qquad (9.79)
$$

and

$$
c_{0,r}^f + \sum_{i=1}^{n} c_{i,r}^f \cdot \bar{x}_i = f_r(\bar{\mathbf{x}}), \qquad (9.80)
$$

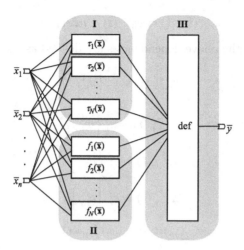

FIGURE 9.5. Network structure of the Takagi-Sugeno system

we get

$$\overline{y} = \frac{\sum_{r=1}^{N} w_r^{\text{def}} \cdot \tau_r\left(\overline{\mathbf{x}}\right) \cdot f_r\left(\overline{\mathbf{x}}\right)}{\sum_{r=1}^{N} w_r^{\text{def}} \cdot \tau_r\left(\overline{\mathbf{x}}\right)} = \text{def} \begin{pmatrix} \tau_1\left(\overline{\mathbf{x}}\right), \ldots, \tau_N\left(\overline{\mathbf{x}}\right), \\ f_1\left(\overline{\mathbf{x}}\right), \ldots, f_N\left(\overline{\mathbf{x}}\right); \\ w_1^{\text{def}}, \ldots, w_N^{\text{def}} \end{pmatrix}. \quad (9.81)$$

The network structure of the Takagi-Sugeno system is presented in Fig. 9.5.

In the Takagi-Sugeno system, the following parameters are subject to learning:

- $p_{u,i,r}^A$, $u = 1, 2, \ldots, P^A$, parameters of input membership functions of the fuzzy sets,

- $c_{i,r}^f$, $i = 0, 1, \ldots, n$, $r = 1, 2, \ldots, N$, parameters of the functional blocks,

- $w_{i,r}^\tau$, $i = 1, 2, \ldots, n$, $r = 1, 2, \ldots, N$, weights of antecedents,

- w_r^{def}, $r = 1, 2, \ldots, N$, weights of rules.

The Takagi-Sugeno system parameters are modified by iteration according to the dependencies below:

$$p_{u,i,r}^A\left(t+1\right) = p_{u,i,r}^A\left(t\right) - \eta \Delta p_{u,i,r}^A\left(t\right), \quad (9.82)$$

$$w_{i,r}^\tau\left(t+1\right) = w_{i,r}^\tau\left(t\right) - \eta \Delta w_{i,r}^\tau\left(t\right), \quad (9.83)$$

$$c_{0,r}^f\left(t+1\right) = c_{0,r}^f\left(t\right) - \eta \Delta c_{0,r}^f\left(t\right), \quad (9.84)$$

$$c_{i,r}^f\left(t+1\right) = c_{i,r}^f\left(t\right) - \eta \Delta c_{i,r}^f\left(t\right), \quad i = 1, \ldots, n, \quad (9.85)$$

$$w_r^{\text{def}}(t+1) = w_r^{\text{def}}(t) - \eta \Delta w_r^{\text{def}}(t).$$ (9.86)

The terms Δ in the above dependencies are defined as follows:

$$\Delta p_{u,i,r}^A(t) = \varepsilon_r^\tau \left\{ p_{u,i,r}^A \right\},$$ (9.87)

$$\Delta w_{i,r}^\tau(t) = \varepsilon_r^\tau \left\{ w_{i,r}^\tau \right\}$$ (9.88)

$$\Delta c_{0,r}^f(t) = \varepsilon_r^f \left\{ c_{0,r}^f \right\},$$ (9.89)

$$\Delta c_{i,r}^f(t) = \varepsilon_r^f \left\{ c_{i,r}^f \right\},$$ (9.90)

$$\Delta w_r^{\text{def}}(t) = \varepsilon^{\text{def}} \left\{ w_r^{\text{def}} \right\}.$$ (9.91)

The errors propagated by individual layers of the Takagi-Sugeno system are defined as follows (Fig. 9.6):

$$\varepsilon_r^\tau = \varepsilon^{\text{def}} \left\{ \tau_r(\overline{\mathbf{x}}) \right\},$$ (9.92)

$$\varepsilon_r^f = \varepsilon^{\text{def}} \left\{ f_r(\overline{\mathbf{x}}) \right\},$$ (9.93)

$$\varepsilon^{\text{def}} = \varepsilon = \overline{y} - d.$$ (9.94)

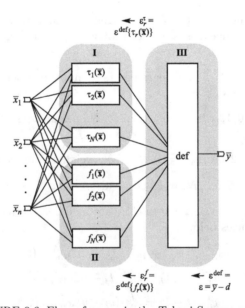

FIGURE 9.6. Flow of errors in the Takagi-Sugeno system

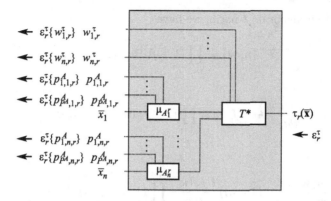

FIGURE 9.7. Block of rules activation of the Takagi-Sugeno system

The errors propagated by blocks of rules activation of the Takagi-Sugeno system are defined as follows (Fig. 9.7):

$$\varepsilon_r^\tau \left\{ p_{u,i,r}^A \right\} = \varepsilon_r^\tau \frac{\partial T^* \left\{ \begin{matrix} \mu_{A_1^r}(\overline{x}_1), \mu_{A_2^r}(\overline{x}_2), \ldots, \mu_{A_n^r}(\overline{x}_n); \\ w_{1,r}^\tau, w_{2,r}^\tau, \ldots, w_{n,r}^\tau \end{matrix} \right\}}{\partial \mu_{A_i^r}(\overline{x}_i)} \tag{9.95}$$

$$\cdot \frac{\partial \mu_{A_i^r}(\overline{x}_i)}{\partial p_{u,i,r}^A},$$

$$\varepsilon_r^\tau \left\{ w_{i,r}^\tau \right\} = \varepsilon_r^\tau \frac{\partial T^* \left\{ \begin{matrix} \mu_{A_1^r}(\overline{x}_1), \mu_{A_2^r}(\overline{x}_2), \ldots, \mu_{A_n^r}(\overline{x}_n); \\ w_{1,r}^\tau, w_{2,r}^\tau, \ldots, w_{n,r}^\tau \end{matrix} \right\}}{\partial w_{i,r}^\tau}. \tag{9.96}$$

We should notice that we are solving an optimization problem with constraints. That is why further in our considerations, we will apply the so-called *constraint function* $f_z(\cdot)$ given by dependency

$$f_z(x) = \frac{1}{1 + \exp\left(-\left(p_1 x - p_2\right)\right)}, \tag{9.97}$$

while

$$\frac{\partial f_z(x)}{\partial x} = p_1 \left(1 - f_z(x)\right) f_z(x). \tag{9.98}$$

In the simulations, it has been assumed that $p_1 = 10$ and $p_2 = 5$.

Example 9.1

We will show the method for the determination of partial derivatives in formula (9.95) and (9.96). Using the notation of the constraint function in the definition of weighted t-norm, we get

$$T^* \left\{ \begin{matrix} a_1, a_2, \ldots, a_n; \\ w_1, w_2, \ldots, w_n \end{matrix} \right\} = T^* \left\{ \mathbf{a}; \mathbf{w} \right\} = T_{i=1}^n \left\{ 1 - f_z(w_i)(1 - a_i) \right\}. \tag{9.99}$$

In case of an algebraic t-norm, we have

$$T^* \{\mathbf{a}; \mathbf{w}\} = \prod_{i=1}^{n} (1 - f_z (w_i) (1 - a_i)). \tag{9.100}$$

Then

$$\frac{\partial T^* \{\mathbf{a}; \mathbf{w}\}}{\partial a_i} = f_z (w_i) \prod_{\substack{u=1 \\ u \neq i}}^{n} (1 - f_z (w_u) (1 - a_u)) \tag{9.101}$$

and

$$\frac{\partial T^* \{\mathbf{a}; \mathbf{w}\}}{\partial w_i} = - (1 - a_i) \frac{\partial f_z (w_i)}{\partial w_i} \prod_{\substack{u=1 \\ u \neq i}}^{n} (1 - f_z (w_u) (1 - a_u)). \tag{9.102}$$

Example 9.2

We will determine the partial derivatives of the Gaussian membership function of the input fuzzy set A (in order to have a clear notation, we will omit appropriate indexes)

$$\mu_A (x) = \exp \left(- \left(\frac{x - \overline{x}}{\sigma} \right)^2 \right). \tag{9.103}$$

Let us notice that:

$$P^A = 2, \quad p^A_{1,i,r} = x, \quad p^A_{2,i,r} = \sigma. \tag{9.104}$$

Appropriate derivatives take the form

$$\frac{\partial \mu_A (x)}{\partial x} = -\mu_A (x) \frac{2 (x - \overline{x})}{\sigma^2}, \tag{9.105}$$

$$\frac{\partial \mu_A (x)}{\partial \overline{x}} = \mu_A (x) \frac{2 (x - \overline{x})}{\sigma^2}, \tag{9.106}$$

$$\frac{\partial \mu_A (x)}{\partial \sigma} = \mu_A (x) \frac{2 (x - \overline{x})^2}{\sigma^3}. \tag{9.107}$$

The errors propagated by functional blocks of the Takagi-Sugeno system are determined as follows (Fig. 9.8):

$$\varepsilon^f_r \left\{ c^f_{0,r} \right\} = \varepsilon^f_r, \tag{9.108}$$

$$\varepsilon^f_r \left\{ c^f_{i,r} \right\} = \varepsilon^f_r \overline{x}_i. \tag{9.109}$$

The errors propagated by the defuzzification block of the Takagi-Sugeno system are determined as follows (Fig. 9.9):

$$\varepsilon^{\text{def}} \{\tau_r (\overline{\mathbf{x}})\} = \varepsilon^{\text{def}} \frac{\partial}{\partial \tau_r (\overline{\mathbf{x}})} \text{def} \left(\begin{array}{c} \tau_1 (\overline{\mathbf{x}}), \ldots, \tau_N (\overline{\mathbf{x}}), \\ f_1 (\overline{\mathbf{x}}), \ldots, f_N (\overline{\mathbf{x}}); \\ w_1^{\text{def}}, \ldots, w_N^{\text{def}} \end{array} \right), \tag{9.110}$$

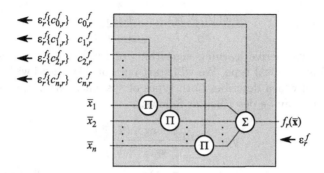

FIGURE 9.8. Functional block of the Takagi-Sugeno system

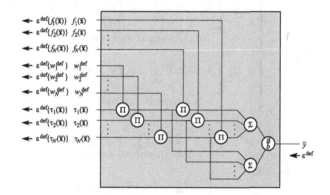

FIGURE 9.9. Defuzzification block of the Takagi-Sugeno system

$$\varepsilon^{\text{def}} \left\{ f_r(\overline{\mathbf{x}}) \right\} = \varepsilon^{\text{def}} \frac{\partial}{\partial f_r(\overline{\mathbf{x}})} \text{def} \left(\begin{array}{c} \tau_1(\overline{\mathbf{x}}), \ldots, \tau_N(\overline{\mathbf{x}}), \\ f_1(\overline{\mathbf{x}}), \ldots, f_N(\overline{\mathbf{x}}); \\ w_1^{\text{def}}, \ldots, w_N^{\text{def}} \end{array} \right), \qquad (9.111)$$

$$\varepsilon^{\text{def}} \left\{ w_r^{\text{def}} \right\} = \varepsilon^{\text{def}} \frac{\partial}{\partial w_r^{\text{def}}} \text{def} \left(\begin{array}{c} \tau_1(\overline{\mathbf{x}}), \ldots, \tau_N(\overline{\mathbf{x}}), \\ f_1(\overline{\mathbf{x}}), \ldots, f_N(\overline{\mathbf{x}}); \\ w_1^{\text{def}}, \ldots, w_N^{\text{def}} \end{array} \right), \qquad (9.112)$$

where

$$\text{def} \left(\begin{array}{c} a_1, a_2, \ldots, a_n, \\ b_1, b_2, \ldots, b_n; \\ w_1, w_2, \ldots, w_n \end{array} \right) = \text{def}(\mathbf{a}, \mathbf{b}; \mathbf{w}) = \frac{\sum_{i=1}^n w_i a_i b_i}{\sum_{i=1}^n w_i a_i}, \qquad (9.113)$$

$$\frac{\partial \text{def}(\mathbf{a}, \mathbf{b}; \mathbf{w})}{\partial a_j} = (b_j - \text{def}(\mathbf{a}, \mathbf{b}; \mathbf{w})) \frac{w_j}{\sum_{i=1}^n w_i a_i}, \qquad (9.114)$$

$$\frac{\partial \text{def}(\mathbf{a}, \mathbf{b}; \mathbf{w})}{\partial w_j} = (b_j - \text{def}(\mathbf{a}, \mathbf{b}; \mathbf{w})) \frac{a_j}{\sum_{i=1}^n w_i a_i}, \qquad (9.115)$$

$$\frac{\partial \text{def} (\mathbf{a}, \mathbf{b}; \mathbf{w})}{\partial b_j} = \frac{w_j a_j}{\sum_{i=1}^{n} w_i a_i}. \tag{9.116}$$

Now we will derive learning algorithms of the neuro-fuzzy systems of Mamdani and logical type. We will start our considerations with a generalized model which describes both types of systems. The output signal of such system may be described as follows:

$$\bar{y} = \frac{\sum_{r=1}^{N} \bar{y}^r \cdot \text{agr}_r (\mathbf{\bar{x}}, \bar{y}^r)}{\sum_{r=1}^{N} \text{agr}_r (\mathbf{\bar{x}}, \bar{y}^r)}. \tag{9.117}$$

The operation of operators $\text{agr}_r (\mathbf{\bar{x}}, \bar{y}^r)$, $r = 1, 2, \ldots, N$, depends on the type of inference applied in a given system, i.e.

$$\text{agr}_r (\mathbf{\bar{x}}, \bar{y}^r) = \begin{cases} S^* \left\{ \begin{array}{c} I_{1,r} (\mathbf{\bar{x}}, \bar{y}^r), \ldots, I_{N,r} (\mathbf{\bar{x}}, \bar{y}^r); \\ w_1^{\text{agr}}, \ldots, w_N^{\text{agr}} \end{array} \right\} \\ \text{for } Mamdani \text{ inference,} \\ \\ T^* \left\{ \begin{array}{c} I_{1,r} (\mathbf{\bar{x}}, \bar{y}^r), \ldots, I_{N,r} (\mathbf{\bar{x}}, \bar{y}^r); \\ w_1^{\text{agr}}, \ldots, w_N^{\text{agr}} \end{array} \right\} \\ \text{for } logical \text{ inference,} \end{cases} \tag{9.118}$$

where

$$I_{k,r} (\mathbf{\bar{x}}, \bar{y}^r) = \begin{cases} T \left\{ \tau_k (\mathbf{\bar{x}}), \mu_{B^k} (\bar{y}^r) \right\} \\ \text{for } Mamdani \text{ inference} \\ \\ I_{\text{fuzzy}} (\tau_k (\mathbf{\bar{x}}), \mu_{B^k} (\bar{y}^r)) \\ \text{for } logical \text{ inference} \end{cases} \tag{9.119}$$

and

$$I_{\text{fuzzy}} (\tau_k (\mathbf{\bar{x}}), \mu_{B^k} (\bar{y}^r)) = \begin{cases} S \left\{ N (\tau_k (\mathbf{\bar{x}})), \mu_{B^k} (\bar{y}^r) \right\} \\ \text{for } S - implication, \\ \\ t_{\text{mul}}^{-1} \left(\min \left\{ 1, \dfrac{t_{\text{mul}} (\mu_{B^k} (\bar{y}^r))}{t_{\text{mul}} (\tau_k (\mathbf{\bar{x}}))} \right\} \right) \\ \text{for } R - implication, \\ \\ S \left\{ N (\tau_k (\mathbf{\bar{x}})), T \left\{ \tau_k (\mathbf{\bar{x}}), \mu_{B^k} (\bar{y}^r) \right\} \right\} \\ \text{for } Q - implication, \end{cases} \tag{9.120}$$

In formula (9.120), the definition of R-implication has been used, taking into consideration the multiplicative generators $t_{\text{mul}}(\cdot)$ of Archimedean t-norm. The rules activation operator $\tau_k (\mathbf{\bar{x}})$, $k = 1, 2, \ldots, N$, has been described similarly to the Takagi-Sugeno system considered earlier. Figure 9.10 presents the network structure of a generalized neuro-fuzzy system.

In the considered neuro-fuzzy system, the following parameters are subject to learning:

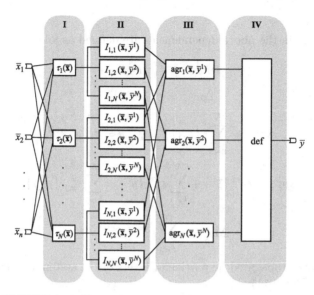

FIGURE 9.10. Network structure of the neuro-fuzzy system

- $p^A_{u,i,k}$, $u = 1, 2, \ldots, P^A$, $i = 1, 2, \ldots, n$, $k = 1, 2, \ldots, N$, parameters of input membership functions of the fuzzy sets,

- $p^B_{1,k} = \overline{y}^k$, $k = 1, 2, \ldots, N$, centers of membership functions of output fuzzy sets,

- $p^B_{u,k}$, $u = 2, 3, \ldots, P^B$, $k = 1, 2, \ldots, N$, other parameters of membership functions of output fuzzy sets,

- $w^\tau_{i,k}$, $i = 1, 2, \ldots, n$, $k = 1, 2, \ldots, N$, weights of antecedents,

- w^{agr}_k, $k = 1, 2, \ldots, N$, weights of rules.

The system parameters are modified by iteration according to the dependencies below:

$$p^A_{u,i,k}(t+1) = p^A_{u,i,k}(t) - \eta \Delta p^A_{u,i,k}(t), \tag{9.121}$$

$$w^\tau_{i,k}(t+1) = w^\tau_{i,k}(t) - \eta \Delta w^\tau_{i,k}(t), \tag{9.122}$$

$$p^B_{u,k}(t+1) = p^B_{u,k}(t) - \eta \Delta p^B_{u,k}(t), \quad u = 2, \ldots, P^B, \tag{9.123}$$

$$\overline{y}^r(t+1) = p^B_{1,r}(t+1) = \overline{y}^r(t) - \eta \Delta \overline{y}^r(t), \tag{9.124}$$

$$w^{\mathrm{agr}}_k(t+1) = w^{\mathrm{agr}}_k(t) - \eta \Delta w^{\mathrm{agr}}_k(t). \tag{9.125}$$

The terms Δ in the above dependencies are defined as follows:

$$\Delta p_{u,i,k}^A = \varepsilon_k^\tau \left\{ p_{u,i,k}^A \right\}, \tag{9.126}$$

$$\Delta w_{i,k}^\tau = \varepsilon_k^\tau \left\{ w_{i,k}^\tau \right\}, \tag{9.127}$$

$$\Delta p_{u,k}^B = \sum_{r=1}^N \varepsilon_{k,r}^I \left\{ p_{u,k}^B \right\}, \quad u = 2, \ldots, P^B, \tag{9.128}$$

$$\Delta \overline{y}^r = \Delta p_{1,r}^B = \varepsilon^{\mathrm{def}} \left\{ \overline{y}^r \right\} + \sum_{k=1}^N \varepsilon_{k,r}^I \left\{ \overline{y}^r \right\} + \sum_{k=1}^N \varepsilon_{r,k}^I \left\{ p_{1,r}^B \right\}, \tag{9.129}$$

$$\Delta w_k^{\mathrm{agr}} = \sum_{r=1}^N \varepsilon_r^{\mathrm{agr}} \left\{ w_k^{\mathrm{agr}} \right\}. \tag{9.130}$$

The errors propagated by particular layers of the system are determined as follows (Fig. 9.11):

$$\varepsilon_k^\tau = \sum_{r=1}^N \varepsilon_{k,r}^I \left\{ \tau_k \left(\overline{\mathbf{x}} \right) \right\}, \tag{9.131}$$

$$\varepsilon_{k,r}^I = \varepsilon_r^{\mathrm{agr}} \left\{ I_{k,r} \left(\overline{\mathbf{x}}, \overline{y}^r \right) \right\}, \tag{9.132}$$

$$\varepsilon_r^{\mathrm{agr}} = \varepsilon^{\mathrm{def}} \left\{ \mathrm{agr}_r \left(\overline{\mathbf{x}}, \overline{y}^r \right) \right\}, \tag{9.133}$$

$$\varepsilon^{\mathrm{def}} = \varepsilon = \overline{y} - d. \tag{9.134}$$

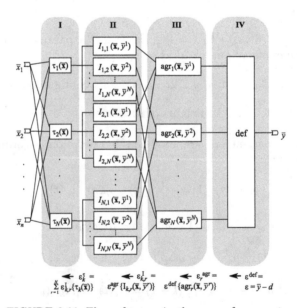

FIGURE 9.11. Flow of errors in the neuro-fuzzy system

The errors propagated by blocks of rules activation of the system are determined similarly as in Takagi-Sugeno system. The method of determination of errors propagated by implication blocks of the system depends of the chosen inference model (Mamdani or logical) as well as type of applied fuzzy implication (S, R, Q-implication) in case of logical inference.

The errors propagated by implication blocks of the system with Mamdani type inference are determined as follows (Fig. 9.12):

$$\varepsilon_{k,r}^{I}\left\{p_{u,k}^{B}\right\} = \varepsilon_{k,r}^{I}\frac{\partial T\left\{\tau_{k}\left(\overline{\mathbf{x}}\right),\mu_{B^{k}}\left(\overline{y}^{r}\right)\right\}}{\partial \mu_{B^{k}}\left(\overline{y}^{r}\right)}\frac{\partial \mu_{B^{k}}\left(\overline{y}^{r}\right)}{\partial p_{u,k}^{B}}, \tag{9.135}$$

$$\varepsilon_{k,r}^{I}\left\{\overline{y}^{r}\right\} = \varepsilon_{k,r}^{I}\frac{\partial T\left\{\tau_{k}\left(\overline{\mathbf{x}}\right),\mu_{B^{k}}\left(\overline{y}^{r}\right)\right\}}{\partial \mu_{B^{k}}\left(\overline{y}^{r}\right)}\frac{\partial \mu_{B^{k}}\left(\overline{y}^{r}\right)}{\partial \overline{y}^{r}}, \tag{9.136}$$

$$\varepsilon_{k,r}^{I}\left\{\tau_{k}\left(\overline{\mathbf{x}}\right)\right\} = \varepsilon_{k,r}^{I}\frac{\partial T\left\{\tau_{k}\left(\overline{\mathbf{x}}\right),\mu_{B^{k}}\left(\overline{y}^{r}\right)\right\}}{\partial \tau_{k}\left(\overline{\mathbf{x}}\right)}, \tag{9.137}$$

whereas the derivatives $\frac{\partial \mu_{B^{k}}(\overline{y}^{r})}{\partial p_{u,k}^{B}}$, $\frac{\partial \mu_{B^{k}}(\overline{y}^{r})}{\partial \overline{y}^{r}}$, $\frac{\partial T\left\{\tau_{k}(\overline{\mathbf{x}}),\mu_{B^{k}}(\overline{y}^{r})\right\}}{\partial \tau_{k}(\overline{\mathbf{x}})}$ and $\frac{\partial T\left\{\tau_{k}(\overline{\mathbf{x}}),\mu_{B^{k}}(\overline{y}^{r})\right\}}{\partial \mu_{B^{k}}(\overline{y}^{r})}$ are determined using the dependencies provided with the description of the learning method of the Takagi-Sugeno system.

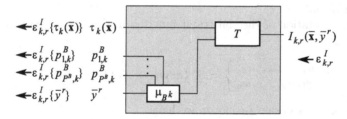

FIGURE 9.12. Implication block of the system with Mamdani type inference

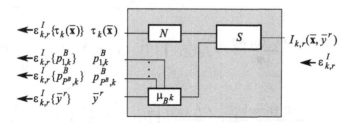

FIGURE 9.13. Implication block of the neuro-fuzzy system with inference of logical type (S-implication)

The errors propagated by implication blocks of the system with inference of logical type using the S-implication are determined as follows (Fig. 9.13):

$$\varepsilon_{k,r}^I \left\{ p_{u,k}^B \right\} = \varepsilon_{k,r}^I \frac{\partial S \left\{ N \left(\tau_k \left(\overline{\mathbf{x}} \right) \right), \mu_{B^k} \left(\overline{y}^r \right) \right\}}{\partial \mu_{B^k} \left(\overline{y}^r \right)} \frac{\partial \mu_{B^k} \left(\overline{y}^r \right)}{\partial p_{u,k}^B}, \tag{9.138}$$

$$\varepsilon_{k,r}^I \left\{ \overline{y}^r \right\} = \varepsilon_{k,r}^I \frac{\partial S \left\{ N \left(\tau_k \left(\overline{\mathbf{x}} \right) \right), \mu_{B^k} \left(\overline{y}^r \right) \right\}}{\partial \mu_{B^k} \left(\overline{y}^r \right)} \frac{\partial \mu_{B^k} \left(\overline{y}^r \right)}{\partial \overline{y}^r}, \tag{9.139}$$

$$\varepsilon_{k,r}^I \left\{ \tau_k \left(\overline{\mathbf{x}} \right) \right\} = \varepsilon_{k,r}^I \frac{\partial S \left\{ N \left(\tau_k \left(\overline{\mathbf{x}} \right) \right), \mu_{B^k} \left(\overline{y}^r \right) \right\}}{\partial N \left(\tau_k \left(\overline{\mathbf{x}} \right) \right)} \frac{\partial N \left(\tau_k \left(\overline{\mathbf{x}} \right) \right)}{\partial \tau_k \left(\overline{\mathbf{x}} \right)}, \tag{9.140}$$

while

$$N \left(a \right) = 1 - a, \tag{9.141}$$

$$\frac{\partial N \left(a \right)}{\partial a} = -1 \tag{9.142}$$

and the derivatives $\frac{\partial \mu_{B^k} \left(\overline{y} \right)}{\partial p_{u,k}^B}$ and $\frac{\partial \mu_{B^k} \left(\overline{y}^r \right)}{\partial \overline{y}^r}$ are determined using the dependencies provided with the description of the learning method of the Takagi-Sugeno system.

The method of determination of partial derivatives in formulas (9.138) - (9.140) will be shown in Example 9.3. This example relates to a more general case, taking into account any number of arguments and their weights in the definition of the t-conorm.

Example 9.3

Using the notation of the constraint function in the definition of weighted t-conorm, we get

$$S^* \left\{ \begin{array}{l} a_1, a_2, \ldots, a_n; \\ w_1, w_2, \ldots, w_n \end{array} \right\} = S^* \left\{ \mathbf{a}; \mathbf{w} \right\} = \mathop{S}_{i=1}^{n} \left\{ f_z \left(w_i \right) a_i \right\}. \tag{9.143}$$

In case of an algebraic t-conorm, we have

$$S^* \left\{ \mathbf{a}; \mathbf{w} \right\} = 1 - \prod_{i=1}^{n} \left(1 - f_z \left(w_i \right) a_i \right). \tag{9.144}$$

Then

$$\frac{\partial S^* \left\{ \mathbf{a}; \mathbf{w} \right\}}{\partial a_i} = f_z \left(w_i \right) \prod_{\substack{u=1 \\ u \neq i}}^{n} \left(1 - f_z \left(w_u \right) a_u \right) \tag{9.145}$$

and

$$\frac{\partial S^* \left\{ \mathbf{a}; \mathbf{w} \right\}}{\partial w_i} = a_i \frac{\partial f_z \left(w_i \right)}{\partial w_i} \prod_{\substack{u=1 \\ u \neq i}}^{n} \left(1 - f_z \left(w_u \right) a_u \right). \tag{9.146}$$

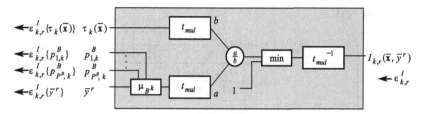

FIGURE 9.14. Implication block of the neuro-fuzzy system with inference of logical type (R-implication)

The errors propagated by implication blocks of the system with inference of logical type using the R-implication are determined as follows (Fig. 9.14):

$$
\varepsilon_{k,r}^{I}\left\{p_{u,k}^{B}\right\} =
$$

$$
= \varepsilon_{k,r}^{I}\left(
\begin{array}{c}
\dfrac{\partial t_{\mathrm{mul}}^{-1}\left(\min\left\{1, \dfrac{t_{\mathrm{mul}}\left(\mu_{B^{k}}\left(\overline{y}^{r}\right)\right)}{t_{\mathrm{mul}}\left(\tau_{k}\left(\overline{\mathbf{x}}\right)\right)}\right\}\right)}{\partial \min\left\{1, \dfrac{t_{\mathrm{mul}}\left(\mu_{B^{k}}\left(\overline{y}^{r}\right)\right)}{t_{\mathrm{mul}}\left(\tau_{k}\left(\overline{\mathbf{x}}\right)\right)}\right\}} \\[4ex]
\dfrac{\partial \min\left\{1, \dfrac{t_{\mathrm{mul}}\left(\mu_{B^{k}}\left(\overline{y}^{r}\right)\right)}{t_{\mathrm{mul}}\left(\tau_{k}\left(\overline{\mathbf{x}}\right)\right)}\right\}}{\partial \dfrac{t_{\mathrm{mul}}\left(\mu_{B^{k}}\left(\overline{y}^{r}\right)\right)}{t_{\mathrm{mul}}\left(\tau_{k}\left(\overline{\mathbf{x}}\right)\right)}} \cdot \dfrac{\partial \dfrac{t_{\mathrm{mul}}\left(\mu_{B^{k}}\left(\overline{y}^{r}\right)\right)}{t_{\mathrm{mul}}\left(\tau_{k}\left(\overline{\mathbf{x}}\right)\right)}}{\partial t_{\mathrm{mul}}\left(\mu_{B^{k}}\left(\overline{y}^{r}\right)\right)} \\[4ex]
\dfrac{\partial t_{\mathrm{mul}}\left(\mu_{B^{k}}\left(\overline{y}^{r}\right)\right)}{\partial \mu_{B^{k}}\left(\overline{y}^{r}\right)} \dfrac{\partial \mu_{B^{k}}\left(\overline{y}^{r}\right)}{\partial p_{u,k}^{B}}
\end{array}
\right), \qquad (9.147)
$$

$$
\varepsilon_{k,r}^{I}\left\{\overline{y}^{r}\right\} =
$$

$$
= \varepsilon_{k,r}^{I}\left(
\begin{array}{c}
\dfrac{\partial t_{\mathrm{mul}}^{-1}\left(\min\left\{1, \dfrac{t_{\mathrm{mul}}\left(\mu_{B^{k}}\left(\overline{y}^{r}\right)\right)}{t_{\mathrm{mul}}\left(\tau_{k}\left(\overline{\mathbf{x}}\right)\right)}\right\}\right)}{\partial \min\left\{1, \dfrac{t_{\mathrm{mul}}\left(\mu_{B^{k}}\left(\overline{y}^{r}\right)\right)}{t_{\mathrm{mul}}\left(\tau_{k}\left(\overline{\mathbf{x}}\right)\right)}\right\}} \\[4ex]
\dfrac{\partial \min\left\{1, \dfrac{t_{\mathrm{mul}}\left(\mu_{B^{k}}\left(\overline{y}^{r}\right)\right)}{t_{\mathrm{mul}}\left(\tau_{k}\left(\overline{\mathbf{x}}\right)\right)}\right\}}{\partial \dfrac{t_{\mathrm{mul}}\left(\mu_{B^{k}}\left(\overline{y}^{r}\right)\right)}{t_{\mathrm{mul}}\left(\tau_{k}\left(\overline{\mathbf{x}}\right)\right)}} \cdot \dfrac{\partial \dfrac{t_{\mathrm{mul}}\left(\mu_{B^{k}}\left(\overline{y}^{r}\right)\right)}{t_{\mathrm{mul}}\left(\tau_{k}\left(\overline{\mathbf{x}}\right)\right)}}{\partial t_{\mathrm{mul}}\left(\mu_{B^{k}}\left(\overline{y}^{r}\right)\right)} \\[4ex]
\dfrac{\partial t_{\mathrm{mul}}\left(\mu_{B^{k}}\left(\overline{y}^{r}\right)\right)}{\partial \mu_{B^{k}}\left(\overline{y}^{r}\right)} \dfrac{\partial \mu_{B^{k}}\left(\overline{y}^{r}\right)}{\partial \overline{y}^{r}}
\end{array}
\right), \qquad (9.148)
$$

$$\varepsilon_{k,r}^{I}\left\{\tau_{k}\left(\overline{\mathbf{x}}\right)\right\} =$$

$$= \varepsilon_{k,r}^{I} \left(\frac{\dfrac{\partial t_{mul}^{-1}\left(\min\left\{1, \dfrac{t_{mul}\left(\mu_{B^{k}}\left(\overline{y}^{r}\right)\right)}{t_{mul}\left(\tau_{k}\left(\overline{\mathbf{x}}\right)\right)}\right\}\right)}{\partial \min\left\{1, \dfrac{t_{mul}\left(\mu_{B^{k}}\left(\overline{y}^{r}\right)\right)}{t_{mul}\left(\tau_{k}\left(\overline{\mathbf{x}}\right)\right)}\right\}} }{ \dfrac{\partial \min\left\{1, \dfrac{t_{mul}\left(\mu_{B^{k}}\left(\overline{y}^{r}\right)\right)}{t_{mul}\left(\tau_{k}\left(\overline{\mathbf{x}}\right)\right)}\right\}}{\partial \dfrac{t_{mul}\left(\mu_{B^{k}}\left(\overline{y}^{r}\right)\right)}{t_{mul}\left(\tau_{k}\left(\overline{\mathbf{x}}\right)\right)}} \cdot \dfrac{\partial \dfrac{t_{mul}\left(\mu_{B^{k}}\left(\overline{y}^{r}\right)\right)}{t_{mul}\left(\tau_{k}\left(\overline{\mathbf{x}}\right)\right)}}{\partial t_{mul}\left(\tau_{k}\left(\overline{\mathbf{x}}\right)\right)} } \cdot \dfrac{\partial t_{mul}\left(\tau_{k}\left(\overline{\mathbf{x}}\right)\right)}{\partial \tau_{k}\left(\overline{\mathbf{x}}\right)} \right). \qquad (9.149)$$

In the above formulas, there are derivatives of the division operator and the minimum operator. The method of their determination has been provided in Subchapter 10.6 of the following chapter.

Example 9.4

To generate the Goguen R-implication, the following multiplicative generator of the t-norm may be used:

$$t_{mul}\left(a\right) = a^{p}, \quad p > 0. \qquad (9.150)$$

Then in formulas (9.147) - (9.149) the following dependencies are used:

$$\frac{\partial t_{mul}\left(a\right)}{\partial a} = pa^{p-1}, \qquad (9.151)$$

$$t_{mul}^{-1}\left(a\right) = a^{\frac{1}{p}}, \qquad (9.152)$$

$$\frac{\partial t_{mul}^{-1}\left(a\right)}{\partial a} = \frac{1}{p}a^{\frac{1}{p}-1}. \qquad (9.153)$$

The errors propagated by implication blocks of the system with inference of logical type using the Q-implication are determined as follows (Fig. 9.15):

$$\varepsilon_{k,r}^{I}\left\{p_{u,k}^{B}\right\} = \varepsilon_{k,r}^{I}\frac{\partial S\left\{N\left(\tau_{k}\left(\overline{\mathbf{x}}\right)\right), T\left\{\tau_{k}\left(\overline{\mathbf{x}}\right), \mu_{B^{k}}\left(\overline{y}^{r}\right)\right\}\right\}}{\partial T\left\{\tau_{k}\left(\overline{\mathbf{x}}\right), \mu_{B^{k}}\left(\overline{y}^{r}\right)\right\}} \qquad (9.154)$$

$$\cdot \frac{\partial T\left\{\tau_{k}\left(\overline{\mathbf{x}}\right), \mu_{B^{k}}\left(\overline{y}^{r}\right)\right\}}{\partial \mu_{B^{k}}\left(\overline{y}^{r}\right)} \frac{\partial \mu_{B^{k}}\left(\overline{y}^{r}\right)}{\partial p_{u,k}^{B}},$$

$$\varepsilon_{k,r}^{I}\left\{\overline{y}^{r}\right\} = \varepsilon_{k,r}^{I}\frac{\partial S\left\{N\left(\tau_{k}\left(\overline{\mathbf{x}}\right)\right), T\left\{\tau_{k}\left(\overline{\mathbf{x}}\right), \mu_{B^{k}}\left(\overline{y}^{r}\right)\right\}\right\}}{\partial T\left\{\tau_{k}\left(\overline{\mathbf{x}}\right), \mu_{B^{k}}\left(\overline{y}^{r}\right)\right\}} \qquad (9.155)$$

$$\cdot \frac{\partial T\left\{\tau_{k}\left(\overline{\mathbf{x}}\right), \mu_{B^{k}}\left(\overline{y}^{r}\right)\right\}}{\partial \mu_{B^{k}}\left(\overline{y}^{r}\right)} \frac{\partial \mu_{B^{k}}\left(\overline{y}^{r}\right)}{\partial \overline{y}^{r}},$$

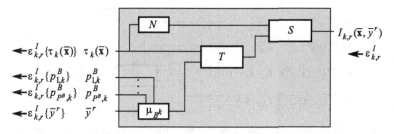

FIGURE 9.15. Implication block of the neuro-fuzzy system with inference of logical type (Q-implication)

$$\varepsilon_{k,r}^{I} \left\{ \tau_k \left(\overline{\mathbf{x}} \right) \right\}$$

$$= \varepsilon_{k,r}^{I} \left(\frac{\partial N \left(\tau_k \left(\overline{\mathbf{x}} \right) \right)}{\partial \tau_k \left(\overline{\mathbf{x}} \right)} \cdot \frac{\partial S \left\{ N \left(\tau_k \left(\overline{\mathbf{x}} \right) \right), T \left\{ \tau_k \left(\overline{\mathbf{x}} \right), \mu_{B^k} \left(\overline{y}^r \right) \right\} \right\}}{\partial N \left(\tau_k \left(\overline{\mathbf{x}} \right) \right)} + \frac{\partial S \left\{ N \left(\tau_k \left(\overline{\mathbf{x}} \right) \right), T \left\{ \tau_k \left(\overline{\mathbf{x}} \right), \mu_{B^k} \left(\overline{y}^r \right) \right\} \right\}}{\partial T \left\{ \tau_k \left(\overline{\mathbf{x}} \right), \mu_{B^k} \left(\overline{y}^r \right) \right\}} \cdot \frac{\partial T \left\{ \tau_k \left(\overline{\mathbf{x}} \right), \mu_{B^k} \left(\overline{y}^r \right) \right\}}{\partial \tau_k \left(\overline{\mathbf{x}} \right)} \right), \quad (9.156)$$

and the derivatives

$$\frac{\partial \mu_{B^k} \left(\overline{y}^r \right)}{\partial p_{u,k}^{B}}, \frac{\partial \mu_{B^k} \left(\overline{y}^r \right)}{\partial \overline{y}^r}, \frac{\partial T \left\{ \tau_k \left(\overline{\mathbf{x}} \right), \mu_{B^k} \left(\overline{y}^r \right) \right\}}{\partial \tau_k \left(\overline{\mathbf{x}} \right)}, \frac{\partial T \left\{ \tau_k \left(\overline{\mathbf{x}} \right), \mu_{B^k} \left(\overline{y}^r \right) \right\}}{\partial \mu_{B^k} \left(\overline{y}^r \right)},$$

$$\frac{\partial S \left\{ N \left(\tau_k \left(\overline{\mathbf{x}} \right) \right), T \left\{ \tau_k \left(\overline{\mathbf{x}} \right), \mu_{B^k} \left(\overline{y}^r \right) \right\} \right\}}{\partial N \left(\tau_k \left(\overline{\mathbf{x}} \right) \right)}, \frac{\partial S \left\{ N \left(\tau_k \left(\overline{\mathbf{x}} \right) \right), T \left\{ \tau_k \left(\overline{\mathbf{x}} \right), \mu_{B^k} \left(\overline{y}^r \right) \right\} \right\}}{\partial T \left\{ \tau_k \left(\overline{\mathbf{x}} \right), \mu_{B^k} \left(\overline{y}^r \right) \right\}},$$

$$\text{and} \frac{\partial N \left(\tau_k \left(\overline{\mathbf{x}} \right) \right)}{\partial \tau_k \left(\overline{\mathbf{x}} \right)}$$

are determined using the dependencies provided earlier and with the description of the learning method of the Takagi-Sugeno system.

Errors propagated by aggregation blocks of the system are determined depending on the chosen inference method. The errors propagated by aggregation blocks of the system with Mamdani type inference are determined as follows (Fig. 9.16):

$$\varepsilon_r^{\text{agr}} \left\{ w_r^{\text{agr}} \right\} = \varepsilon_r^{\text{agr}} \frac{\partial S^* \left\{ \begin{array}{c} I_{1,r} \left(\overline{\mathbf{x}}, \overline{y}^r \right), \ldots, I_{N,r} \left(\overline{\mathbf{x}}, \overline{y}^r \right); \\ w_1^{\text{agr}}, \ldots, w_N^{\text{agr}} \end{array} \right\}}{\partial w_r^{\text{agr}}}, \quad (9.157)$$

$$\varepsilon_r^{\text{agr}} \left\{ I_{k,r} \left(\overline{\mathbf{x}}, \overline{y}^r \right) \right\} = \varepsilon_r^{\text{agr}} \frac{\partial S^* \left\{ \begin{array}{c} I_{1,r} \left(\overline{\mathbf{x}}, \overline{y}^r \right), \ldots, I_{N,r} \left(\overline{\mathbf{x}}, \overline{y}^r \right); \\ w_1^{\text{agr}}, \ldots, w_N^{\text{agr}} \end{array} \right\}}{\partial I_{k,r} \left(\overline{\mathbf{x}}, \overline{y}^r \right)}, \quad (9.158)$$

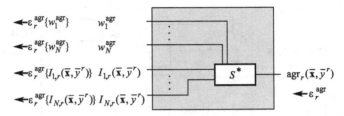

FIGURE 9.16. Aggregation block of the neuro-fuzzy system with Mamdani type inference

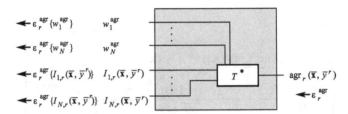

FIGURE 9.17. Aggregation block of the neuro-fuzzy system with inference of logical type

while the derivatives

$$\frac{\partial S^* \left\{ \begin{smallmatrix} I_{1,r}(\overline{\mathbf{x}},\overline{y}^r),\dots,I_{N,r}(\overline{\mathbf{x}},\overline{y}^r); \\ w_1^{\mathrm{agr}},\dots,w_N^{\mathrm{agr}} \end{smallmatrix} \right\}}{\partial I_{k,r}(\overline{\mathbf{x}},\overline{y}^r)} \quad \text{and} \quad \frac{\partial S^* \left\{ \begin{smallmatrix} I_{1,r}(\overline{\mathbf{x}},\overline{y}^r),\dots,I_{N,r}(\overline{\mathbf{x}},\overline{y}^r); \\ w_1^{\mathrm{agr}},\dots,w_N^{\mathrm{agr}} \end{smallmatrix} \right\}}{\partial w_k^{\mathrm{agr}}}$$

are determined based on the dependencies presented above.

The errors propagated by aggregation blocks of the system with inference of logical type are determined as follows (Fig. 9.17):

$$\varepsilon_r^{\mathrm{agr}}\{w_k^{\mathrm{agr}}\} = \varepsilon_r^{\mathrm{agr}} \frac{\partial T^* \left\{ \begin{smallmatrix} I_{1,r}(\overline{\mathbf{x}},\overline{y}^r),\dots,I_{N,r}(\overline{\mathbf{x}},\overline{y}^r); \\ w_1^{\mathrm{agr}},\dots,w_N^{\mathrm{agr}} \end{smallmatrix} \right\}}{\partial w_k^{\mathrm{agr}}}, \tag{9.159}$$

$$\varepsilon_r^{\mathrm{agr}}\{I_{k,r}(\overline{\mathbf{x}},\overline{y}^r)\} = \varepsilon_r^{\mathrm{agr}} \frac{\partial T^* \left\{ \begin{smallmatrix} I_{1,r}(\overline{\mathbf{x}},\overline{y}^r),\dots,I_{N,r}(\overline{\mathbf{x}},\overline{y}^r); \\ w_1^{\mathrm{agr}},\dots,w_N^{\mathrm{agr}} \end{smallmatrix} \right\}}{\partial I_{k,r}(\overline{\mathbf{x}},\overline{y}^r)}, \tag{9.160}$$

while the derivatives

$$\frac{\partial T^* \left\{ \begin{smallmatrix} I_{1,r}(\overline{\mathbf{x}},\overline{y}^r),\dots,I_{N,r}(\overline{\mathbf{x}},\overline{y}^r); \\ w_1^{\mathrm{agr}},\dots,w_N^{\mathrm{agr}} \end{smallmatrix} \right\}}{\partial I_{k,r}(\overline{\mathbf{x}},\overline{y}^r)} \quad \text{and} \quad \frac{\partial T^* \left\{ \begin{smallmatrix} I_{1,r}(\overline{\mathbf{x}},\overline{y}^r),\dots,I_{N,r}(\overline{\mathbf{x}},\overline{y}^r); \\ w_1^{\mathrm{agr}},\dots,w_N^{\mathrm{agr}} \end{smallmatrix} \right\}}{\partial w_k^{\mathrm{agr}}}$$

are determined using the dependencies provided with the description of the learning method of the Takagi-Sugeno system.

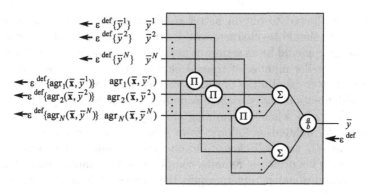

FIGURE 9.18. Defuzzification block of the neuro-fuzzy system

The errors propagated by the defuzzification block of the system are determined as follows (Fig. 9.18):

$$\varepsilon^{\text{def}}\left\{\overline{y}^r\right\} = \varepsilon^{\text{def}} \frac{\partial \det \left(\begin{array}{c} \text{agr}_1\left(\overline{\mathbf{x}}, \overline{y}^1\right), \ldots, \text{agr}_N\left(\overline{\mathbf{x}}, \overline{y}^N\right) ; \\ \overline{y}^1, \ldots, \overline{y}^N \end{array} \right)}{\partial \overline{y}^r}, \qquad (9.161)$$

$$\varepsilon^{\text{def}}\left\{\text{agr}_r\left(\overline{\mathbf{x}}, \overline{y}^r\right)\right\} = \varepsilon^{\text{def}} \frac{\partial \det \left(\begin{array}{c} \text{agr}_1\left(\overline{\mathbf{x}}, \overline{y}^1\right), \ldots, \text{agr}_N\left(\overline{\mathbf{x}}, \overline{y}^N\right) ; \\ \overline{y}^1, \ldots, \overline{y}^N \end{array} \right)}{\partial \, \text{agr}_r\left(\overline{\mathbf{x}}, \overline{y}^r\right)}, \qquad (9.162)$$

while

$$\det\left(a_1, a_2, \ldots, a_n; w_1, w_2, \ldots, w_n\right) = \det\left(\mathbf{a}; \mathbf{w}\right) = \frac{\sum_{i=1}^n w_i a_i}{\sum_{i=1}^n a_i}, \qquad (9.163)$$

$$\frac{\partial \det\left(\mathbf{a}; \mathbf{w}\right)}{\partial a_j} = \left(w_j - \det\left(\mathbf{a}; \mathbf{w}\right)\right) \frac{1}{\sum_{i=1}^n a_i}, \qquad (9.164)$$

$$\frac{\partial \det\left(\mathbf{a}; \mathbf{w}\right)}{\partial w_j} = \left(a_j - \det\left(\mathbf{a}; \mathbf{w}\right) \frac{\partial a_j}{\partial w_j}\right) \frac{1}{\sum_{i=1}^n a_i}. \qquad (9.165)$$

Let us notice that dependencies (9.163) - (9.165) are special cases of dependencies – (9.113) - (9.116).

9.7 Comparison of neuro-fuzzy systems

Simulation analyses in Subchapters 9.3 – 9.5 and an attempt to evaluate the studied neuro-fuzzy systems were based on the mean squared error as the criterion used to compare these systems. These considerations allow to conclude that usually the systems containing a higher number of trained parameters allowed to obtain better results. However, the desired neuro-fuzzy system should be characterized by the smallest possible error but at the same time should be as simple as possible. It should remember that systems with smaller number of trained parameters are characterized among others by better capabilities of generalization of the results obtained. Here, we should mention the so-called *parsimony principle* [235]. This principle is very useful when determining the appropriate order of the model. It may be formulated as follows: *from between two alternative and satisfactory models, we shall choose the one which contains less independent parameters.* This principle remains compliant with common sense: "do not enter any additional parameters into the process description unless they are necessary".

Estimation methods of the system order have been best developed for autoregression processes [107, 132, 202]. Time series $u(n), u(n-1), ..,$ $u(n-p)$ is an autoregression process of order p, if the difference equation is satisfied

$$u(n) + \alpha_1 u(n-1) + ... + \alpha_p u(n-p) = e(n) \tag{9.166}$$

or equivalently

$$u(n) = -\sum_{k=1}^{p} \alpha_k u(n-k) + e(n), \tag{9.167}$$

where $\alpha_1, ..., \alpha_p$ are process coefficient, while $e(n)$ is the white noise

$$\mathbf{E}[e(n)] = 0, \quad \mathbf{E}[e(n)e(m)] = \begin{cases} \sigma^2 & \text{for} \quad n = m, \\ 0 & \text{for} \quad n \neq m. \end{cases} \tag{9.168}$$

In the autoregression theory, criteria allowing to estimate the order of predictor p, determining first the prediction error $\widehat{Q}_p$ based on the learning sequence of the length M, are well known. The most important is the Akaike information criterion (AIC), Schwarz method and the final prediction error (FPE) method.

In the following point, we will first present the basic models evaluation criteria (taking into account their complexity), initially applied to the estimation of orders of autoregression processes, and next they will be adapted to evaluate the effectiveness of neuro-fuzzy systems. By *the effectiveness of operation* of a neuro-fuzzy system, we shall understand the precision (accuracy) of operation achieved by such a system, (expressed by mean squared error or by the number of erroneously classified samples) in the context of its size. By *the system size* we shall understand the number of all parameters that are subject to learning. We shall also present the concept of the so-called *criteria isolines*, which allow to solve the problem of the compromise between the system accuracy and the number of parameters describing this system.

9.7.1 Models evaluation criteria taking into account their complexity

Two general criteria taking into account the complexity of the model, the dependencies between those criteria as well as their special forms are presented below.

9.7.1.1. Criterion A

The general form of criterion A, taking into account the complexity of the model, is given by formula

$$W\left(p\right) = \widehat{Q}_p\left[1 + \beta\left(M, p\right)\right], \qquad (9.169)$$

where $\widehat{Q}_p$ is the mean square error, and $\beta\left(M, p\right)$ is the function of the length of the learning sequence M and the number of parameters p of the model. To eliminate too complex structures (according to the economy principle), we assume that

$$\lim_{p \to \infty} \beta\left(M, p\right) = \infty. \qquad (9.170)$$

At the same time, in order to avoid the situation where the presence of the penalizing term in expression (9.169) hampers the observation of the decreasing of the mean square error $\widehat{Q}_p$ value with the increase of model complexity, we shall assume that

$$\lim_{M \to \infty} \beta\left(M, p\right) = 0. \qquad (9.171)$$

The typical choice is $\beta\left(M, p\right) = 2p/M$ and then

$$W\left(p\right) = \widehat{Q}_p\left[1 + \frac{2p}{M}\right]. \qquad (9.172)$$

9.7.1.2. Criterion B

An alternative criterion to formula (9.169) may be the following dependence:

$$W\left(p\right) = M \log \widehat{Q}_p + \gamma\left(M, p\right), \qquad (9.173)$$

where the additional term $\gamma\left(M, p\right)$ should take into account the penalty for accepting models of an order which is too high. It is easy to check that if

$$\gamma\left(M, p\right) = M\beta\left(M, p\right), \qquad (9.174)$$

then criteria (9.169) and (9.173) are asymptotically equivalent.

Below, we shall present the basic methods of the compromise selection of the model order. Most of these methods are the special cases of criterion A or B presented above.

9.7.1.3. Akaike information criterion (AIC) method

The assumption that $\gamma(M,p) = 2p$ in criterion (9.173) results in the so-called *Akaike Information Criterion*. The complexity of system p may be found by searching for the smallest value of the following expression

$$\text{AIC}(p) = M \ln \widehat{Q}_p + 2p \qquad (9.175)$$

9.7.1.4. Final prediction error (FPE) method

The FPE criterion was also proposed by Akaike. In *the Final Prediction Error* method, which does not result from any general formulas (9.169) and (9.173), the complexity of system p may be found by searching for the smallest value of the expression

$$\text{FPE}(p) = \frac{M+p}{M-p}\widehat{Q}_p. \qquad (9.176)$$

In expression (9.176) together with the increase of parameter p, the factor $\frac{M+p}{M-p}$ increases and the value of the mean square error $\widehat{Q}_p$ decreases. We shall notice that for high values of M, the following approximation may be used:

$$\text{FPE}(p) = \widehat{Q}_p \left[1 + \frac{2p/M}{1-p/M} \right] \approx \widehat{Q}_p \left[1 + \frac{2p}{M} \right], \qquad (9.177)$$

which is of type (9.169), i.e.

$$\beta(M,p) = \frac{2p}{M}. \qquad (9.178)$$

Expressions (9.169) and (9.173) are asymptotically equivalent, if condition (9.174) is satisfied, and hence

$$\gamma(M,p) = 2p. \qquad (9.179)$$

In consequence:

$$\text{FPE}(p) \approx \text{AIC}(p) = M \ln \widehat{Q}_p + 2p. \qquad (9.180)$$

FPE and AIC criteria show a tendency to select a model of a too small order. That is why literature [235] proposes three other methods described below:

9.7.1.5. Schwarz method

Assuming $\gamma\,(M,p) = p \log M$ in criterion (9.173) gives the so-called *Schwarz criterion*. In this method, the complexity of system p may be found by searching for the smallest value of the expression

$$S(p) = M \ln \widehat{Q}_p + p \ln M. \tag{9.181}$$

9.7.1.6. Södeström and Stoica method

Assuming $\gamma\,(M,p) = 2pc \log\,(\log M)$, where $c \geq 1$, in criterion (9.173) gives the so-called *Södeström and Stoica criterion*. In this method, the complexity of system p is found by searching for the smallest value of the expression

$$H(p) = M \ln \widehat{Q}_p + 2pc \log\,(\log M)\,. \tag{9.182}$$

9.7.1.7. CAT method

In the CAT (*Criterion Autoregressive Transfer Function*) method, the complexity of system p may be found by searching for the smallest value of the expression

$$\mathrm{CAT}\,(p) = \frac{1}{M} \sum_{i=1}^{p} \frac{1}{\overline{Q}_i} - \frac{1}{\overline{Q}_p}, \tag{9.183}$$

where $\overline{Q}_i = \frac{m}{M-i} \widehat{Q}_i$.

The methods described above for determination of the order of the model have been first proposed for the analysis of data autoregression processes using formula (9.166). However, it should be stated that these methods allow to determine the appropriate order of the model regardless whether the system belongs to the class of the model structures or not [235].

9.7.2 *Criteria isolines method*

The estimation methods of the prediction order described in the previous point will be adapted now to the evaluation of fuzzy systems. Thanks to this, search for the desired fuzzy system based on two criteria (number of parameters and mean square error) will come down to one selected criterion, i.e. AIC, Schwarz or FPE. They have been adapted for the needs of evaluation of neuro-fuzzy systems in the following form:

$$\mathrm{AIC}(p, \widehat{Q}_p) = M \ln \widehat{Q}_p + 2p, \tag{9.184}$$

$$S(p, \widehat{Q}_p) = M \ln \widehat{Q}_p + p \ln M, \tag{9.185}$$

$$\mathrm{FPE}(p, \widehat{Q}_p) = \frac{Mn + p}{Mn - p} \widehat{Q}_p, \tag{9.186}$$

where p is the number of system parameters subject to learning (number of parameters of all membership functions and number of all weights if they occur in a given system), $\widehat{Q}_p$ is the measure of error used in simulations described in Subchapters 9.3 - 9.5, M is the number of samples in a learning sequence, and n is the number of system inputs. The product $M \cdot n$ may therefore be treated as a measure of size of the problem being solved. Tables 9.62a, 9.62b, 9.63a and 9.63b contain the computed values of criteria for particular tested structures in case of the learning and testing sequence used in the polymerization problem. Figures 9.19 - 9.24 illustrate the coordinates of the points corresponding to particular neuro-fuzzy systems tested. The coordinate p defines the number of parameters of a given system, coordinate Q defines the error with which the system realized the problem to be solved. The criteria isolines present constant values of the AIC, Schwarz and FPE criteria, with different values of the error and the number of parameters. Such an approach allows to solve the problem of the compromise between the system operation error and the number of parameters describing this system. Points located on the criteria isolines with the same values of AIC, Schwarz or FPE criterion characterize the neuro-fuzzy systems making up the Pareto set. In the Pareto set, none of the two values of contradictory criteria may be improved (mean square error *versus* system size), without worsening the other one. Points located on the criteria isolines with the smallest values of AIC, Schwarz or FPE criterion characterize the neuro-fuzzy systems which have been called suboptimal ones. The suboptimal neuro-fuzzy systems presented in graphs ensure the smallest value of criteria within tested structures (the terminology "optimum systems" is not used as all possible structures have not been tested).

Tables 9.62a, 9.62b, 9.63a and 9.63b and figures indicate that both for the learning sequence and for the testing sequence, the AIC criterion evaluates as the best system 1 (simplified Larsen structure), and next, system 29 (Zadeh structure with weights of rules), the FPE criterion – system 29, the Schwarz criterion – definitely system 1.

Analogically, the criteria isolines may be easily drawn for HANG, NDP and modeling the taste of rice problems. Having drawn these lines, it may be checked that in case of the HANG problem, both for the learning and the testing sequence, the AIC and FPE criteria indicate system 23 (Reichenbach structure with weights of rules), and the Schwarz criterion – system 1 (simplified Larsen structure). In case of the NDP problem all three criteria, both for the learning and the testing sequence indicate the selection of system 3 (simplified Larsen structure with weights of inputs and rules). In case of modeling the taste of rice, both for the learning and the testing sequence, the AIC and Schwarz criteria indicate system 1 (simplified Larsen structure) as the best one, and the FPE criterion indicates system 20 (Mamdani structure with weight of rules).

TABLE 9.62a. Values of criteria for the learning sequence

No.	Structure	Polymerization				
		error	p	AIC	FPE	Schwarz
1	Larsen simplified	0.0042	42	−299.09	0.0063	−204.65
2	Larsen simplified with weights of rules	0.0039	48	−292.27	0.0062	−184.35
3	Larsen simplified with weights of inputs and rules	0.0031	66	−272.34	0.0059	−123.94
4	Łukasiewicz simplified	0.0059	42	−275.30	0.0089	−180.86
5	Łukasiewicz simplified with weights of rules	0.0039	48	−292.27	0.0062	−184.35
6	Łukasiewicz simplified with weights of inputs and rules	0.0037	66	−259.96	0.0071	−111.56
7	Zadeh simplified	0.0049	42	−288.30	0.0074	−193.86
8	Zadeh simplified with weights of rules	0.0041	48	−288.77	0.0065	−180.85
9	Zadeh simplified with weights of inputs and rules	0.0038	66	−258.09	0.0073	−109.69
10	Binary	0.0063	48	−258.70	0.01	−150.78
11	Binary with weights of rules	0.0054	54	−257.49	0.0091	−136.08
12	Binary with weights of inputs and rules	0.0036	72	−249.88	0.0074	−87.99
13	Larsen	0.0049	48	−276.30	0.0078	−168.37
14	Larsen with weights of rules	0.0043	54	−273.44	0.0073	−152.02
15	Larsen with weights of inputs and rules	0.0035	72	−251.85	0.0072	−89.96
16	Łukasiewicz	0.0065	48	−256.52	0.0104	−148.59
17	Łukasiewicz with weights of rules	0.0041	54	−276.77	0.0069	−155.36
18	Łukasiewicz with weights of inputs and rules	0.0038	72	−246.09	0.0078	−84.20
19	Mamdani	0.0041	48	−288.77	0.0065	−180.85
20	Mamdani with weights of rules	0.0039	54	−280.27	0.0066	−158.86
21	Mamdani with weights of inputs and rules	0.0034	72	−253.88	0.0069	−91.99
22	Reichenbach	0.0040	48	−290.50	0.0064	−182.57
23	Reichenbach with weights of rules	0.0037	54	−283.96	0.0063	−162.54

TABLE 9.62b. Values of criteria for the learning sequence

No.	Structure	Polymerization				
		error	p	AIC	FPE	Schwarz
24	Reichenbach with weights of inputs and rules	0.0034	72	−253.88	0.0069	−91.990
25	Willmott	0.0056	48	−266.95	0.0089	−159.02
26	Willmott with weights of rules	0.0047	54	−267.21	0.008	−145.79
27	Willmott with weights of inputs and rules	0.0039	72	−244.27	0.008	−82.38
28	Zadeh	0.0038	48	−294.09	0.0061	−186.17
29	Zadeh with weights of rules	0.0030	54	−298.64	0.0051	−177.22
30	Zadeh with weights of inputs and rules	0.0028	72	−267.47	0.0057	−105.58
31	Takagi-Sugeno	0.0034	60	−277.88	0.0061	−142.97
32	Takagi-Sugeno with weights of rules	0.0031	66	−272.34	0.0059	−123.94
33	Takagi-Sugeno with weights of inputs and rules	0.0030	84	−238.64	0.007	−49.77

In all simulations performed so far, the effectiveness of operation of the neuro-fuzzy systems, assuming the defined number of rules to solve a specific problem has been analised. Thanks to it, it was possible to compare 33 different neuro-fuzzy systems. It should be stressed that the method of criteria isolines may be also applied to the appropriate designing of each of these systems. If we concentrate on a single specific neuro-fuzzy system, we may select the number of rules which will ensure the smallest value of one of the criteria listed in point 9.7.1. For example, for the problem of modeling the taste of rice, we have applied the simplified Larsen structure, changing gradually the number of rules from 10 to 2. Individual systems are characterized by the following number of parameters: 110, 99, 88, 77, 66, 55, 44, 33 and 22. Figures 9.25 and 9.26 show the function of dependency of the AIC and Schwarz criteria versus the number of parameters. As it may be inferred from the graphs, the AIC and Schwarz criteria suggest that four rules should be assumed. In the simulations performed, the problem of modeling the taste of rice was analyzed, assuming 5 rules, according to the principle of caution.

TABLE 9.63a. Value of criteria for the testing sequence

No.	Structure	Polymerization				
		error	p	AIC	FPE	Schwarz
1	Larsen simplified	0.0045	42	−294.26	0.0068	−199.82
2	Larsen simplified with weights of rules	0.0041	48	−288.77	0.0065	−180.85
3	Larsen simplified with weights of inputs and rules	0.0033	66	−267.97	0.0063	−119.57
4	Łukasiewicz simplified	0.0063	42	−270.70	0.0095	−176.27
5	Łukasiewicz simplified with weights of rules	0.0041	48	−288.77	0.0065	−180.85
6	Łukasiewicz simplified with weights of inputs and rules	0.0039	66	−256.27	0.0075	−107.87
7	Zadeh simplified	0.0053	42	−282.80	0.0080	−188.37
8	Zadeh simplified with weights of rules	0.0044	48	−283.83	0.0070	−175.90
9	Zadeh simplified with weights of inputs and rules	0.0040	66	−254.50	0.0077	−106.10
10	Binary	0.0067	48	−254.40	0.0107	−146.47
11	Binary with weights of rules	0.0057	54	−253.71	0.0096	−132.29
12	Binary with weights of inputs and rules	0.0038	72	−246.09	0.0078	−84.20
13	Larsen	0.0052	48	−272.14	0.0083	−164.21
14	Larsen with weights of rules	0.0045	54	−270.26	0.0076	−148.84
15	Larsen with weights of inputs and rules	0.0038	72	−246.09	0.0078	−84.20
16	Łukasiewicz	0.0069	48	−252.34	0.0110	−144.41
17	Łukasiewicz with weights of rules	0.0045	54	−270.26	0.0076	−148.84
18	Łukasiewicz with weights of inputs and rules	0.0041	72	−240.77	0.0084	−78.88
19	Mamdani	0.0043	48	−285.44	0.0068	−177.51
20	Mamdani with weights of rules	0.0042	54	−275.09	0.0071	−153.67
21	Mamdani with weights of inputs and rules	0.0037	72	−247.96	0.0076	−86.07
22	Reichenbach	0.0044	48	−283.83	0.0070	−175.90

TABLE 9.63b. Value of criteria for the testing sequence

No.	Structure	Polymerization				
		error	p	AIC	FPE	Schwarz
23	Reichenbach with weights of rules	0.0039	54	−280.27	0.0066	−158.86
24	Reichenbach with weights of inputs and rules	0.0037	72	−247.96	0.0076	−86.07
25	Willmott	0.0060	48	−262.12	0.0096	−154.19
26	Willmott with weights of rules	0.0049	54	−264.30	0.0083	−142.88
27	Willmott with weights of inputs and rules	0.0043	72	−237.44	0.0088	−75.55
28	Zadeh	0.0043	48	−285.44	0.0068	−177.51
29	Zadeh with weights of rules	0.0033	54	−291.97	0.0056	−170.55
30	Zadeh with weights of inputs and rules	0.0031	72	−260.34	0.0063	−98.45
31	Takagi-Sugeno	0.0036	60	−273.88	0.0065	−138.97
32	Takagi-Sugeno with weights of rules	0.0034	66	−265.88	0.0065	−117.48
33	Takagi-Sugeno with weights of inputs and rules	0.0033	84	−231.97	0.0077	−43.09

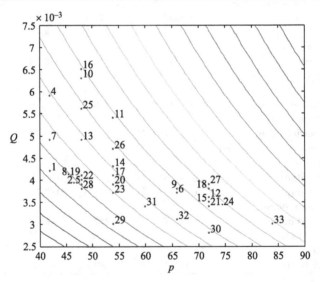

FIGURE 9.19. Criteria isolines: results obtained by particular systems for the Akaike criterion for the learning sequence – polymerization problem

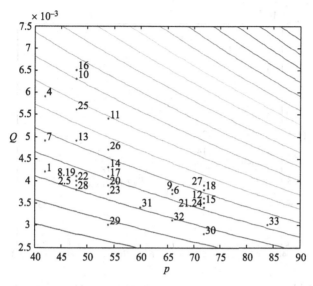

FIGURE 9.20. Criteria isolines: results obtained by particular systems for the FPE criterion for the learning sequence – polymerization problem

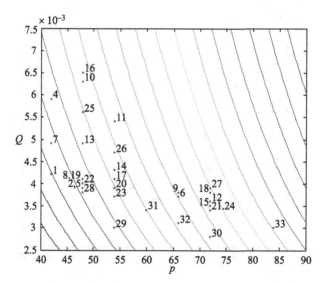

FIGURE 9.21. Criteria isolines: results obtained by particular systems for the Schwarz criterion for the learning sequence – polymerization problem

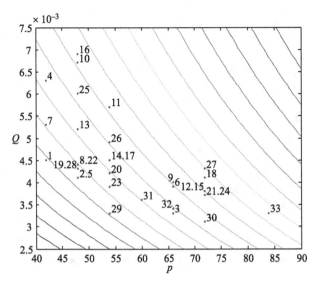

FIGURE 9.22. Criteria isolines: results obtained by particular systems for the Akaike criterion for the testing sequence – polymerization problem

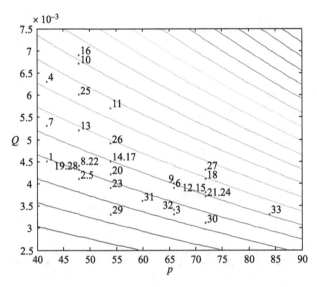

FIGURE 9.23. Criteria isolines: results obtained by particular systems for the FPE criterion for the testing sequence – polymerization problem

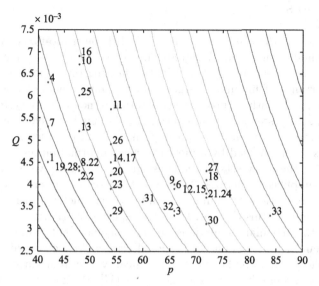

FIGURE 9.24. Criteria isolines: results obtained by particular systems for the
Schwarz criterion for the testing sequence – polymerization problem

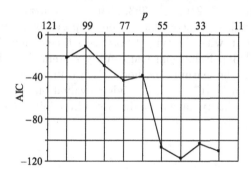

FIGURE 9.25. Values of the Akaike criterion

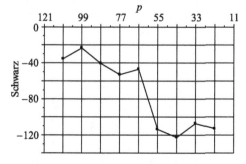

FIGURE 9.26. Values of the Schwarz criterion

9.8 Notes

In this chapter, the object of studies included the neuro-fuzzy systems of the Mamdani, logical and Takagi-Sugeno type. From the simulations performed we may conclude that if weights reflecting the importance of rules and importance of linguistic variables in the antecedents of rules are included, it significantly improves the operation of neuro-fuzzy systems. The Takagi-Sugeno systems are characterized by the smallest mean square error, but this result is obtained with a large number of parameters. Extended structures (characterized by a more extensive information on membership functions of the fuzzy sets in the consequents of rules) give better results than the simplified structures. Moreover, the issue of compromise between the system operation error and the number of parameters describing it has been presented in this chapter. From the analysis of criteria isolines corresponding to particular simulations we may conclude that in most cases the best system, in the meaning of proposed criteria, is the simplified Larsen structure given by formula (9.25). The logical type systems have been studied in monographs by Czogała and Łęski [34], Rutkowska [187] as well as Rutkowski [225]. Different approaches to the issue of designing neuro-fuzzy networks have been presented in works [65, 126, 142, 145, 148, 149, 176, 185, 186, 213, 214, 216, 239, 253, 254]. Neuro-fuzzy structures associated with the rough sets theory have been proposed by Nowicki [151, 152], while in association with the type-2 fuzzy sets theory have been proposed by Starczewski [238]. Relational neuro-fuzzy systems have been analyzed by Scherer [231]. The learning method of neuro-fuzzy structures has been developed by Piliński [172 - 174]. Models evaluation criteria taking into account their complexity have been discussed in detail in monograph [235].

10

Flexible neuro-fuzzy systems

10.1 Introduction

In the previous chapter we considered Mamdani and logical neuro-fuzzy systems. In the present chapter we will build a neuro-fuzzy system, the inference method (Mamdani or logical) of which will be found as a result of the learning process. The structure of such a system will be changing during the learning process. Its operation will be possible thanks to specially constructed adjustable triangular norms. Adjustable triangular norms, applied to aggregate particular rules, take the form of a classic t-norm or t-conorm after the learning process is finished. Adjustable implications, which finally take the form of a "correlation function" between premises and consequents (Mamdani approach) or fuzzy S-implication (logical approach), will be constructed in analogical way. Moreover, the following concepts will be used for construction of the neuro-fuzzy systems: the concept of soft triangular norms, parameterized triangular norms as well as weights which describe the importance of particular rules and antecedents in those rules.

10.2 Soft triangular norms

Soft equivalents of triangular norms shall be defined in the following way:

$$\tilde{T}\{\mathbf{a}; \alpha\} = (1 - \alpha)\frac{1}{n}\sum_{i=1}^{n} a_i + \alpha T\{a_1, \ldots, a_n\} \qquad (10.1)$$

and

$$\tilde{S}\{\mathbf{a};\alpha\} = (1-\alpha)\frac{1}{n}\sum_{i=1}^{n}a_i + \alpha S\{a_1,\ldots,a_n\}, \qquad (10.2)$$

where $\mathbf{a} = [a_1,\ldots,a_n]$ and $\alpha \in [0,1]$. The above operators allow smooth balancing between the arithmetic average of arguments $a_1,\ldots,a_n$ and a classic t-norm or t-conorm operator.

Example 10.1

The soft Zadeh t-norm (of the min type) shall be defined as follows:

$$\tilde{T}\{a_1,a_2;\alpha\} = (1-\alpha)\frac{1}{2}(a_1+a_2) + \alpha \min\{a_1,a_2\}. \qquad (10.3)$$

Its operation is illustrated by Fig. 10.1.

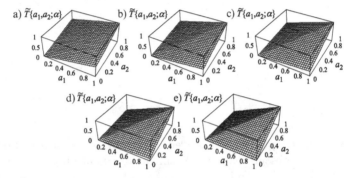

FIGURE 10.1. Hyperplanes of function (10.3) for a) $\alpha = 0.00$, b) $\alpha = 0.25$, c) $\alpha = 0.50$, d) $\alpha = 0.75$, e) $\alpha = 1.00$

The soft Zadeh t-conorm takes the following form

$$\tilde{S}\{a_1,a_2;\alpha\} = (1-\alpha)\frac{1}{2}(a_1+a_2) + \alpha \max\{a_1,a_2\}. \qquad (10.4)$$

Its operation is illustrated by Fig. 10.2.

As we remember, the "correlation function" in the Mamdani approach shall be defined through the t-norm. A soft equivalent of this function shall be notated as follows:

$$\tilde{I}(a,b;\beta) = (1-\beta)\frac{1}{2}(a+b) + \beta T\{a,b\}. \qquad (10.5)$$

The soft S-implication takes the form

$$\tilde{I}(a,b;\beta) = (1-\beta)\frac{1}{2}(1-a+b) + \beta S\{1-a,b\}, \qquad (10.6)$$

where $\beta \in [0,1]$ in both cases.

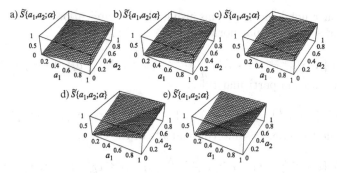

FIGURE 10.2. Hyperplanes of function (10.4) for a) $\alpha = 0.00$, b) $\alpha = 0.25$, c) $\alpha = 0.50$, d) $\alpha = 0.75$, e) $\alpha = 1.00$

Example 10.2

The soft binary S-implication is given by the following formula

$$\widetilde{I}(a, b; \beta) = (1 - \beta) \frac{1}{2} (1 - a + b) + \beta \max \{1 - a, b\}. \tag{10.7}$$

Its operation is illustrated by Fig. 10.3.

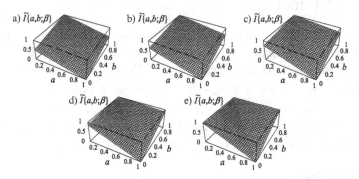

FIGURE 10.3. Hyperplanes of function (10.7) for a) $\beta = 0.00$, b) $\beta = 0.25$, c) $\beta = 0.50$, d) $\beta = 0.75$, e) $\beta = 1.00$

To construct Mamdani systems, we can use the following soft triangular norms:

- $\widetilde{T}_1 \{\mathbf{a}; \alpha^\tau\} = (1 - \alpha^\tau) \frac{1}{n} \sum_{i=1}^n a_i + \alpha^\tau T_{i=1}^n \{a_i\}$ to aggregate the premises in particular rules;

- $\widetilde{T}_2 \{b_1, b_2; \alpha^I\} = (1 - \alpha^I) \frac{1}{2} (b_1 + b_2) + \alpha^I T \{b_1, b_2\}$ to combine the premises and consequents of the rules;

- $\widetilde{S} \{\mathbf{c}; \alpha^{\text{agr}}\} = (1 - \alpha^{\text{agr}}) \frac{1}{N} \sum_{k=1}^N c_k + \alpha^{\text{agr}} S_{k=1}^N \{c_k\}$ to aggregate the rules,

where n is the number of inputs while N is the number of rules.

To construct logical systems using the S-implication, we can use the following soft triangular norms:

- $\tilde{T}_1 \{\mathbf{a}; \alpha^\tau\} = (1 - \alpha^\tau) \frac{1}{n} \sum_{i=1}^{n} a_i + \alpha^\tau T_{i=1}^{n} \{a_i\}$ to aggregate the premises in particular rules;

- $\tilde{S} \{b_1, b_2; \alpha^I\} = (1 - \alpha^I) \frac{1}{2} (1 - b_1 + b_2) + \alpha^I S \{1 - b_1, b_2\}$ to combine the premises and consequents of the rules;

- $\tilde{T}_2 \{\mathbf{c}; \alpha^{\text{agr}}\} = (1 - \alpha^{\text{agr}}) \frac{1}{N} \sum_{k=1}^{N} c_k + \alpha^{\text{agr}} T_{k=1}^{N} \{c_k\}$ to aggregate the rules,

where n is the number of inputs while N is the number of rules. It should be emphasized that parameters α^τ, α^I and α^{agr} can be found as a result of learning.

10.3   Parameterized triangular norms

In order to construct flexible systems, we can also use parameterized variations of triangular norms. These include among other things Dombi, Hamacher, Yager, Frank, Weber, Dubois and Prade, Schweizer and Mizumoto triangular norms. The notations $\overset{\leftrightarrow}{T} \{a_1, a_2, \ldots, a_n; p\}$ and $\overset{\leftrightarrow}{S}\{a_1, a_2, \ldots, a_n; p\}$ will be used to notate them. Parameterized triangular norms are characterized by the fact that their corresponding hyperplanes can be modified as a result of learning the parameter p.

Example 10.3
Parameterized Dombi t-norm is defined as follows:

$$\overset{\leftrightarrow}{T} \{\mathbf{a}; p\} = \begin{cases} \textit{Łukasiewicz t-norm} & \text{for} \quad p = 0, \\ \left(1 + \left(\sum_{i=1}^{n} \left(\frac{1 - a_i}{a_i}\right)^p\right)^{\frac{1}{p}}\right)^{-1} & \text{for} \quad p \in (0, \infty), \\ \textit{Zadeh t-norm} & \text{for} \quad p = \infty. \end{cases} \quad (10.8)$$

Its operation for $n = 2$ is illustrated by Fig. 10.4.
Parameterized Dombi t-conorm is defined as follows:

$$\overset{\leftrightarrow}{S} \{a; p\} = \begin{cases} \textit{Łukasiewicz t} - \textit{conorm} & \text{for} \quad p = 0, \\ 1 - \left(1 + \left(\sum_{i=1}^{n} \left(\frac{a_i}{1 - a_i}\right)^p\right)^{\frac{1}{p}}\right)^{-1} & \text{for} \quad p \in (0, \infty), \\ \textit{Zadeh t} - \textit{conorm} & \text{for} \quad p = \infty. \end{cases} \quad (10.9)$$

Figure 10.5 illustrates its operation.

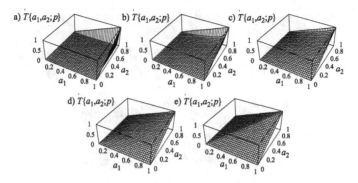

FIGURE 10.4. Hyperplanes of function (10.8) for a) $p = 0.10$, b) $p = 0.25$, c) $p = 0.50$, d) $p = 1.00$, e) $p = 10.00$

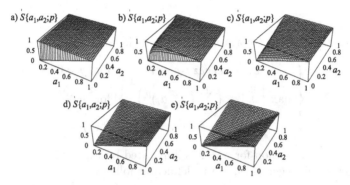

FIGURE 10.5. Hyperplanes of function (10.9) for a) $p = 0.10$, b) $p = 0.25$, c) $p = 0.50$, d) $p = 1.00$, e) $p = 10.00$

The additive generator of parameterized Dombi t-norm takes the form

$$t_{\text{add}}(x) = \left(\frac{1-x}{x}\right)^p,\tag{10.10}$$

while the additive generator of parameterized Dombi t-conorm is defined as follows:

$$s_{\text{add}}(x) = \left(\frac{x}{1-x}\right)^p.\tag{10.11}$$

Parameterized Dombi t-norm for $n = 2$ may play the role of a "correlation function". By combining the concept of parameterized Dombi t-conorm with the concept of S-implication we obtain the parameterized Dombi S-implication which is notated as follows:

$$\overleftrightarrow{I}(a, b; p) = 1 - \left(1 + \left(\left(\frac{1-a}{a}\right)^p + \left(\frac{b}{1-b}\right)^p\right)^{\frac{1}{p}}\right)^{-1}\tag{10.12}$$

for $p \in (0, \infty)$. The operation of parameterized Dombi S-implication is illustrated by Fig. 10.6.

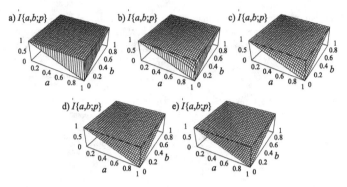

FIGURE 10.6. Hyperplanes of function (10.12) for a) $p = 0.10$, b) $p = 0.25$, c) $p = 0.50$, d) $p = 1.00$, e) $p = 10.00$

Example 10.4

Parameterized Yager t-norm is defined as follows:

$$\overset{\leftrightarrow}{T}\{\mathbf{a};p\} = \begin{cases} \text{Łukasiewicz } t\text{-norm} & \text{for } p = 0 \\ \max\left\{0, 1 - \left(\sum_{i=1}^{n}(1 - a_i)^p\right)^{\frac{1}{p}}\right\} & \text{for } p \in (0,\infty) \\ \text{Zadeh } t\text{-norm} & \text{for } p = \infty \end{cases} \quad (10.13)$$

for $p > 0$. Its operation for $n = 2$ is illustrated by Fig. 10.7. Parameterized Yager t-conorm is defined as follows:

$$\overset{\leftrightarrow}{S}\{\mathbf{a};p\} = \begin{cases} \text{boundary } t\text{-conorm} & \text{for } p = 0, \\ \min\left\{1, \left(\sum_{i=1}^{n}(a_i)^p\right)^{\frac{1}{p}}\right\} & \text{for } p \in (0,\infty), \\ \text{Zadeh } t\text{-conorm} & \text{for } p = \infty. \end{cases} \quad (10.14)$$

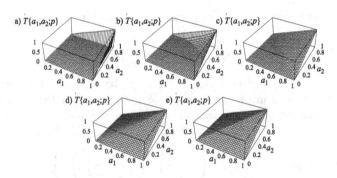

FIGURE 10.7. Hyperplanes of function (10.13) for a) $p = 0.1$, b) $p = 0.5$, c) $p = 1.0$, d) $p = 10.0$, e) $p = 100.0$

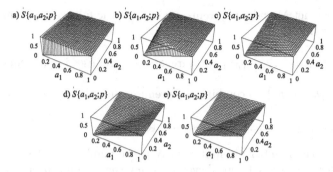

FIGURE 10.8. Hyperplanes of function (10.14) for a) $p = 0.1$, b) $p = 0.5$, c) $p = 1.0$, d) $p = 10.0$, e) $p = 100.0$

Figure 10.8 illustrates its operation.

The additive generator of parameterized Yager t-norm takes the form

$$t_{\text{add}}(x) = (1 - x)^p, \tag{10.15}$$

while the additive generator of parameterized Yager t-conorm is defined as follows:

$$s_{\text{add}}(x) = x^p. \tag{10.16}$$

Parameterized Yager t-norm for $n = 2$ can be used as "correlation function". By combining the concept of parameterized Yager t-conorm with the concept of S-implication we obtain parameterized Yager S-implication which is notated as follows:

$$\overleftrightarrow{I}(a, b; p) = \min\left\{1, ((1 - a)^p + b^p)^{\frac{1}{p}}\right\}. \tag{10.17}$$

The operation of parameterized Yager S-implication is illustrated by Fig. 10.9.

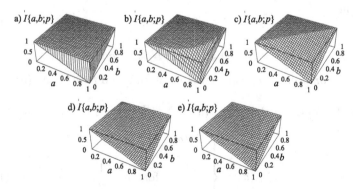

FIGURE 10.9. Hyperplanes of function (10.17) for a) $p = 0.1$, b) $p = 0.5$, c) $p = 1.0$, d) $p = 10.0$, e) $p = 100.0$

To construct Mamdani systems, we can use the following parameterized triangular norms:

- $\overleftrightarrow{T}_1 \{a_1, a_2, \ldots, a_n; p^\tau\}$ to aggregate the premises in particular rules;

- $\overleftrightarrow{T}_2 \{b_1, b_2; p^I\}$ to combine the premises and consequents of the rules;

- $\overleftrightarrow{S} \{c_1, c_2, \ldots, c_N; p^{\mathrm{agr}}\}$ to aggregate the rules,

where n is the number of inputs and N is the number of rules.

In order to construct logical systems using the S-implication, we can use the following parameterized triangular norms:

- $\overleftrightarrow{T}_1 \{a_1, a_2, \ldots, a_n; p^\tau\}$ to aggregate the premises in particular rules;

- $\overleftrightarrow{S} \{1 - b_1, b_2; p^I\}$ to combine the premises and consequents of the rules;

- $\overleftrightarrow{T}_2 \{c_1, c_2, \ldots, c_N; p^{\mathrm{agr}}\}$ to aggregate the rules,

where n is the number of inputs and N is the number of rules.

It should be emphasized that parameters p^τ, p^I and p^{agr} can be found in the process of learning.

10.4   Adjustable triangular norms

We will build the function $H(\mathbf{a}; \nu)$ which, depending on the value of the parameter ν, takes the form of t-norm or t-conorm. To construct this function we will use the compromise operator defined below.

Definition 10.1
Function
$$\widetilde{N}_\nu : [0, 1] \rightarrow [0, 1] \tag{10.18}$$
defined as
$$\widetilde{N}_\nu(a) = (1 - \nu) N(a) + \nu a \tag{10.19}$$
is called a *compromise operator*, where $\nu \in [0, 1]$ and $N(a) = \widetilde{N}_0(a) = 1 - a$.

It could be observed that $\widetilde{N}_{1-\nu}(a) = \widetilde{N}_\nu(1 - a) = 1 - \widetilde{N}_\nu(a)$ and

$$\widetilde{N}_\nu(a) = \begin{cases} N(a) & \text{for } \nu = 0, \\ \dfrac{1}{2} & \text{for } \nu = \dfrac{1}{2}, \\ a & \text{for } \nu = 1. \end{cases} \tag{10.20}$$

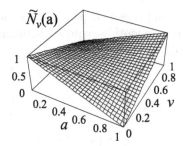

FIGURE 10.10. Illustration of the operation of the compromise operator (10.19)

Function $\widetilde{N}_\nu$ for $\nu = 0$ is *a strong type negation*. Its operation is illustrated by Fig. 10.10.

Definition 10.2
Function

$$H : [0, 1]^n \to [0, 1] \tag{10.21}$$

defined as

$$H(\mathbf{a}; \nu) = \widetilde{N}_\nu \left(\overset{n}{\underset{i=1}{S}} \left\{ \widetilde{N}_\nu (a_i) \right\} \right) = \widetilde{N}_{1-\nu} \left(\overset{n}{\underset{i=1}{T}} \left\{ \widetilde{N}_{1-\nu} (a_i) \right\} \right) \tag{10.22}$$

is called *H-function*, where $\nu \in [0, 1]$.

Theorem 10.1
Let T and S be dual triangular norms. Then function H, defined by formula (10.22), changes its shape from the t-norm to the t-conorm, when ν changes from 0 to 1.

Proof. The assumption says that

$$T\{\mathbf{a}\} = N\left(S\{N(a_1), N(a_2), \dots, N(a_n)\}\right). \tag{10.23}$$

For $\nu = 0$ formula (10.23) can be notated as follows:

$$T\{\mathbf{a}\} = \widetilde{N}_0 \left(S\left\{ \widetilde{N}_0(a_1), \widetilde{N}_0(a_2), \dots, \widetilde{N}_0(a_n) \right\} \right). \tag{10.24}$$

At the same time

$$S\{\mathbf{a}\} = \widetilde{N}_1 \left(S\left\{ \widetilde{N}_1(a_1), \widetilde{N}_1(a_2), \dots, \widetilde{N}_1(a_n) \right\} \right) \tag{10.25}$$

for $\nu = 1$. The right sides of formulas (10.24) and (10.25) can be notated as follows:

$$H(\mathbf{a}; \nu) = \widetilde{N}_\nu \left(\overset{n}{\underset{i=1}{S}} \left\{ \widetilde{N}_\nu(a_i) \right\} \right) \tag{10.26}$$

for, respectively, $\nu = 0$ and $\nu = 1$. If parameter ν changes its value from 0 to 1, then function H is smoothly switched between the t-norm and the t-conorm. It could easily be observed that:

$$H(\mathbf{a};\nu) = \begin{cases} T\{\mathbf{a}\} & \text{for } \nu = 0, \\[2mm] \dfrac{1}{2} & \text{for } \nu = \dfrac{1}{2}, \\[2mm] S\{\mathbf{a}\} & \text{for } \nu = 1. \end{cases} \tag{10.27}$$

Example 10.5
The adjustable H-function constructed with the use of Zadeh t-norm or t-conorm takes the form

$$H(a_1, a_2; \nu) = \tilde{N}_{1-\nu}\left(\min\left\{\tilde{N}_{1-\nu}(a_1), \tilde{N}_{1-\nu}(a_2)\right\}\right) \tag{10.28}$$

$$= \tilde{N}_\nu\left(\max\left\{\tilde{N}_\nu(a_1), \tilde{N}_\nu(a_2)\right\}\right),$$

while ν changes from value 0 to 1. It could easily be observed that:

$$H(a_1, a_2; 0) = T\{a_1, a_2\} = \min\{a_1, a_2\}, \tag{10.29}$$

$$H(a_1, a_2; 1) = S\{a_1, a_2\} = \max\{a_1, a_2\}. \tag{10.30}$$

The operation of Zadeh H-function is illustrated by Fig. 10.11.

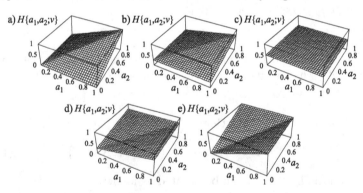

FIGURE 10.11. Hyperplanes of function (10.28) for a) $\nu = 0.00$, b) $\nu = 0.15$, c) $\nu = 0.50$, d) $\nu = 0.85$, e) $\nu = 1.00$

Example 10.6
The adjustable H-function constructed with the use of algebraic t-norm or t-conorm takes the form:

$$H(a_1, a_2; \nu) = \tilde{N}_{1-\nu}\left(\tilde{N}_{1-\nu}(a_1)\,\tilde{N}_{1-\nu}(a_1)\right) \tag{10.31}$$

$$= \tilde{N}_\nu\left(1 - \left(1 - \tilde{N}_\nu(a_1)\right)\left(1 - \tilde{N}_\nu(a_1)\right)\right),$$

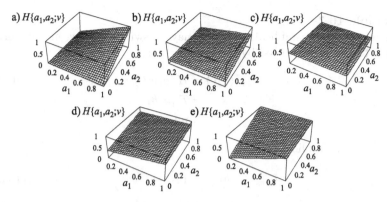

FIGURE 10.12. Hyperplanes of function (10.31) for a) $v = 0.00$, b) $v = 0.15$, c) $v = 0.50$, d) $v = 0.85$, e) $v = 1.00$

while ν changes from value 0 to 1. It could easily be observed that:

$$T\{a_1, a_2\} = H(a_1, a_2; 0) = a_1 a_2, \tag{10.32}$$

$$S\{a_1, a_2\} = H(a_1, a_2; 1) = a_1 + a_2 - a_1 a_2. \tag{10.33}$$

The operation of algebraic H-function is illustrated by Fig. 10.12.

Now we will construct the so-called H-implication which may be switched between the "correlation function" (t-norm) and fuzzy implication (S-implication).

Theorem 10.2

Let T and S be dual triangular norms. Then the H-implication defined as follows:

$$I(a, b; \nu) = H\left(\tilde{N}_{1-\nu}(a), b; \nu\right) \tag{10.34}$$

changes from the "engineering implication"

$$I_{\text{cor}}(a, b) = I(a, b; 0) = T\{a, b\} \tag{10.35}$$

to the fuzzy implication

$$I_{\text{fuzzy}}(a, b) = I(a, b; 1) = S\{1 - a, b\} \tag{10.36}$$

when parameter ν changes its value from 0 to 1.

Proof. Theorem 10.2 is a direct consequence of Theorem 10.1.

Example 10.7

The adjustable H-implication which may be switched between the "correlation function" expressed by the Zadeh t-norm

$$I_{\text{eng}}(a, b) = H(a, b; 0) \tag{10.37}$$
$$= T\{a, b\}$$
$$= \min\{a, b\}$$

and binary S-implication

$$I_{\text{fuzzy}}(a, b) = H\left(\tilde{N}_0(a), b; 1\right) \tag{10.38}$$
$$= S\{N(a), b\}$$
$$= \max\{N(a), b\}$$

may be expressed as follows:

$$I(a, b; \nu) = H\left(\tilde{N}_{1-\nu}(a), b; \nu\right), \tag{10.39}$$

while ν changes from 0 to 1. The operation of H-implication given by formula (10.39) is illustrated by Fig. 10.13.

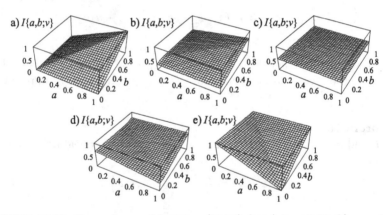

FIGURE 10.13. Hyperplanes of function (10.39) for a) $\nu = 0.00$, b) $\nu = 0.15$, c) $\nu = 0.50$, d) $\nu = 0.85$, e) $\nu = 1.00$

Example 10.8

The adjustable H-implication which may be switched between the "correlation function" expressed by algebraic t-norm

$$I_{\text{eng}}(a, b) = H(a, b; 0) \tag{10.40}$$
$$= T\{a, b\}$$
$$= ab,$$

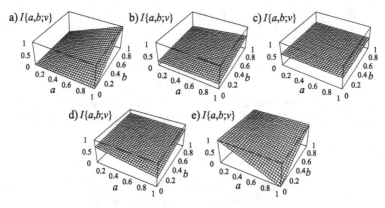

FIGURE 10.14. Hyperplanes of function (10.42) for a) $\nu = 0.00$, b) $\nu = 0.15$, c) $\nu = 0.50$, d) $\nu = 0.85$, e) $\nu = 1.00$

and binary S-implication

$$I_{\text{fuzzy}}(a, b) = H\left(\widetilde{N}_0(a), b; 1\right) \tag{10.41}$$
$$= S\{N(a), b\}$$
$$= 1 - a + ab,$$

may be expressed as follows:

$$I(a, b; \nu) = H\left(\widetilde{N}_{1-\nu}(a), b; \nu\right), \tag{10.42}$$

while ν changes from 0 to 1. The operation of H-implication given by formula (10.42) is illustrated by Fig. 10.14.

10.5   Flexible systems

Using the concept of adjustable triangular norms and adjustable implications, we will build a neuro-fuzzy system the structure of which can change between the system of Mamdani type and the logical type system.

Theorem 10.3
Let T and S be dual triangular norms. Then the neuro-fuzzy system

$$\tau_k(\overline{\mathbf{x}}) = H\left(\begin{array}{c} \mu_{A_1^k}(\overline{x}_1), \ldots, \mu_{A_n^k}(\overline{x}_n); \\ 0 \end{array}\right), \tag{10.43}$$

$$I_{k,r}(\overline{\mathbf{x}}, \overline{y}^r) = H\left(\begin{array}{c} \widetilde{N}_{1-\nu}(\tau_k(\overline{\mathbf{x}})), \mu_{B^k}(\overline{y}^r); \\ \nu \end{array}\right), \tag{10.44}$$

$$\mathrm{agr}_r \left(\overline{\mathbf{x}}, \overline{y}^r \right) = H \left(\frac{I_{1,r} \left(\overline{\mathbf{x}}, \overline{y}^r \right), \dots, I_{N,r} \left(\overline{\mathbf{x}}, \overline{y}^r \right);}{1 - \nu} \right), \qquad (10.45)$$

$$\overline{y} = \frac{\sum_{r=1}^{N} \overline{y}^r \cdot \mathrm{agr}_r \left(\overline{\mathbf{x}}, \overline{y}^r \right)}{\sum_{r=1}^{N} \mathrm{agr}_r \left(\overline{\mathbf{x}}, \overline{y}^r \right)} \qquad (10.46)$$

changes between the Mamdani type system ($\nu = 0$) and the logical type system ($\nu = 1$) together with the change of parameter ν from 0 to 1.

Proof. For $\nu = 0$ formula (10.46) takes the form

$$\overline{y} = \frac{\sum_{r=1}^{N} \overline{y}^r \cdot S_{k=1}^{N} \left\{ T \left\{ \tau_k \left(\overline{\mathbf{x}} \right), \mu_{B^k} \left(\overline{y}^r \right) \right\} \right\}}{\sum_{r=1}^{N} S_{k=1}^{N} \left\{ T \left\{ \tau_k \left(\overline{\mathbf{x}} \right), \mu_{B^k} \left(\overline{y}^r \right) \right\} \right\}}. \qquad (10.47)$$

It could easily be observed that the above formula describes the Mamdani type system. For $\nu = 1$ we have

$$\overline{y} = \frac{\sum_{r=1}^{N} \overline{y}^r \cdot T_{k=1}^{N} \left\{ S \left\{ N \left(\tau_k \left(\overline{\mathbf{x}} \right) \right), \mu_{B^k} \left(\overline{y}^r \right) \right\} \right\}}{\sum_{r=1}^{N} T_{k=1}^{N} \left\{ S \left\{ N \left(\tau_k \left(\overline{\mathbf{x}} \right) \right), \mu_{B^k} \left(\overline{y}^r \right) \right\} \right\}}. \qquad (10.48)$$

Dependency (10.48) describes a logical system using the S-implication. For the value of parameter $\nu \in (0, 1)$ the inference is performed according to the definition of the H-implication, which ends the proof.

Table 10.1 presents implication and aggregation operators for changing parameter ν. The system described by means of dependencies (10.43) - (10.46) is a flexible system as it enables the choice of inference model as a result of the learning process. However, that system does not include the other flexibility aspects described in Subchapters 10.2 and 10.3.

At present the concept of soft triangular norms, parameterized triangular norms, weights of rules and weights of rules premises will be introduced to system (10.46) given in Theorem 10.3. Then the flexible neuro-fuzzy system takes the following form:

$$\tau_k \left(\overline{\mathbf{x}} \right) = \left(\begin{array}{c} \left(1 - \alpha^\tau \right) \mathrm{avg} \left(\mu_{A_1^k} \left(\overline{x}_1 \right), \dots, \mu_{A_n^k} \left(\overline{x}_n \right) \right) + \\ + \alpha^\tau \overset{\leftrightarrow^*}{H} \left(\begin{array}{c} \mu_{A_1^k} \left(\overline{x}_1 \right), \dots, \mu_{A_n^k} \left(\overline{x}_n \right); \\ w_{1,k}^\tau, \dots, w_{n,k}^\tau, p^\tau, 0 \end{array} \right) \end{array} \right), \qquad (10.49)$$

TABLE 10.1. Implication and aggregation operators for changing parameter ν

Parameter ν	Implication	Aggregation
$\nu = 0$	$T \{a, b\}$	t-conorma
$\nu = 1$	$S \{1 - a, b\}$	t-norma
$0 < \nu < 1$	$H \left(\tilde{N}_{1-\nu} \left(a \right), b; \nu \right)$	$H \left(a, b; 1 - \nu \right)$
$\nu = 0.5$	$H \left(a, b; 0.5 \right) = 0.5$	$H \left(a, b; 0.5 \right) = 0.5$

$$I_{k,r}\left(\overline{\mathbf{x}}, \overline{y}^{r}\right) = \left(\begin{array}{c} \left(1 - \alpha^{I}\right) \operatorname{avg}\left(\tilde{N}_{1-\nu}\left(\tau_{k}\left(\overline{\mathbf{x}}\right)\right), \mu_{B^{k}}\left(\overline{y}^{r}\right)\right) + \\ +\alpha^{I} \overleftrightarrow{H}\left(\begin{array}{c} \tilde{N}_{1-\nu}\left(\tau_{k}\left(\overline{\mathbf{x}}\right)\right), \mu_{B^{k}}\left(\overline{y}^{r}\right); \\ p^{I}, \nu \end{array}\right) \end{array}\right), \quad (10.50)$$

$$\operatorname{agr}_{r}\left(\overline{\mathbf{x}}, \overline{y}^{r}\right) = \left(\begin{array}{c} \left(1 - \alpha^{\mathrm{agr}}\right) \operatorname{avg}\left(I_{1,r}\left(\overline{\mathbf{x}}, \overline{y}^{r}\right), \dots, I_{N,r}\left(\overline{\mathbf{x}}, \overline{y}^{r}\right)\right) + \\ +\alpha^{\mathrm{agr}} \overleftrightarrow{H}^{*}\left(\begin{array}{c} I_{1,r}\left(\overline{\mathbf{x}}, \overline{y}^{r}\right), \dots, I_{N,r}\left(\overline{\mathbf{x}}, \overline{y}^{r}\right); \\ w_{1}^{\mathrm{agr}}, \dots, w_{N}^{\mathrm{agr}}, p^{\mathrm{agr}}, 1 - \nu \end{array}\right) \end{array}\right). \quad (10.51)$$

In the system described by means of dependencies (10.46) and (10.49) - (10.51) we can distinguish the following parameters:

- $\nu \in [0, 1]$, parameter of the type of inference model,

- $\alpha^{\tau} \in [0, 1]$, $\alpha^{I} \in [0, 1]$, $\alpha^{\mathrm{agr}} \in [0, 1]$, flexibility parameters (in the sense of Yager and Filev) in operators of premises aggregation, operators of inference and operators of rules aggregation,

- $p^{\tau} \in [0, \infty)$, $p^{I} \in [0, \infty)$, $p^{\mathrm{agr}} \in [0, \infty)$, parameters of the hyperplanes shape of premises aggregation operators, operators of inference and operators of rules aggregation,

- $w_{i,k}^{\tau} \in [0, 1]$, $i = 1, \dots, n$, $k = 1, \dots, N$, weights of rules premises,

- $w_{k}^{\mathrm{agr}} \in [0, 1]$, $k = 1, \dots, N$, weights of rules,

- $p_{u,i,k}^{A}$, $u = 1, 2, \dots, P^{A}$, $i = 1, 2, \dots, n$, parameters of the shape of membership function of input fuzzy sets,

- $p_{1,k}^{B} = \overline{y}^{k}$, $k = 1, 2, \dots, N$, centers of membership functions of output fuzzy sets,

- $p_{u,k}^{B}$, $u = 2, 3, \dots, P^{B}$, $k = 1, 2, \dots, N$, parameters of the shape of membership functions of output fuzzy sets.

The above mentioned parameters will be subject to learning in the following subchapter.

10.6   Learning algorithms

Now we will derive gradient learning algorithms of the system described by means of dependencies (10.46) and (10.49) - (10.51). Those parameters are modified by iteration according to the dependencies below:

$$\nu(t + 1) = \nu(t) - \eta \Delta \nu(t), \quad (10.52)$$

$$\alpha^{\tau}(t + 1) = \alpha^{\tau}(t) - \eta \Delta \alpha^{\tau}(t) \quad (10.53)$$

$$\alpha^I \left(t+1 \right) = \alpha^I \left(t \right) - \eta \Delta \alpha^I \left(t \right), \tag{10.54}$$

$$\alpha^{\mathrm{agr}} \left(t+1 \right) = \alpha^{\mathrm{agr}} \left(t \right) - \eta \Delta \alpha^{\mathrm{agr}} \left(t \right) \tag{10.55}$$

$$p^\tau \left(t+1 \right) = p^\tau \left(t \right) - \eta \Delta p^\tau \left(t \right), \tag{10.56}$$

$$p^I \left(t+1 \right) = p^I \left(t \right) - \eta \Delta p^I \left(t \right), \tag{10.57}$$

$$p^{\mathrm{agr}} \left(t+1 \right) = p^{\mathrm{agr}} \left(t \right) - \eta \Delta p^{\mathrm{agr}} \left(t \right), \tag{10.58}$$

$$w^\tau_{i,k} \left(t+1 \right) = w^\tau_{i,k} \left(t \right) - \eta \Delta w^\tau_{i,k} \left(t \right), \tag{10.59}$$

$$w^{\mathrm{agr}}_k \left(t+1 \right) = w^{\mathrm{agr}}_k \left(t \right) - \eta \Delta w^{\mathrm{agr}}_k \left(t \right), \tag{10.60}$$

$$p^A_{u,i,k} \left(t+1 \right) = p^A_{u,i,k} \left(t \right) - \eta \Delta p^A_{u,i,k} \left(t \right), \tag{10.61}$$

$$p^B_{u,k} \left(t+1 \right) = p^B_{u,k} \left(t \right) - \eta \Delta p^B_{u,k} \left(t \right); \quad u = 2, \ldots, P^B, \tag{10.62}$$

$$\overline{y}^r \left(t+1 \right) = p^B_{1,r} \left(t+1 \right) = \overline{y}^r \left(t \right) - \eta \Delta \overline{y}^r \left(t \right). \tag{10.63}$$

The terms Δ in the above dependencies are defined as follows:

$$\Delta \nu = \sum_{k=1}^{N} \sum_{r=1}^{N} \varepsilon^I_{k,r} \left\{ \nu \right\} + \sum_{r=1}^{N} \varepsilon^{\mathrm{agr}}_r \left\{ \nu \right\}, \tag{10.64}$$

$$\Delta \alpha^\tau = \sum_{k=1}^{N} \varepsilon^\tau_k \left\{ \alpha^\tau \right\}, \tag{10.65}$$

$$\Delta \alpha^I = \sum_{k=1}^{N} \sum_{r=1}^{N} \varepsilon^I_{k,r} \left\{ \alpha^I \right\}, \tag{10.66}$$

$$\Delta \alpha^{\mathrm{agr}} = \sum_{r=1}^{N} \varepsilon^{\mathrm{agr}}_r \left\{ \alpha^{\mathrm{agr}} \right\}, \tag{10.67}$$

$$\Delta p^\tau = \sum_{k=1}^{N} \varepsilon^\tau_k \left\{ p^\tau \right\}, \tag{10.68}$$

$$\Delta p^I = \sum_{k=1}^{N} \sum_{r=1}^{N} \varepsilon^I_{k,r} \left\{ p^I \right\}, \tag{10.69}$$

$$\Delta p^{\mathrm{agr}} = \sum_{r=1}^{N} \varepsilon^{\mathrm{agr}}_r \left\{ p^{\mathrm{agr}} \right\}, \tag{10.70}$$

$$\Delta w^\tau_{i,k} = \varepsilon^\tau_k \left\{ w^\tau_{i,k} \right\}, \tag{10.71}$$

$$\Delta w^{\mathrm{agr}}_k = \sum_{r=1}^{N} \varepsilon^{\mathrm{agr}}_r \left\{ w^{\mathrm{agr}}_k \right\}, \tag{10.72}$$

$$\Delta p_{u,i,k}^A = \varepsilon_k^\tau \left\{ p_{u,i,k}^A \right\}, \tag{10.73}$$

$$\Delta p_{u,k}^B = \sum_{r=1}^N \varepsilon_{k,r}^I \left\{ p_{u,k}^B \right\}; \quad u = 2, \ldots, P^B, \tag{10.74}$$

$$\Delta \overline{y}^r = \Delta p_{1,r}^B = \varepsilon^{\mathrm{def}} \left\{ \overline{y}^r \right\} + \sum_{k=1}^N \varepsilon_{k,r}^I \left\{ \overline{y}^r \right\} + \sum_{k=1}^N \varepsilon_{r,k}^I \left\{ p_{1,r}^B \right\}. \tag{10.75}$$

The errors propagated through particular system layers are defined similarly to the learning algorithms related to non-flexible systems which have been described in point 9.6. The method of error propagation is illustrated in Fig. 9.11.

The errors propagated by blocks of rules activation are defined as follows (Fig. 10.15):

$$\varepsilon_k^\tau \{\alpha^\tau\} = \varepsilon_k^\tau \frac{\partial \tau_k(\overline{\mathbf{x}})}{\partial \alpha^\tau}, \tag{10.76}$$

$$\varepsilon_k^\tau \{p^\tau\} = \varepsilon_k^\tau \frac{\partial \tau_k(\overline{\mathbf{x}})}{\partial b_k^\tau(\overline{\mathbf{x}})} \frac{\partial b_k^\tau(\overline{\mathbf{x}})}{\partial p^\tau}, \tag{10.77}$$

$$\varepsilon_k^\tau \{w_{i,k}^\tau\} = \varepsilon_k^\tau \frac{\partial \tau_k(\overline{\mathbf{x}})}{\partial b_k^\tau(\overline{\mathbf{x}})} \frac{\partial b_k^\tau(\overline{\mathbf{x}})}{\partial w_{i,k}^\tau}, \tag{10.78}$$

$$\varepsilon_k^\tau \left\{ p_{u,i,k}^A \right\} = \varepsilon_k^\tau \left(\begin{array}{c} \dfrac{\partial \tau_k(\overline{\mathbf{x}})}{\partial b_k^\tau(\overline{\mathbf{x}})} \dfrac{\partial b_k^\tau(\overline{\mathbf{x}})}{\partial \mu_{A_i^k}(\overline{x}_i)} + \\[2mm] + \dfrac{\partial \tau_k(\overline{\mathbf{x}})}{\partial a_k^\tau(\overline{\mathbf{x}})} \dfrac{\partial a_k^\tau(\overline{\mathbf{x}})}{\partial \mu_{A_i^k}(\overline{x}_i)} \end{array} \right) \frac{\partial \mu_{A_i^k}(\overline{x}_i)}{\partial p_{u,i,k}^A}, \tag{10.79}$$

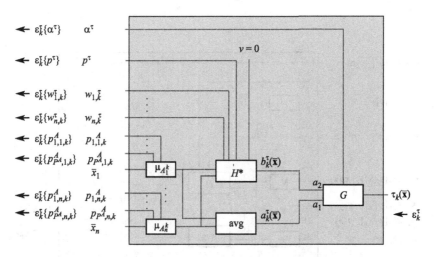

FIGURE 10.15. Block of rules activation of a flexible system

while

$$\frac{\partial \tau_k\left(\overline{\mathbf{x}}\right)}{\partial a_k^\tau\left(\overline{\mathbf{x}}\right)} = \frac{\partial}{\partial a_k^\tau\left(\overline{\mathbf{x}}\right)} G\left(\begin{array}{c} a_k^\tau\left(\overline{\mathbf{x}}\right), b_k^\tau\left(\overline{\mathbf{x}}\right); \\ \alpha^\tau \end{array}\right),\tag{10.80}$$

$$\frac{\partial \tau_k\left(\overline{\mathbf{x}}\right)}{\partial b_k^\tau\left(\overline{\mathbf{x}}\right)} = \frac{\partial}{\partial b_k^\tau\left(\overline{\mathbf{x}}\right)} G\left(\begin{array}{c} a_k^\tau\left(\overline{\mathbf{x}}\right), b_k^\tau\left(\overline{\mathbf{x}}\right); \\ \alpha^\tau \end{array}\right),\tag{10.81}$$

$$\frac{\partial \tau_k\left(\overline{\mathbf{x}}\right)}{\partial \alpha^\tau} = \frac{\partial}{\partial \alpha^\tau} G\left(\begin{array}{c} a_k^\tau\left(\overline{\mathbf{x}}\right), b_k^\tau\left(\overline{\mathbf{x}}\right); \\ \alpha^\tau \end{array}\right),\tag{10.82}$$

$$\frac{\partial a_k^\tau\left(\overline{\mathbf{x}}\right)}{\partial \mu_{A_i^k}\left(\overline{x}_i\right)} = \frac{\partial}{\partial \mu_{A_i^k}\left(\overline{x}_i\right)} \operatorname{avg}\left(\mu_{A_1^k}\left(\overline{x}_1\right), \ldots, \mu_{A_n^k}\left(\overline{x}_n\right)\right),\tag{10.83}$$

$$\frac{\partial b_k^\tau\left(\overline{\mathbf{x}}\right)}{\partial p^\tau} = \frac{\partial}{\partial p^\tau} \overset{\leftrightarrow *}{H}\left(\begin{array}{c} \mu_{A_1^k}\left(\overline{x}_1\right), \ldots, \mu_{A_n^k}\left(\overline{x}_n\right); \\ w_{1,k}^\tau, \ldots, w_{n,k}^\tau, p^\tau, 0 \end{array}\right),\tag{10.84}$$

$$\frac{\partial b_k^\tau\left(\overline{\mathbf{x}}\right)}{\partial w_{i,k}^\tau} = \frac{\partial}{\partial w_{i,k}^\tau} \overset{\leftrightarrow *}{H}\left(\begin{array}{c} \mu_{A_1^k}\left(\overline{x}_1\right), \ldots, \mu_{A_n^k}\left(\overline{x}_n\right); \\ w_{1,k}^\tau, \ldots, w_{n,k}^\tau, p^\tau, 0 \end{array}\right),\tag{10.85}$$

$$\frac{\partial b_k^\tau\left(\overline{\mathbf{x}}\right)}{\partial \mu_{A_i^k}\left(\overline{x}_i\right)} = \frac{\partial}{\partial \mu_{A_i^k}\left(\overline{x}_i\right)} \overset{\leftrightarrow *}{H}\left(\begin{array}{c} \mu_{A_1^k}\left(\overline{x}_1\right), \ldots, \mu_{A_n^k}\left(\overline{x}_n\right); \\ w_{1,k}^\tau, \ldots, w_{n,k}^\tau, p^\tau, 0 \end{array}\right).\tag{10.86}$$

The derivatives in the above dependencies are determined with use of the formulas specified in the further part of this subchapter.

The errors propagated by blocks of implications are defined as follows (Fig. 10.16):

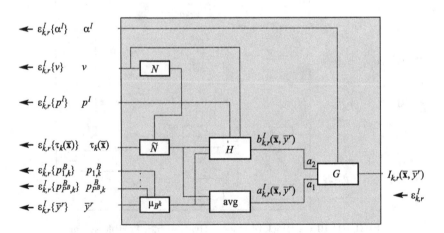

FIGURE 10.16. Block of implications of a flexible system

$$
\varepsilon_{k,r}^{I}\left\{\nu\right\} = \varepsilon_{k,r}^{I}\left(
\begin{array}{c}
\dfrac{\partial I_{k,r}\left(\overline{\mathbf{x}},\overline{y}^{r}\right)}{\partial b_{k,r}^{I}\left(\overline{\mathbf{x}},\overline{y}^{r}\right)}\dfrac{\partial b_{k,r}^{I}\left(\overline{\mathbf{x}},\overline{y}^{r}\right)}{\partial \nu} + \\[2mm]
+\left(
\begin{array}{c}
\dfrac{\partial I_{k,r}\left(\overline{\mathbf{x}},\overline{y}^{r}\right)}{\partial b_{k,r}^{I}\left(\overline{\mathbf{x}},\overline{y}^{r}\right)}\dfrac{\partial b_{k,r}^{I}\left(\overline{\mathbf{x}},\overline{y}^{r}\right)}{\partial \widetilde{N}_{1-\nu}\left(\tau_{k}\left(\overline{\mathbf{x}}\right)\right)} + \\[2mm]
+\dfrac{\partial I_{k,r}\left(\overline{\mathbf{x}},\overline{y}^{r}\right)}{\partial a_{k,r}^{I}\left(\overline{\mathbf{x}},\overline{y}^{r}\right)}\dfrac{\partial a_{k,r}^{I}\left(\overline{\mathbf{x}},\overline{y}^{r}\right)}{\partial \widetilde{N}_{1-\nu}\left(\tau_{k}\left(\overline{\mathbf{x}}\right)\right)}
\end{array}
\right)\cdot \\[2mm]
\cdot\dfrac{\partial \widetilde{N}_{1-\nu}\left(\tau_{k}\left(\overline{\mathbf{x}}\right)\right)}{\partial\left(1-\nu\right)}\dfrac{\partial N\left(\nu\right)}{\partial \nu}
\end{array}
\right), \tag{10.87}
$$

$$
\varepsilon_{k,r}^{I}\left\{\alpha^{I}\right\} = \varepsilon_{k,r}^{I}\frac{\partial I_{k,r}\left(\overline{\mathbf{x}},\overline{y}^{r}\right)}{\partial \alpha^{I}}, \tag{10.88}
$$

$$
\varepsilon_{k,r}^{I}\left\{p^{I}\right\} = \varepsilon_{k,r}^{I}\frac{\partial I_{k,r}\left(\overline{\mathbf{x}},\overline{y}^{r}\right)}{\partial b_{k,r}^{I}\left(\overline{\mathbf{x}},\overline{y}^{r}\right)}\frac{\partial b_{k,r}^{I}\left(\overline{\mathbf{x}},\overline{y}^{r}\right)}{\partial p^{I}}, \tag{10.89}
$$

$$
\varepsilon_{k,r}^{I}\left\{p_{u,k}^{B}\right\} = \varepsilon_{k,r}^{I}\left(
\begin{array}{c}
\dfrac{\partial I_{k,r}\left(\overline{\mathbf{x}},\overline{y}^{r}\right)}{\partial b_{k,r}^{I}\left(\overline{\mathbf{x}},\overline{y}^{r}\right)}\dfrac{\partial b_{k,r}^{I}\left(\overline{\mathbf{x}},\overline{y}^{r}\right)}{\partial \mu_{B^{k}}\left(\overline{y}^{r}\right)} + \\[2mm]
+\dfrac{\partial I_{k,r}\left(\overline{\mathbf{x}},\overline{y}^{r}\right)}{\partial a_{k,r}^{I}\left(\overline{\mathbf{x}},\overline{y}^{r}\right)}\dfrac{\partial a_{k,r}^{I}\left(\overline{\mathbf{x}},\overline{y}^{r}\right)}{\partial \mu_{B^{k}}\left(\overline{y}^{r}\right)}
\end{array}
\right)\frac{\partial \mu_{B^{k}}\left(\overline{y}^{r}\right)}{\partial p_{u,k}^{B}}, \tag{10.90}
$$

$$
\varepsilon_{k,r}^{I}\left\{\overline{y}^{r}\right\} = \varepsilon_{k,r}^{I}\left(
\begin{array}{c}
\dfrac{\partial I_{k,r}\left(\overline{\mathbf{x}},\overline{y}^{r}\right)}{\partial b_{k,r}^{I}\left(\overline{\mathbf{x}},\overline{y}^{r}\right)}\dfrac{\partial b_{k,r}^{I}\left(\overline{\mathbf{x}},\overline{y}^{r}\right)}{\partial \mu_{B^{k}}\left(\overline{y}^{r}\right)} + \\[2mm]
+\dfrac{\partial I_{k,r}\left(\overline{\mathbf{x}},\overline{y}^{r}\right)}{\partial a_{k,r}^{I}\left(\overline{\mathbf{x}},\overline{y}^{r}\right)}\dfrac{\partial a_{k,r}^{I}\left(\overline{\mathbf{x}},\overline{y}^{r}\right)}{\partial \mu_{B^{k}}\left(\overline{y}^{r}\right)}
\end{array}
\right)\frac{\partial \mu_{B^{k}}\left(\overline{y}^{r}\right)}{\partial \overline{y}^{r}}, \tag{10.91}
$$

$$
\varepsilon_{k,r}^{I}\left\{\tau_{k}(\overline{\mathbf{x}})\right\} = \varepsilon_{k,r}^{I}\left(
\begin{array}{c}
\dfrac{\partial I_{k,r}(\overline{\mathbf{x}},\overline{y}^{r})}{\partial b_{k,r}^{I}\left(\overline{\mathbf{x}},\overline{y}^{r}\right)}\dfrac{\partial b_{k,r}^{I}(\overline{\mathbf{x}},\overline{y}^{r})}{\partial \widetilde{N}_{1-\nu}\left(\tau_{k}\left(\overline{\mathbf{x}}\right)\right)} + \\[2mm]
+\dfrac{\partial I_{k,r}(\overline{\mathbf{x}},\overline{y}^{r})}{\partial a_{k,r}^{I}(\overline{\mathbf{x}},\overline{y}^{r})}\dfrac{\partial a_{k,r}^{I}(\overline{\mathbf{x}},\overline{y}^{r})}{\partial \widetilde{N}_{1-\nu}(\tau_{k}\left(\overline{\mathbf{x}}\right))}
\end{array}
\right)\frac{\partial \widetilde{N}_{1-\nu}(\tau_{k}\left(\overline{\mathbf{x}}\right))}{\partial \tau_{k}(\overline{\mathbf{x}})}, \tag{10.92}
$$

while

$$
\frac{\partial I_{k,r}\left(\overline{\mathbf{x}},\overline{y}^{r}\right)}{\partial a_{k,r}^{I}\left(\overline{\mathbf{x}},\overline{y}^{r}\right)} = \frac{\partial}{\partial a_{k,r}^{I}\left(\overline{\mathbf{x}},\overline{y}^{r}\right)}G\left(\begin{array}{c} a_{k,r}^{I}\left(\overline{\mathbf{x}},\overline{y}^{r}\right),b_{k,r}^{I}\left(\overline{\mathbf{x}},\overline{y}^{r}\right); \\ \alpha^{I} \end{array}\right), \tag{10.93}
$$

$$
\frac{\partial I_{k,r}\left(\overline{\mathbf{x}},\overline{y}^{r}\right)}{\partial b_{k,r}^{I}\left(\overline{\mathbf{x}},\overline{y}^{r}\right)} = \frac{\partial}{\partial b_{k,r}^{I}\left(\overline{\mathbf{x}},\overline{y}^{r}\right)}G\left(\begin{array}{c} a_{k,r}^{I}\left(\overline{\mathbf{x}},\overline{y}^{r}\right),b_{k,r}^{I}\left(\overline{\mathbf{x}},\overline{y}^{r}\right); \\ \alpha^{I} \end{array}\right), \tag{10.94}
$$

$$\frac{\partial I_{k,r}\left(\overline{\mathbf{x}},\overline{y}^{r}\right)}{\partial\alpha^{I}}=\frac{\partial}{\partial\alpha^{I}}G\left(\begin{array}{c}a^{I}_{k,r}\left(\overline{\mathbf{x}},\overline{y}^{r}\right),b^{I}_{k,r}\left(\overline{\mathbf{x}},\overline{y}^{r}\right);\\\alpha^{I}\end{array}\right),\qquad(10.95)$$

$$\frac{\partial a^{I}_{k,r}\left(\overline{\mathbf{x}},\overline{y}^{r}\right)}{\partial\widetilde{N}_{1-\nu}\left(\tau_{k}\left(\overline{\mathbf{x}}\right)\right)}=\frac{\partial}{\partial\widetilde{N}_{1-\nu}\left(\tau_{k}\left(\overline{\mathbf{x}}\right)\right)}\operatorname{avg}\left(\widetilde{N}_{1-\nu}\left(\tau_{k}\left(\overline{\mathbf{x}}\right)\right),\mu_{B^{k}}\left(\overline{y}^{r}\right)\right),\quad(10.96)$$

$$\frac{\partial a^{I}_{k,r}\left(\overline{\mathbf{x}},\overline{y}^{r}\right)}{\partial\mu_{B^{k}}\left(\overline{y}^{r}\right)}=\frac{\partial}{\partial\mu_{B^{k}}\left(\overline{y}^{r}\right)}\operatorname{avg}\left(\widetilde{N}_{1-\nu}\left(\tau_{k}\left(\overline{\mathbf{x}}\right)\right),\mu_{B^{k}}\left(\overline{y}^{r}\right)\right),\qquad(10.97)$$

$$\frac{\partial b^{I}_{k,r}\left(\overline{\mathbf{x}},\overline{y}^{r}\right)}{\partial\nu}=\frac{\partial}{\partial\nu}\overset{\leftrightarrow}{H}\left(\begin{array}{c}\widetilde{N}_{1-\nu}\left(\tau_{k}\left(\overline{\mathbf{x}}\right)\right),\mu_{B^{k}}\left(\overline{y}^{r}\right);\\p^{I},\nu\end{array}\right),\qquad(10.98)$$

$$\frac{\partial b^{I}_{k,r}\left(\overline{\mathbf{x}},\overline{y}^{r}\right)}{\partial p^{I}}=\frac{\partial}{\partial p^{I}}\overset{\leftrightarrow}{H}\left(\begin{array}{c}\widetilde{N}_{1-\nu}\left(\tau_{k}\left(\overline{\mathbf{x}}\right)\right),\mu_{B^{k}}\left(\overline{y}^{r}\right);\\p^{I},\nu\end{array}\right),\qquad(10.99)$$

$$\frac{\partial b^{I}_{k,r}\left(\overline{\mathbf{x}},\overline{y}^{r}\right)}{\partial\widetilde{N}_{1-\nu}\left(\tau_{k}\left(\overline{\mathbf{x}}\right)\right)}=\frac{\partial}{\partial\widetilde{N}_{1-\nu}\left(\tau_{k}\left(\overline{\mathbf{x}}\right)\right)}\overset{\leftrightarrow}{H}\left(\begin{array}{c}\widetilde{N}_{1-\nu}\left(\tau_{k}\left(\overline{\mathbf{x}}\right)\right),\mu_{B^{k}}\left(\overline{y}^{r}\right);\\p^{I},\nu\end{array}\right),\quad(10.100)$$

$$\frac{\partial b^{I}_{k,r}\left(\overline{\mathbf{x}},\overline{y}^{r}\right)}{\partial\mu_{B^{k}}\left(\overline{y}^{r}\right)}=\frac{\partial}{\partial\mu_{B^{k}}\left(\overline{y}^{r}\right)}\overset{\leftrightarrow}{H}\left(\begin{array}{c}\widetilde{N}_{1-\nu}\left(\tau_{k}\left(\overline{\mathbf{x}}\right)\right),\mu_{B^{k}}\left(\overline{y}^{r}\right);\\p^{I},\nu\end{array}\right).\qquad(10.101)$$

The derivatives in the above dependencies are determined with use of the formulas specified in the further part of this subchapter.

The errors propagated by blocks of aggregation are defined as follows (Fig. 10.17):

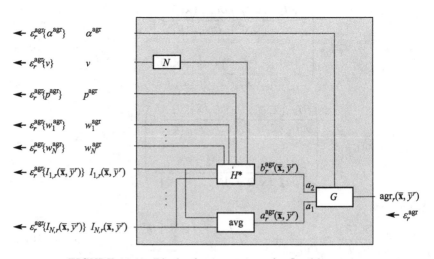

FIGURE 10.17. Block of aggregation of a flexible system

$$\varepsilon_r^{\mathrm{agr}}\{\nu\} = \varepsilon_r^{\mathrm{agr}} \frac{\partial \mathrm{agr}_r(\overline{\mathbf{x}}, \overline{y}^r)}{\partial b_r^{\mathrm{agr}}(\overline{\mathbf{x}}, \overline{y}^r)} \frac{\partial b_r^{\mathrm{agr}}(\overline{\mathbf{x}}, \overline{y}^r)}{\partial (1-\nu)} \frac{\partial N(\nu)}{\partial \nu}, \qquad (10.102)$$

$$\varepsilon_r^{\mathrm{agr}}\{\alpha^{\mathrm{agr}}\} = \varepsilon_r^{\mathrm{agr}} \frac{\partial \mathrm{agr}_r(\overline{\mathbf{x}}, \overline{y}^r)}{\partial \alpha^{\mathrm{agr}}}, \qquad (10.103)$$

$$\varepsilon_r^{\mathrm{agr}}\{p^{\mathrm{agr}}\} = \varepsilon_r^{\mathrm{agr}} \frac{\partial \mathrm{agr}_r(\overline{\mathbf{x}}, \overline{y}^r)}{\partial b_r^{\mathrm{agr}}(\overline{\mathbf{x}}, \overline{y}^r)} \frac{\partial b_r^{\mathrm{agr}}(\overline{\mathbf{x}}, \overline{y}^r)}{\partial p^{\mathrm{agr}}}, \qquad (10.104)$$

$$\varepsilon_r^{\mathrm{agr}}\{w_k^{\mathrm{agr}}\} = \varepsilon_r^{\mathrm{agr}} \frac{\partial \mathrm{agr}_r(\overline{\mathbf{x}}, \overline{y}^r)}{\partial b_r^{\mathrm{agr}}(\overline{\mathbf{x}}, \overline{y}^r)} \frac{\partial b_r^{\mathrm{agr}}(\overline{\mathbf{x}}, \overline{y}^r)}{\partial w_k^{\mathrm{agr}}}, \qquad (10.105)$$

$$\varepsilon_r^{\mathrm{agr}}\{I_{k,r}(\overline{\mathbf{x}}, \overline{y}^r)\} = \varepsilon_r^{\mathrm{agr}} \left(\begin{array}{c} \dfrac{\partial \mathrm{agr}_r(\overline{\mathbf{x}}, \overline{y}^r)}{\partial b_r^{\mathrm{agr}}(\overline{\mathbf{x}}, \overline{y}^r)} \dfrac{\partial b_r^{\mathrm{agr}}(\overline{\mathbf{x}}, \overline{y}^r)}{\partial I_{k,r}(\overline{\mathbf{x}}, \overline{y}^r)} + \\[2ex] + \dfrac{\partial \mathrm{agr}_r(\overline{\mathbf{x}}, \overline{y}^r)}{\partial a_r^{\mathrm{agr}}(\overline{\mathbf{x}}, \overline{y}^r)} \dfrac{\partial a_r^{\mathrm{agr}}(\overline{\mathbf{x}}, \overline{y}^r)}{\partial I_{k,r}(\overline{\mathbf{x}}, \overline{y}^r)} \end{array} \right), \qquad (10.106)$$

while

$$\frac{\partial \mathrm{agr}_r(\overline{\mathbf{x}}, \overline{y}^r)}{\partial a_r^{\mathrm{agr}}(\overline{\mathbf{x}}, \overline{y}^r)} = \frac{\partial}{\partial a_r^{\mathrm{agr}}(\overline{\mathbf{x}}, \overline{y}^r)} G \left(\begin{array}{c} a_r^{\mathrm{agr}}(\overline{\mathbf{x}}, \overline{y}^r), b_r^{\mathrm{agr}}(\overline{\mathbf{x}}, \overline{y}^r); \\ \alpha^{\mathrm{agr}} \end{array} \right), \qquad (10.107)$$

$$\frac{\partial \mathrm{agr}_r(\overline{\mathbf{x}}, \overline{y}^r)}{\partial b_r^{\mathrm{agr}}(\overline{\mathbf{x}}, \overline{y}^r)} = \frac{\partial}{\partial b_r^{\mathrm{agr}}(\overline{\mathbf{x}}, \overline{y}^r)} G \left(\begin{array}{c} a_r^{\mathrm{agr}}(\overline{\mathbf{x}}, \overline{y}^r), b_r^{\mathrm{agr}}(\overline{\mathbf{x}}, \overline{y}^r); \\ \alpha^{\mathrm{agr}} \end{array} \right), \qquad (10.108)$$

$$\frac{\partial \mathrm{agr}_r(\overline{\mathbf{x}}, \overline{y}^r)}{\partial \alpha^{\mathrm{agr}}} = \frac{\partial}{\partial \alpha^{\mathrm{agr}}} G \left(\begin{array}{c} a_r^{\mathrm{agr}}(\overline{\mathbf{x}}, \overline{y}^r), b_r^{\mathrm{agr}}(\overline{\mathbf{x}}, \overline{y}^r); \\ \alpha^{\mathrm{agr}} \end{array} \right), \qquad (10.109)$$

$$\frac{\partial a_r^{\mathrm{agr}}(\overline{\mathbf{x}}, \overline{y}^r)}{\partial I_{k,r}(\overline{\mathbf{x}}, \overline{y}^r)} = \frac{\partial}{\partial I_{k,r}(\overline{\mathbf{x}}, \overline{y}^r)} \mathrm{avg}(I_{1,r}(\overline{\mathbf{x}}, \overline{y}^r), \dots, I_{N,r}(\overline{\mathbf{x}}, \overline{y}^r)), \qquad (10.110)$$

$$\frac{\partial b_r^{\mathrm{agr}}(\overline{\mathbf{x}}, \overline{y}^r)}{\partial (1-\nu)} = \frac{\partial}{\partial (1-\nu)} \overset{\leftrightarrow}{H}{}^* \left(\begin{array}{c} I_{1,r}(\overline{\mathbf{x}}, \overline{y}^r), \dots, I_{N,r}(\overline{\mathbf{x}}, \overline{y}^r); \\ w_1^{\mathrm{agr}}, \dots, w_N^{\mathrm{agr}}, p^{\mathrm{agr}}, 1-\nu \end{array} \right), \qquad (10.111)$$

$$\frac{\partial b_r^{\mathrm{agr}}(\overline{\mathbf{x}}, \overline{y}^r)}{\partial p^{\mathrm{agr}}} = \frac{\partial}{\partial p^{\mathrm{agr}}} \overset{\leftrightarrow}{H}{}^* \left(\begin{array}{c} I_{1,r}(\overline{\mathbf{x}}, \overline{y}^r), \dots, I_{N,r}(\overline{\mathbf{x}}, \overline{y}^r); \\ w_1^{\mathrm{agr}}, \dots, w_N^{\mathrm{agr}}, p^{\mathrm{agr}}, 1-\nu \end{array} \right), \qquad (10.112)$$

$$\frac{\partial b_r^{\mathrm{agr}}(\overline{\mathbf{x}}, \overline{y}^r)}{\partial w_k^{\mathrm{agr}}} = \frac{\partial}{\partial w_k^{\mathrm{agr}}} \overset{\leftrightarrow}{H}{}^* \left(\begin{array}{c} I_{1,r}(\overline{\mathbf{x}}, \overline{y}^r), \dots, I_{N,r}(\overline{\mathbf{x}}, \overline{y}^r); \\ w_1^{\mathrm{agr}}, \dots, w_N^{\mathrm{agr}}, p^{\mathrm{agr}}, 1-\nu \end{array} \right), \qquad (10.113)$$

$$\frac{\partial b_r^{\mathrm{agr}}(\overline{\mathbf{x}}, \overline{y}^r)}{\partial I_{k,r}(\overline{\mathbf{x}}, \overline{y}^r)} = \frac{\partial}{\partial I_{k,r}(\overline{\mathbf{x}}, \overline{y}^r)} \overset{\leftrightarrow}{H}{}^* \left(\begin{array}{c} I_{1,r}(\overline{\mathbf{x}}, \overline{y}^r), \dots, I_{N,r}(\overline{\mathbf{x}}, \overline{y}^r); \\ w_1^{\mathrm{agr}}, \dots, w_N^{\mathrm{agr}}, p^{\mathrm{agr}}, 1-\nu \end{array} \right). \qquad (10.114)$$

The errors propagated by defuzzification block are defined similarly to the learning algorithms related to non-flexible systems which have been described in Subchapters 9.3 - 9.5.

The learning algorithms derived above of a flexible neuro-fuzzy system require determining derivatives for different types of operators. Below a computation method of those derivatives is presented.

10.6.1   Basic operators

Summation operator

$$y = \sum_{i=1}^{n} x_i \qquad (10.115)$$

$$\frac{\partial y}{\partial x_i} = 1 \qquad (10.116)$$

Multiplication operator

$$y = \prod_{i=1}^{n} x_i \qquad (10.117)$$

$$\frac{\partial y}{\partial x_i} = \prod_{\substack{j=1 \\ j \neq i}}^{n} x_j \qquad (10.118)$$

Division operator

$$y = \frac{a}{b} \qquad (10.119)$$

$$\frac{\partial y}{\partial a} = \frac{1}{b} \qquad (10.120)$$

$$\frac{\partial y}{\partial b} = -\frac{a}{b^2} \qquad (10.121)$$

Minimum operator

$$y = \min_{i=1...n} \{x_i\} \qquad (10.122)$$

$$\frac{\partial y}{\partial x_i} = \begin{cases} 1 & \text{for} \quad x_i = y \\ 0 & \text{for} \quad x_i \neq y \end{cases} \qquad (10.123)$$

Maximum operator

$$y = \max_{i=1...n} \{x_i\} \qquad (10.124)$$

$$\frac{\partial y}{\partial x_i} = \begin{cases} 1 & \text{for} \quad x_i = y \\ 0 & \text{for} \quad x_i \neq y \end{cases} \qquad (10.125)$$

Compromise operator

$$\widetilde{N}_\nu(a) = (1 - f_z(\nu))(1 - a) + f_z(\nu) a \qquad (10.126)$$

$$\frac{\partial \widetilde{N}_\nu(a)}{\partial a} = 2f_z(\nu) - 1 \qquad (10.127)$$

$$\frac{\partial \widetilde{N}_\nu (a)}{\partial \nu} = (2a - 1) \frac{\partial f_z (\nu)}{\partial \nu} \qquad (10.128)$$

Arithmetic average operator

$$\text{avg}(a_1, a_2, ..., a_n) = \frac{1}{n} \sum_{i=1}^{n} a_i \qquad (10.129)$$

$$\frac{\partial \text{avg}(a_1, a_2, \ldots, a_n)}{\partial a_i} = \frac{1}{n} \qquad (10.130)$$

Aggregation operator

$$G(a_1, a_2; \phi) = (1 - f_z(\phi)) a_1 + f_z(\phi) a_2 \qquad (10.131)$$

$$\frac{\partial G(a_1, a_2; \phi)}{\partial a_1} = 1 - f_z(\phi) \qquad (10.132)$$

$$\frac{\partial G(a_1, a_2; \phi)}{\partial a_2} = f_z(\phi) \qquad (10.133)$$

$$\frac{\partial G(a_1, a_2; \phi)}{\partial \phi} = -(a_1 - a_2) \frac{\partial f_z(\phi)}{\partial \phi} \qquad (10.134)$$

Defuzzification operator

$$\text{def}(a_1, a_2, \ldots, a_n; w_1, w_2, \ldots, w_n) = \text{def}(\mathbf{a}; \mathbf{w}) = \frac{\sum_{i=1}^{n} w_i a_i}{\sum_{i=1}^{n} a_i} \qquad (10.135)$$

$$\frac{\partial \text{def}(\mathbf{a}; \mathbf{w})}{\partial a_j} = (w_j - \text{def}(\mathbf{a}; \mathbf{w})) \frac{1}{\sum_{i=1}^{n} a_i} \qquad (10.136)$$

$$\frac{\partial \text{def}(\mathbf{a}; \mathbf{w})}{\partial w_j} = \left(a_j - \text{def}(\mathbf{a}; \mathbf{w}) \frac{\partial a_j}{\partial w_j} \right) \frac{1}{\sum_{i=1}^{n} a_i} \qquad (10.137)$$

10.6.2 Membership functions

Gaussian membership function

$$\mu_A(x) = \exp\left(-\left(\frac{x - \overline{x}}{\sigma} \right)^2 \right) \qquad (10.138)$$

$$\frac{\partial \mu_A(x)}{\partial x} = -\mu_A(x) \frac{2(x - \overline{x})}{\sigma^2} \qquad (10.139)$$

$$\frac{\partial \mu_A(x)}{\partial \overline{x}} = \mu_A(x)\frac{2(x-\overline{x})}{\sigma^2} \qquad (10.140)$$

$$\frac{\partial \mu_A(x)}{\partial \sigma} = \mu_A(x)\frac{2(x-\overline{x})^2}{\sigma^3} \qquad (10.141)$$

Triangular membership function

$$\mu_A(x) = \begin{cases} 0 & \text{for} \quad x \le a \text{ or } x \ge c \\ \dfrac{x-a}{b-a} & \text{for} \quad a \le x \le b \\ \dfrac{c-x}{c-b} & \text{for} \quad b \le x \le c \end{cases} \qquad (10.142)$$

$$\frac{\partial \mu_A(x)}{\partial x} = \begin{cases} 0 & \text{for} \quad x < a \text{ or } x > c \\ \dfrac{1}{2(b-a)} & \text{for} \quad x = a \\ \dfrac{1}{b-a} & \text{for} \quad a < x < b \\ \dfrac{c-2b+a}{2(c-b)(b-a)} & \text{for} \quad x = b \\ -\dfrac{1}{c-b} & \text{for} \quad b < x < c \\ -\dfrac{1}{2(c-b)} & \text{for} \quad x = c \end{cases} \qquad (10.143)$$

$$\frac{\partial \mu_A(x)}{\partial a} = \begin{cases} 0 & \text{for} \quad x \le a \text{ or } x > b \\ \dfrac{1}{2(b-a)} & \text{for} \quad x = b \\ \dfrac{x-a}{(b-a)^2} & \text{for} \quad a \le x < b \end{cases} \qquad (10.144)$$

$$\frac{\partial \mu_A(x)}{\partial b} = \begin{cases} 0 & \text{for} \quad x \le a \text{ or } x \ge c \\ \dfrac{a-x}{(b-a)^2} & \text{for} \quad a \le x < b \\ \dfrac{a-2b+c}{2(c-b)(b-a)} & \text{for} \quad x = b \\ \dfrac{c-x}{(c-b)^2} & \text{for} \quad b < x \le c \end{cases} \qquad (10.145)$$

$$\frac{\partial \mu_A(x)}{\partial c} = \begin{cases} 0 & \text{for} \quad x \le b \text{ or } x > c \\ \dfrac{1}{2(c-b)} & \text{for} \quad x = c \\ \dfrac{x-b}{(c-b)^2} & \text{for} \quad b < x < c \end{cases} \qquad (10.146)$$

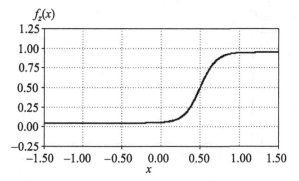

FIGURE 10.18. Plot of function (10.147) for $p_{z1} = 10$, $p_{z2} = 5$, $p_{z3} = 0.9$, $p_{z4} = 0.05$

10.6.3   Constraints

Constraints for parameters $\nu \in [0, 1]$, $\lambda \in [0, 1]$, $\alpha^\tau \in [0, 1]$, $\alpha^I \in [0, 1]$, $\alpha^{\mathrm{agr}} \in [0, 1]$, $w_{i,k}^\tau \in [0, 1]$, $i = 1, \ldots, n$, $k = 1, \ldots, N$, $w_k^{\mathrm{agr}} \in [0, 1]$, $k = 1, \ldots, N$

$$f_z(x) = \frac{p_{z3}}{1 + \exp(-(p_{z1}x - p_{z2}))} + p_{z4} \qquad (10.147)$$

$$\frac{\partial f_z(x)}{\partial x} = -\frac{p_{z1}}{p_{z3}}(p_{z3} + p_{z4} - f_z(x))(p_{z4} - f_z(x)) \qquad (10.148)$$

Constraints for parameters $p^\tau \in [0, \infty)$, $p^I \in [0, \infty)$, $p^{\mathrm{agr}} \in [0, \infty)$

$$f_z(x) = \frac{x}{1 + \exp(-(p_{z1}x - p_{z2}))} + p_{z3} \qquad (10.149)$$

$$\frac{\partial f_z(x)}{\partial x} = \frac{-p_{z3} + f_z(x)}{x}(1 + p_{z1}(p_{z3} + x - f_z(x))) \qquad (10.150)$$

In Figures 10.18 and 10.19 we show plots of functions 10.147 and 10.149, respectively.

10.6.4   H-functions

Argument of H-functions

$$\arg_i(a_i, w_i, \nu) = G\left(\frac{N(f_z(w_i)N(a_i)), f_z(w_i)a_i;}{\nu}\right) \qquad (10.151)$$

$$\frac{\partial \arg_i(a_i, w_i, \nu)}{\partial a_i} = f_z(w_i) \qquad (10.152)$$

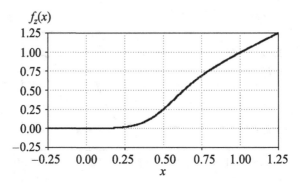

FIGURE 10.19. Plot of function (10.149) for $p_{z1} = 10$, $p_{z2} = 5$, $p_{z3} = 0$

$$\frac{\partial \arg_i (a_i, w_i, \nu)}{\partial w_i} = (a + \nu - 1) \frac{\partial f_z (w_i)}{\partial w_i} \tag{10.153}$$

$$\frac{\partial \arg_i (a_i, w_i, \nu)}{\partial \nu} = f_z (w_i) - 1 \tag{10.154}$$

Zadeh H-function

$$H^* (\mathbf{a}; \mathbf{w}, \nu) = \widetilde{N}_\nu \left(\max_{i=1,\dots,n} \left\{ \widetilde{N}_\nu (\arg_i (a_i, w_i, \nu)) \right\} \right) \tag{10.155}$$

$$H^* (\mathbf{a}; \mathbf{w}, \nu) = \widetilde{N}_\nu (h^* (\mathbf{a}; \mathbf{w}, \nu)) \tag{10.156}$$

where

$$h^* (\mathbf{a}; \mathbf{w}, \nu) = \max_{i=1,\dots,n} \left\{ \widetilde{N}_\nu (\arg_i (a_i, w_i, \nu)) \right\} \tag{10.157}$$

$$\frac{\partial H^* (\mathbf{a}; \mathbf{w}, \nu)}{\partial a_i} = \begin{cases} (2 f_z (\nu) - 1)^2 \cdot \dfrac{\partial \arg_i (a_i, w_i, \nu)}{\partial a_i} \\ \quad \text{for} \quad h^* (\mathbf{a}; \mathbf{w}, \nu) = \widetilde{N}_\nu (\arg_i (a_i, w_i, \nu)) \\ 0 \quad \text{for} \quad h^* (\mathbf{a}; \mathbf{w}, \nu) \neq \widetilde{N}_\nu (\arg_i (a_i, w_i, \nu)) \end{cases} \tag{10.158}$$

$$\frac{\partial H^* (\mathbf{a}; \mathbf{w}, \nu)}{\partial w_i} = \begin{cases} (2 f_z (\nu) - 1)^2 \cdot \dfrac{\partial \arg_i (a_i, w_i, \nu)}{\partial w_i} \\ \quad \text{for} \quad h^* (\mathbf{a}; \mathbf{w}, \nu) = \widetilde{N}_\nu (\arg_i (a_i, w_i, \nu)) \\ 0 \quad \text{for} \quad h^* (\mathbf{a}; \mathbf{w}, \nu) \neq \widetilde{N}_\nu (\arg_i (a_i, w_i, \nu)) \end{cases} \tag{10.159}$$

$$\frac{\partial H^*\left(\mathbf{a};\mathbf{w},\nu\right)}{\partial \nu} = \frac{\partial f_z\left(\nu\right)}{\partial \nu}\left(2h^*\left(\mathbf{a};\mathbf{w},\nu\right)-1\right)$$

$$+\left(2f_z\left(\nu\right)-1\right)\max_{i=1,\ldots,n}\left\{\begin{array}{l}\left(2f_z\left(\nu\right)-1\right)\dfrac{\partial \arg_i\left(a_i,w_i,\nu\right)}{\partial \nu}+ \\[2ex] +\left(2\arg_i\left(a_i,w_i,\nu\right)-1\right)\dfrac{\partial f_z\left(\nu\right)}{\partial \nu}\end{array}\right\} \qquad (10.160)$$

Algebraic H-function

$$H^*\left(\mathbf{a};\mathbf{w},\nu\right) = \widetilde{N}_\nu\left(1-\prod_{i=1}^{n}\left(1-\widetilde{N}_\nu\left(\arg_i\left(a_i,w_i,\nu\right)\right)\right)\right) \qquad (10.161)$$

$$H^*\left(\mathbf{a};\mathbf{w},\nu\right) = \widetilde{N}_\nu\left(h^*\left(\mathbf{a};\mathbf{w},\nu\right)\right) \qquad (10.162)$$

where

$$h^*\left(\mathbf{a};\mathbf{w},\nu\right) = 1-\prod_{i=1}^{n}\left(1-\widetilde{N}_\nu\left(\arg_i\left(a_i,w_i,\nu\right)\right)\right) \qquad (10.163)$$

$$\frac{\partial H^*\left(\mathbf{a};\mathbf{w},\nu\right)}{\partial a_i} = \left(2f_z\left(\nu\right)-1\right)^2\frac{\partial \arg_i\left(a_i,w_i,\nu\right)}{\partial a_i}$$
$$\cdot \prod_{\substack{u=1 \\ u\neq i}}^{n}\left(1-\widetilde{N}_\nu\left(\arg_u\left(a_u,w_u,\nu\right)\right)\right) \qquad (10.164)$$

$$\frac{\partial H^*\left(\mathbf{a};\mathbf{w},\nu\right)}{\partial w_i} = \left(2f_z\left(\nu\right)-1\right)^2\frac{\partial \arg_i\left(a_i,w_i,\nu\right)}{\partial w_i}\cdot$$
$$\cdot \prod_{\substack{u=1 \\ u\neq i}}^{n}\left(1-\widetilde{N}_\nu\left(\arg_u\left(a_u,w_u,\nu\right)\right)\right) \qquad (10.165)$$

$$\frac{\partial H^*\left(\mathbf{a};\mathbf{w},\nu\right)}{\partial \nu} = \left(2h^*\left(\mathbf{a};\mathbf{w},\nu\right)-1\right)\frac{\partial f_z\left(\nu\right)}{\partial \nu}+$$
$$+\left(2f_z\left(\nu\right)-1\right)\sum_{i=1}^{n}\left(\left(\begin{array}{l}\left(2\arg_i\left(a_i,w_i,\nu\right)-1\right)\dfrac{\partial f_z\left(\nu\right)}{\partial \nu}+ \\[2ex] +\left(2f_z\left(\nu\right)-1\right)\dfrac{\partial \arg_i\left(a_i,w_i,\nu\right)}{\partial \nu}\end{array}\right)\cdot\right.$$
$$\left.\cdot\prod_{\substack{u=1 \\ u\neq i}}^{n}\left(1-\widetilde{N}_\nu\left(\arg_u\left(a_u,w_u,\nu\right)\right)\right)\right) \qquad (10.166)$$

Dombi H-function

$$\overset{\leftrightarrow *}{H}(\mathbf{a};\mathbf{w},p,\nu)$$

$$=\tilde{N}_\nu\!\left(1-\left(1+\left(\sum_{i=1}^{n}\left(\tilde{N}_\nu\left(\arg_i\left(a_i,w_i,\nu\right)\right)^{-1}-1\right)^{-f_{z1}(p)}\right)^{\frac{1}{f_{z1}(p)}}\right)^{-1}\right) \tag{10.167}$$

$$\overset{\leftrightarrow *}{H}(\mathbf{a};\mathbf{w},p,\nu)=\tilde{N}_\nu\left(1-\overset{\leftrightarrow *}{h}(\mathbf{a};\mathbf{w},p,\nu)\right) \tag{10.168}$$

$$p\in(0,\infty) \tag{10.169}$$

where

$$\overset{\leftrightarrow *}{h}(\mathbf{a};\mathbf{w},p,\nu)$$

$$=\left(1+\left(\sum_{i=1}^{n}\left(\tilde{N}_\nu\left(\arg_i\left(a_i,w_i,\nu\right)\right)^{-1}-1\right)^{-f_{z1}(p)}\right)^{\frac{1}{f_{z1}(p)}}\right)^{-1} \tag{10.170}$$

$$\frac{\partial\overset{\leftrightarrow *}{H}(\mathbf{a};\mathbf{w},p,\nu)}{\partial a_i}=\left(2f_z\left(\nu\right)-1\right)^2\cdot$$

$$\cdot\frac{\left(\overset{\leftrightarrow *}{h}(\mathbf{a};\mathbf{w},p,\nu)^{-1}-1\right)^{1-f_{z1}(p)}}{\overset{\leftrightarrow *}{h}(\mathbf{a};\mathbf{w},p,\nu)^{-2}}\cdot$$

$$\tag{10.171}$$

$$\cdot\frac{\left(\tilde{N}_\nu\left(\arg_i\left(a_i,w_i,\nu\right)\right)^{-1}-1\right)^{-f_{z1}(p)-1}}{\tilde{N}_\nu\left(\arg_i\left(a_i,w_i,\nu\right)\right)^2}\cdot$$

$$\cdot\frac{\partial\arg_i\left(a_i,w_i,\nu\right)}{\partial a_i}$$

$$\frac{\partial\overset{\leftrightarrow *}{H}(\mathbf{a};\mathbf{w},p,\nu)}{\partial w_i}=\left(2f_z\left(\nu\right)-1\right)^2\cdot$$

$$\cdot\frac{\left(\overset{\leftrightarrow *}{h}(\mathbf{a};\mathbf{w},p,\nu)^{-1}-1\right)^{1-f_{z1}(p)}}{\overset{\leftrightarrow *}{h}(\mathbf{a};\mathbf{w},p,\nu)^{-2}}\cdot$$

$$\cdot\frac{\left(\tilde{N}_\nu\left(\arg_i\left(a_i,w_i,\nu\right)\right)^{-1}-1\right)^{-f_{z1}(p)-1}}{\tilde{N}_\nu\left(\arg_i\left(a_i,w_i,\nu\right)\right)^2}\cdot$$

$$\tag{10.172}$$

$$\cdot\frac{\partial\arg_i\left(a_i,w_i,\nu\right)}{\partial w_i}$$

$$\frac{\partial \overset{\leftrightarrow*}{H}(\mathbf{a};\mathbf{w},p,\nu)}{\partial p} = \frac{2f_z(\nu)-1}{f_{z1}(p)} \frac{\left(\overset{\leftrightarrow*}{h}(\mathbf{a};\mathbf{w},p,\nu)^{-1}-1\right)^{1-f_{z1}(p)}}{\overset{\leftrightarrow*}{h}(\mathbf{a};\mathbf{w},p,\nu)^{-2}} \cdot$$

$$\cdot \left(\sum_{i=1}^{n} \frac{-\ln\left(\tilde{N}_\nu(\arg_i(a_i,w_i,\nu))^{-1}-1\right)}{\left(\tilde{N}_\nu(\arg_i(a_i,w_i,\nu))^{-1}-1\right)^{f_{z1}(p)}} + \right.$$

$$\left. + \frac{\ln\left(\overset{\leftrightarrow*}{h}(\mathbf{a};\mathbf{w},p,\nu)^{-1}-1\right)}{\left(\overset{\leftrightarrow*}{h}(\mathbf{a};\mathbf{w},p,\nu)^{-1}-1\right)^{-f_{z1}(p)}} \right) \tag{10.173}$$

$$\cdot \frac{\partial f_{z1}(p)}{\partial p}$$

$$\frac{\partial \overset{\leftrightarrow*}{H}(\mathbf{a};\mathbf{w},p,\nu)}{\partial \nu} = \left(1 - 2\overset{\leftrightarrow*}{h}(\mathbf{a};\mathbf{w},p,\nu)\right)\frac{\partial f_z(\nu)}{\partial \nu} +$$

$$+ (2f_z(\nu)-1) \frac{\left(\overset{\leftrightarrow*}{h}(\mathbf{a};\mathbf{w},p,\nu)^{-1}-1\right)^{1-f_{z1}(p)}}{\overset{\leftrightarrow*}{h}(\mathbf{a};\mathbf{w},p,\nu)^{-2}} \cdot$$

$$\cdot \sum_{i=1}^{n} \left(\frac{\left(\tilde{N}_\nu(\arg_i(a_i,w_i,\nu))^{-1}-1\right)^{-f_{z1}(p)-1}}{\tilde{N}_\nu(\arg_i(a_i,w_i,\nu))^2} \cdot \right. \tag{10.174}$$

$$\left. \cdot \left((2f_z(\nu)-1)\frac{\partial \arg_i(a_i,w_i,\nu)}{\partial \nu} + \right.\right.$$

$$\left.\left. + (2\arg_i(a_i,w_i,\nu)-1)\frac{\partial f_z(\nu)}{\partial \nu} \right) \right)$$

Yager H-function

$$\overset{\leftrightarrow*}{H}(\mathbf{a};\mathbf{w},p,\nu)$$
$$= \tilde{N}_\nu\left(\min\left\{1,\left(\sum_{i=1}^{n}\tilde{N}_\nu(\arg_i(a_i,w_i,\nu))^{f_{z1}(p)}\right)^{\frac{1}{f_{z1}(p)}}\right\}\right) \tag{10.175}$$

$$\overset{\leftrightarrow*}{H}(\mathbf{a};\mathbf{w},p,\nu) = \tilde{N}_\nu\left(\min\left\{1,\overset{\leftrightarrow*}{h}(\mathbf{a};\mathbf{w},p,\nu)\right\}\right) \tag{10.176}$$

$$\overset{\leftrightarrow*}{H}(\mathbf{a};\mathbf{w},p,\nu) = \begin{cases} \tilde{N}_\nu\left(\overset{\leftrightarrow*}{h}(\mathbf{a};\mathbf{w},p,\nu)\right) & \text{for } \overset{\leftrightarrow*}{h}(\mathbf{a};\mathbf{w},p,\nu) \leq 1 \\ \tilde{N}_\nu(1) & \text{for } \overset{\leftrightarrow*}{h}(\mathbf{a};\mathbf{w},p,\nu) > 1 \end{cases} \tag{10.177}$$

$$p \in (0,\infty) \tag{10.178}$$

where

$$
\overset{\leftrightarrow}{h}{}^{*}\left(\mathbf{a};\mathbf{w},p,\nu\right)=\left(\sum_{i=1}^{n}\widetilde{N}_{\nu}\left(\arg_{i}\left(a_{i},w_{i},\nu\right)\right)^{f_{z1}(p)}\right)^{\frac{1}{f_{z1}(p)}} \tag{10.179}
$$

$$
\frac{\partial \overset{\leftrightarrow}{H}{}^{*}\left(\mathbf{a};\mathbf{w},p,\nu\right)}{\partial a_{i}}=
\begin{cases}
\left(2f_{z}\left(\nu\right)-1\right)^{2}\cdot \\
\quad\cdot \overset{\leftrightarrow}{h}{}^{*}\left(\mathbf{a};\mathbf{w},p,\nu\right)^{1-f_{z1}(p)}\cdot \\
\quad\cdot \widetilde{N}_{\nu}\left(\arg_{i}\left(a_{i},w_{i},\nu\right)\right)^{f_{z1}(p)-1}\cdot \\
\quad\cdot \dfrac{\partial \arg_{i}\left(a_{i},w_{i},\nu\right)}{\partial a_{i}} \\
\quad\text{for }\ \overset{\leftrightarrow}{h}{}^{*}\left(\mathbf{a};\mathbf{w},p,\nu\right)\le 1 \\
0\quad\text{for }\ \overset{\leftrightarrow}{h}{}^{*}\left(\mathbf{a};\mathbf{w},p,\nu\right)>1
\end{cases} \tag{10.180}
$$

$$
\frac{\partial \overset{\leftrightarrow}{H}{}^{*}\left(\mathbf{a};\mathbf{w},p,\nu\right)}{\partial w_{i}}=
\begin{cases}
\left(2f_{z}\left(\nu\right)-1\right)^{2}\cdot \\
\quad\cdot \overset{\leftrightarrow}{h}{}^{*}\left(\mathbf{a};\mathbf{w},p,\nu\right)^{1-f_{z1}(p)}\cdot \\
\quad\cdot \widetilde{N}_{\nu}\left(\arg_{i}\left(a_{i},w_{i},\nu\right)\right)^{f_{z1}(p)-1}\cdot \\
\quad\cdot \dfrac{\partial \arg_{i}\left(a_{i},w_{i},\nu\right)}{\partial w_{i}} \\
\quad\text{for }\ \overset{\leftrightarrow}{h}{}^{*}\left(\mathbf{a};\mathbf{w},p,\nu\right)\le 1 \\
0\quad\text{for }\ \overset{\leftrightarrow}{h}{}^{*}\left(\mathbf{a};\mathbf{w},p,\nu\right)>1
\end{cases} \tag{10.181}
$$

$$
\frac{\partial \overset{\leftrightarrow}{H}{}^{*}\left(\mathbf{a};\mathbf{w},p,\nu\right)}{\partial p}=
\begin{cases}
\dfrac{2f_{z}\left(\nu\right)-1}{f_{z1}\left(p\right)}\overset{\leftrightarrow}{h}{}^{*}\left(\mathbf{a};\mathbf{w},p,\nu\right)^{1-f_{z1}(p)}\cdot \\
\left(\sum\limits_{i=1}^{n}\dfrac{\ln\left(\widetilde{N}_{\nu}\left(\arg_{i}\left(a_{i},w_{i},\nu\right)\right)\right)}{\widetilde{N}_{\nu}\left(\arg_{i}\left(a_{i},w_{i},\nu\right)\right)^{-f_{z1}(p)}}+\right. \\
\left.\quad -\dfrac{\ln\left(\overset{\leftrightarrow}{h}{}^{*}\left(\mathbf{a};\mathbf{w},p,\nu\right)\right)}{\overset{\leftrightarrow}{h}{}^{*}\left(\mathbf{a};\mathbf{w},p,\nu\right)^{-f_{z1}(p)}}\right)\cdot \\
\quad\cdot \dfrac{\partial f_{z1}\left(p\right)}{\partial p} \\
\quad\text{for }\ \overset{\leftrightarrow}{h}{}^{*}\left(\mathbf{a};\mathbf{w},p,\nu\right)\le 1 \\
0\quad\text{for }\ \overset{\leftrightarrow}{h}{}^{*}\left(\mathbf{a};\mathbf{w},p,\nu\right)>1
\end{cases} \tag{10.182}
$$

$$\frac{\partial \overset{\leftrightarrow}{H}{}^{*}(\mathbf{a};\mathbf{w},p,\nu)}{\partial \nu}=\begin{cases}\begin{aligned}&\left(2\overset{\leftrightarrow}{h}{}^{*}(\mathbf{a};\mathbf{w},p,\nu)-1\right)\frac{\partial f_{z}(\nu)}{\partial \nu}+\\&+\left(2f_{z}(\nu)-1\right)\frac{1}{\overset{\leftrightarrow}{h}{}^{*}(\mathbf{a};\mathbf{w},p,\nu)^{f_{z1}(p)-1}}\cdot\\&\begin{pmatrix}\tilde{N}_{\nu}\left(\arg_{i}(a_{i},w_{i},\nu)\right)^{f_{z1}(p)-1}\cdot\\\sum_{i=1}^{n}\begin{pmatrix}\left(2f_{z}(\nu)-1\right)\dfrac{\partial \arg_{i}(a_{i},w_{i},\nu)}{\partial \nu}+\\+\left(2\arg_{i}(a_{i},w_{i},\nu)-1\right)\dfrac{\partial f_{z}(\nu)}{\partial \nu}\end{pmatrix}\end{pmatrix}\\&\qquad\text{for}\quad \overset{\leftrightarrow}{h}{}^{*}(\mathbf{a};\mathbf{w},p,\nu)\le 1\\&\frac{\partial f_{z}(\nu)}{\partial \nu}\quad\text{for}\quad \overset{\leftrightarrow}{h}{}^{*}(\mathbf{a};\mathbf{w},p,\nu)>1\end{aligned}\end{cases}$$

(10.183)

10.7 Simulation examples

We will present the results of simulation for previously described flexible neuro-fuzzy systems. Simulations concern problems of polymerization, modeling the taste of rice, classification of iris flower and classification of wine presented in Subchapter 9.2. To remind of those problems, they have been listed in Table 10.2. Two simulation series have been conducted for each simulation example. Each series has been conducted and described in analogic way:

- In the first experiment only the parameters of membership function of input and output fuzzy sets and the parameter of inference model $\nu \in [0,1]$ were learnt. The value of this parameter after completion of the learning process belongs to the set $\nu \in \{0,1\}$.

- In the second experiment the parameters of membership function of input and output fuzzy sets were also learnt, whereas the value of

TABLE 10.2. Simulation examples used

Simulation problem	Type of the problem	Number of inputs	Length of the learning sequence	Length of the testing sequence
Polymerization	approximation	3	70	–
Modeling the taste of rice	approximation	5	75	30
Classification of iris flowers	classification	4	105	45
Classification of wine	classification	13	125	53

the parameter of inference model ν was chosen as a opposite value (0 or 1) to the one obtained in the first experiment. As we will see, the accuracy obtained in this experiment is worse than the one obtained in the first experiment.

- In the third experiment the same parameters as in the first experiment and also flexibility parameters $\alpha^\tau \in [0,1]$, $\alpha^I \in [0,1]$, $\alpha^{\mathrm{agr}} \in [0,1]$ and parameters of the shape of the applied operators $p^\tau \in [0,\infty)$, $p^I \in [0,\infty)$, $p^{\mathrm{agr}} \in [0,\infty)$ were learnt. The latter parameters occur in case of applying adjustable Dombi and Yager H-functions (in the second series of experiments).

- In the fourth experiment the same parameters as in the third experiment, as well as weights of rules premises $w_{i,k}^\tau \in [0,1]$, $i = 1,\ldots,n$, $k = 1,\ldots,N$ and weights of particular rules $w_k^{\mathrm{agr}} \in [0,1]$, $k = 1,\ldots,N$ were learnt. The values of weights, after completion of the learning process, are illustrated in the diagrams in which the weights of premises and the weights of rules are separated with a vertical dotted line. In the diagrams we assume that the grayer the field which symbolizes a given weight, the value of the weight is closer to zero.

In the first simulation series, in each of the four experiments described above, non-adjustable Zadeh and algebraic H-functions and H-implications were applied. In the second series of experiments adjustable H-functions and Dombi and Yager H-implications were applied, instead of non- adjustable operators.

10.7.1 Polymerization

The results of simulation for the polymerization problem are presented in Tables 10.3a and 10.3b for non-adjustable H-functions (Zadeh and algebraic) and in Tables 10.4a and 10.4b for adjustable H-functions (Dombi and Yager). Moreover, for the experiment (iv) the values of weights of rules premises $w_{i,k}^\tau \in [0,1]$ and values of weights of rules $w_k^{\mathrm{agr}} \in [0,1]$ of the considered systems with non-adjustable H-functions are symbolically presented in Fig. 10.20, while the values of weight of systems with adjustable H-functions are presented in Fig. 10.21.

10.7.2 Modeling the taste of rice

The results of simulation for the problem of modeling the taste of rice are presented in Tables 10.5a and 10.5b for non-adjustable H-functions (Zadeh and algebraic) and in Tables 10.6a and 10.6b for adjustable H-functions (Dombi and Yager). Moreover, for the experiment (iv) the values of weights of rules premises $w_{i,k}^\tau \in [0,1]$ and values of weights of rules

TABLE 10.3a. The results of simulation of a flexible system with non-parame-terized H-functions – the problem of polymerization

Flexible system with non-parameterized H-functions (Polymerization)		
Simulation number	Name of flexibility parameter	Initial value
i	ν	0.5
ii	ν	1
iii	ν	0.5
	α^τ	1
	α^I	1
	α^{agr}	1
iv	ν	0.5
	α^τ	1
	α^I	1
	α^{agr}	1
	$\mathbf{w}^\tau$	1
	$\mathbf{w}^{\mathrm{arg}}$	1

TABLE 10.3b. The results of simulation of a flexible system with non-parame-terized H-functions – the problem of polymerization

Flexible system with non-parametrized H-functions (Polymerization)				
Simulation number	Final value after learning		RMSE (learning sequence)	
	Zadeh H-function	Algebraic H-function	Zadeh H-function	Algebraic H-function
i	0.0000	0.0000	0.0096	0.0060
ii	–	–	0.0115	0.0063
iii	0.0000	0.0000	0.0059	0.0056
	0.7158	0.9678		
	0.7613	0.9992		
	0.7277	0.9930		
iv	0.0000	0.0000	0.0056	0.0044
	0.6941	0.9987		
	0.7783	0.9992		
	0.6713	0.9334		
	Fig. 10.20a	Fig. 10.20b		
	Fig. 10.20a	Fig. 10.20b		

TABLE 10.4a. The results of simulation of a flexible system with parameterized H-functions – the problem of polymerization

Flexible system with parameterized H-functions (Polymerization)		
Simulation number	Name of flexibility parameter	Initial value
i	ν	0.5
ii	ν	1
iii	ν	0.5
	p^τ	10
	p^I	10
	p^{agr}	10
	α^τ	1
	α^I	1
	α^{agr}	1
iv	ν	0.5
	p^τ	10
	p^I	10
	p^{agr}	10
	α^τ	1
	α^I	1
	α^{agr}	1
	$\mathbf{w}^\tau$	1
	$\mathbf{w}^{\mathrm{arg}}$	1

$w_k^{\mathrm{agr}} \in [0,1]$ of considered systems with non-adjustable H-functions are symbolically presented in Fig. 10.22, while the values of weights of systems with adjustable H-functions are presented in Fig. 10.23.

10.7.3   Classification of iris flower

The results of simulation for the problem of classification of iris flower are presented in Tables 10.7a and 10.7b for non-adjustable H-functions (Zadeh and algebraic) and in Tables 10.8a and 10.8b for adjustable H-functions (Dombi and Yager). Moreover, for the experiment (iv) the values of weights of rules premises $w_{i,k}^\tau \in [0,1]$ and values of weights of rules $w_k^{\mathrm{agr}} \in [0,1]$ of the considered systems with non-adjustable H-functions are symbolically presented in Fig. 10.24, while the values of weight of systems with adjustable H-functions are presented in Fig. 10.25.

TABLE 10.4b. The results of simulation of a flexible system with parameterized H-functions – the problem of polymerization

Flexible system with parameterized H-functions (Polymerization)				
Simulation number	Final value after learning		RMSE (learning sequence)	
	Dombi H-function	Yager H-function	Dombi H-function	Yager H-function
i	0.0000	0.0000	0.0117	0.0110
ii	–	–	0.0133	0.0113
iii	0.0000 9.9714 10.0042 9.9835 0.6996 0.7743 0.9941	0.0000 10.2089 10.2594 9.3991 0.1624 0.5344 0.9942	0.0077	0.0061
iv	0.0000 13.1310 15.3619 3.4720 0.7127 0.7148 0.9335 Fig. 10.21a Fig. 10.21a	0.0000 7.5714 11.7834 13.9273 0.1375 0.4742 0.9910 Fig. 10.21b Fig. 10.21b	0.0069	0.0053

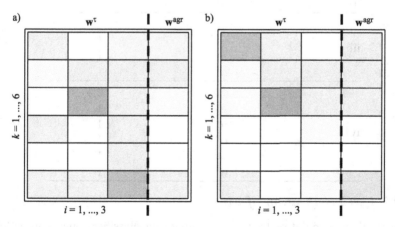

FIGURE 10.20. Weights of rules premises and weights of rules for a flexible system which solves the problem of polymerization in case of a) Zadeh H-function, b) algebraic H-function

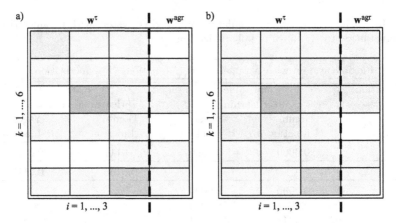

FIGURE 10.21. Weights of rules premises and weights of rules for a flexible system which solves the problem of polymerization in case of a) Dombi H-function, b) Yager H-function

TABLE 10.5a. The results of simulation of a flexible system with non-parameterized H-functions – the problem of modeling the taste of rice

Flexible system with non-parameterized H-functions (Modeling the taste of rice)		
Simulation number	Name of flexibility parameter	Initial value
i	ν	0.5
ii	ν	1
iii	ν	0.5
	α^τ	1
	α^I	1
	α^{agr}	1
iv	ν	0.5
	α^τ	1
	α^I	1
	α^{agr}	1
	$\mathbf{w}^\tau$	1
	$\mathbf{w}^{\mathrm{arg}}$	1

10.7.4 Classification of wine

The results of simulation for the problem of wine classification are presented in Tables 10.9a and 10.9b for non-adjustable H-functions (Zadeh and algebraic) and in Tables 10.10a and 10.10b for adjustable H-functions (Dombi

TABLE 10.5b. The results of simulation of a flexible system with non-parameterized H-functions – the problem of modeling the taste of rice

Flexible system with non-parameterized H-functions (Modeling the taste of rice)				
Simulation number	Final value after learning		RMSE (learning sequence)	
	Zadeh H-function	Algebraic H-function	Zadeh H-function	Algebraic H-function
i	0.0000	0.0000	0.0184	0.0185
ii	–	–	0.0186	0.0192
iii	0.0000 0.2954 0.9843 0.4658	0.0000 0.9972 0.9979 0.9958	0.0163	0.0173
iv	0.0000 0.3101 0.9575 0.5496 Fig. 10.22a Fig. 10.22a	0.0000 0.9519 0.9512 0.9085 Fig. 10.22b Fig. 10.22b	0.0140	0.0159

TABLE 10.6a. The results of simulation of a flexible system with parameterized H-functions – the problem of modeling the taste of rice

Flexible system with parametrized H-functions (Modeling the taste of rice)		
Simulation number	Name of flexibility parameter	Initial value
i	ν	0.5
ii	ν	1
iii	ν p^τ p^I p^{agr} α^τ α^I α^{agr}	0.5 10 10 10 1 1 1
iv	ν p^τ p^I p^{agr} α^τ α^I α^{agr} $\mathbf{w}^\tau$ $\mathbf{w}^{\mathrm{arg}}$	0.5 10 10 10 1 1 1 1 1

TABLE 10.6b. The results of simulation of a flexible system with parameterized H-functions – the problem of modeling the taste of rice

	Flexible system with parametrized H-functions (Modeling the taste of rice)			
Simulation number	Final value after learning		RMSE (learning sequence)	
	Dombi H-function	Yager H-function	Dombi H-function	Yager H-function
i	0.0000	0.0000	0.0186	0.0187
ii	–	–	0.0192	0.0197
iii	0.0000	0.0000		
	9.9268	10.7365		
	10.0026	10.1154		
	9.7692	10.8200	0.0181	0.0184
	0.4606	0.6895		
	0.9943	0.9993		
	0.9865	0.9728		
iv	0.0000	0.0000		
	10.1449	10.9117		
	9.9448	10.0472		
	9.2063	10.0148		
	0.4380	0.6763	0.0160	0.0169
	0.9201	0.9263		
	0.8967	0.9927		
	Fig. 10.23a	Fig. 10.23b		
	Fig. 10.23a	Fig. 10.23b		

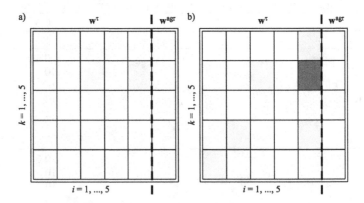

FIGURE 10.22. Weights of rules premises and weights of rules for a flexible system which solves the problem of modeling the taste of rice in case of a) Zadeh H-function, b) algebraic H-function

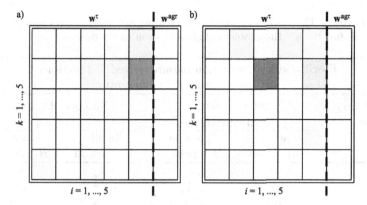

a) $\mathbf{w}^\tau$ $\mathbf{w}^{agr}$ b) $\mathbf{w}^\tau$ $\mathbf{w}^{agr}$

$k = 1, ..., 5$ $k = 1, ..., 5$

$i = 1, ..., 5$ $i = 1, ..., 5$

FIGURE 10.23. Weights of rules premises and weights of rules for a flexible system which solves the problem of modeling the taste of rice in case of a) Dombi H-function, b) Yager H-function

TABLE 10.7a. The results of simulation of a flexible system with non-parameterized H-functions – the problem of classification of iris flowers

Flexible system with non-parameterized H-functions (Classification of iris flowers)				
Simulation number	Name of flexibility parameter	Initial value	Final value after learning	
			Zadeh H-function	Algebraic H-function
i	ν	0.5	1.0000	1.0000
ii	ν	0	–	–
iii	ν	0.5	1.0000	1.0000
	α^τ	1	0.2032	0.9922
	α^I	1	0.9891	0.6082
	α^{agr}	1	0.9994	
iv	ν	0.5	1.0000	1.0000
	α^τ	1	0.2442	0.9592
	α^I	1	0.9845	0.5753
	α^{agr}	1	0.9650	0.9937
	$\mathbf{w}^\tau$	1	Fig. 10.24a	Fig. 10.24b
	$\mathbf{w}^{arg}$	1	Fig. 10.24a	Fig. 10.24b

and Yager). Moreover, for the experiment (iv) the values of weights of rules premises $w_{i,k}^\tau \in [0,1]$ and values of weights of rules $w_k^{agr} \in [0,1]$ of considered systems with non-adjustable H-functions are symbolically presented in Fig. 10.26, while the values of weights of systems with adjustable H-functions are presented in Fig. 10.27.

TABLE 10.7b. The results of simulation of a flexible system with non-parameterized H-functions – the problem of classification of iris flowers

Flexible system with non-parameterized H-functions (Classification of iris flowers)				
Simulation number	Number of errors [%] (learning sequence)		Number of errors [%] (testing sequence)	
	Zadeh H-function	Algebraic H-function	Zadeh H-function	Algebraic H-function
i	0.95	0.95	4.44	4.44
ii	0.95	0.95	6.67	6.67
iii	0.00	0.95	4.44	4.44
iv	0.00	0.00	4.44	4.44

TABLE 10.8a. The results of simulation of a flexible system with parameterized H-functions – the problem of classification of iris flowers

Flexible system with parameterized H-functions (Classification of iris flowers)				
Simulation number	Name of flexibility parameter	Initial value	Final value after learning	
			Dombi H-function	Yager H-function
i	ν	0.5	1.0000	1.0000
ii	ν	0	–	–
iii	ν	0.5	1.0000	1.0000
	p^τ	10	13.2031	4.3306
	p^I	10	10.0001	7.5741
	p^{agr}	10	9.9974	10.1209
	α^τ	1	0.8259	0.7846
	α^I	1	0.9924	0.9931
	α^{agr}	1	0.9985	0.9985
iv	ν	0.5	1.0000	1.0000
	p^τ	10	13.5253	4.3621
	p^I	10	10.8610	8.0120
	p^{agr}	10	9.4218	9.3590
	α^τ	1	0.8739	0.8068
	α^I	1	0.9871	0.9731
	α^{agr}	1	0.9698	0.9661
	$\mathbf{w}^\tau$	1	Fig. 10.25a	Fig. 10.25b
	$\mathbf{w}^{\mathrm{arg}}$	1	Fig. 10.25a	Fig. 10.25b

TABLE 10.8b. The results of simulation of a flexible system with parameterized H-functions – the problem of classification of iris flowers

| Simulation number | Flexible system with parameterized H-functions (Classification of iris flowers) | | | |
| | Number of errors [%] (learning sequence) | | Number of errors [%] (testing sequence) | |
	Dombi H-function	Yager H-function	Dombi H-function	Yager H-function
i	0.00	0.95	4.44	4.44
ii	0.95	0.95	6.67	6.67
iii	0.00	0.00	4.44	4.44
iv	0.00	0.00	2.22	2.22

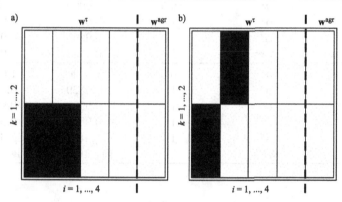

FIGURE 10.24. Weights of rules premises and weights of rules for a flexible system which solves the problem of classification of iris flowers in case of a) Zadeh H-function, b) algebraic H-function

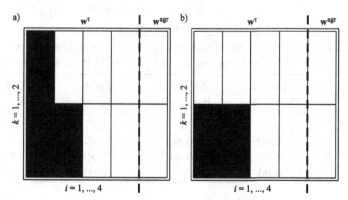

FIGURE 10.25. Weights of rules premises and weights of rules for a flexible system which solves the problem of classification of iris flowers in case of a) Dombi H-function, b) Yager H-function

TABLE 10.9a. The results of simulation of a flexible system with non-parameterized H-functions – the problem of wine classification

Flexible system with non-parameterized H-functions (Classification of wine)				
Simulation number	Name of flexibility parameter	Initial value	Final value after learning	
			Zadeh H-function	Algebraic H-function
i	ν	0.5	1.0000	1.0000
ii	ν	0	–	–
iii	ν	0.5	1.0000	1.0000
	α^τ	1	0.0004	0.0036
	α^I	1	0.9907	0.9986
	α^{agr}	1	0.9938	0.9908
iv	ν	0.5	1.0000	1.0000
	α^τ	1	0.0329	0.0180
	α^I	1	0.9987	0.9756
	α^{agr}	1	0.9896	0.9861
	$\mathbf{w}^\tau$	1	Fig. 10.26a	Fig. 10.26b
	$\mathbf{w}^{arg}$	1	Fig. 10.26a	Fig. 10.26b

TABLE 10.9b. The results of simulation of a flexible system with non-parameterized H-functions – the problem of wine classification

Flexible system with non-parameterized H-functions (Classification of wine)				
Simulation number	Number of errors [%] (learning sequence)		Number of errors [%] (testing sequence)	
	Zadeh H-function	Algebraic H-function	Zadeh H-function	Algebraic H-function
i	0.00	0.00	3.77	1.89
ii	0.80	0.80	3.77	3.77
iii	0.00	0.00	1.89	1.89
iv	0.00	0.00	0.00	0.00

TABLE 10.10a. The results of simulation of a flexible system with parameterized H-functions – the problem of wine classification

	Flexible system with parameterized H-functions (Classification of wine)			
Simulation number	Name of flexibility parameter	Initial value	Final value after learning	
			Dombi H-function	Yager H-function
i	ν	0.5	1.0000	1.0000
ii	ν	0	–	–
iii	ν	0.5	1.0000	1.0000
	p^τ	10	9.9999	10.0498
	p^I	10	10.0005	9.9936
	p^{agr}	10	9.9991	10.0014
	α^τ	1	0.0032	0.0029
	α^I	1	0.9911	0.9917
	α^{agr}	1	0.9919	0.9920
iv	ν	0.5	1.0000	1.0000
	p^τ	10	7.8330	6.9528
	p^I	10	11.7084	13.3122
	p^{agr}	10	14.3699	12.1427
	α^τ	1	0.0028	0.0389
	α^I	1	0.9826	0.9740
	α^{agr}	1	0.9914	0.9599
	$\mathbf{w}^\tau$	1	Fig. 10.27a	Fig. 10.27b
	$\mathbf{w}^{\mathrm{arg}}$	1	Fig. 10.27a	Fig. 10.27b

TABLE 10.10b. The results of simulation of a flexible system with parameterized H-functions – the problem of wine classification

	Flexible system with parameterized H-functions (Classification of wine)			
Simulation number	Number of errors [%] (learning sequence)		Number of errors [%] (testing sequence)	
	Dombi H-function	Yager H-function	Dombi H-function	Yager H-function
i	0.00	0.00	1.89	1.89
ii	0.00	0.00	3.77	3.77
iii	0.00	0.00	1.89	1.89
iv	0.00	0.00	0.00	0.00

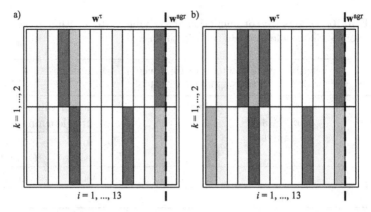

FIGURE 10.26. Weights of rules premises and weights of rules for a flexible system which solves the problem of distinguishing the brand of wine in case of a) Zadeh H-function, b) algebraic H-function

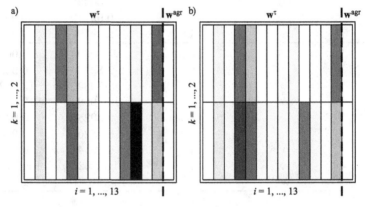

FIGURE 10.27. Weights of rules premises and weights of rules for a flexible system which solves the problem of distinguishing the brand of wine in case of a) Dombi H-function, b) Yager H-function

10.8 Notes

The concept of flexible neuro-fuzzy systems presented in this chapter allows us to determine the type of system (Mamdani or logical) as a result of the learning process. It could be inferred from the simulation examples presented in Subchapter 10.7 that a flexible system becomes a Mamdani system after completion of the learning process (parameter $\nu = 0$) for the problems of approximation or identification. In contrast, for the problems of classification a flexible system becomes a logical system (parameter $\nu = 1$) as a result of learning. The above results could be treated as a recommendation of the Mamdani system to solve the problems of approximation or identification and the logical system to solve the problems of classification. It should be mentioned that the concept of soft triangular norms was presented by Yager and Filev [262], while Klement [111] and Lowen [128] presented in detail various types of parameterized triangular norms. Various types of flexible neuro-fuzzy structures were proposed by Cpałka [30]. The subject of those systems is discussed in more detail in monograph [225]. We refer the interested Reader to the following works [210–212, 215, 217, 218, 220, 223, 227].

References

[1] Aliev R.A., Aliev R.R., *Soft Computing and its Applications*, World Scientific, Singapore 2001.

[2] Arabas J., *Lectures on Evolutionary Algorithms*, Scientific-Technical Publishing House WNT, Warsaw 2001 (in Polish).

[3] Babuška R., *Fuzzy Modeling for Control*, Kluwer Academic Publishers, Boston 1998.

[4] Backer E., *Computer-Assisted Reasoning in Cluster Analysis*, Prentice Hall, New York 1995.

[5] Badźmirowski K., Kubiś M., *Expert Systems*, Industrial Institute of Electronics, Warsaw 1991.

[6] Bartlett P., Downs T., *Training a Neural Networks with a Genetic Algorithm*, Technical Report, Dept. of Elec. Eng., Univ. of Queensland, 1990.

[7] Belew R.K., McInerney J., Schraudolph N.N., *Evolving Networks: Using Genetic Algorithms with Connectionist Learning*, CSE technical report CS90-174, La Jolla, CA: University of California at Dan Diego 1990.

[8] Bellman R.E., Giertz M., On the analytical formalism of fuzzy sets, *Information Sciences*, 5, 149–156 (1975).

496   References

[9] Bezdek J.C., Pal S.K., *Fuzzy Models for Pattern Recognition*, IEEE Press, New York 1992.

[10] Bezdek J., Keller J., Krisnapuram R., Pal N.R., *Fuzzy Models and Algorithms for Pattern Recognition and Image Processing*, Kluwer Academic Press, 1999.

[11] Bilski J., Rutkowski L., A fast training algorithm for neural networks, *IEEE Trans. on Circuits and Systems II*, June 1998, 749–753 (1998).

[12] Bishop Ch.M., *Neural Networks for Pattern Recognition*, Oxford University Press, Oxford, New York 1995.

[13] Bojadziev G., Bojadziev M., *Fuzzy Logic for Business, Finance and Management*, World Scientific, Singapore 1999.

[14] Bonabeau E., Theraulaz G., Swarm smarts. *Scientific American*, 282 (3) 2000, pp. 72-79.

[15] Branke J., *Evolutionary Algorithm for Neural Network Design and Training*, University of Karlsruhe, 1995.

[16] Brindle M., *Genetic Algorithms for Function Optimization*, Ph.D. dissertation, University of Alberta, 1981.

[17] Brown M., Harris C., *Neuro-fuzzy Adaptive Modelling and Control*, Prentice Hall PTR, Upper Saddle River, NJ 1994.

[18] Bubnicki Z., *Control theory and algorithms*, Polish Scientific Publishers PWN, Warsaw 2002 (in Polish).

[19] Bubnicki Z., *Analysis and Decision Making in Uncertain Systems*, Series: Communications and Control Engineering, Springer-Verlag, Heidelberg 2004.

[20] Cacoullos T., Estimation of a multivariate density, *Ann. Inst. Statist. Math.*, 18, 179–189 (1965).

[21] Calvo T., Mayor G., Mesiar R. (Eds.), *Aggregation Operators. New Trends and Applications*, Physica-Verlag. A Springer-Verlag Company, New York 2002.

[22] Capcarrère M., Tettamazi A., Tomassini M., Sipper M., A statistical study of a class of cellular evolutionary algorithms, *Evolutionary Computation*, 7(3), 255–274 (1999).

[23] Chi Z., Yan H., Pham T., *Fuzzy Algorithms: With Applications to Image Processing and Pattern Recognition*, World Scientific, Singapore 1996.

[24] Chong E.K.P., Żak S.H., *An Introduction to Optimization*, John Wiley, 1996.

[25] Chromiec J., Strzemieczna E., *Artificial intelligence. Methods of Construction and Analysis of Expert Systems*, Academic Publishing House PLJ, Warsaw 1994 (in Polish).

[26] Cichocki A., Unbehauen R., *Neural Networks for Optimization and Signal Processing*, John Wiley & Sons — Interscience, New York 1993.

[27] Cichosz P., *Self-Learning Systems*, Scientific-Technical Publishing House WNT, Warsaw 2000 (in Polish).

[28] Cordón O., Herrera F., Evolutionary design of TSK fuzzy ruled based systems using (μ, λ)-evolutionary strategies, *Proceedings of the Sixth IEEE International Conference on Fuzzy Systems* (FUZZ–IEEE'97), vol. 1, 509–514, Barcelona 1997.

[29] Cordón O., Herrera F., Hoffman F., Magdalena L., *Genetic Fuzzy Systems. Evolutionary Tuning and Learning of Fuzzy Knowledge Bases*, World Scientific, Singapore 2001.

[30] Cpałka K., *Flexible Neuro-Fuzzy Inference Systems*, PhD Thesis, Częstochowa University of Technology, Częstochowa 2001 (in Polish).

[31] Cytowski J., *Genetic Algorithms*, Academic Publiching House, Warsaw 1996 (in Polish).

[32] Czabański R., *Automatic Determining of Fuzzy IF-THEN Rules from Numerical Data*, PhD Thesis, Silesian University of Technology, Gliwice 2002 (in Polish).

[33] Czogała E., Pedrycz W., *Elements and Methods of Fuzzy Sets Theory*, Polish Scientific Publishers PWN, Warsaw 1985 (in Polish).

[34] Czogała E., Łęski J., *Fuzzy and Neuro-Fuzzy Intelligent Systems*, Physica-Verlag, Heidelberg, New York 2000.

[35] Davis L., *Handbook of Genetic Algorithms*, Van Nostrand Reinhold, New York 1991.

[36] Dąbrowski M., Rudeński A., Application of Evolutionary Procedures in Electric Machines Optimization, *X Scientific-Technical Conference: Applications of Computers in Electrotechnics*, 11–12, Poznań/Kiekrz 2005 (in Polish).

498   References

[37] Dąbrowski M., Rudeński A., Discrete Independent Variables in Nondeterministic Electric Machine Optimization, *Proc. of XLI International Symposium on Electrical Machines* SME-2005, Opole–Jarnołtówek 2005 (in Polish).

[38] De Jong K.A., *An Analysis of the Behavior of a Class of Genetic Adaptive Systems*, Doctoral Dissertation, University of Michigan, 1975.

[39] De Silva C.W., *Intelligent Control: Fuzzy Logic Applications*, CRC Press, Boca Raton 1995.

[40] Driankov D., Hellendoorn H., Reinfrank M., *An Introduction to Fuzzy Control*, Springer-Verlag New York, Inc., New York 1993.

[41] Dubois D., Prade H., Operations on fuzzy numbers, *Intern. Journal System Science*, **9**, 613–626 (1978).

[42] Dubois D., Prade H., Fuzzy Sets and Systems: Theory and Applications, *Mathematics in Science Engineering*, Academic Press, Inc., vol. 144, San Diego 1980.

[43] Dubois D., Prade H., Rough fuzzy sets and fuzzy rough sets, *International J. General Systems*, 17(2–3), 191–209 (1990).

[44] Dubois D., Prade H., Putting Rough Sets and Fuzzy Sets Together, in: Słowiński R. (Ed.) *Intelligent Decision Support: Handbook of Applications and Advances of the Rough Sets Theory*, Kluwer Academic Publishers, 203–232, Dordrecht 1992.

[45] Duch W., *Fascinating World of Computer Programs*, NAKOM Publishing House, Poznań 1997 (In Polish).

[46] Duch W., What is Cognitivism?, *Cognitivism and Media in Education* 1/1998, Adam Marszałek Publishing House, Toruń 1998 (In Polish).

[47] Duch W., Korbicz J., Rutkowski L., Tadeusiewicz R., *Biocybernetics and Biomedical Engineering*, vol 6: *Neural Networks*, Academic Publishing House EXIT, Warsaw 2000 (In Polish).

[48] Duda R.O., Hart P.E., Stork D.G., *Pattern Classification*, John Wiley & Sons, Inc., Toronto 2001.

[49] Elman J.L., Finding structure in time, *Cognitive Science*, 14, 179–211 (1990).

[50] Evolver — the Genetic Algorithm Problem Solver, Axcelis, Inc., 4668 Eastern Avenue N., Seattle, WA 98103, USA.

[51] Fausett L., *Fundamentals of Neural Networks. Architectures, Algorithms, and Applications*, Prentice Hall, New Jersey 1994.

[52] Ferber J., *Multi-Agent systems*, Addison-Wesley, New York 1999.

[53] Findeisen W., Szymanowski W., Wierzbicki A., *Theory and Computational Methods in Optimization*, Polish Scientific Publishers PWN, Warsaw 1977 (In Polish).

[54] FlexTool (GA) M2.1, Flexible Intelligence Group, L.L.C., Tuscaloosa, AL35486-1477, USA.

[55] Fodor J.C., On fuzzy implication, *Fuzzy Sets and Systems*, vol. 42, 293–300 (1991).

[56] Fogel D.B., *Evolutionary Computation. Towards a New Philosophy of Machine Intelligence*, IEEE Press, New York 1995.

[57] Fraser A.S., Simulation of genetic systems by automatic digital computers. I. Introduction, Part I, II, *Aust. J. Biol. Sci.*, vol. 10, 484–499 (1957).

[58] Fukami S., Mizumoto M., Tanaka K., Some considerations of fuzzy conditional inference, *Fuzzy Sets and Systems*, vol. 4, 243–273 (1980).

[59] Galar R., *Soft Selection in Random Global Adaptation in R^n*, Wrocław University of Technology Publishing House, Wrocław 1990 (in Polish).

[60] Gen M., Cheng R., *Genetic Algorithms and Engineering Design*, John Wiley & Sons, Inc., New York 1997.

[61] Giergiel M.J., Hendzel Z., Żylski W., *Modelling and Control of Mobile Wheel Robots*, Polish Scientific Publishers PWN, Warsaw 2002 (In Polish).

[62] Głowiński C., Artificial Intelligence, *PC Kurier*, 14–27, February 1999 (In Polish).

[63] Goldberg D.E., *Genetic Algorithms in Search, Optimization and Machine Learning*, Kluwer Academic Publishers, Boston, MA. 1989.

[64] Gorzałczany M.B., A interval-valued fuzzy inference method — some basic properties, *Fuzzy Sets and Systems*, vol. 31, 243–251 (1989).

[65] Gorzałczany M.B., *Computational Intelligence Systems and Applications, Neuro-Fuzzy and Fuzzy Neural Synergisms*, Springer-Verlag, Heidelberg 2002.

[66] Greco S., Matarazzo B., Słowiński R., Rough set processing of vague information using fuzzy similarity relations, *Finite Versus Infinite — Contributions to an Eternal Dilemma*, C.S. Calude and G. Paun (Eds.), Springer-Verlag, 149–173, London 2000.

[67] Grzymała-Busse J.W., LERS — A system for learning from examples based on rough sets, in: Słowiński R. (Ed.), *Intelligent Decision Support: Handbook of Applications and Advences of the Rough Sets Theory*, Kluwer Academic Publishers, 3–18, Dordrecht 1992.

[68] Hagan M., Menhaj M.B., Training feed forward networks with the Marquardt algorithm, *IEEE Trans. on Neural Networks*, vol. 5, 989–993 (1994).

[69] Härdle W., *Applied Nonparametric Regression*, Cambridge University Press, London 1990.

[70] Harel D., *Algorithmics: The Spirit of Computing*, Addison-Wesley, Reading, MA, 1987.

[71] Harris C.J., Moore C.G., Brown M., *Intelligent Control: Aspects of fuzzy logic and neural nets*, World Scientific, 1993.

[72] Hassoun M.H., *Fundamentals of Artificial Neural Networks*, The MIT Press, Cambridge, Massachusetts 1995.

[73] Haykin S., *Neural Networks: A Comprehensive Foundation*, Macmillan College Publishing Company, New York 1994.

[74] Hebb D., *Organization of behaviour*, John Wiley, New York 1949.

[75] Helt P., Parol M., Piotrowski P., *Artificial Intelligence Methods in Electroenergetics*, Warsaw University of Technology Publishing House, Warsaw 2000 (In Polish).

[76] Hertz J., Krogh A., Palmer R.G., *Introduction to the Theory of Neural Computation*, Westview Press, 1991.

[77] Herrera F., Herrera E., Lozano M., Verdegay J.L., Fuzzy tools to improve genetic algorithms, in: *Proc. First European Congress on Fuzzy and Intelligent Technologies* (EUFIT'94), 1532–1539, Aachen 1994.

[78] Herrera F., Lozano M., Verdegay J.L., Fuzzy connectives based crossover operators to model genetic algorithms population diversity, *Fuzzy Sets and Systems*, 92(1), 21–30 (1997).

[79] Hirota K. (Ed.), *Industrial Applications of Fuzzy Technology*, Springer-Verlag, Heidelberg, New York 1993.

[80] Hisdal E., The IF THEN ELSE statement and interval-valued fuzzy sets of higher type, *Int. J. Man-Machine Studies*, 15, Academic Press Inc., 385–455, London 1981.

[81] Hoffman P., Check enter. Garri Kasparow vs Deep Junior, *Wprost*, vol. 57, February 9, 2003 (In Polish).

[82] Holland J.H., *Adaptation in Natural and Artificial Systems*, Cambridge, The MIT Press, Cambridge, Massachusetts 1992.

[83] Höppner F., Klawonn F., Kruse R., Runkler T., *Fuzzy cluster analysis. Methods for Classification, Data Analysis and Image Recognition*, John Wiley & Sons, Chichester 1999.

[84] Hornik K., Approximation capabilities of multilayer feedforward networks, *Neural Networks*, vol. 4, 251–257 (1991).

[85] Hornik K., Some new results on neural network approximation, *Neural Networks*, vol. 6, 1069–1072 (1993).

[86] Hu Y.H., Hwang J. (Eds.), *Handbook of Neural Network Signal Processing*, CRC Press, Boca Raton 2002.

[87] Inuiguchi M., Tanino T., *New Fuzzy Rough Sets Based on Certainty Qualification*, in: Pal S.K., Polkowski L., Skowron A. (Eds.), Rough-Neural Computing: Techniques for Computing with Words, Springer-Verlag, 277–296, Heidelberg, New York 2004.

[88] Ishibuchi H., Nakashima T., Murata T., *40 Techniques and application of genetic algorithm-based methods for design compact fuzzy classification systems*, Fuzzy Theory Systems, Cornelius T. Leondes (Ed.), Academic Press, 1999.

[89] Jagielski J., *Knowledge Engineering in Expert Systems*, Lubusky Publishing House, Zielona Góra, Poland 2001 (In Polish).

[90] Jang J.-S.R., ANFIS: Adaptive-network-based fuzzy inference systems, *IEEE Trans. Syst., Man, Cybern.*, 23(3), 665–685 (1993).

[91] Jankowski N., *Ontogenic Neural Networks. On Networks Changing their Structure*, Academic Publishing House EXIT, Warsaw 2003 (In Polish).

[92] Jordan M.I., Attractor dynamics and parallelism in a connectionist sequential machine, *Proc. of the Eight Annual Conference of the Cognitive Science Society*, 531–546, Hillsdale, NJ: Erlbaum 1986.

[93] Kacprzak T., Ślot K., *Cellular Neural Networks*, Polish Scientific Publishers PWN, Warsaw 1994 (In Polish).

502   References

[94] Kacprzyk J., *Fuzzy Sets in System Analysis*, Polish Scientific Publishers PWN, Warszawa 1986 (In Polish).

[95] Kacprzyk J., *Multistage Fuzzy Control*, Wiley, 1997.

[96] Karayiannis N.B., Venetsanopoulos A.N., *Artificial Neural Networks*, Kluwer Academic Publishers, Boston, Dordrecht 1993.

[97] Karnik N.N., Mendel J.M., *An Introduction to Type-2 Fuzzy Logic Systems*, University of Southern California, Los Angeles 1998.

[98] Karnik N.N., Mendel J.M., Applications of type-2 fuzzy logic systems to forecasting of time-series, *Information Sciences* 120, Elsevier, 89–111, 1999.

[99] Karnik N.N., Mendel J.M., Liang Q., Type-2 fuzzy logic systems, *IEEE Trans. on Fuzzy Systems*, 7(6), 643–658 (1999).

[100] Karnik N.N., Mendel J.M, Operations on type-2 fuzzy sets, *Fuzzy sets and systems*, vol. 122, 327–348 (2000).

[101] Karnik N.N., Mendel J.M., Centroid of a type-2 fuzzy set, *Information Sciences* 132, Elsevier, 195–220, 2001.

[102] Karr C.L., Design of an adaptive fuzzy logic controller using a genetic algorithm, *Proc. of the Fourth International Conference on Genetic Algorithms*, Morgan Kaufmann, San Mateo, CA 1991.

[103] Karr C.L., Genetic algorithms for fuzzy controllers, *AI Expert*, vol. 1991, 26–33.

[104] Kasabov N., *Foundations of Neural Networks*, Fuzzy Systems and Knowledge Engineering, The MIT Press, Cambridge, Massachusetts 1996.

[105] Kasperski M.J., *Artificial Intelligence*, Helion Publishing House, Gliwice 2003 (In Polish).

[106] Kawa K., Artificial Intelligence Language, *PC Kurier*, 118–120, June 18, 1998 (In Polish).

[107] Kay S.M., *Modern Spectral Estimation. Theory and application*, Prentice Hall, Englewood Cliffs, New Jersey 1988.

[108] Kecman V., *Learning and Soft Computing*, The MIT Press, Cambridge, Massachusetts 2001.

[109] Khosla R., Dillon T., *Engineering intelligent hybrid multi-agent systems*, Kluwer Academic Publishers, Boston 1997.

[110] Kitano H., Designing neural networks by genetic algorithms using graph generation systems, *Complex Systems*, vol. 4, 461–476 (1990).

[111] Klement E.P., Mesiar R., Pap E., *Triangular Norms*, Kluwer Academic Publishers, Netherlands 2000.

[112] Köhn P., *Genetic Encoding Strategies for Neural Networks*, University of Tennessee — Universität Erlangen-Nümberg, Dreibergstr. 5, 91056 Erlangen, Germany 1996.

[113] Kohonen T., *Self-Organization and Associative Memory*; Springer-Verlag, Heidelberg, New York 1988.

[114] Komosiński M., *Artificial Life*, Software 2.0 Magazine, nr 2(86), 32–38, February 2002 (In Polish).

[115] Korbicz J., Obuchowicz A., Uciński D., *Artificial Neural Networks. Theory and applications*, Academic Publishing House PLJ, Warsaw 1994 (In Polish).

[116] Korbicz J., Koscielny J.M., Kowalczuk Z., Cholewa W., (Eds.), *Fault Diagnosis. Models, Artificial Intelligence, Applications.*, Springer-Verlag, Berlin, Heidelberg 2004.

[117] Kosiński R.A., *Artificial Neural Networks – nonlinear dynamics and chaos*, Scientific-Technical Publishing House WNT, Warsaw 2002 (In Polish).

[118] Kosko B., *Neural Networks and Fuzzy Systems*, Prentice Hall, Englewood Cliffs, NJ 1992.

[119] Kościelny J.M., *Automated Industrial Process Diagnostics*, Academic Publishing House EXIT, Warsaw 2000 (in Polish).

[120] Koza J.R., *Genetic Programming II. Automatic Discovery of Reusable Programs*, The MIT Press, 1994.

[121] Kröse B., Smagt P., *An introduction to Neural Networks*, Eight Ed., 1996.

[122] Kruse R., Gebhardt J., Klawonn F., *Foundations of Fuzzy Systems*, John Wiley & Sons – Interscience, New York 1994.

[123] Krawiec K., Stefanowski J., *Machine Learning and Neural Networks*, Poznan University of Technology Publishing House, Poznań 2003 (in Polish).

[124] Kurowski J., Man vs Computer — Deep Blue — facts, myths, what next, *Software* 2.0 Magazine, nr 2(86), 26–31, February 2002 (in Polish).

504   References

[125] Lee M.A. Takagi H., *Dynamic control of genetic algorithms using fuzzy logic techniques*, Fifth International Conference on Genetic Algorithms (ICGA' 93), 76–83, Urbana-Champaign 1993.

[126] Li H., Chen C.L.P., Huang H., *Fuzzy Neural Intelligent Systems. Mathematical Foundation and the Applications in Engineering*, CRC Press, Boca Raton 2001.

[127] Liu J., *Autonomous agents and multi-agent systems*, World Scientific, Singapore 2001.

[128] Lowen R., *Fuzzy Set Theory Basic Concepts, Techniques and Bibliography*, Kluwer Academic Publishers, Dordrecht 1996.

[129] Luger G.F., *Artificial Intelligence: Structures and Strategies for Complex Problem Solving*, Fifth Ed., Addison-Wesley, London 2005.

[130] Łachwa A., *Fuzzy World of Sets, Numbers, Relations, Facts, Rules and Decisions*, Academic Publishing House EXIT, Warsaw 2001 (in Polish).

[131] Mańdziuk J., *Hopfield-type Neural Networks. Theory and applications*, Academic Publishing House EXIT, Warsaw 2000 (in Polish).

[132] Marple S.L. Jr., *Digital Spectral Analysis with applications*, Prentice Hall, Englewood Cliffs, New Jersey 1987.

[133] Mc Culloch W.S., A logical calculus of the ideas immanent in nervous activity, *Bulletin of Mathematical Biophysics*, vol. 5, 115–133 (1943).

[134] Mendel J.M., *Uncertain Rule-Based Fuzzy Logic Systems: Introduction and New Directions*, Prentice Hall PTR, London 2001.

[135] Menzel P., D'Aluisio F., *Robo Sapiens*, Gruner + Jahr Poland Publishing House, Warsaw 2002 (in Polish).

[136] Michalewicz Z., *Genetic Algorithms + Data Structures = Evolution Programs*, Springer-Verlag, London 1998.

[137] Miller G.F, Todd P.M., Hagde S.U., Designing neural networks using genetic algorithms, *Proceedings of the Third International Conference on Genetic Algorithms and Their Applications*, Schaffer J.D. (Ed.), Morgan Kauffmann, 379–384, San Mateo, CA 1989.

[138] Minsky M., Papert S., *Perceptrons: an introduction to computational geometry*, Cambridge, MA, 1988.

[139] Montana D., Davis L., Training feedforward neural networks using genetic algorithms, *Proceedings of 11th International Conference on Artificial Intelligence*, 762–767, Morgan Kauffman, 1989.

[140] Mrózek A., Płonka L., *Data Analysis with Rough Sets. Applications in Economics, Medicine and Control*, Academic Publishing House PLJ, Warsaw 1999 (in Polish).

[141] Mulawka J., *Expert Systems*, Scientific-Technical Publishing House WNT, Warszawa 1996 (in Polish).

[142] Nauck D., Klawon F., Kruse R., *Foundations of Neuro-Fuzzy Systems*, John Wiley & Sons – Interscience, New York 1997.

[143] Nęcka E., *Intelligence. Genesis, Structure, Functions*, Gdańsk Psychological Publishing House, Gdańsk 2003 (in Polish).

[144] Nguyen H.T., Walker E.A., *A First Course in Fuzzy Logic*, Second Ed., Chapman & Hall/CRC, Boca Raton 2000.

[145] Nie J., Linkens D., *Fuzzy-Neural Control*, Prentice Hall, Englewood Cliffs, NJ 1995.

[146] Niederliński A., *Rule Expert Systems*, Jacek Skalmierski Publishing House, Gliwice 2000 (in Polish).

[147] Nilsson N.J., *Artificial Intelligence: A New Synthesis*, Morgan Kaufmann, San Francisco 1998.

[148] Nowicki R., *Neuro-Fuzzy Systems with Various Inference Methods*, PhD Thesis, AGH, Kraków 1999 (in Polish).

[149] Nowicki R., Rutkowski L., Rough-Neuro-Fuzzy System for Classification, *Proc. of Fuzzy Systems and Knowledge Discovery*, 149, Singapore 2002.

[150] Nowicki R., Rutkowski L., *Rough Fuzzy Sets in Diagnostics*, VI National Conference Industrial Process Diagnostics, PWNT, 67–72, Gdańsk 2003 (in Polish).

[151] Nowicki R., *Rough Sets in the Neuro-Fuzzy Architectures Based on Monotonic Fuzzy Implications*, Lecture Notes in Artificial Intelligence, nr 3070, Artificial Intelligence and Soft Computing, Rutkowski L., Siekmann J., Tadeusiewicz R., Zadeh L.A. (Eds.), Springer-Verlag, 510–517, Berlin, Heidelberg 2004.

[152] Nowicki R., *Rough Sets in the Neuro-Fuzzy Architectures Based on Non-monotonic Fuzzy Implications*, Lecture Notes in Artificial Intelligence, nr 3070, Artificial Intelligence and Soft Computing, Rutkowski L., Siekmann J., Tadeusiewicz R., Zadeh L.A. (Eds.), Springer-Verlag, 518–525, Berlin, Heidelberg 2004.

506   References

[153] Obuchowicz A., *Evolutionary Algorithms for Global Optimization and Dynamic System Diagnosis*, Lubusky Scientific Society in Zielona Góra, Zielona Góra 2003.

[154] Osowski Ł., Building Speech Synthesizer, *Software* 2.0 Magazine, nr 2(98), 26–35, February 2003 (in Polish).

[155] Osowski S., *Neural Networks in Algorithmic Approach*, Scientific-Technical Publishing House WNT, Warsaw 1996 (in Polish).

[156] Osowski S., *Neural Networks for Information Processing*, Warsaw University of Technology Publishing House, Warsaw 2000 (in Polish).

[157] Pagan A., Ullah A., *Nonparametric Econometrics*, Cambridge University Press, London 1999.

[158] Pal S.K., Polkowski L., Skowron A., *Rough-Neural Computing, Techniques for Computing with Words*, Springer-Verlag, Berlin 2004.

[159] Parzen E., On estimation of probability density function and mode, *Annals Mathematics of Statistics*, vol. 33, 1065–1076 (1962).

[160] Pawlak M., *Evolutionary Algorithms for Production Scheduling*, Polish Scientific Publishers PWN, Warsaw 1999 (in Polish).

[161] Pawlak Z., Rough sets, *Proc. on Int. Journal of Information and Computer Science*, vol. 11, 341 (1982).

[162] Pawlak Z., *Information Systems. Theoretical Bases*, Scientific-Technical Publishing House WNT, Warsaw 1983 (in Polish).

[163] Pawlak Z., *Rough Sets — Theoretical Aspect of Reasoning About Data*, Kluwer Academic Publishers, Dordrecht 1991.

[164] Pawlak Z., Rough sets, decision algorithms and Bayes' theorem, *European Journal of Operational Research*, vol. 136, 181–189 (2002).

[165] Pedrycz W., *Fuzzy Control and Fuzzy Systems*, John Wiley & Sons – Interscience, New York 1993.

[166] Pedrycz W., *Fuzzy Sets Engineering*, CRC Press, Inc., Boca Raton 1995.

[167] Pedrycz W., Gomide F., *An Introduction to Fuzzy Sets: Analysis and Design*, A Bradford Book, The MIT Press, Cambridge, Massachusetts 1998.

[168] Penrose R., *The Emperor's New Mind*, Oxford University Press, 1989.

[169] Perta R., Mutant Invasion, *Chip* Magazine, vol. 9, 106–110, 2003 (in Polish).

[170] Pieczyński A., *Knowledge Representation in Diagnostic Expert System*, Lubusky Publishing House, Zielona Góra 2003 (in Polish).

[171] Piegat A., *Fuzzy Modeling and Control*, Springer-Verlag, New York 2001.

[172] Piliński M., *FLiNN — User Manual*, Polish Neural Network Society, Częstochowa 1996 (in Polish).

[173] Piliński M., *Fuzzy Controllers Using Neural Networks*, PhD Thesis, Wrocław University of Technology, 1996 (in Polish).

[174] Piliński M., Universal network trainer, *Proc. of the Second Conference, Neural Networks and Their Applications*, vol. 2, 383–391, Częstochowa 1996.

[175] Podsiadło M., Remarks to Schema Theorem, *National Conference: Evolutionary Algorithms*, Warsaw University of Technology, 1996 (in Polish).

[176] Pokropińska A., *Investigation of Effectiveness of Neuro-Fuzzy systems*, PhD Thesis, Częstochowa University of Technology, Częstochowa 2005 (in Polish).

[177] Polkowski L., *Rough Sets, Mathematical Foundation*, Springer-Verlag, Heidelberg, New York 2002.

[178] Principe J.C., Euliano N.R., Lefebvre W.C., *Neural and Adaptive Systems*, John Wiley & Sons, New York 2000.

[179] Radosiński E., *Information Systems in Dynamic Decision Analysis*, Polish Scientific Publishers PWN, Warszawa–Wrocław 2001 (in Polish).

[180] Radzikowska A.M., Kerre E.E., A comparative study of fuzzy rough sets, *Fuzzy sets and systems*, vol. 126, 137–155 (2002).

[181] Rosenblatt F., *On the Convergence of Reinforcement Procedures In Simple Perceptrons*, Cornell Aeronautical Laboratory Report VG-1196-G-4, Buffalo, NY 1960.

[182] Rudeński A., Application of Deterministic and Evolutionary Methods in Detailed and General Optimization of Electrical Machines. *Proc. of XL International Symposium on Electrical Machines* SME-2004, 200–204, 2004 (in Polish).

[183] Rumelhart D.E., Hinton G.E., Williams R.J., Learning internal representations by error propagation, in: *Parallel Distributed Processing*, vol. 1, ch. 8, Rumelhart D.E. and McClelland J.L. (Eds.), The MIT Press, Cambridge, Massachusetts 1986.

508   References

[184] Russell S.J., Norvig P., *Artificial Intelligence: A Modern Approach (2nd Ed.)*, Prentice Hall, 2002.

[185] Rutkowska D., *Intelligent Computing Systems. Genetic Algorithms and Neural Networks in Fuzzy Systems*, Academic Publishing House PLJ, Warsaw 1997 (in Polish).

[186] Rutkowska D., Piliński M., Rutkowski L., *Genetic Algorithms, Neural Networks and Fuzzy Systems*, Polish Scientific Publishers PWN, Warsaw 1999 (in Polish).

[187] Rutkowska D., *Neuro-Fuzzy Architectures and Hybrid Learning*, Physica-Verlag, Springer-Verlag Company, Heidelberg, New York 2002.

[188] Rutkowski L., Sequential estimates of probability densities by orthogonal series and their application in pattern classification, *IEEE Trans. on Systems, Man, and Cybernetics*, 10(12), 918–920 (1980).

[189] Rutkowski L., Sequential estimates of a regression function by orthogonal series with applications in discrimination, in: *Lectures Notes in Statistics*, Springer-Verlag, vol. 8, 236–244, Heidelberg, New York 1981.

[190] Rutkowski L., On Bayes risk consistent pattern recognition procedures in a quasi-stationary environment, *IEEE Trans. on Pattern Analysis and Machine Intelligence*, PAMI-4(1), 84–87 (1982).

[191] Rutkowski L., On system identification by nonparametric function fitting, *IEEE Trans. on Automatic Control*, vol. AC-27, 225–227 (1982).

[192] Rutkowski L., On-line identification of time-varying systems by nonparametric techniques, *IEEE Trans. on Automatic Control*, vol. AC-27, 228–230 (1982).

[193] Rutkowski L., On nonparametric identification with prediction of time-varying systems, *IEEE Trans. on Automatic Control*, vol. AC-29, 58–60 (1984).

[194] Rutkowski L., Nonparametric identification of quasi-stationary systems, *Systems and Control Letters*, vol. 6, 33–35 (1985).

[195] Rutkowski L., The real-time identification of time-varying systems by nonparametric algorithms based on the Parzen kernels, *International Journal of Systems Science*, vol. 16, 1123–1130 (1985).

[196] Rutkowski L., Sequential pattern recognition procedures derived from multiple Fourier series, *Pattern Recognition Letters*, vol. 8, 213–216 (1988).

[197] Rutkowski L., Rafajłowicz E., On global rate of convergence of some nonparametric identification procedures, *IEEE Trans. on Automatic Control*, vol. AC-34, nr 10, 1089–1091 (1989).

[198] Rutkowski L., Nonparametric learning algorithms in the time-varying environments, *Signal Processing*, vol. 18, 129–137 (1989).

[199] Rutkowski L., An application of multiple Fourier series to identification of multivariable nonstationary systems, *Int. Journal of Systems Science*, 20(10), 1993–2002 (1989).

[200] Rutkowski L., Identification of MISO nonlinear regressions in the presence of a wide class of disturbances, *IEEE Trans. on Information Theory*, vol. IT-37, 214–216 (1991).

[201] Rutkowski L., Multiple Fourier series procedures for extraction of nonlinear regressions from noisy data, *IEEE Trans. on Signal Processing*, 41(10), 3062–3065 (1993).

[202] Rutkowski L., *Adaptive Filters and Adaptive Signal Processing: Theory and Applications*, Scientific-Technical Publishing House WNT, Warsaw 1994 (in Polish).

[203] Rutkowski L., Gałkowski T., On pattern classification and system identification by probabilistic neural networks, *Appl. Math. and Comp. Sci.*, 4(3), 413–422 (1994).

[204] Rutkowski L., *Neural Networks and Neurocomputers*, Częstochowa University of Technology Publishing House, Częstochowa 1996 (in Polish).

[205] Rutkowski L., Cierniak R., Image compression by competitive learning neural networks and predictive vector quantization, *Appl. Math. and Comp. Science*, 6(3), 431–445 (1996).

[206] Rutkowski L., Tadeusiewicz R. (Eds.), *Proc. of the Fourth Conference: Neural Networks and Their Applications*, Zakopane 1999.

[207] Rutkowski L., Cierniak R., On image compression by competitive neural networks and optimal linear predictors, *Signal Processing: Image Communication — a Eurosip Journal*, Elsevier Science B.V., February, vol. 15, nr 6, 559–565, Amsterdam 2000.

[208] Rutkowski L., Cpałka K., Flexible structures of neuro-fuzzy systems, *Quo Vadis Computational Intelligence, Studies in Fuzziness and Soft Computing*, Springer-Verlag, vol. 54, 479–484, Heidelberg, New York 2000.

510   References

[209] Rutkowski L., Tadeusiewicz R. (Eds.), *Proc. of the Fifth Conference: Neural Networks and Soft Computing*, Zakopane 2000.

[210] Rutkowski L., Cpałka K., A neuro-fuzzy controller with a compromise fuzzy reasoning, *Control and Cybernetics*, 31(2), 297–308, 2002.

[211] Rutkowski L., Cpałka K., Compromise approach to neuro-fuzzy systems, *Proc. of the 2nd Euro-International Symposium on Computational Intelligence*, vol. 76, 85–90, Koszyce 2002.

[212] Rutkowski L., Cpałka K., Elastyczne systemy rozmyto-neuronowe, *XIV Krajowa Konferencja Automatyki*, 813–818, 24–27 czerwca, Zielona Góra 2002.

[213] Rutkowski L., Scherer R., New Neuro-Fuzzy Structures, *XIV National Conference on Automatics*, 809–812, June 24–27, Zielona Góra 2002 (in Polish).

[214] Rutkowski L., Starczewski J., Nowe metody modelowania niepewności w systemach rozmytych, *XIV Krajowa Konferencja Automatyki*, 885–888, 24–27 czerwca, Zielona Góra 2002.

[215] Rutkowski L., Cpałka K., Flexible weighted neuro-fuzzy systems, in: *Proc. 9th Int. Conf. on Neural Information Processing* (ICONIP'02), Orchid Country Club, November 18–22, Singapore 2002.

[216] Rutkowski L., Nowicki R., Hybrid soft computing techniques and their applications, *Proc. of the Symposium on Methods of Artificial Intelligence* AI-METH, 55–60, November 13–15, Gliwice 2002.

[217] Rutkowski L., Cpałka K., Compromise weighted neuro-fuzzy systems, in: Rutkowski L., Kacprzyk J. (Eds.), *Neural Networks and Soft Computing*, Heidelberg, Springer-Verlag, 557–562, Heidelberg, New York 2003.

[218] Rutkowski L., Cpałka K., Flexible neuro-fuzzy systems, *IEEE Trans. on Neural Networks*, vol. 14, 554–574 (2003).

[219] Rutkowski L., Kacprzyk J. (Eds.), *Neural Networks and Soft Computing*, Springer-Verlag, Heidelberg, New York 2003.

[220] Rutkowski L., Cpałka K., Neuro-fuzzy systems derived from quasi-triangular norms, *Proc. of the IEEE Int. Conf. on Fuzzy Systems*, 1031–1036, July 26–29, Budapest 2004.

[221] Rutkowski L., Adaptive probabilistic neural-networks for pattern classification in time-varying environment, *IEEE Trans. Neural Networks*, vol. 15, 811–827 (2004).

[222] Rutkowski L., Generalized regression neural networks in time-varying environment, *IEEE Trans. Neural Networks*, vol. 15, 576–596 (2004).

[223] Rutkowski L., A New Method for System Modelling and Pattern Classification, *Bulletin of the Polish Academy of Sciences*, 52(1), 11–24 (2004).

[224] Rutkowski L., New Soft Computing Techniques for System Modelling, in: *Pattern Classification and Image Processing*, Springer-Verlag, Heidelberg and New York, 2004.

[225] Rutkowski L., *Flexible Neuro-Fuzzy Systems: Structures, Learning and Performance Evaluation*, Kluwer Academic Publishers, Boston, Dordrecht 2004.

[226] Rutkowski L., Siekmann J., Tadeusiewicz R., Zadeh L. A., (Eds.): *Proc. of the Seventh Int. Conf. Artificial Intelligence and Soft Computing*, ICAISC-2004, Zakopane, Springer-Verlag, Heidelberg, New York 2004.

[227] Rutkowski L., Cpałka K., Designing and learning of adjustable quasi-triangular norms with applications to neuro-fuzzy systems, *IEEE Trans. on Fuzzy Systems*, vol. 14, 140–151 (2005).

[228] Rzewuski M., Playing Creator, *PC Kurier* Magazine, vol. 19, 28–37, 2002 (in Polish).

[229] Sadowski W., Atomic Warhead, *Polityka* Magazine, Nr 16 (2397), April 19, 2003 (in Polish).

[230] Schaffer J.D., Whitley L., Eshelman J., Combinations of genetic algorithms and neural networks, a survey of the state of the art, *Proc. of Int. Workshop on Combinations of Genetic Algorithms and Neural Networks*, COGANN-92, 1992.

[231] Scherer R., *Methods of Classification Using Neuro-Fuzzy Systems*, PhD Thesis, Częstochowa University of Technology, Częstochowa 2002 (in Polish).

[232] Shonkwiler R., Miller K.R., Genetic algorithm, neural network synergy for nonlinearly constrained optimization problems, *Proc. of Int. Workshop on Combinations of Genetic Algorithms and Neural Networks*, COGANN-92, 248–257, 1992.

[233] Słowiński R. (Ed.), *Intelligent Decision Support, Handbook of Applications and Advances of the Rough Sets Theory*, Kluwer Academic Publishers, Boston, Dordrecht 1992.

512   References

[234] Słupecki J., Hałkowska K., Piróg-Rzepecka K., *Logic and Set Theory*, Polish Scientific Publishers PWN, Warsaw 1994 (in Polish).

[235] Söderström T., Stoica P., *System Identification*, Prentice-Hall, London, United Kingdom, 1989.

[236] Specht D.F., Probabilistic neural networks, *Neural Networks*, vol. 3, 109–118 (1990).

[237] Specht D.F., A general regression neural network, *IEEE Trans. on Neural Networks*, vol. 2, 568–576 (1991).

[238] Starczewski J., *Fuzzy Type-2 Inference Systems*, PhD Thesis, Częstochowa University of Technology, Częstochowa 2002 (in Polish).

[239] Starczewski J., Rutkowski L., Neuro-Fuzzy Systems of Type 2, *1st Int'l Conf. on Fuzzy Systems and Knowledge Discovery*, vol. 2, 458–462, Singapore 2002.

[240] Starczewski J., What differs interval type-2 FLS from type-1 FLS, in: Rutkowski L. (Ed.) *Int'l Conf. on Artificial Inteligence and Soft Computing*, Zakopane, June 2004, *Lecture Notes in Computer Science*, Springer, 2004.

[241] Tadeusiewicz R., *Problems of Biocybernetics*, Polish Scientific Publishers PWN, Warsaw 1991 (in Polish).

[242] Tadeusiewicz R., *Neural Networks*, Academic Publishing House RM, Warsaw 1993 (in Polish).

[243] Tadeusiewicz R., Rutkowski L., Chojcan J. (Eds.), *Proc. of the Third Conference: Neural Networks and Their Applications*, Kule 1997.

[244] Tadeusiewicz R., *Elementary Introduction to Neural Networks with Exemplary Programs*, Academic Publishing House, Warsaw 1998 (in Polish).

[245] Tadeusiewicz R., Ogiela M.R., *Medical Image Understanding Technology*, Springer-Verlag, Heidelberg 2004.

[246] Takagi T., Sugeno M., Fuzzy identification of systems and its application to modeling and control, *IEEE Trans. Systems, Man, and Cybernetics*, vol. 15, 116–132 (1985).

[247] Takayama H., Aliens in the Palace!, *Newsweek* Magazine, 92, January 12, 2003 (in Polish).

[248] Tettamanzi A., Tomassini M., *Soft Computing. Integrating Evolutionary, Neural, and Fuzzy Systems*, Springer-Verlag, Berlin, Heidelberg 2001.

[249] Valenzuela-Rendon M., The fuzzy classifier system: A classifier system for continuously varying variables, *Proceedings of the Fourth International Conference on Genetic Algorithms*, 346–353, Morgan Kaufmann, San Mateo 1991.

[250] Volna E., Genetic Search of Optimal Neural Network Topology for problem of Pattern Recognition, *Fifth Conference Neural Networks and Soft Computing*, 657–662, Zakopane 2000.

[251] Volna E., Learning Algorithm of Feedforward Neural Networks, *Fourth Conference Neural Networks and Their Applications*, 611–616, Zakopane 1999.

[252] Wang L.-X., Mendel J.M., Generating fuzzy Rules by Learning from Examples, *IEEE Trans. Systems, Man, and Cybernetics*, **22**(6), 1414–1427, 1992.

[253] Wang L.-X., *Adaptive Fuzzy Systems and Control, Design and Stability Analysis*, Prentice Hall PTR, Upper Saddle River, NJ 1994.

[254] Wang L., Yen J., Application of statistical information criteria for optimal fuzzy model construction, *IEEE Trans. on Fuzzy Systems*, vol. 6, 353–362 (1998).

[255] Weiss G. (Ed.), *Multiagent systems. A modern approach to distributed artificial intelligence*, The MIT Press, Cambridge 1999.

[256] Whitley D., Applying genetic algorithms to neural network learning, *Proceedings of the Seventh Conference of the Society of Artificial Intelligence and Simulation of Behavior*, Sussex, England, Pitman Publishing, 137–144, 1989.

[257] Widrow B., Hoff M.E. Jr., Adaptive switching circuits, *Western Conf. Rec.*, IRE, Part 4, 94–104, 1960.

[258] Widrow B., Stearns S., *Adaptive Signal Processing*, Prentice Hall, Englewood Cliffs, NJ 1985.

[259] Widrow B., Lehr M.A., 30 Years of Adaptive Neural Networks: Perceptron, Madaline and Backpropagation, *Proc. of the IEEE*, 78(9), 1415–1442 (1990).

[260] Winter G., Périaux J., Galán M., Cuesta P., *Genetic algorithms in Engineering and Computer Science*, John Wiley & Sons, Chichester 1995.

[261] Wu H., Mendel J.M., Uncertainty bounds and their use in the design of interval type-2 fuzzy logic systems, *IEEE Trans. Fuzzy Syst.*, 10(5), 622–639 (2002).

[262] Yager R.R., Filev D.P., *Essentials of Fuzzy Modeling and Control*, John Wiley & Sons, New York, 1994.

[263] Yao X., A review of evolutionary artificial neural networks, *Int. Journal of Intelligent Systems*, 1993, 539–567.

[264] Yao X., Liu Y., Towards designing artificial neural networks by evolution, *Applied Mathematics and Computation*, 91(1), 83–90 (1998).

[265] Zadeh L.A., Fuzzy sets, *Information and Control*, vol. 8, 338–353 (1965).

[266] Zadeh L.A., The concept of a linguistic variable and its application to approximate reasoning, Parts 1–3, *Inf. Sci.*, no. 8, 199–249; no. 8, 301–357; no. 9, 43–80; 1975.

[267] Zalewski A., Cegiełka R., *Matlab — numerical calculations and their applications*, NAKOM Publisher, Poznań 2002.

[268] Zieliński J.S. (Ed.), *Intelligent Systems in Management. Theory and Practice*, Polish Scientific Publishers PWN, Warsaw 2000 (in Polish).

[269] Zimmermann H.-J., *Fuzzy Set Theory*, Kluwer Academic Publishers, Boston, Dordrecht 1994.

[270] Żurada J., *Introduction to Artificial Neural Systems*, West Publishing Company, St. Paul, Minnesota, 1992.

[271] Żurada J., Barski M., Jędruch W., *Neural Networks*, Polish Scientific Publishers PWN, Warsaw 1996 (in Polish).